N36.50

CLINICAL ATLAS OF MUSCLE
AND
MUSCULOCUTANEOUS FLAPS

PRINCIPLES OF SAFE MUSCLE FLAP TRANSPOSITION

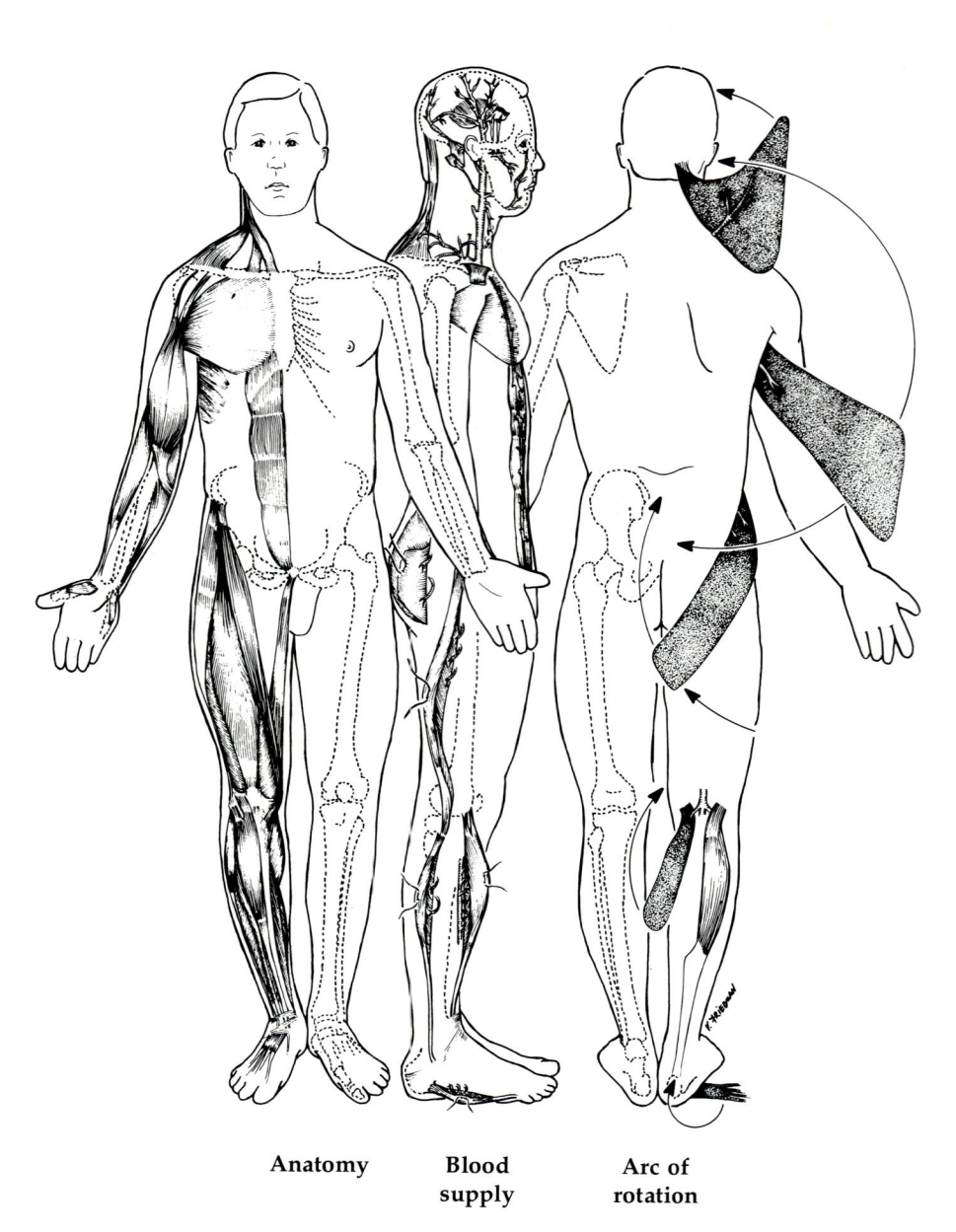

Anatomy   Blood supply   Arc of rotation

CLINICAL ATLAS OF MUSCLE AND MUSCULOCUTANEOUS FLAPS

STEPHEN J. MATHES, M.D.

Associate Professor of Surgery (Plastic and Reconstructive),
University of California at San Francisco,
San Francisco, California;
formerly Assistant Professor of Surgery, Washington University,
St. Louis, Missouri

FOAD NAHAI, M.D.

Assistant Professor of Surgery (Plastic and Reconstructive),
Emory University School of Medicine,
Atlanta, Georgia

with 576 illustrations

Vicki Moses Friedman, Illustrator

The C. V. Mosby Company

ST. LOUIS · TORONTO · LONDON   1979

Copyright © 1979 by The C. V. Mosby Company

All rights reserved. No part of this book may be reproduced in any manner without written permission of the publisher.

Printed in the United States of America

The C. V. Mosby Company
11830 Westline Industrial Drive, St. Louis, Missouri 63141

Library of Congress Cataloging in Publication Data

Mathes, Stephen J
   Clinical atlas of muscle and musculocutaneous flaps.

  Bibliography: p.
  Includes index.
  1. Surgery, Plastic. 2. Muscle. 3. Flaps (Surgery).
I. Nahai, Foad, 1943-   joint author. II. Title.
[DNLM: 1. Muscles—Transplantation—Atlases. 2. Skin—Transplantation—Atlases. 3. Surgery, Plastic—Atlases. WE17 M427c]
RD118.M38   617′.95   79-10739
ISBN 0-8016-3141-6

C/CB/B 9 8 7 6 5 4 3 2 1   01/D/055

To our wives
JENNIFER and SHAHNAZ

FOREWORD

Growth, whether in a biologic system or in an intellectual one, is not a linear phenomenon. All things being equal, it is characterized by periodicity where there is an initial phase of rapid growth, followed by a slowing, and then virtual standstill. If there be provision for the renewal of resources, another period of growth will ensue followed by the same change of events until senescence.

So it is in a profession. In surgery there was a long period of quiescence; indeed, even some regression concomitant with the Dark Ages. During the Renaissance, dogma was questioned and inquiry championed. Not only did art, science, and literature flourish, but also surgery—particularly because of the influence of Ambroise Paré. The discovery of anesthesia and the concept of antisepsis gave rise to the great expansion of surgery in the late nineteenth century, especially in England and on the continent.

In the very recent past, transplantation of the kidney and its impact on the whole field of immunology and genetics has given rise to the entire field of transplant surgery. At the same time a completely new specialty has arisen as a result of the work of Blalock and Taussig, Gross, and others—cardiac surgery. In brief, growth within the field of surgery has been nothing short of remarkable in the past two decades.

Plastic and reconstructive surgery was established as a distinct specialty consequent to the growth and interest in the field occasioned by two world wars. Ideas in reconstructive surgery, however, that were described in 1892—the island flap, the transposition of the latissimus dorsi muscle in breast surgery, or even blood supply to the skin—were short lived or not accepted. Instead, surgeons became intent upon reconstructing defects by means of delayed tubed pedicle flaps as described by Filatov and Gillies, requiring many stages. Growth in the field continued but was slow.

For example, in the 1960s the specialty in this country was quite

FOREWORD

small. From 1960 to 1969, 458 surgeons were trained and certified so that the entire total of practicing board-certified plastic surgeons was but 907. However, from 1969 until the present the number of board-certified plastic surgeons has increased to over 2000 with an equal number of physicians in other specialties of surgery devoting a significant amount of their time to the practice of plastic surgery, largely regional in nature.

At the same time, largely because of refinements in vascular techniques, particularly microvascular, in transplantation, there has been great enthusiasm for immediate reconstruction. Edgerton articulated the idea as particularly applicable to patients with head and neck cancer; Bakamjian and McGregor made the idea practical reality and firmly established the notion of the arterialized flap and vascular territories. Ger reintroduced the idea of transposition of muscle, and Orticochea reintroduced the possibility of the musculocutaneous flap suggested some years before by Neal Owens.

The transposition of muscle and musculocutaneous flaps is then an idea whose time has come. The energy and enthusiasm of young men in the field, for example, Vasconez, McCraw, Bostwick, Brown, Arnold, Dibbell, Mathes, and Nahai, have firmly established not only the anatomic basis but the clinical utility of the transposition of muscle. More importantly, the blood supply to skin is now known, and much of the empiricism, mystery, and tedious delay in reconstructive surgery has been stripped aside.

Because of muscle flaps, musculocutaneous flaps, arterialized flaps, and microsurgical techniques, the entire field of plastic and reconstructive surgery is undergoing a remarkable period of growth and metamorphosis. In certain specific anatomic regions these approaches have completely supplanted all others to all intents and purpose. For example, the anterior tibial defect consequent to an open fracture with wound failure and chronic osteomyelitis can probably be handled best in the distal third of the leg by a free flap. All other areas in the leg are easily closed in a straightforward fashion by muscle flaps and skin graft for integumentary cover or by a musculocutaneous flap. Practically gone are the tedious cross-leg flaps with their requirement for protracted hospitalization.

Perineal wounds with tissue loss are much easier to manage by musculocutaneous flaps or muscle flaps than by any other method. Similarly these techniques have provided additional options in recon-

FOREWORD

struction of the vagina or penis following radical surgery or avulsion injury.

Tissue losses of the abdomen and chest equally can now be approached with confidence because of muscle transposition. For example, the heretofore extremely difficult reconstruction of the anterior chest wall can be managed in a straightforward direct fashion by transposition of the pectoralis major or latissimus dorsi muscle.

The impact that the latissimus dorsi musculocutaneous technique has had and will have on reconstruction of the breast following radical mastectomy is enormous. Clearly it may do for breast reconstruction what the deltopectoral flap has done for head and neck reconstruction.

The need, therefore, for such an atlas of muscle and musculocutaneous flaps is immediate and obvious. The vascular anatomy of muscle is basically the key to knowledge of blood supply to the skin. Armed with this information, the trained surgeon can undertake flap transposition with an enhanced degree of safety and reliability. The book, however, is a guide, and each surgeon interested in the field should use the book as it is intended. The atlas is not a cookbook to clinical practice, but rather is a guide to anatomic dissection so that clinical skills can be enhanced thereby. For example, the latissimus dorsi musculocutaneous flap can be outlined in the anatomy laboratory for a specific clinical problem, and by using this book as a guide the surgeon can proceed to become completely familiar with the technique and its limitations before entering the operating room.

Drs. Mathes and Nahai have firsthand knowledge in the field. They have participated from the start in the development of this exciting aspect of reconstructive surgery. They have spent countless hours in the laboratory and subsequently in the operating rooms at Emory University and at Washington University to detail the precise vascular anatomy of muscle and the overlying skin. To the end of safe clinical practice of reconstructive surgery they have written this book.

Maurice J. Jurkiewicz, M.D.
Professor of Surgery and Chief,
Division of Plastic and Reconstructive Surgery,
Emory University School of Medicine, Atlanta, Georgia

PREFACE

This clinical atlas would not be possible without the contributions and interest of many surgeons. The concept of modern muscle flap transposition was introduced by Ralph Ger for reconstruction of lower extremity defects. Miguel Orticochea expanded Neal Owens' concept of the musculocutaneous flap. John McCraw isolated the musculocutaneous perforating vessels and defined the cutaneous territory of many muscles. Progress continues in this field with contributions being made by many surgeons. Their contributions are cited in the Selected Readings.

Under the able guidance of Maurice Jurkiewicz, at Emory University in Atlanta, the use of the muscle and musculocutaneous flaps for reconstructive surgery has expanded to include all body regions. We especially acknowledge the contributions of Luis Vasconez and his stimulation and encouragement over the past years during the development of these reconstructive techniques. Each of the following surgeons from Emory University has made a significant contribution in the expansion and extension of the muscle and musculocutaneous flap as a reconstructive technique.

Maurice Jurkiewicz	P. G. Arnold
Luis Vasconez	Jim Madden
John McCraw	John Silverton
John Bostwick	William Schneider
Robert Leonard	H. Louis Hill
Paul Silverstein	Rod Hester
Nate Mayl	Berkeley Powell
Robert Brown	

Without the support and encouragement of Paul Weeks, Professor of Surgery, Division of Plastic Surgery at Washington University, this atlas could not have been completed. Roy Peterson, Professor of Anatomy, Washington University School of Medicine, generously supplied his advice, staff, and required cadaver specimens. All dissections for

PREFACE

this text were performed in the Anatomy Laboratory of Washington University School of Medicine in St. Louis.

We thank Ms. Vicki Moses Friedman, our illustrator, and Gerard Huth, our photographer, for their excellent work and willingness to adjust to our irregular schedules.

Stephen J. Mathes
Foad Nahai

CONTENTS

Introduction, 1

SECTION ONE
ANTERIOR THIGH, 7

Gracilis, 13
Sartorius, 33
Rectus femoris, 41
Vastus lateralis, 51
Tensor fascia lata (TFL), 63

SECTION TWO
POSTERIOR THIGH, 87

Gluteus maximus, 91
Biceps femoris, 105
Semitendinosus, 115
Semimembranosus, 123

SECTION THREE
MEDIAL LEG, 133

Gastrocnemius, 141
Soleus, 157
Flexor digitorum longus, 179
Flexor hallucis longus, 189

SECTION FOUR
LATERAL LEG, 199

Anterior group, 199
  Tibialis anterior, 207
  Extensor digitorum longus, 217
  Extensor hallucis longus, 227
Posterior group, 237
  Peroneus longus, 241
  Peroneus brevis, 251

SECTION FIVE
FOOT, 263

Abductor hallucis, 269
Flexor digitorum brevis, 279
Abductor digiti minimi, 291
Extensor digitorum brevis, 301

SECTION SIX
TRUNK, 309

Anterior trunk, 309
  Pectoralis major, 317
  Serratus anterior, 337
  Rectus abdominis, 347
Posterior trunk, 363
  Latissimus dorsi, 369
  Trapezius, 393

SECTION SEVEN
UPPER EXTREMITY, 421

Biceps brachii, 425
Brachioradialis, 433
Flexor carpi ulnaris, 441

SECTION EIGHT
HAND, 449

First dorsal interosseous, 453
Abductor pollicis brevis, 459
Abductor digiti minimi, 465

SECTION NINE
HEAD AND NECK, 471

Sternocleidomastoid, 475
Temporalis, 487

Appendix, 497

Suggested readings, 500

CLINICAL ATLAS OF MUSCLE
AND
MUSCULOCUTANEOUS FLAPS

INTRODUCTION

I profess both to learn and to teach anatomy, not from books but from dissections; not from positions of philosophers but from the fabric of nature.

De Motu Cordis et Sanguinis (1628)
William Harvey

The use of muscle and musculocutaneous flaps as a single-stage reconstructive technique has evolved from an operation for the lower extremity to one applicable to all body regions. The rapid growth of this reconstructive technique has been directly related to increased attention to the precise anatomy of the blood supply of muscle. Every muscle is a potential flap if the surgeon has an accurate knowledge of its functional and vascular anatomy.

The advantages of muscle or musculocutaneous flaps for reconstructive surgery are now well documented. Although the flap donor area is adjacent to the defect, the blood supply of the muscle flap is generally located proximal to the defect and not influenced by local vascular insufficiency whether the cause is traumatic, neoplastic, or radiation injury. Muscle or musculocutaneous flap transposition is accomplished in one operation. The operation, when performed with a thorough knowledge of the anatomy of the muscle chosen for transposition, can be accomplished with safety and relative ease.

Although the external and internal vascular anatomy of muscle is available in a multitude of anatomy texts, these anatomic descriptions have obviously not considered the muscle's anatomic configuration in regard to its application and suitability for transposition in reconstructive surgery. We initially studied the muscles of the lower extremity with colored latex vascular injections. These studies defined the location of the dominant vascular pedicles of these muscles, allowing development of safe guidelines for their use in reconstruction of defects in the lower extremity. Based on this experience and the increasing interest in the principle of muscle transposition as a general technique for reconstructive surgery, these anatomic studies have again been performed in all body regions. The results of these anatomic studies and our experience in their clinical application are presented in this clinical atlas of muscle and musculocutaneous flaps.

Requisites for muscle flaps

The muscles included in this text either fulfill our criteria as a useful muscle for transposition or as a muscle for specialized use as in microsurgical free transfer. All muscles have some potential for transposition. It is our hope that this atlas will prompt reconstructive surgeons to both critically evaluate our anatomic data and assess the many muscles not included here for their application in the continuing development and expansion of this reconstructive technique.

Each muscle must meet certain criteria for suitability for safe and successful transposition. These criteria were developed in our initial anatomic studies and have proven reliable in our clinical experience. These safe guidelines have been applied to each muscle in our text.

Size of muscle belly

The muscle must be adequate in size to cover the defect. The muscle size is demonstrated both in the anatomic drawings and photographs of actual cadaver dissections. Fresh cadaver dissections were used in the photographs illustrating the arc of rotation. The fresh muscle rotated with its dominant vascular pedicle intact accurately simulates the extent of coverage provided by the muscle. The muscles have not been stretched, a source of muscle flap failure when a muscle of inadequate size has been chosen by the reconstructive surgeon. However, the epimysium may be safely removed from muscle, increasing its size without adversely effecting its blood supply. Our clinical experience in patients with upper and lower motor neuron disease has demonstrated no significant alteration in muscle bulk, arc of rotation, or cutaneous territory in the musculocutaneous flap.

Proximity of vascular pedicle to muscle origin

Muscle blood supply may consist of a single pedicle (e.g., tensor fascia lata), two dominant pedicles (e.g., semimembranosus), or segmental blood supply either with a dominant pedicle (e.g., gracilis) or without a dominant pedicle (e.g., sartorius). When the muscle has a single or dominant pedicle close to the origin or point of rotation, the muscle is ideally suited for successful transposition. The muscle that requires division of segmental vascular pedicles will have more limited application as a transposition flap. Both muscle origin and insertion may be divided when the dominant vascular pedicle is long, allowing rotation of either muscle or musculocutaneous unit as an island flap.

The vascular anatomy is demonstrated in illustrations and photographs of actual cadaver dissections. Unlike the majority of anatomic texts, the pedicle is demonstrated as it will be seen during actual distal-to-proximal dissection of the muscle for transposition. In the majority of flap transposition reconstructive procedures, it is rarely necessary to actually visualize the muscle's dominant vascular pedicle except in development of an island or free transfer of muscle or musculocutaneous flaps.

The venous drainage of the muscle is consistently located with the arterial pedicle. In order to simplify the illustrations and photographs, the adjacent veins have not been included by our artist and are purposely excised in the cadaver dissections.

Our observations regarding the use of muscle minor pedicles as the point of rotation require a prior strategic delay of the major pedicle. We still regard this data as experimental and have included this anatomic information in the hope of stimulating further interest in this aspect of muscle transposition.

Accessibility

The majority of the muscles included in this text are superficially located, and the dissection of the muscle is not complicated. A knowledge of muscle origin and insertion is necessary to accurately locate the muscle intended for transposition. Whenever possible, we have attempted to simplify this anatomic data regarding muscle origin and insertion. Hopefully the anatomist will excuse these generalizations, since this data is provided for the surgeon solely as a landmark to locate the proper muscle.

Preservation of function

When a particular muscle is used as a transposition flap, it will no longer serve its original function. When synergistic muscles are available to prevent a functional deficit, this muscle has been designated as an expendable muscle. Certain muscles not considered expendable (e.g., tibialis anterior) may be transposed, leaving the tendon intact with function preservation. Occasionally, transposition of a nonexpendable muscle accepting a functional defect may be warranted in a difficult reconstructive problem. On occasion, several members of either the same muscle group or adjacent groups have been transposed to cover a large defect. Muscles with segmental blood supply (e.g., gluteus maximus) may be split, transposing one half and leaving the other half intact for function preservation.

Proximity of motor nerve entry to muscle origin

The motor nerve generally enters the proximal muscle. However, motor nerve location in relation to the point of rotation is not an important factor in muscle transposition. When the motor nerve limits the arc of rotation, it should be divided. When loss of muscle bulk is desirable following transposition, it may be helpful to deliberately divide the motor nerve. In general the motor nerve is not visualized, since it is closely related to the dominant vascular pedicle of the muscle.

In order to maintain simplicity of the illustrations, the motor nerve has not been included. In the photographs of cadaver dissection the motor nerve has been preserved but generally not identified. However, the location of the motor nerve in relation to the vascular pedicle is included in the anatomic descriptions.

In certain muscles (e.g., extensor digitorum brevis) the motor nerve is illustrated, since these muscles have potential for functional muscle free transfer.

Sensory innervation

When a muscle is elevated in association with its cutaneous territory as a musculocutaneous flap, it may be important to preserve sensation to the skin. In the musculocutaneous free flap the sensory nerve may be included with the flap as a source of innervation by direct repair with intact sensory nerves in the recipient region. For this reason this sensory nerve has been identified when it is available for inclusion in the flap design.

Osseous territory

When the origin or insertion of the muscle unit with bone is not fascial or tendinous, vascular connections can be demonstrated between bone and muscle. Experimental and clinical data have demonstrated that a portion of this bone can be included with the muscle either in flap transposition or free transfer.

Elevation of flap

A description of flap elevation is included for each muscle, stressing pertinent anatomic features. A model has been used to demonstrate the location of both the most direct incision site for the muscle flap and the cutaneous territory of the musculocutaneous flap. Illustrations and photographs of cadaver dissections are also included to assist the surgeon in rapid identification of the proper muscle intended for transposition. Although all the muscles included in this atlas have been successfully used in transposition procedures, specific clinical examples were included only when further clarification of the techniques appeared necessary.

Precautions

With a thorough knowledge of the anatomy of the muscle selected for transposition, technical complications are rare. In distal extremity muscle flaps, the use of the tourniquet will assist the surgeon in flap elevation and avoid injury to important adjacent structures. When a muscle flap is transposed, the muscle should be immediately skin grafted. This prevents muscle desiccation and usually avoids the need for a secondary grafting procedure. When a musculocutaneous flap is transposed, the surgeon is cautioned to suture the muscle edge to the edge of the cutaneous territory as the dissection progresses. This avoids accidental stretching or disruption of the vascular perforators that supply circulation from muscle to skin. When the viability of the muscle or musculocutaneous flap is questioned, circulatory status of the flap should be confirmed by fluorescein injection. This study should be delayed until all flap dissection is completed and the flap is sutured in the transposed position. Each muscle discussion is concluded with specific precautions regarding proximity of the muscle to important structures and potential flap complications.

Summary

We hope this clinical atlas will be useful as an anatomic guide for the reconstructive surgeon contemplating muscle or musculocutaneous flap transposition. The illustrations and anatomic descriptions should be used together with cadaver dissections in the anatomy laboratory as the surgeon evaluates the suitability of this reconstructive modality for the patient.

Muscle and musculocutaneous flap transposition has provided the surgeon a safe, effective technique for reconstruction in one operation. Hopefully this atlas will elucidate the single requirement for successful use of this procedure—a precise knowledge of the vascular and functional anatomy of muscles.

SECTION ONE
ANTERIOR THIGH

Gracilis

Sartorius

Rectus femoris

Vastus lateralis

Tensor fascia lata (TFL)

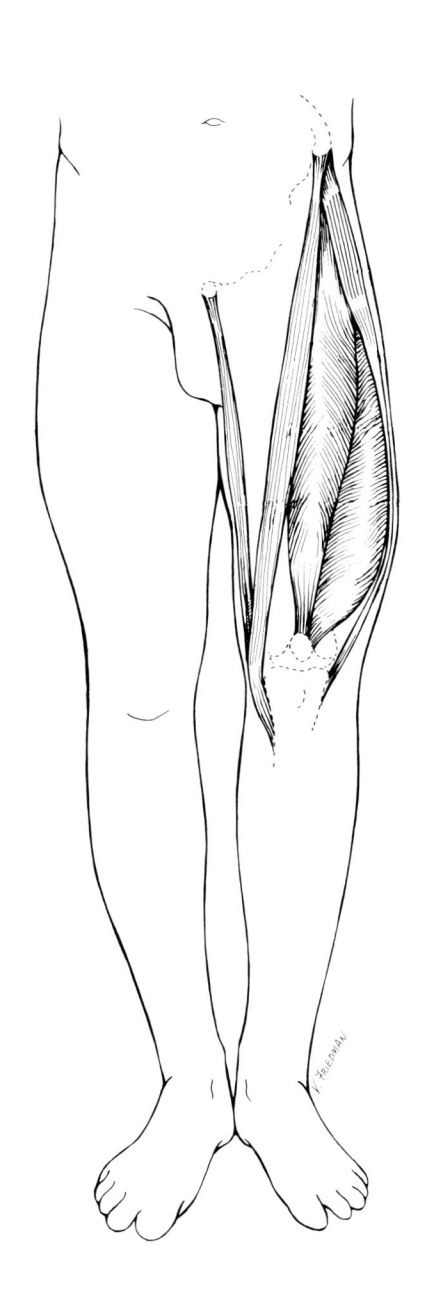

ANTERIOR THIGH

- Tensor fascia lata
- Rectus femoris
- Sartorius
- Gracilis
- Vastus lateralis

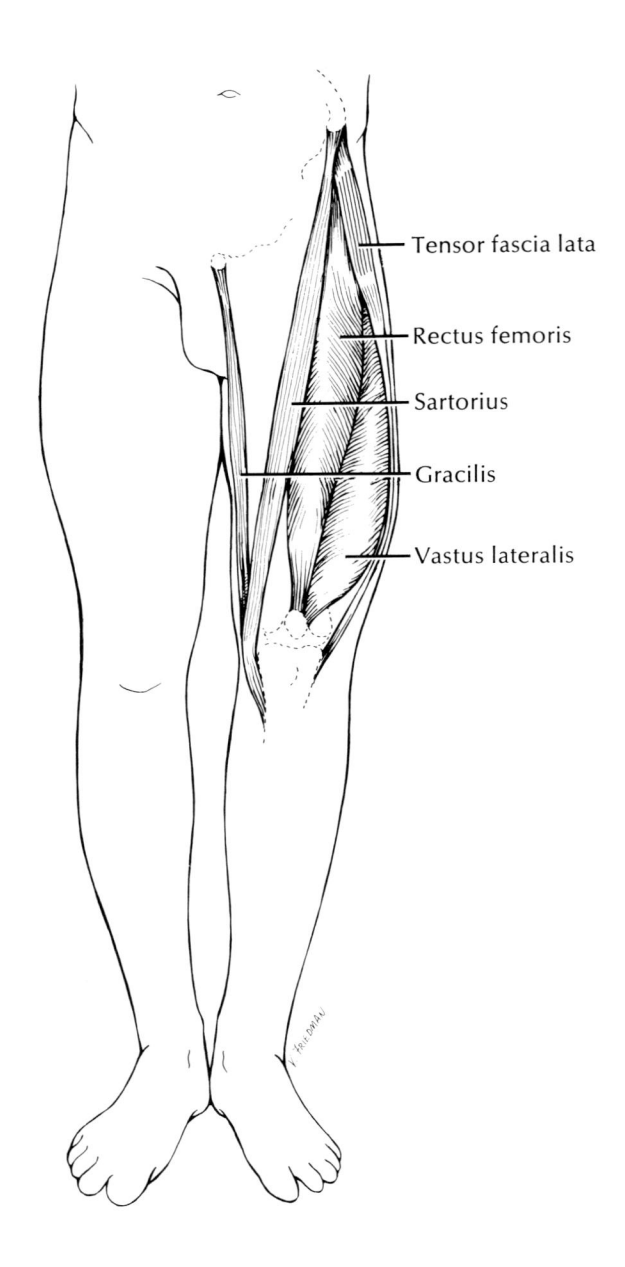

The anterior thigh muscles have great potential as muscle and musculocutaneous flaps for reconstructive surgery of the trunk, perineum, groin, genitalia, anus, buttocks, and free flap transfer.

In general the muscles of this group have a dominant, proximally based blood supply. These pedicles fall within 10 cm of the inguinal ligament, and based on the pedicle, the group has a useful arc of rotation.

ANTERIOR THIGH

- Tensor fascia lata
- Sartorius
- Adductor longus
- Gracilis
- Rectus femoris
- Vastus lateralis

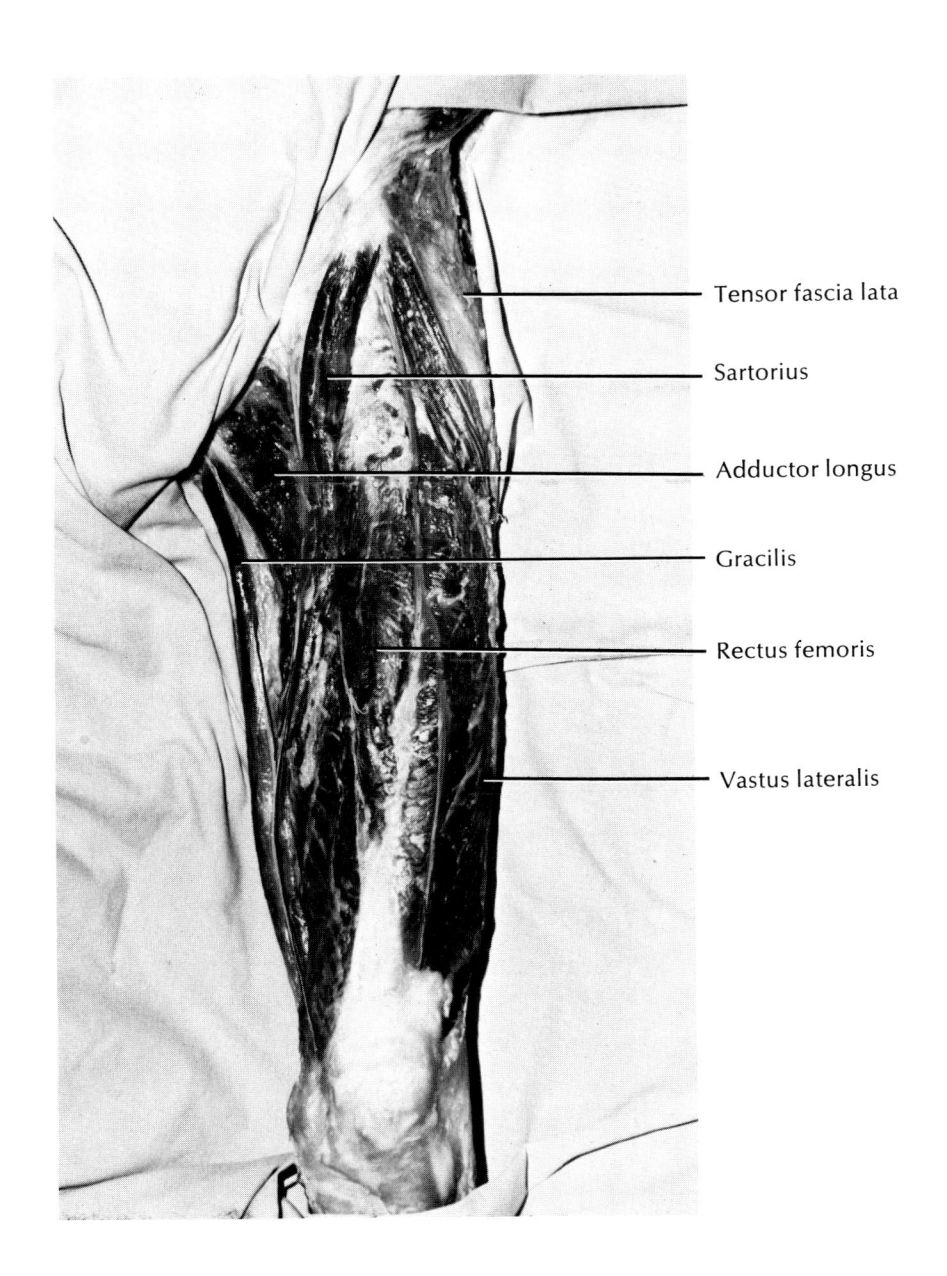

ANTERIOR THIGH

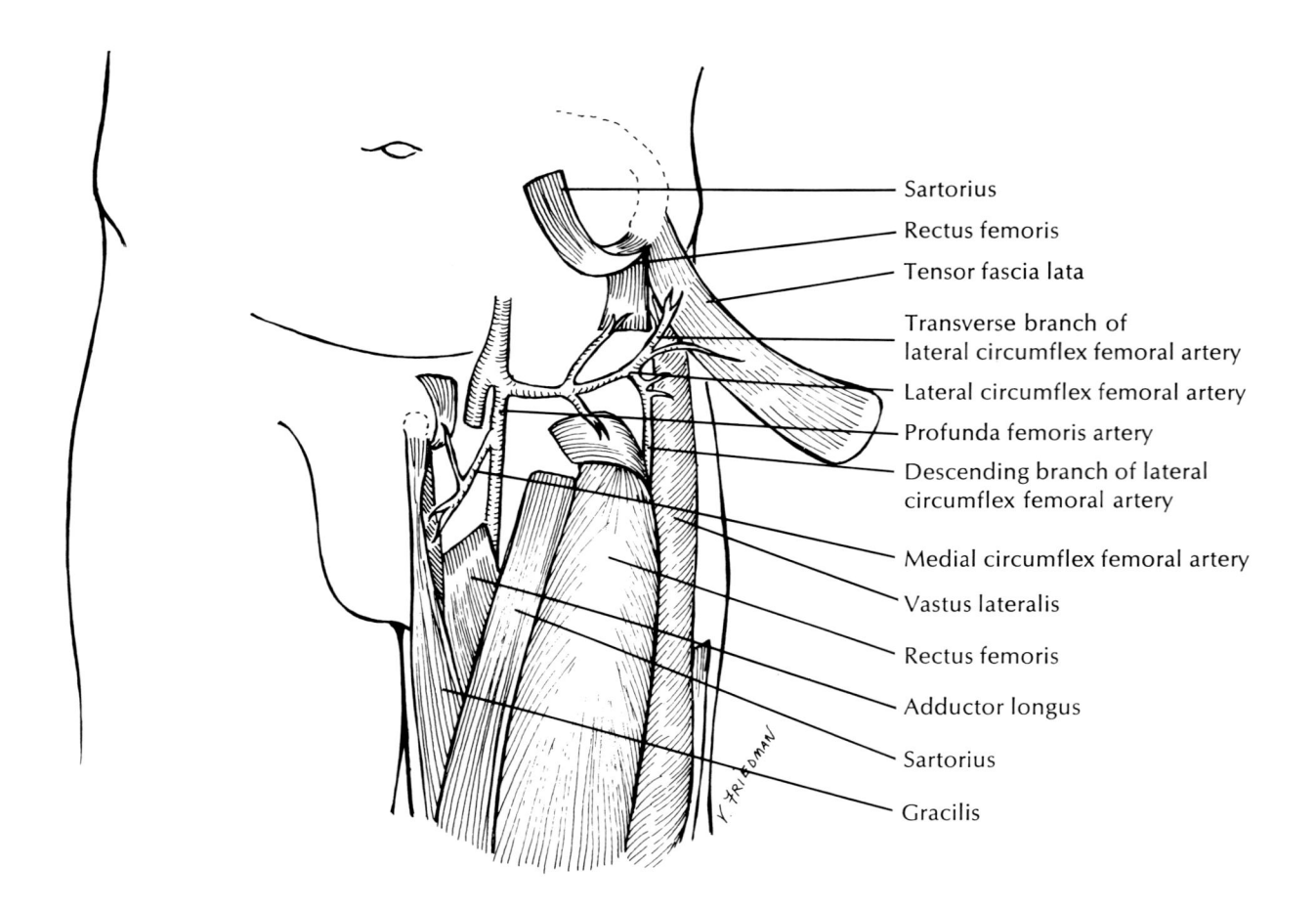

Blood supply

The blood supply of the anterior thigh muscles is based predominantly on the profunda femoris artery through its medial and lateral circumflex femoral branches. The medial circumflex femoral artery, the smaller of these two arteries, courses medially deep to the adductor longus muscle. This artery lies on the adductor magnus muscle and terminates in the gracilis muscle as its major vascular pedicle. The lateral circumflex femoral artery courses laterally deep to the rectus femoris, which it supplies through two branches on the deep surface, emerging at the lateral border of the rectus femoris deep to the sartorius muscle. Here the vessel supplies the vastus lateralis through a descending branch and continues laterally where it has a small branch to the gluteus minimus muscle and two or three large terminal branches into the TFL.

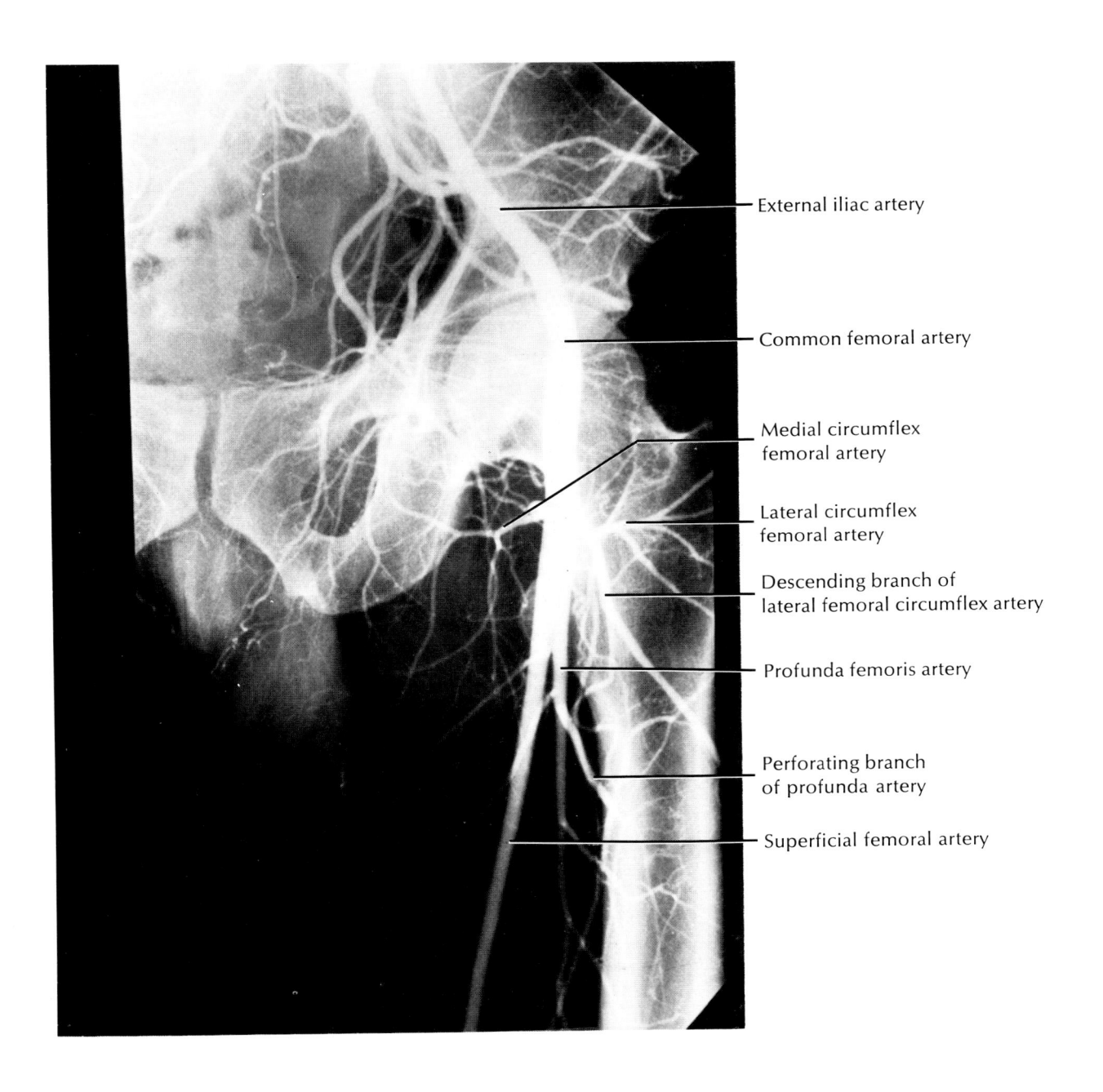

ANTERIOR THIGH

- External iliac artery
- Common femoral artery
- Medial circumflex femoral artery
- Lateral circumflex femoral artery
- Descending branch of lateral femoral circumflex artery
- Profunda femoris artery
- Perforating branch of profunda artery
- Superficial femoral artery

Gracilis

MUSCLE FLAP
MUSCULOCUTANEOUS FLAP
FREE FLAP

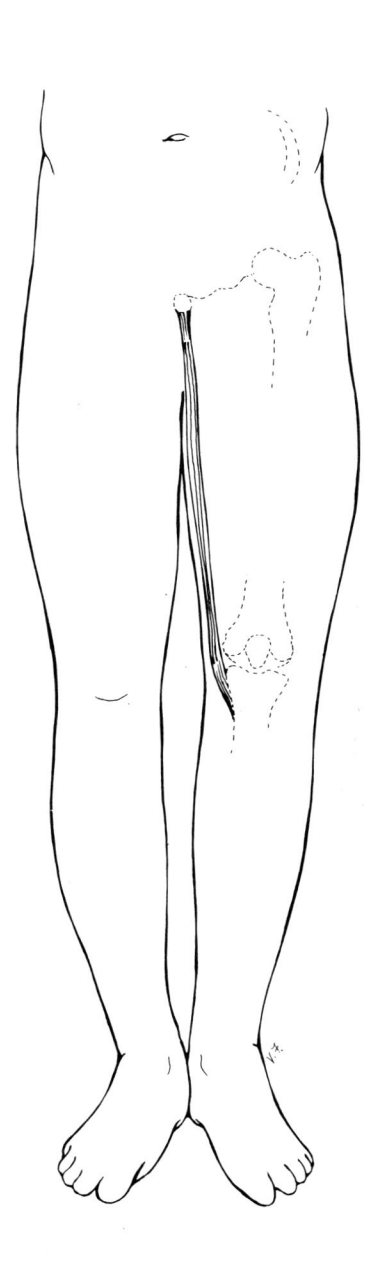

Applications

Reconstruction of:

PENIS
VAGINA
VULVA
ANAL MUSCULATURE

Coverage of:

PUBIS
GROIN
PERINEUM
ABDOMINAL WALL
ISCHIUM

Free flap:

FUNCTIONAL MUSCLE
DISTANT COVERAGE

ANTERIOR THIGH

Gracilis

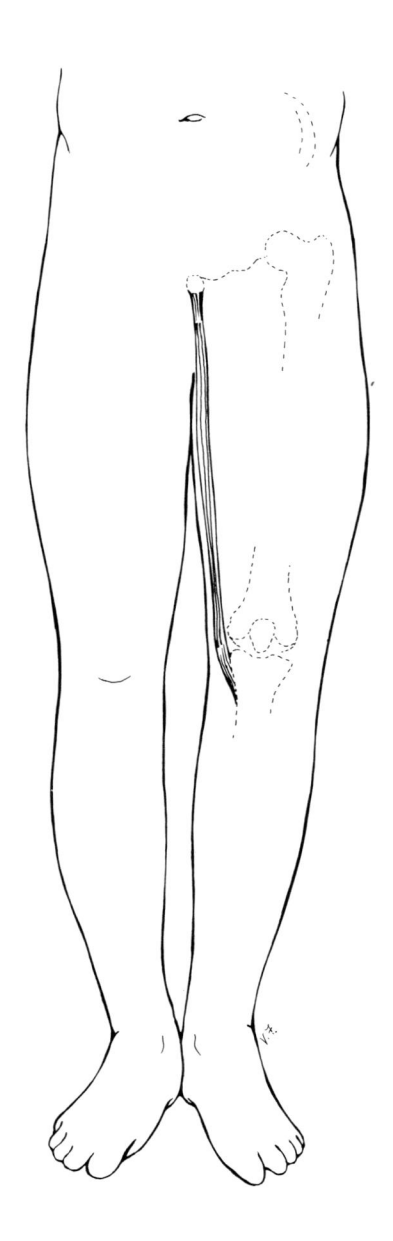

The gracilis is a flat, thin muscle located superficially in the medial thigh between the adductor longus and sartorius muscles anteriorly and the semimembranosus posteriorly. The adductor magnus lies deep to the gracilis muscle. The muscle is broad at its origin and narrows toward its insertion.

Origin: Pubic symphysis.
Insertion: Medial tibial condyle.

ANTERIOR THIGH

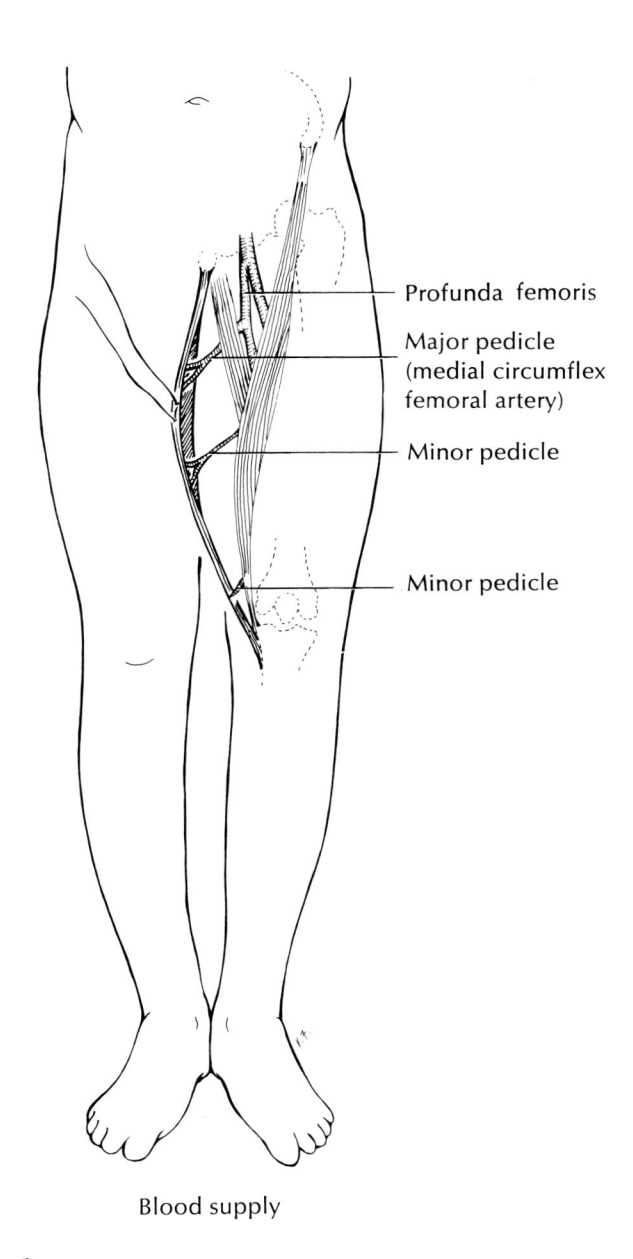

Blood supply

- Profunda femoris
- Major pedicle (medial circumflex femoral artery)
- Minor pedicle
- Minor pedicle

Nerve supply:
  Motor – Anterior branch of the obturator. This motor nerve is located between the adductor longus and magnus muscles. It enters the medial muscle superior and adjacent to the major vascular pedicle. The motor nerve must be preserved when functional muscle transfer is planned.
  Sensory – Obturator nerve.
Function: This is an expendable muscle that serves as an accessory thigh adductor.

ANTERIOR THIGH

Gracilis

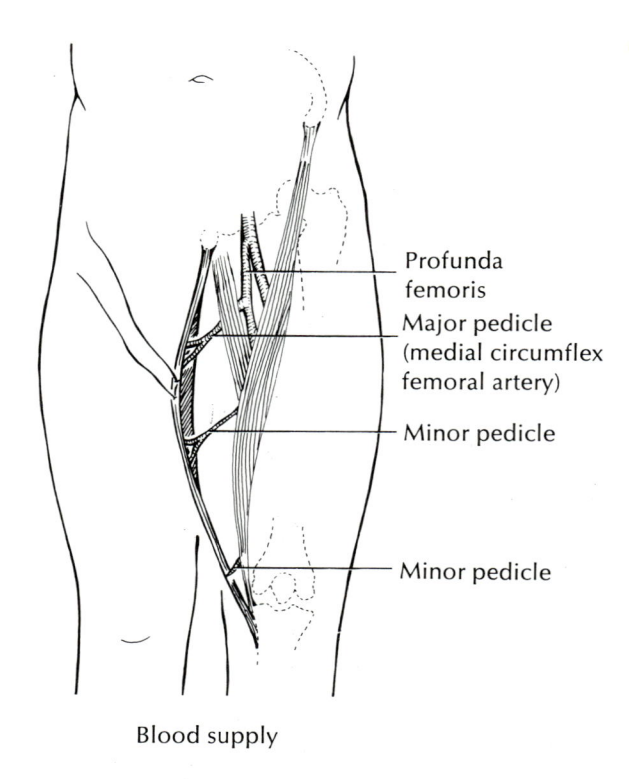

Blood supply

Blood supply

The gracilis has a single dominant vascular pedicle. This major pedicle (medial circumflex femoral artery) is a branch of the profunda femoris artery. The medial femoral circumflex artery enters the upper third of the medial muscle belly approximately 10 cm inferior to the pubic tubercle. Since this dominant pedicle is located beneath the adjacent adductor longus muscle, this pedicle is protected by the adductor muscle during dissection. By medial retraction of the adductor longus muscle, the pedicle may be visualized and preserved. When free flap transfer is considered, the medial circumflex femoral artery can be dissected to its origin from the profunda femoris artery for greater pedicle length and lumen diameter.

The minor pedicles shown are direct branches from the superficial femoral artery. A small branch of the obturator artery enters the muscle adjacent to its origin. All minor pedicles may be safely divided as the entire unit will survive on its dominant major pedicle.

ANTERIOR THIGH

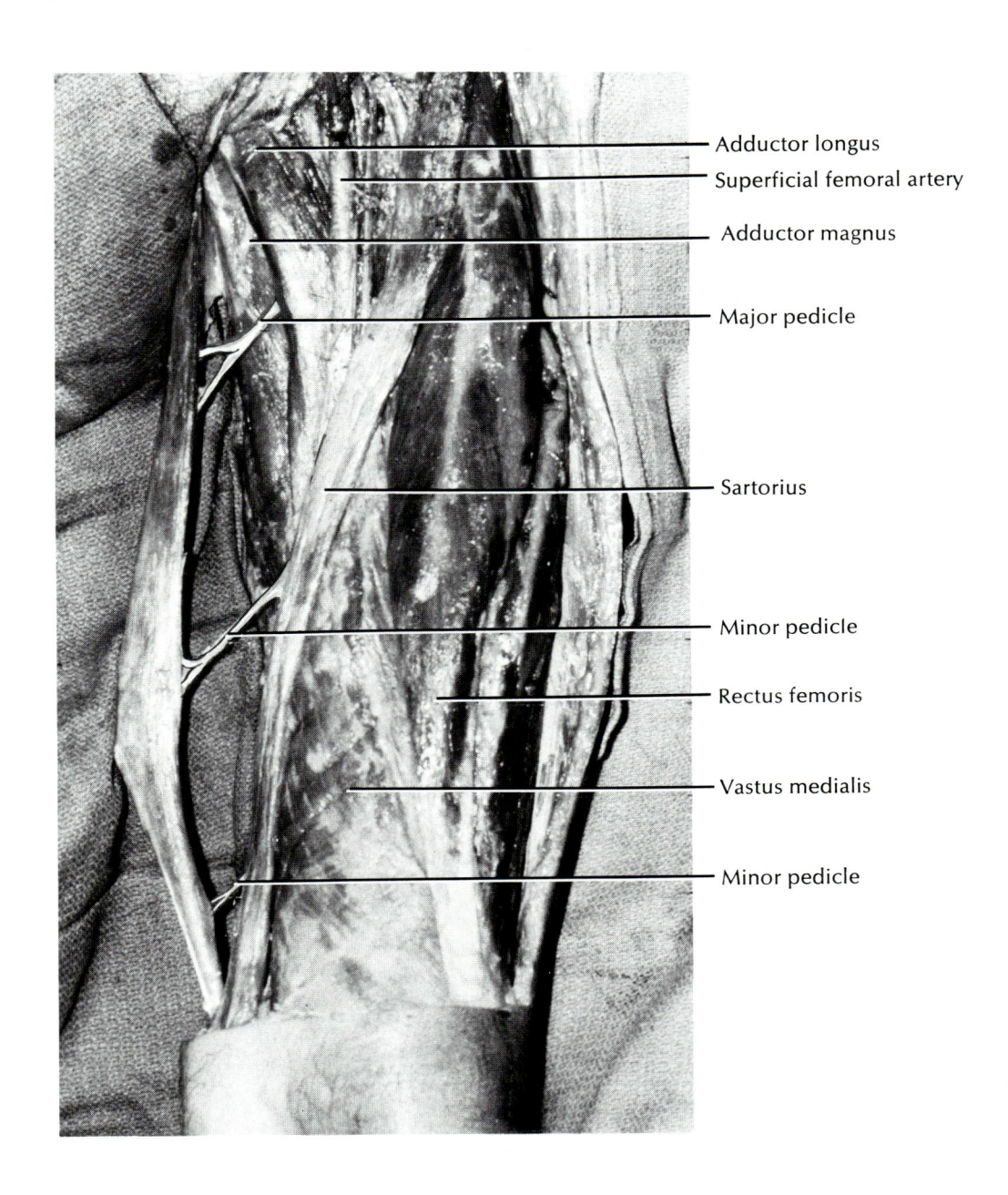

ANTERIOR THIGH

Gracilis

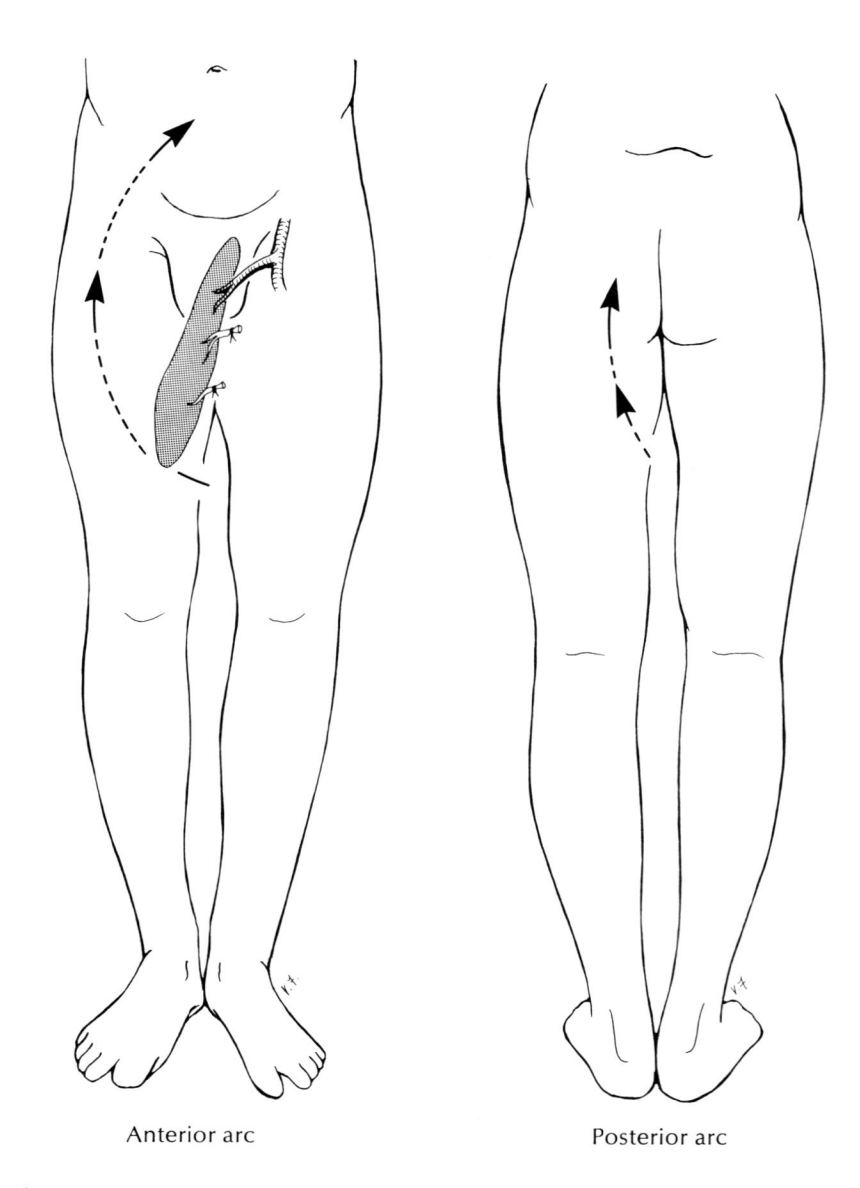

Anterior arc   Posterior arc

Arc of rotation

ANTERIOR ARC

The proximal point of rotation of this unit is the major pedicle, which is located 10 cm inferior to the pubic tubercle. After division of the minor pedicles and muscle insertion, the anterior arc will reach the groin, genitalia, and lower abdomen.

POSTERIOR ARC

The posterior arc of rotation of the gracilis muscle will reach the thigh, perineum, anus, ischium, and buttocks.

ANTERIOR THIGH

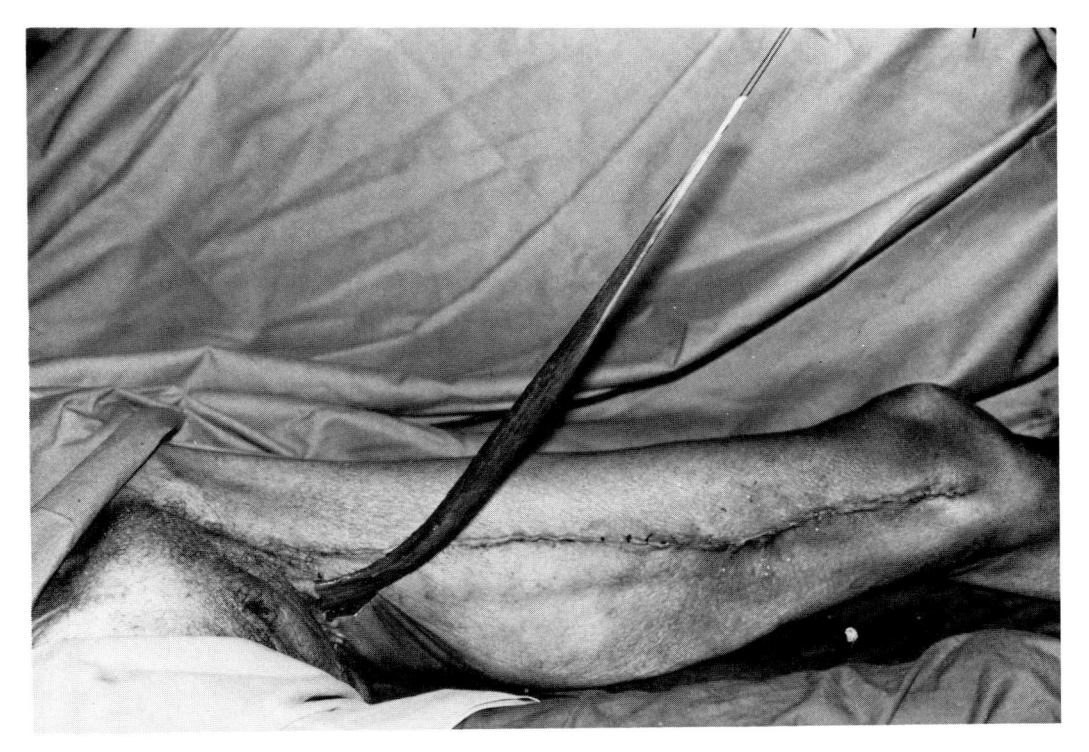

Anterior arc

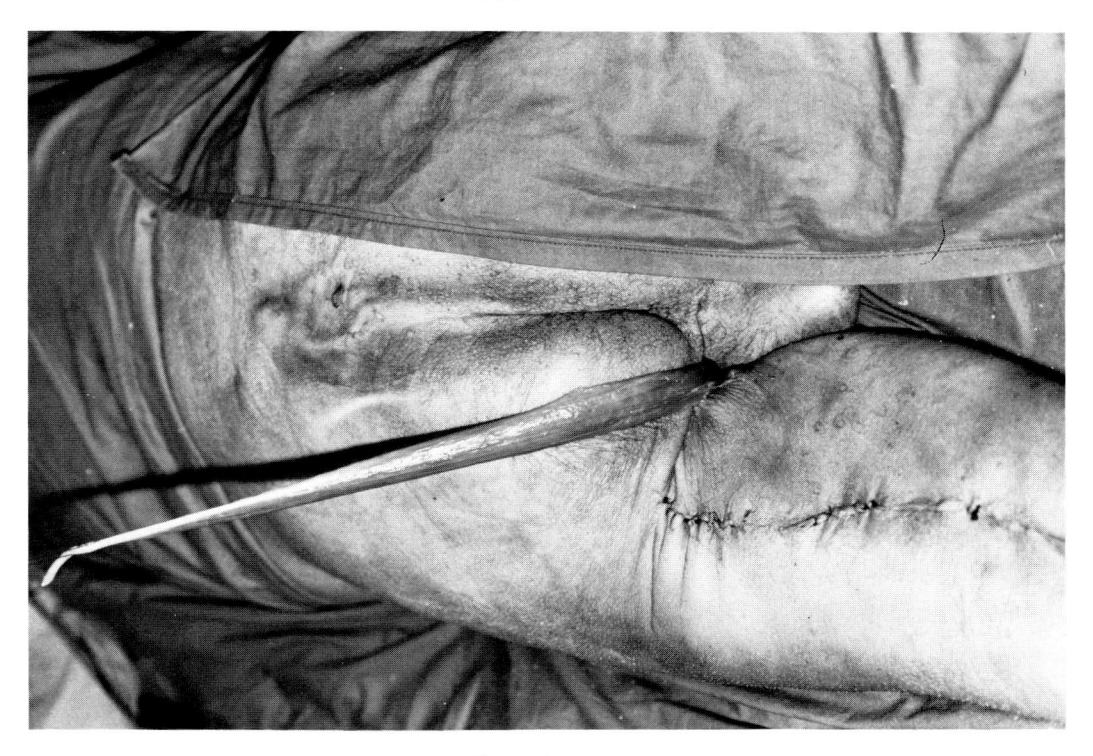

Posterior arc

ANTERIOR THIGH
Gracilis

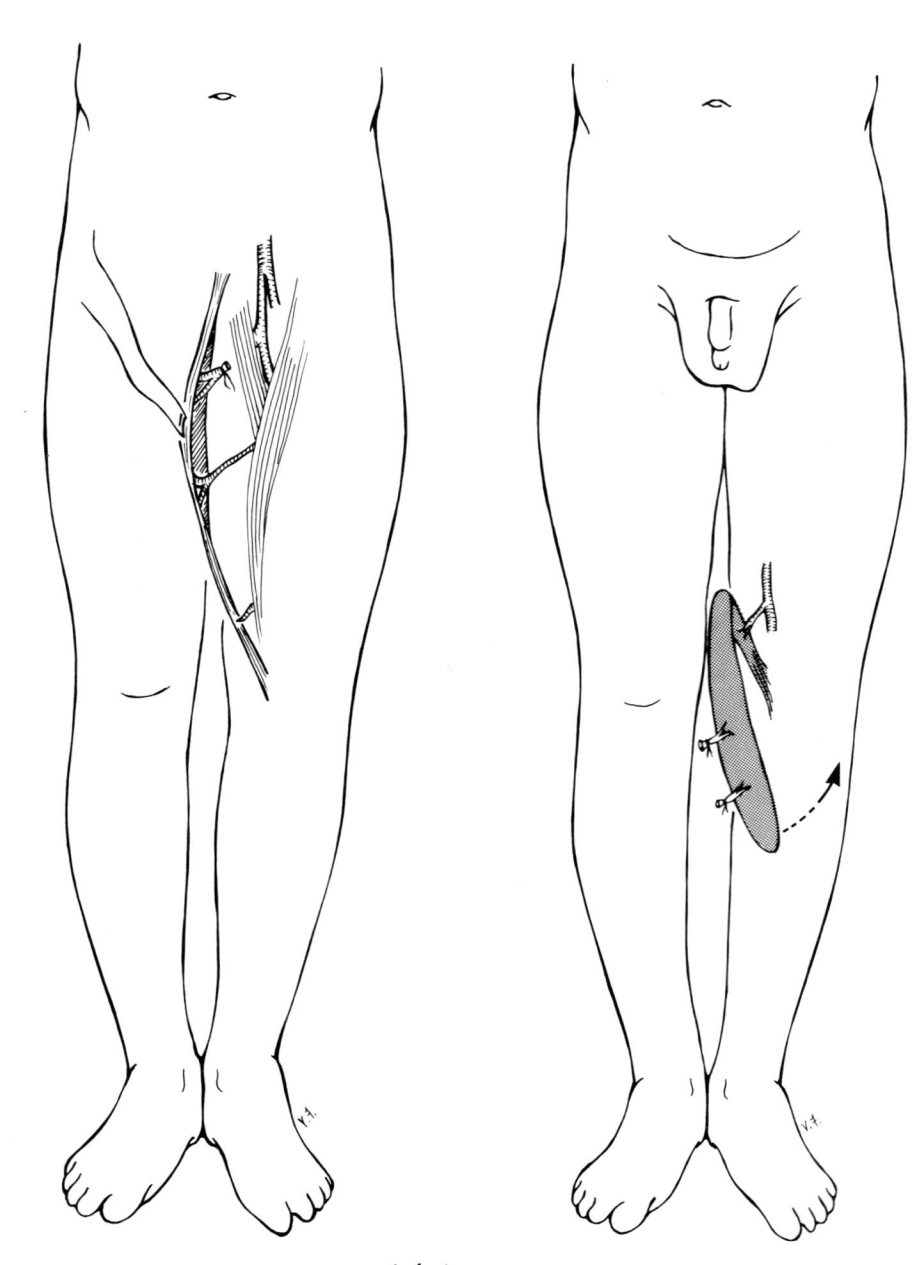

Inferior arc

20

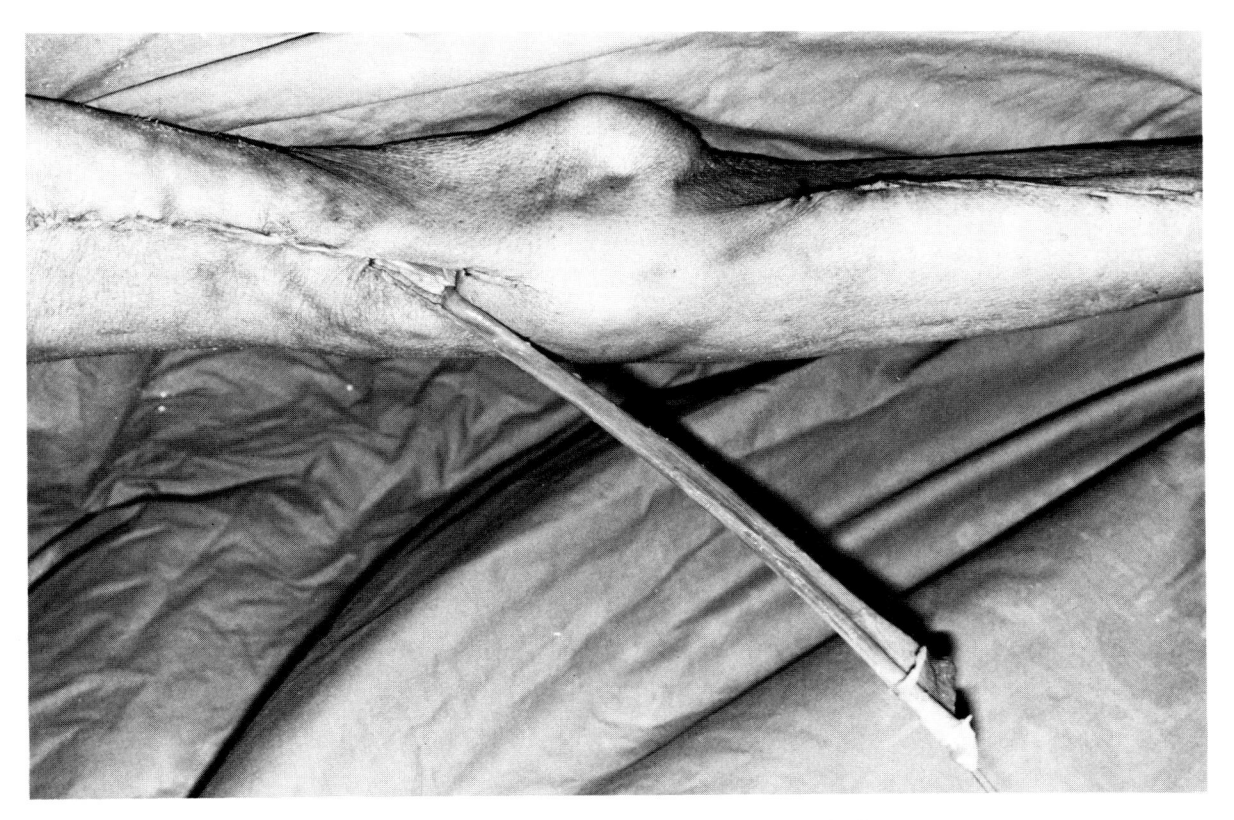

ANTERIOR THIGH

Inferior arc for distally based delayed muscle flap

INFERIOR ARC

Preliminary data indicates the potential use of the inferior minor pedicle as the point of rotation. Under normal circumstances the blood flow via the inferior pedicle to the gracilis is inadequate to support the unit following division of the major pedicle. However, if the major pedicle is divided as shown, 2 or 3 weeks in advance of muscle transposition, an inferior arc of rotation on the minor pedicle may be possible following this strategic delay. The gracilis unit could have application for coverage of defects on the knee and popliteal fossa.

ANTERIOR THIGH

Gracilis

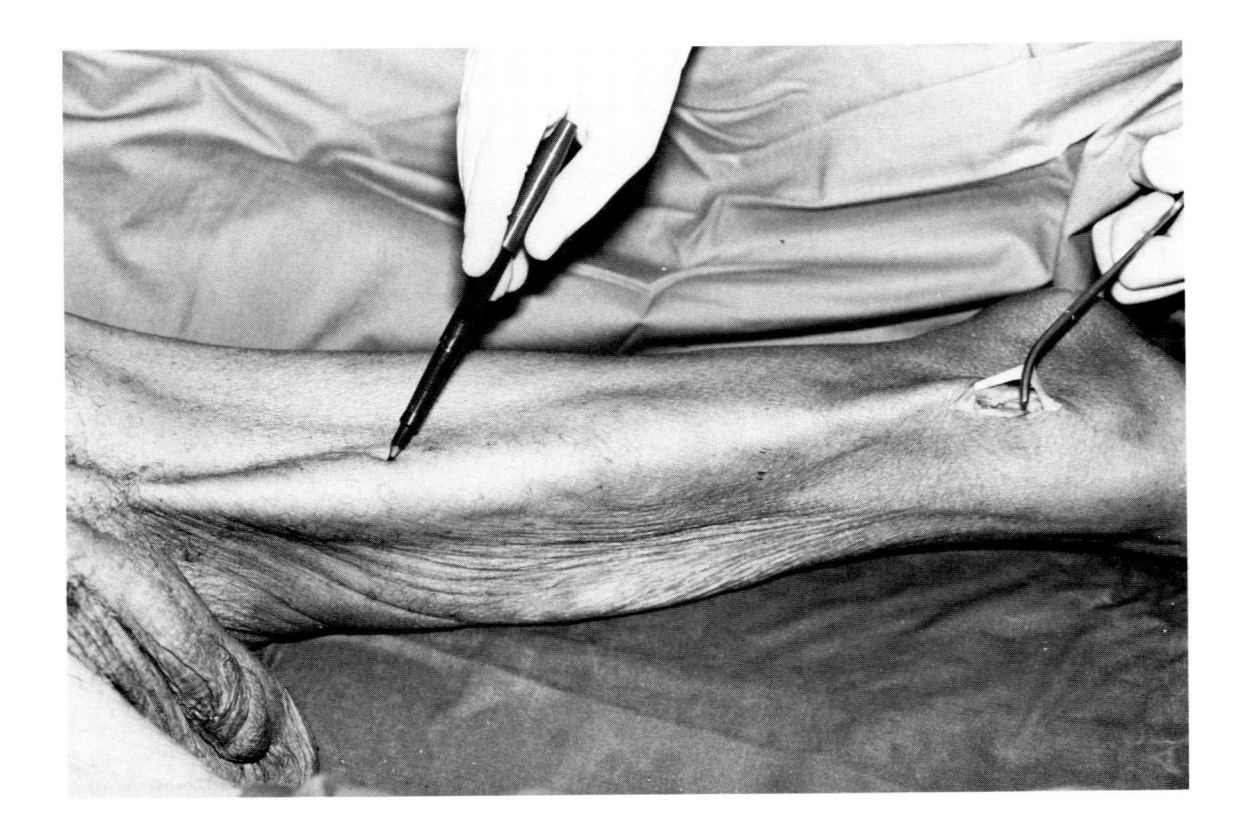

Elevation of flap

The key to successful elevation of the gracilis muscle is accurate localization of the muscle near its insertion. The muscle is located posterior to a line connecting the pubic tubercle with the medial tibial condyle. An incision over the distal portion of the predicted location of the muscle will allow accurate muscle localization. At this level the musculotendinous portion of the gracilis muscle lies between the longitudinally oriented muscular fibers of the sartorius anteriorly and the fascial expanse of the semimembranosus posteriorly. By traction of the distal gracilis muscle the cutaneous territory can then be accurately outlined when this muscle is used as a musculocutaneous flap. This maneuver is essential in the obese patient, since the skin is very mobile in the medial thigh and the gracilis cutaneous territory cannot be safely predicted by external anatomic landmarks.

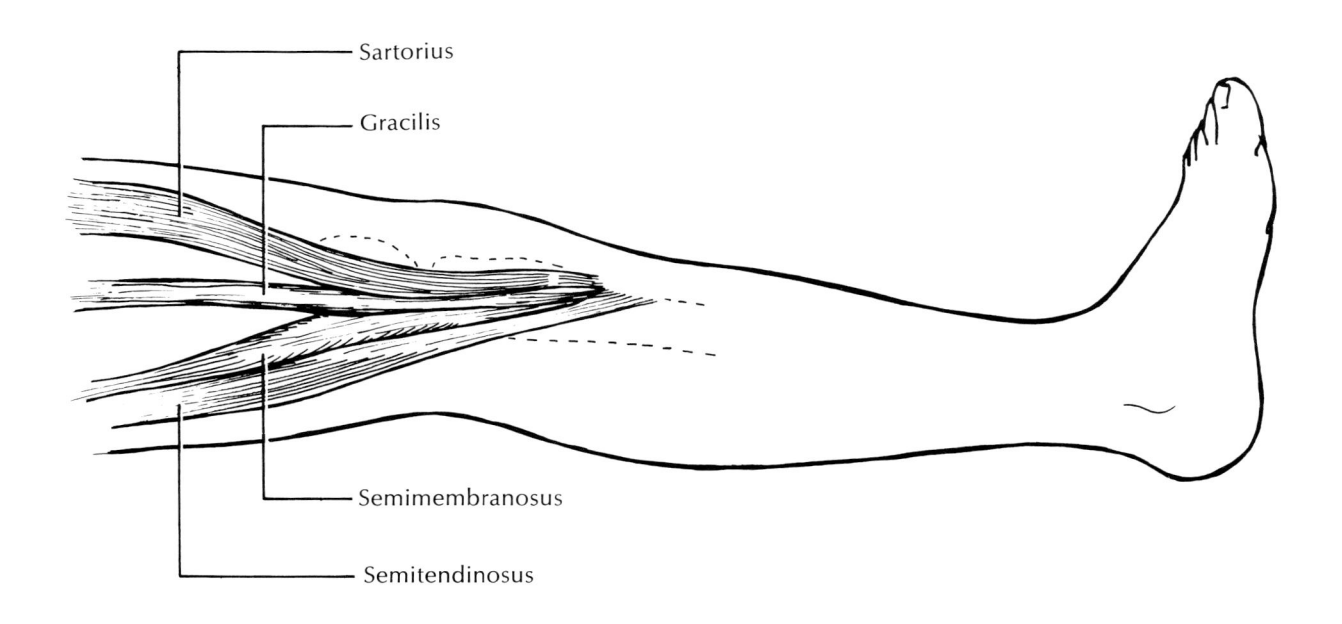

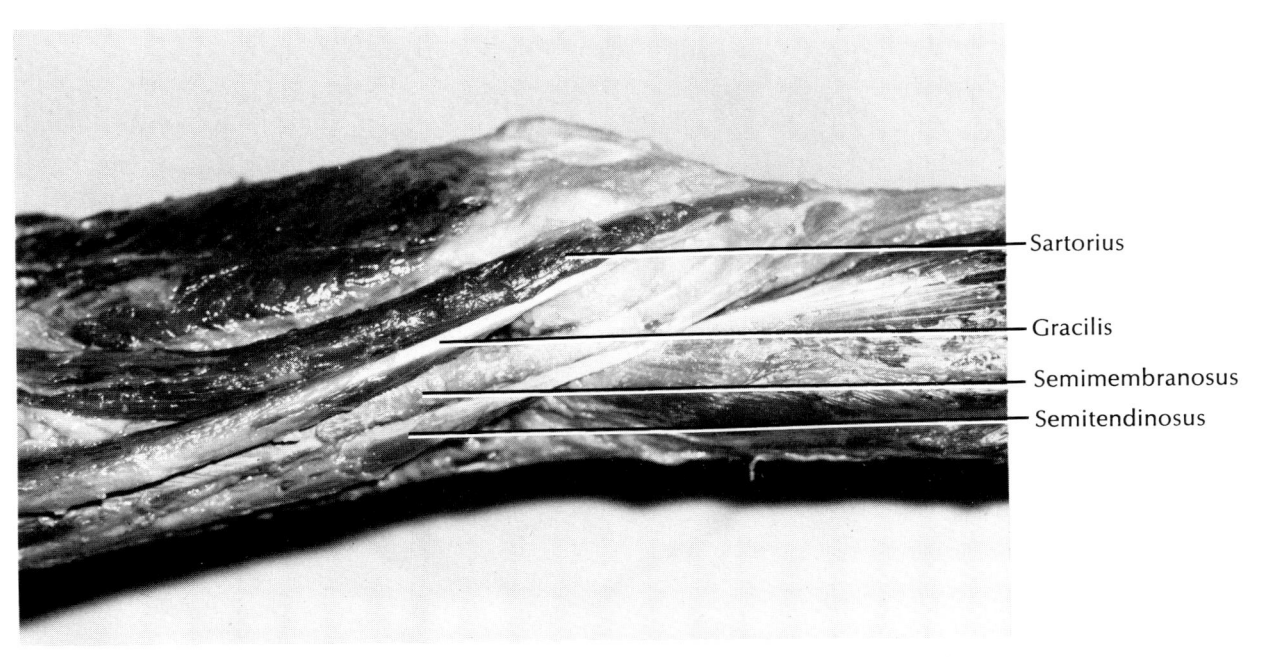

ANTERIOR THIGH

Sartorius
Gracilis
Semimembranosus
Semitendinosus

Sartorius
Gracilis
Semimembranosus
Semitendinosus

Round tendon of gracilis lies between muscular sartorius anteriorly and fascial tendon of semimembranosus posteriorly

23

ANTERIOR THIGH

Gracilis

The distal cutaneous territory of the gracilis muscle is not completely reliable when the flap is rotated superiorly on the dominant vascular pedicle. For this reason the cutaneous territory is based over the proximal two-thirds of the muscle and tapered at the ends to allow primary closure of the donor defect. The flap width extends 2 or 3 cm beyond the medial and lateral borders of the gracilis muscle. The musculocutaneous flap can then either be designed with the skin as an island or with the proximal skin bridge left intact, depending on the planned arc of rotation.

After suture of the dermis of the cutaneous territory to the gracilis muscle edges to prevent disruption of arterial perforators, the muscle is divided at its insertion. The muscle is retracted medially and the minor pedicles divided. In the proximal third of the muscle the adductor longus muscle is retracted in a lateral direction, allowing rapid identification of the major pedicle. For greater arc length the pedicle can be mobilized as necessary. The muscle origin is not divided unless the muscle or musculocutaneous unit is to be used in free transfer by microvascular techniques. The donor defect, following transposition of either a muscle or musculocutaneous gracilis flap, can be closed primarily.

Precautions

- Accurate location of the muscle belly at its insertion is essential for safe elevation.
- Careful determination of the cutaneous territory.
- Suture of cutaneous territory to muscle to avoid disruption of perforators.
- Avoid excessive traction on the dominant pedicle during transposition.

ANTERIOR THIGH

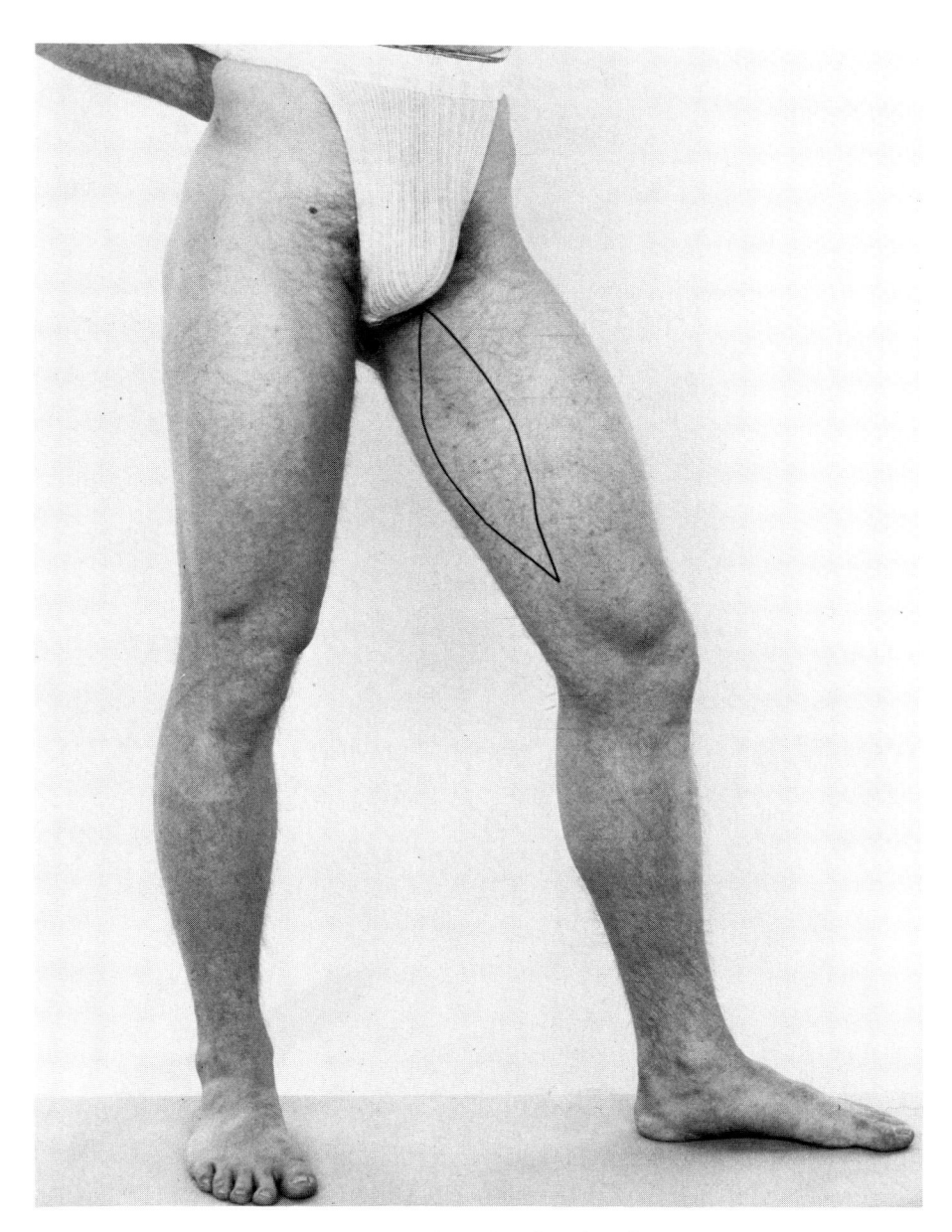

Gracilis skin island outlined

25

ANTERIOR THIGH
Gracilis

Coverage of ischial pressure ulcer

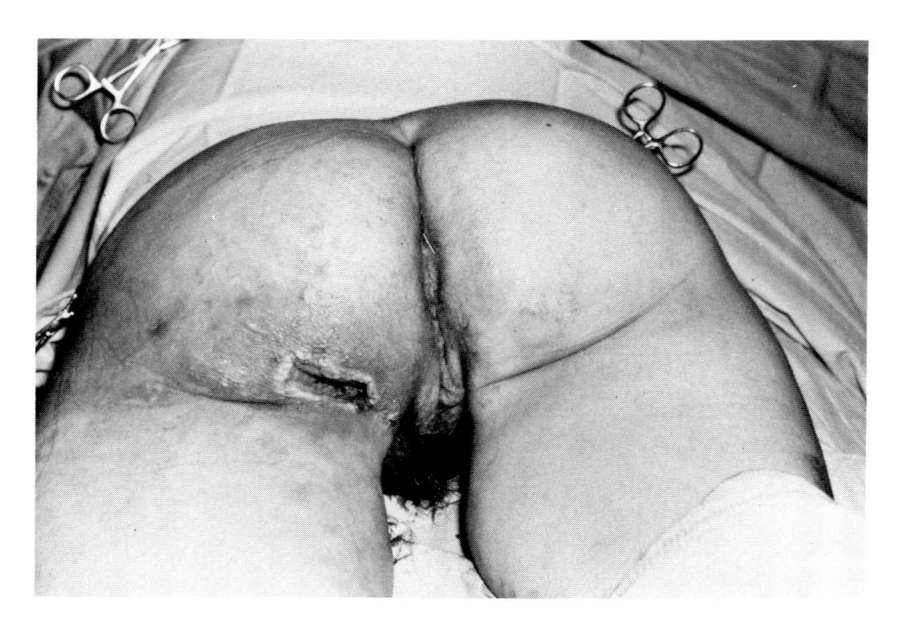

Unilateral ischial pressure ulcer.

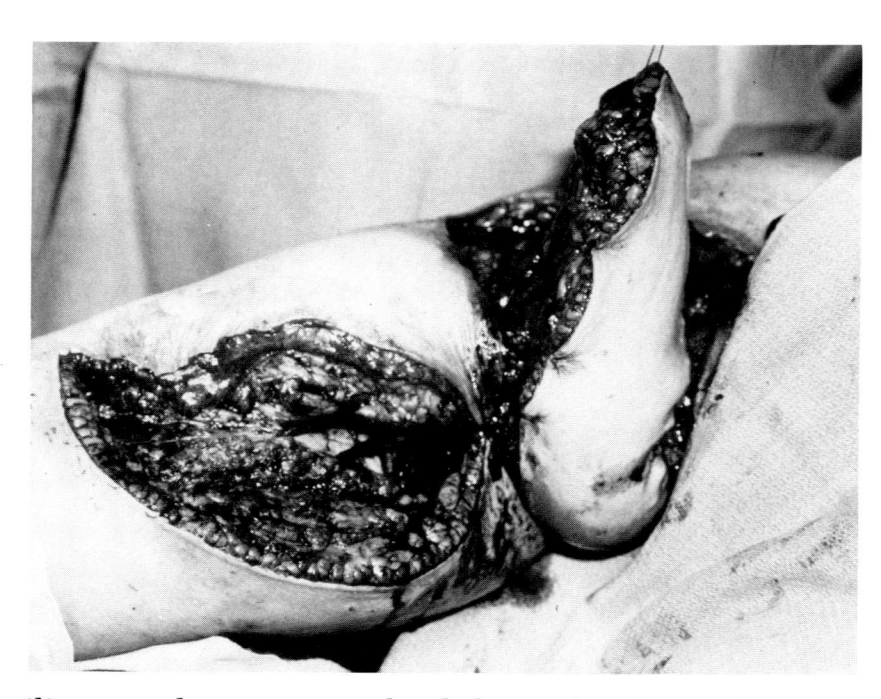

Gracilis musculocutaneous island elevated and rotated posteriorly.

ANTERIOR THIGH

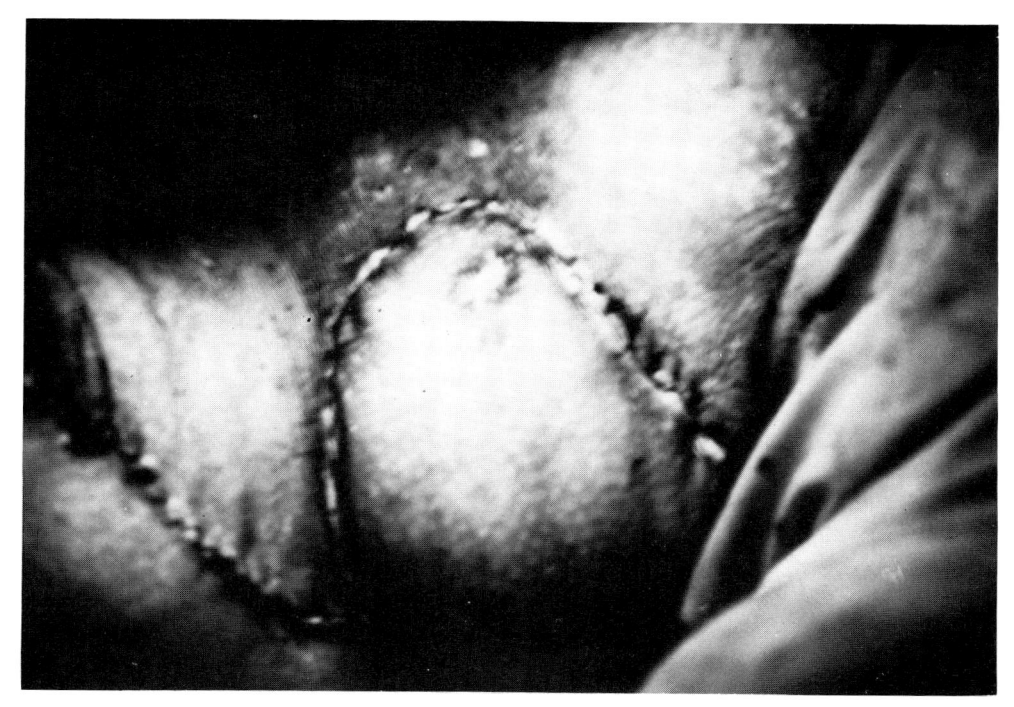

Adequate fluorescence confirms viability of skin island.

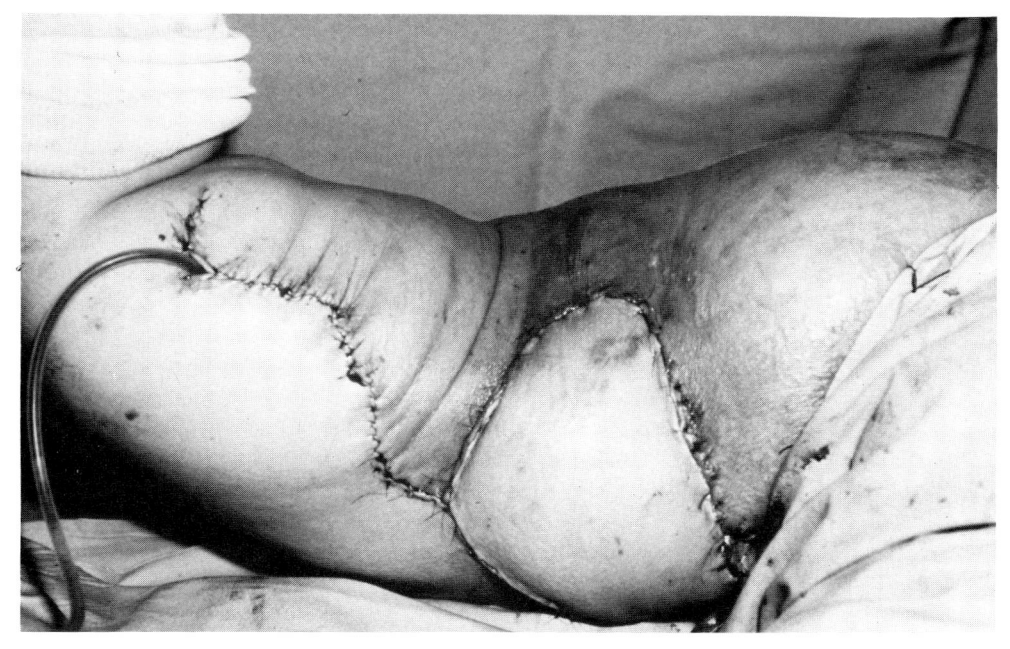

Flap sutured and donor defect closed directly.

ANTERIOR THIGH

Gracilis

Reconstruction of the vagina

Absence of pelvic musculature and vagina following radical hysterectomy.

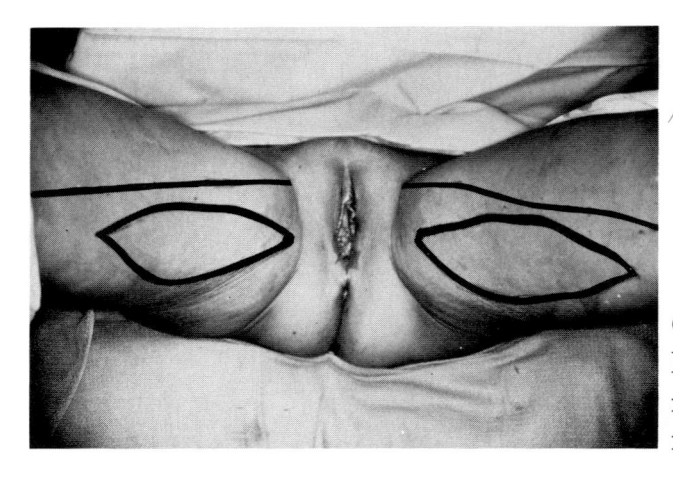

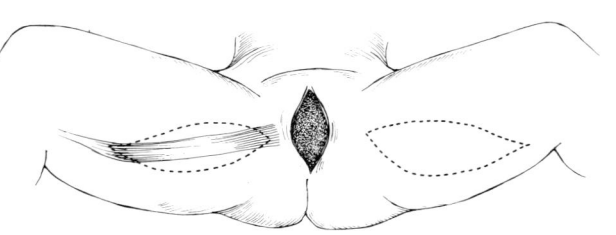

Cutaneous territory marked for bilateral gracilis musculocutaneous flaps. Territory is inferior to line connecting pubis with medial condyle.

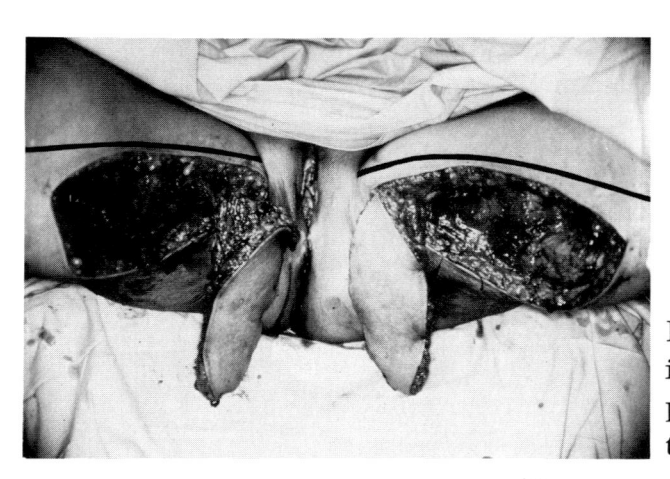

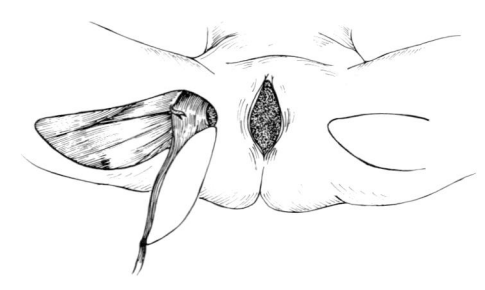

Bilateral gracilis musculocutaneous island flap elevated with proximal pedicle intact and muscle origin intact.

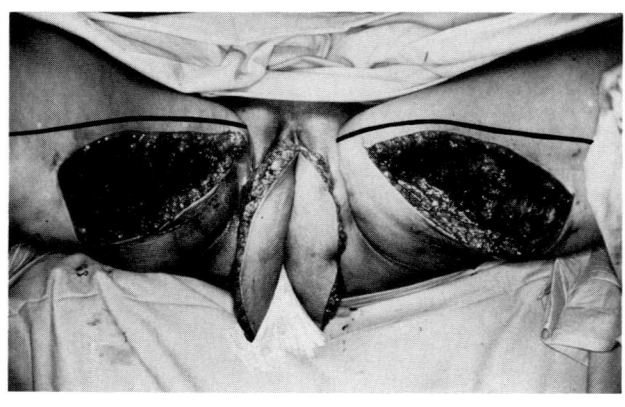

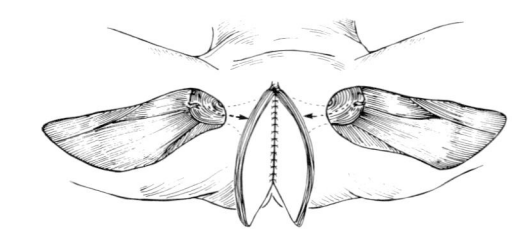

Musculocutaneous flaps passed through subcutaneous tunnel into perineal region.

ANTERIOR THIGH

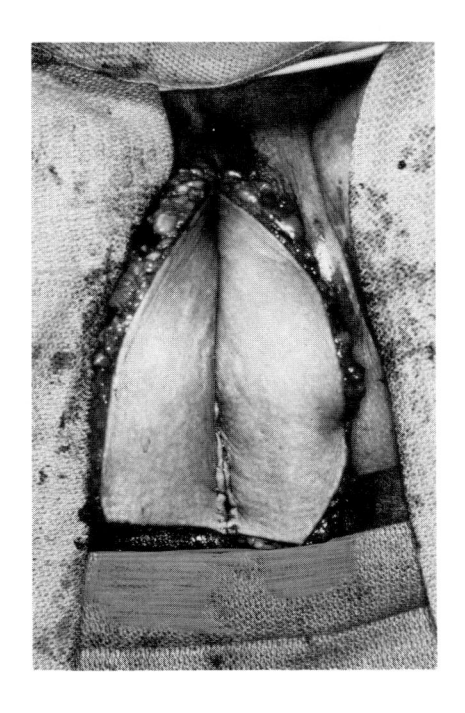

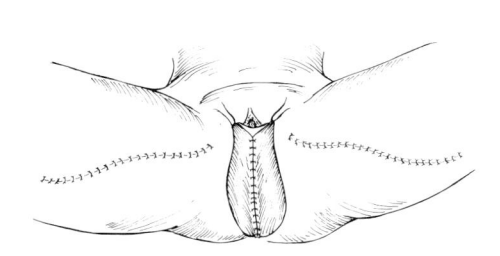

Bilateral cutaneous territories sutured to form reconstructed vagina. Muscle is then sutured to encircle neovagina.

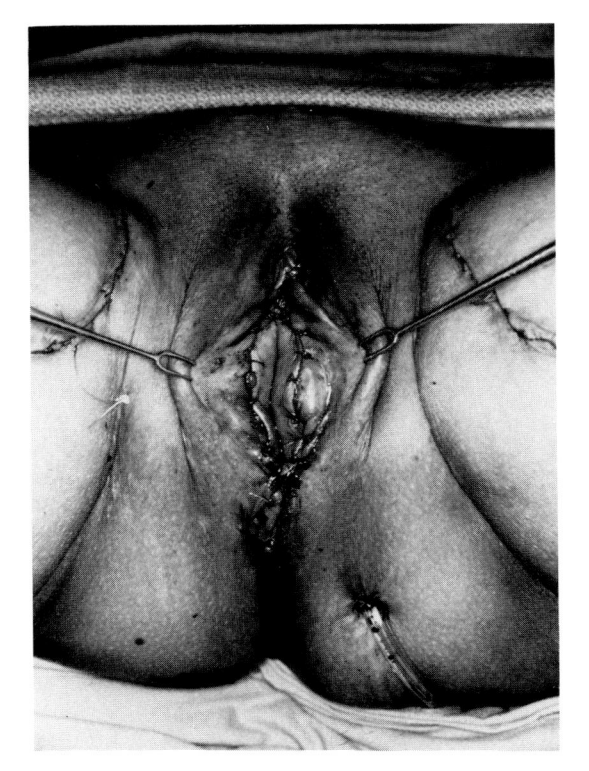

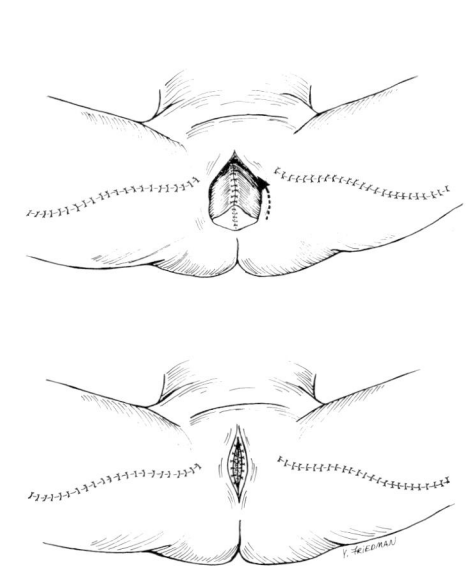

Reconstructed vagina is placed into perineal defect. Skin edges are sutured to labial skin to complete single-stage vaginal reconstruction.

ANTERIOR THIGH

Gracilis

Reconstruction of the penis (bilateral muscle flaps)

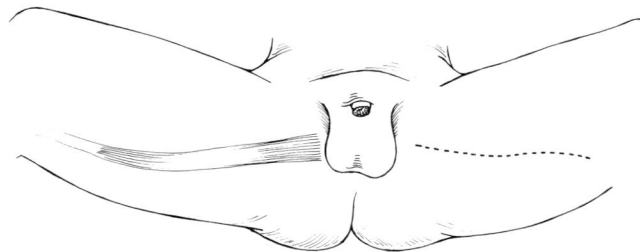

Skin incision.

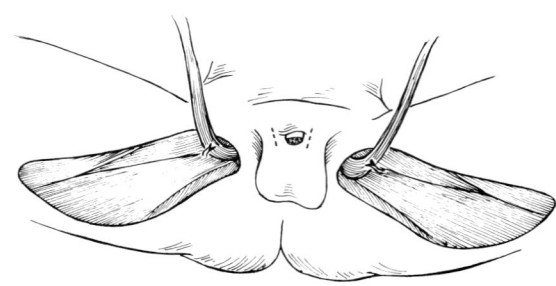

Bilateral muscle flaps elevated.

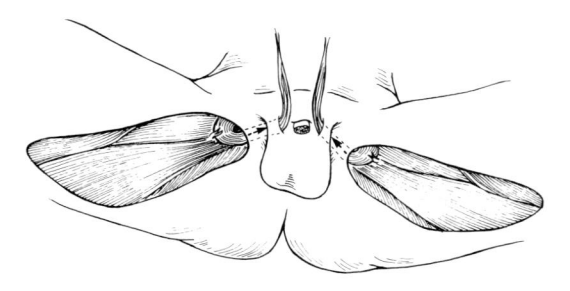

Muscles are tunnelled through.

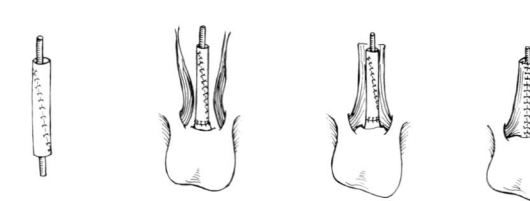

Reverse skin tube for urethra reconstruction is covered with bilateral gracilis muscle flaps.

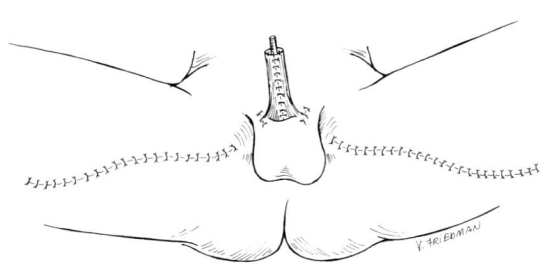

Single-stage penile reconstruction with bilateral gracilis muscle flaps. Skin-graft coverage of muscle completes penile reconstruction.

ANTERIOR THIGH

Reconstruction of the penis (musculocutaneous flap)

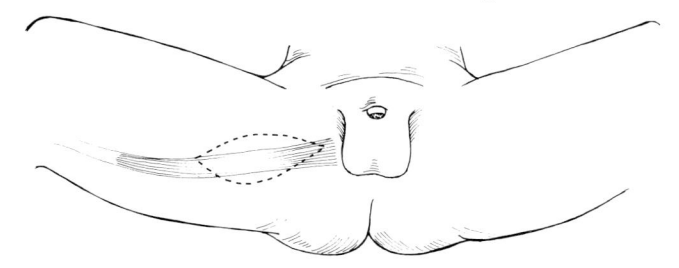

Flap outlined.

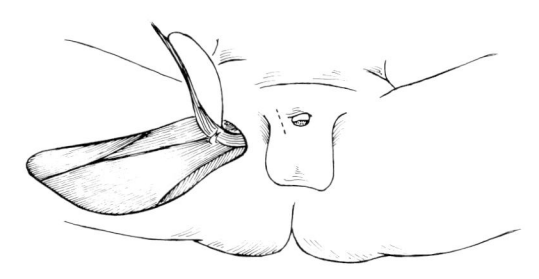

Flap elevated.

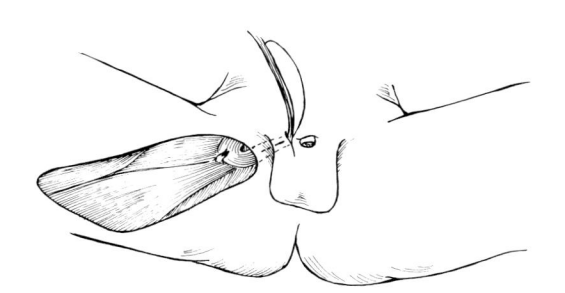

Flap tunnelled through.

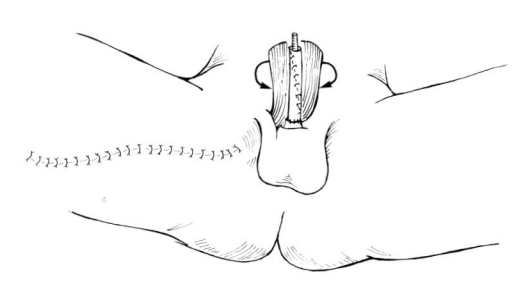

Muscle wrapped around urethral skin graft. Secondary defect closed directly.

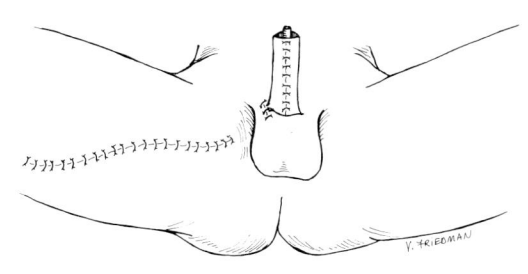

Single-stage penile reconstruction with single gracilis musculocutaneous flap.

ANTERIOR THIGH

Sartorius

MUSCLE FLAP

Applications

    Coverage of:

        GROIN
        FEMORAL VESSELS
        KNEE REGION

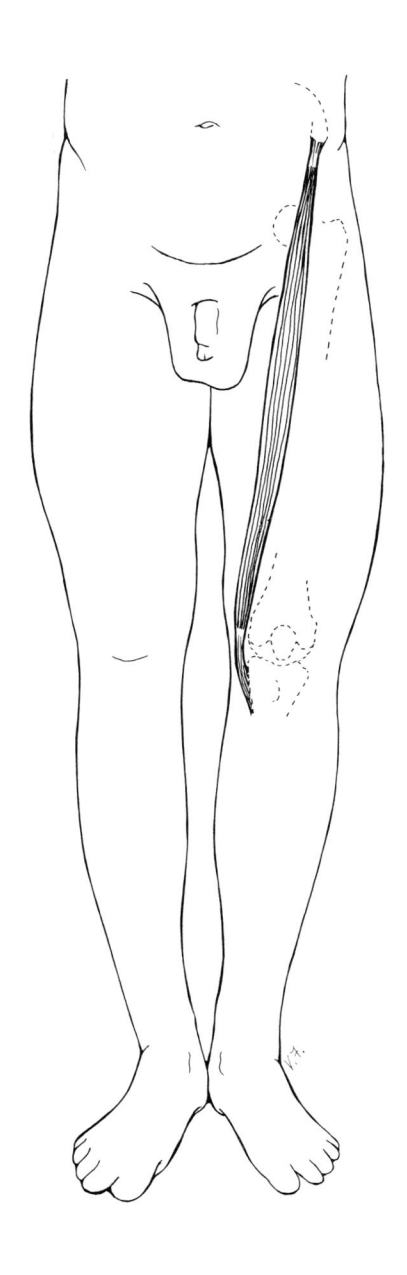

ANTERIOR THIGH

Sartorius

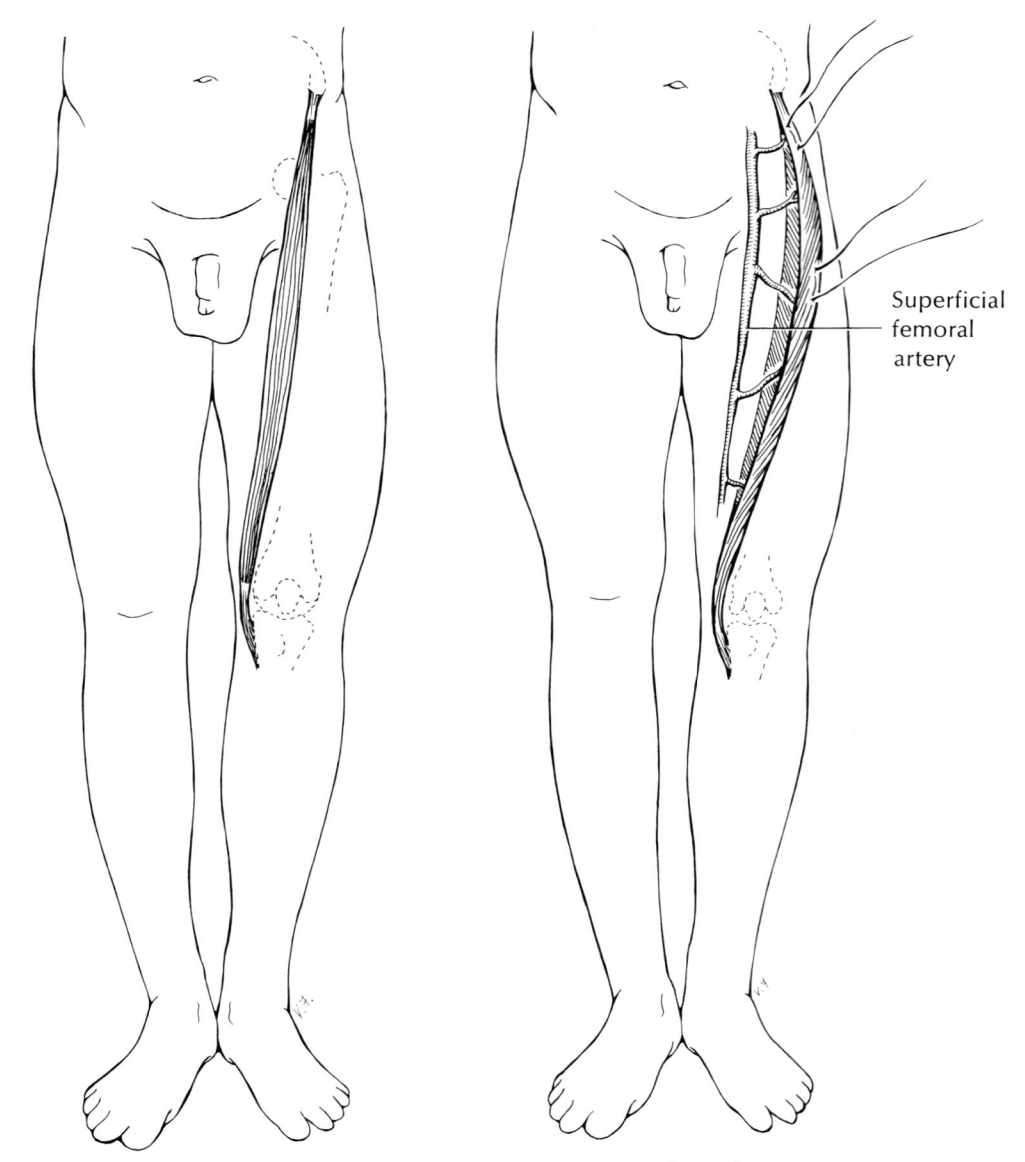

Blood supply

The sartorius is a thin, flat muscle that lies superficially in the thigh and courses from lateral to medial. It is related to the TFL and rectus femoris at its origin and the gracilis at its insertion.

Origin: Anterior superior iliac spine.
Insertion: Medial tibial condyle.
Nerve supply:
   Motor — Branch of femoral nerve.
Function: An expendable muscle that is a lateral rotator and flexor of the thigh.

ANTERIOR THIGH

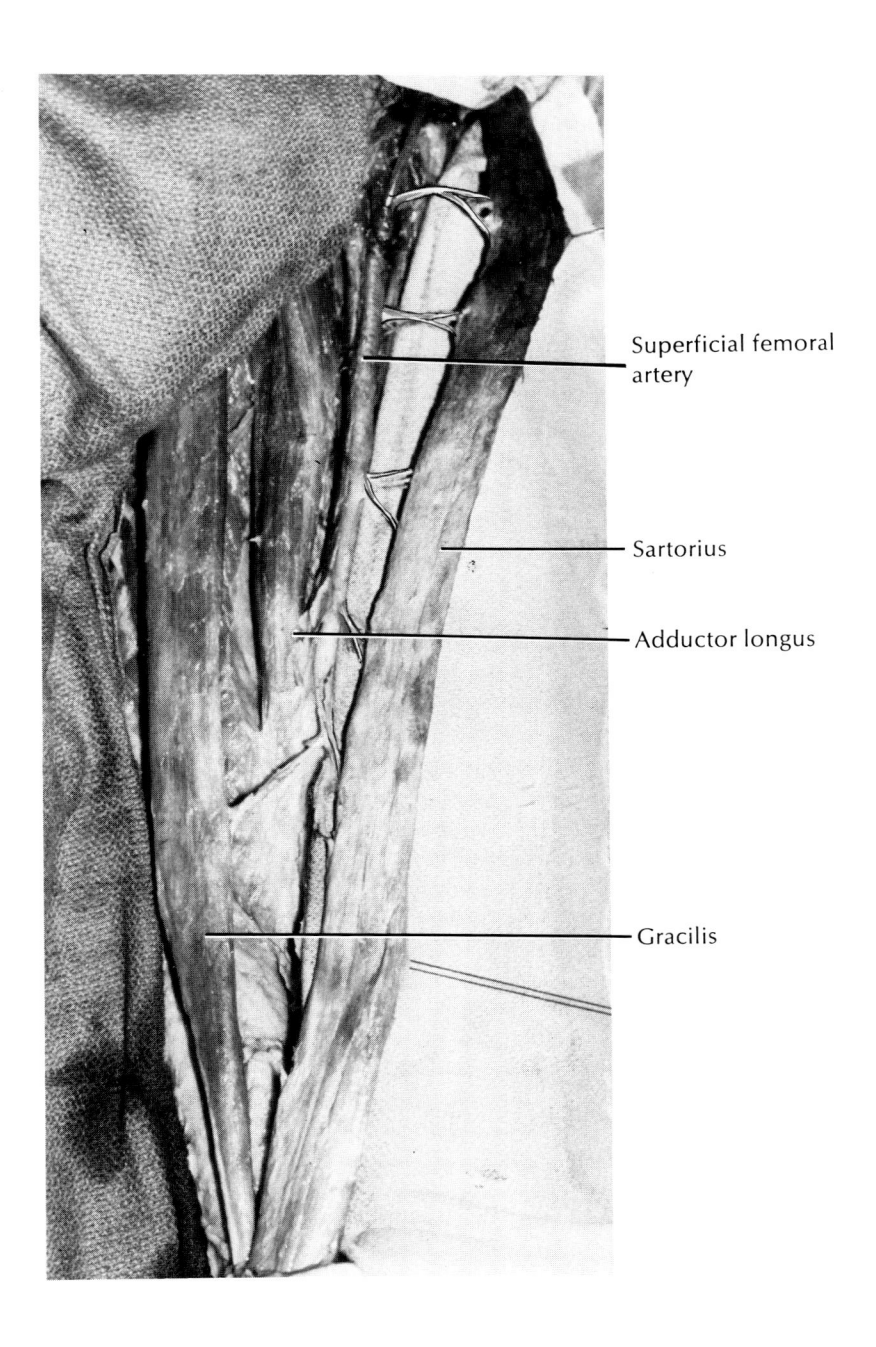

- Superficial femoral artery
- Sartorius
- Adductor longus
- Gracilis

Blood supply

The blood supply is segmental with five to six direct branches from the superficial femoral artery.

ANTERIOR THIGH

Sartorius

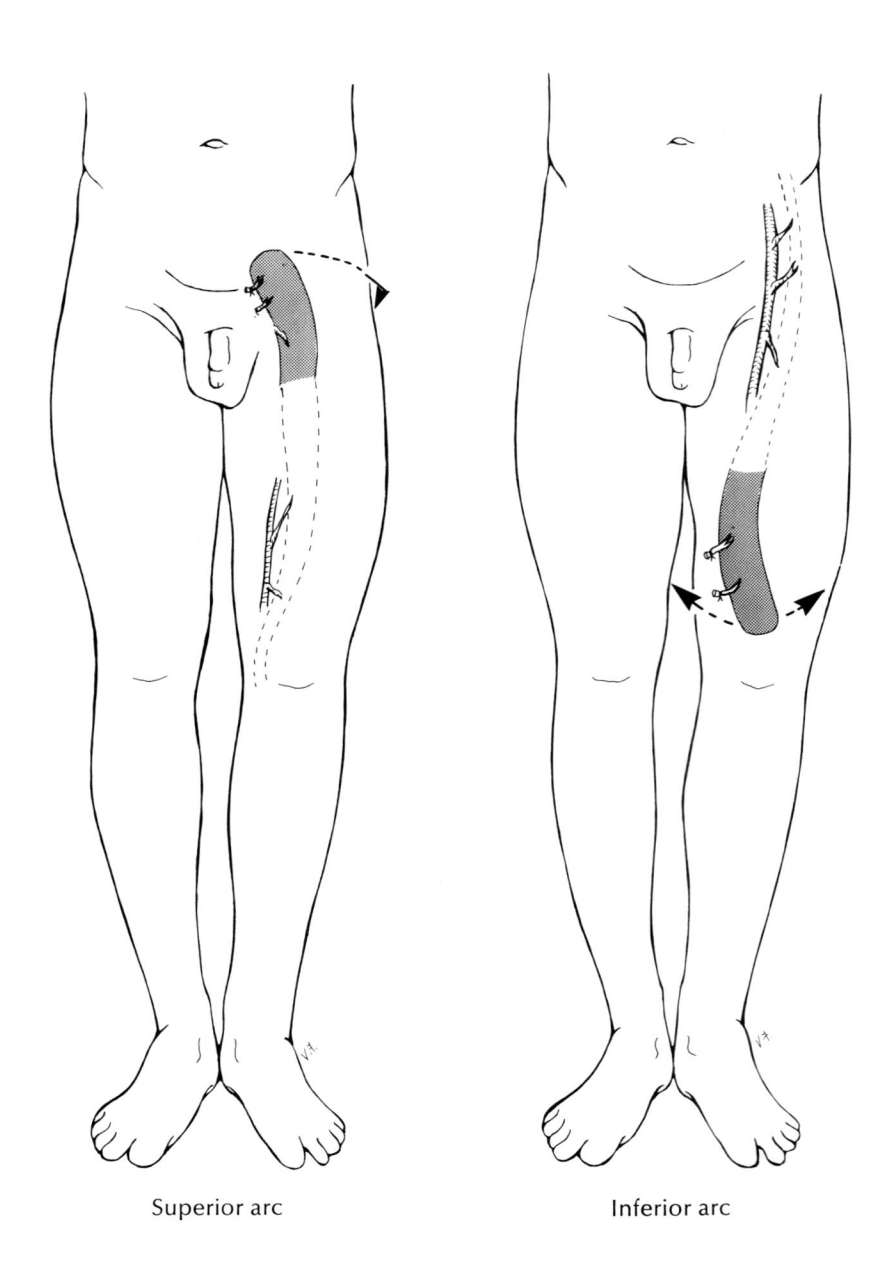

Superior arc

Inferior arc

36

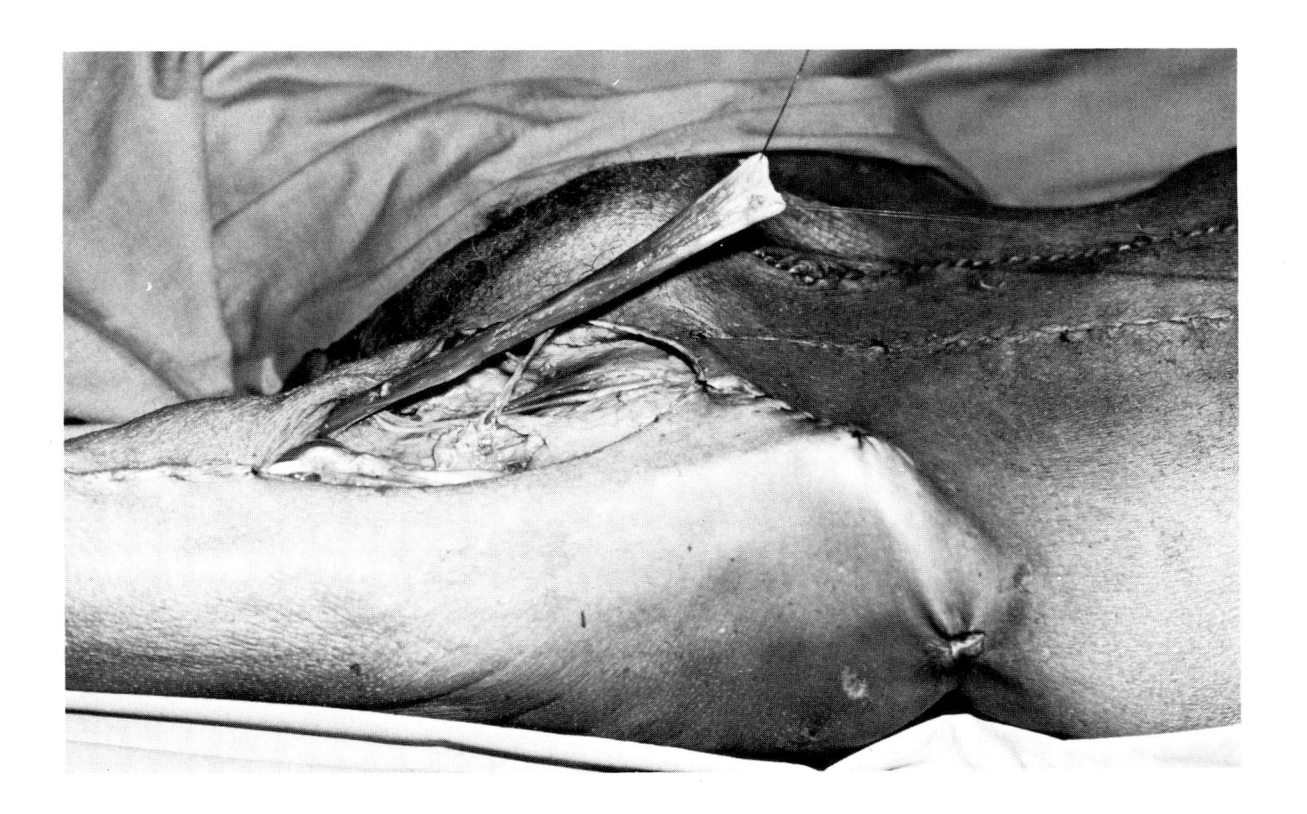

Arc of rotation

The segmental blood supply limits the use of this muscle.

SUPERIOR ARC

With ligation of the proximal one or two pedicles, the origin of the muscle can be transposed into the inguinal region.

INFERIOR ARC

Similarly the insertion of the muscle can be transposed for coverage of small defects in the knee region.

ANTERIOR THIGH

Sartorius

Elevation of flap

Following radical groin lymphadenectomy, the femoral vessels are exposed. The superior origin of the sartorius muscle is generally visualized in the region of the femoral vessels. By division of the origin of the muscle and the two proximal vascular pedicles, the muscle can be transposed and sutured to the inguinal ligament to provide vascularized muscle coverage of the femoral vessels. Using this technique, the femoral vessels can be protected by the superior arc of the sartorius muscle. Likewise, in the region of the knee the insertion of the sartorius muscle may be divided and the muscle rotated following division of one or two of its distal vascular pedicles. In this manner, by an inferior arc of rotation, sartorius muscle transposition will provide coverage of small defects around the knee. The segmental blood supply of the sartorius muscle makes this a poor musculocutaneous unit. Following simple muscle transposition, the donor defect can be closed primarily.

Precautions

- The arc of rotation is short because of the segmental blood supply.
- Division of more than two adjacent pedicles will result in devascularization.
- This is not recommended as a musculocutaneous unit, as the skin territory is small.

ANTERIOR THIGH

Coverage of femoral vessels

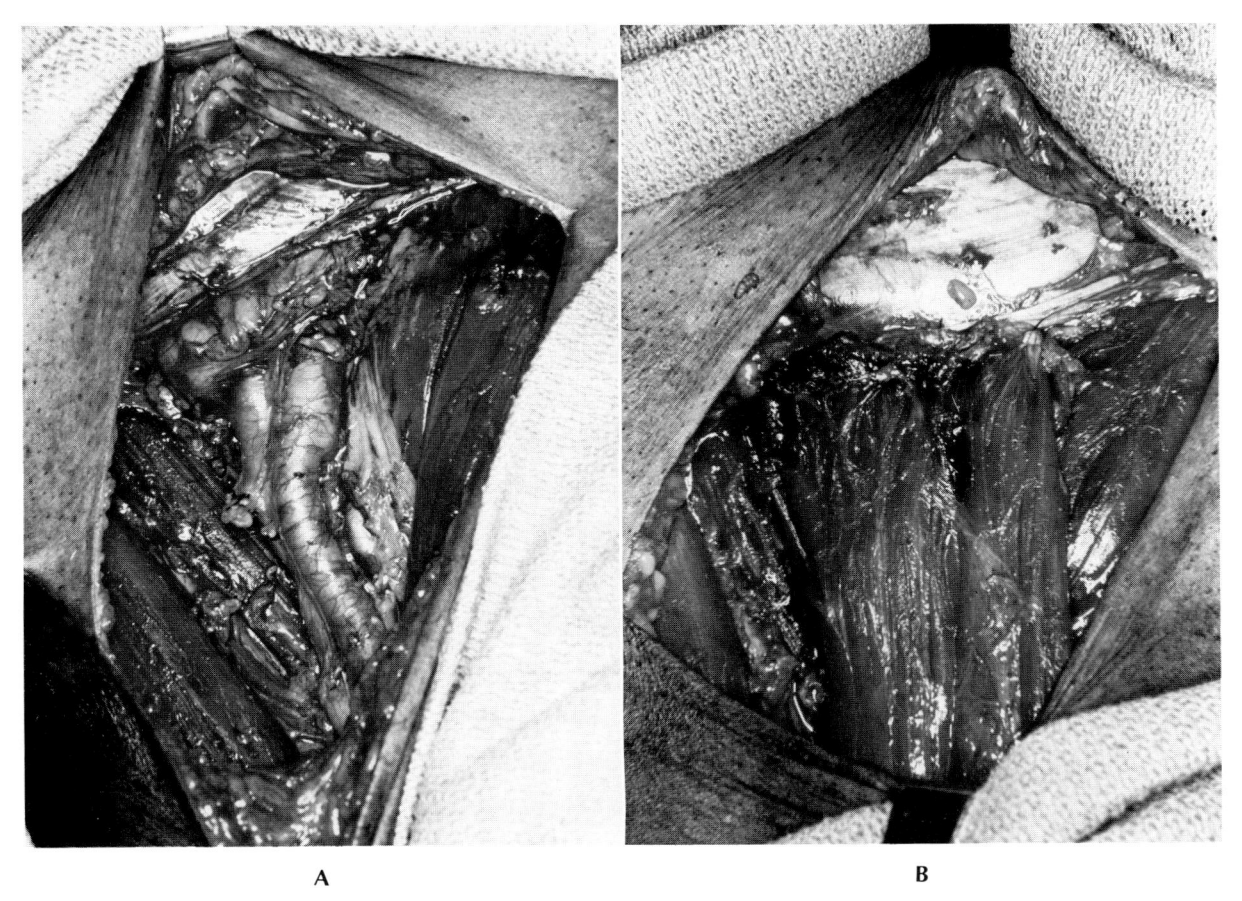

A   Exposed femoral vessels following groin dissection.
B   Vessels covered by transposed sartorius.

ANTERIOR THIGH

Rectus femoris

MUSCLE FLAP
MUSCULOCUTANEOUS FLAP
FREE FLAP

Applications

Reconstruction of:
ABDOMINAL WALL

Coverage of:
ABDOMINAL WALL
GROIN
PERINEUM
TROCHANTER
ISCHIUM

Free flap:
DISTANT COVERAGE
FUNCTIONAL MUSCLE
SENSORY FREE FLAP

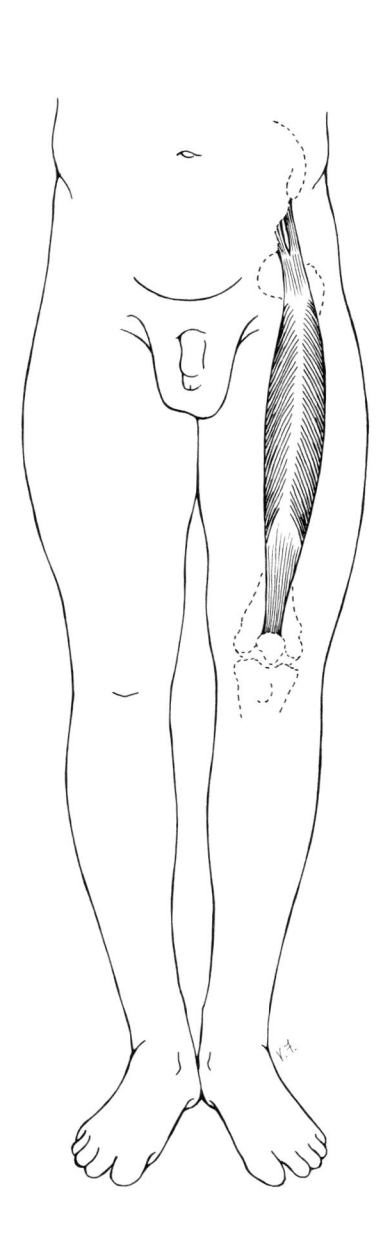

ANTERIOR THIGH

Rectus femoris

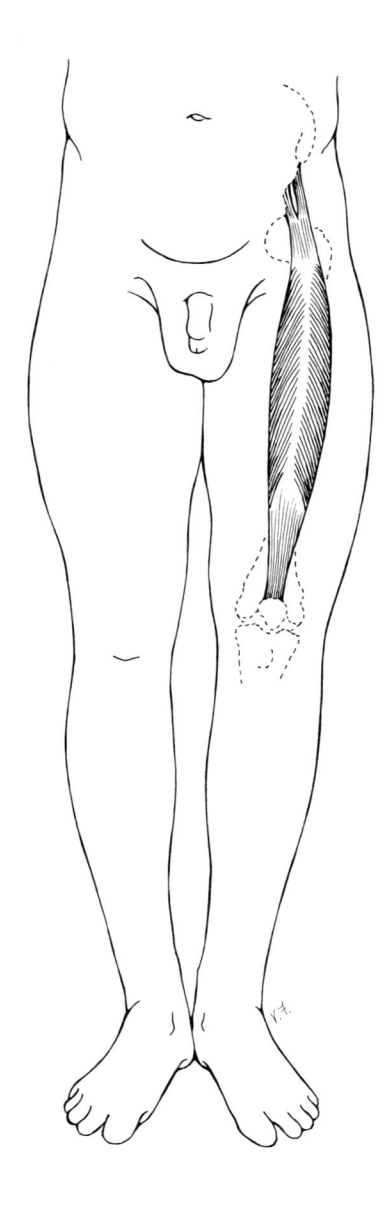

The rectus femoris is a large fusiform muscle located superficially in the central anterior thigh. This muscle lies between the vastus lateralis and medialis and is related to the sartorius and TFL at its origin.

Origin: Inferior iliac spine.
Insertion: Patella.

ANTERIOR THIGH

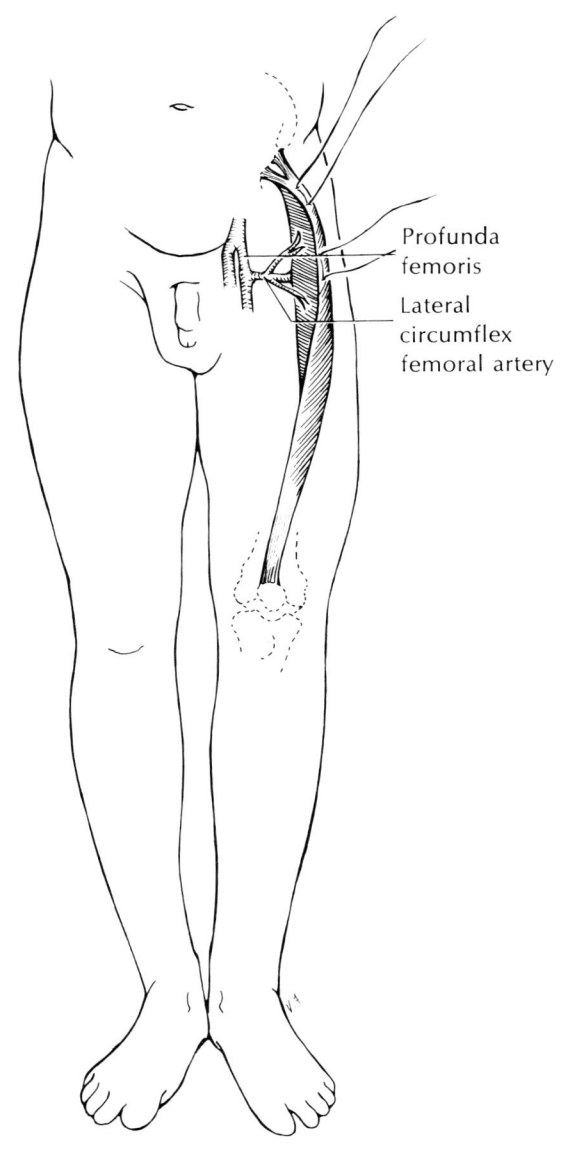

Blood supply

Nerve supply:
    Motor — Muscular branch of the femoral nerve. This motor nerve enters the proximal muscle belly deep to the medial border adjacent to the vascular pedicles.
    Sensory — Anterior cutaneous nerve of thigh.
Function: This is not an expendable muscle. As a strong leg extensor and thigh flexor, the remaining members of the quadriceps femoris should be intact to avoid functional disability.

ANTERIOR THIGH

Rectus femoris

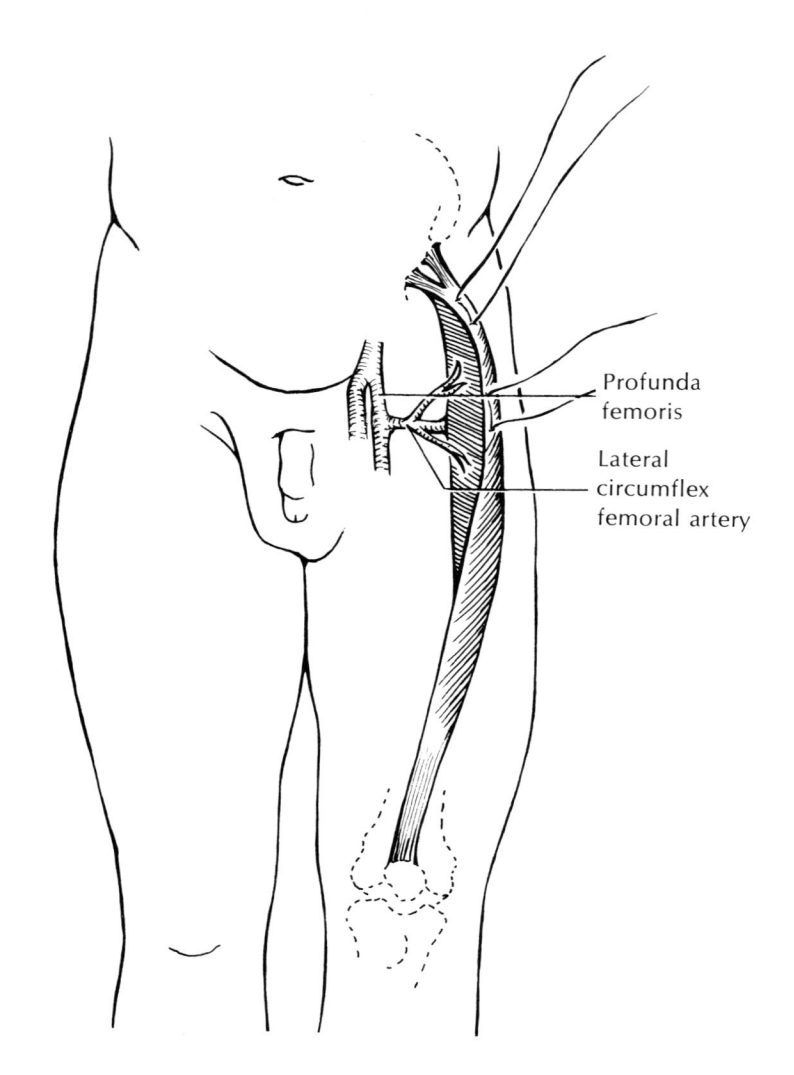

Blood supply

The rectus femoris receives its blood supply from branches of the lateral circumflex femoral artery. Two or three pedicles enter the posterior proximal third of the muscle belly. The major pedicles may be isolated and protected by medial retraction of the proximal sartorius muscle.

ANTERIOR THIGH

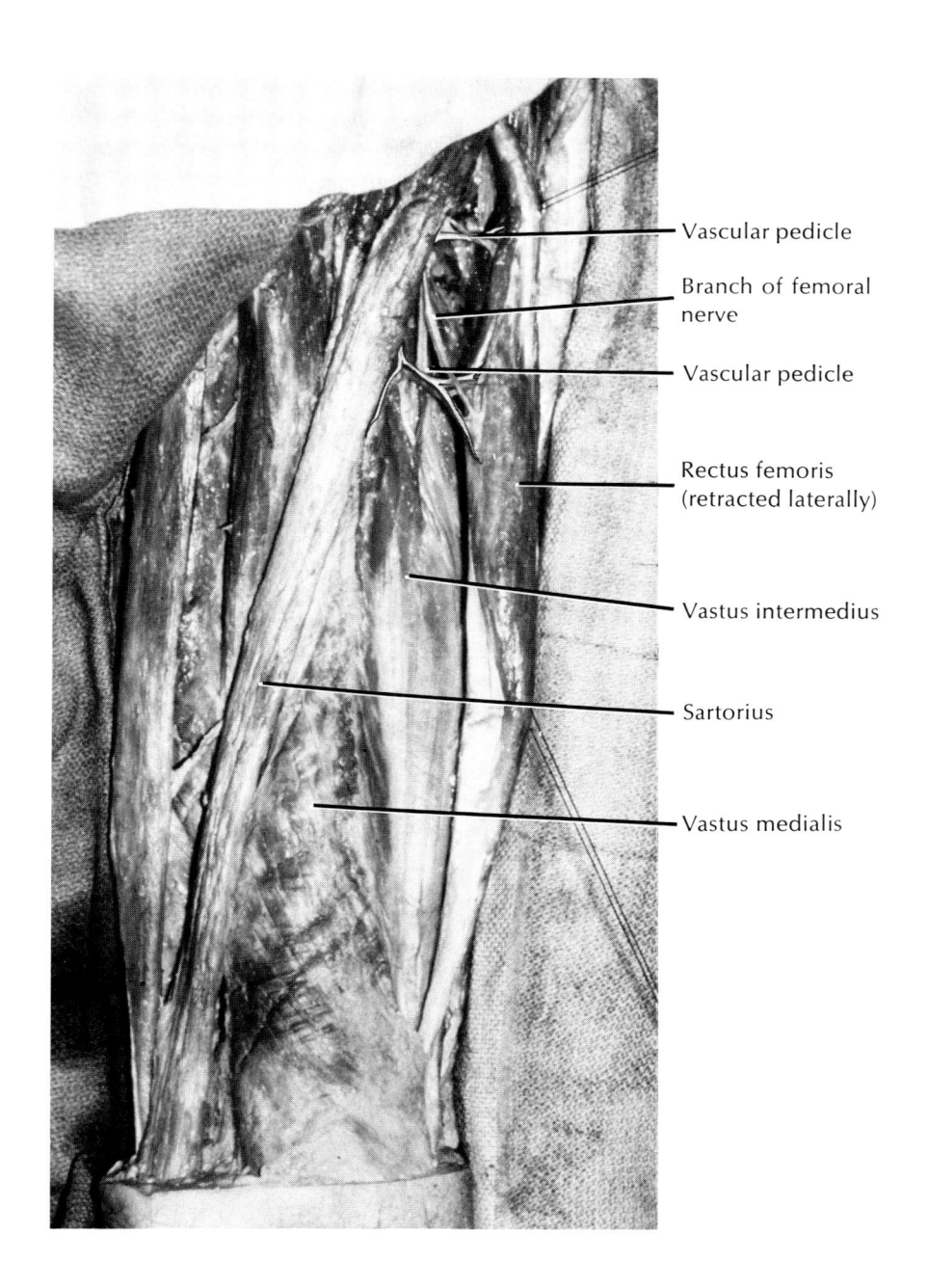

ANTERIOR THIGH

Rectus femoris

Anterior arc

Posterior arc

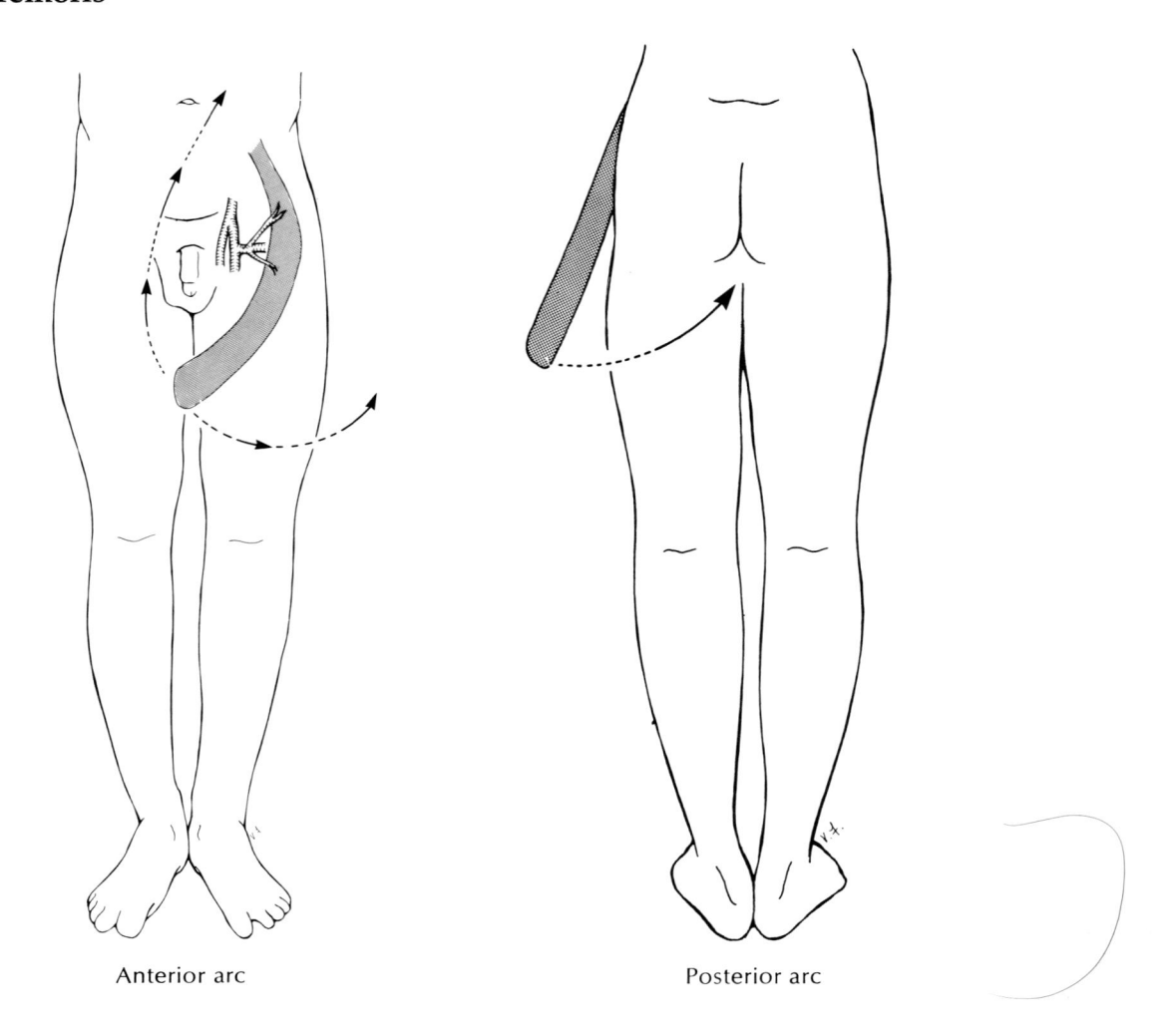

Arc of rotation

ANTERIOR ARC

This long and broad muscle belly has a wide arc of rotation from the point of entrance of its vascular pedicles 10 cm inferior to the inguinal ligament. The muscle or musculocutaneous unit may be used for major abdominal wall reconstruction incorporating the epimysium as fascial replacement. The unit will also cover defects in the abdominal wall, pubic region, perineum, and groin.

POSTERIOR ARC

The posterior arc of rotation of the rectus femoris muscle will cover defects of the trochanter and ischium.

ANTERIOR THIGH

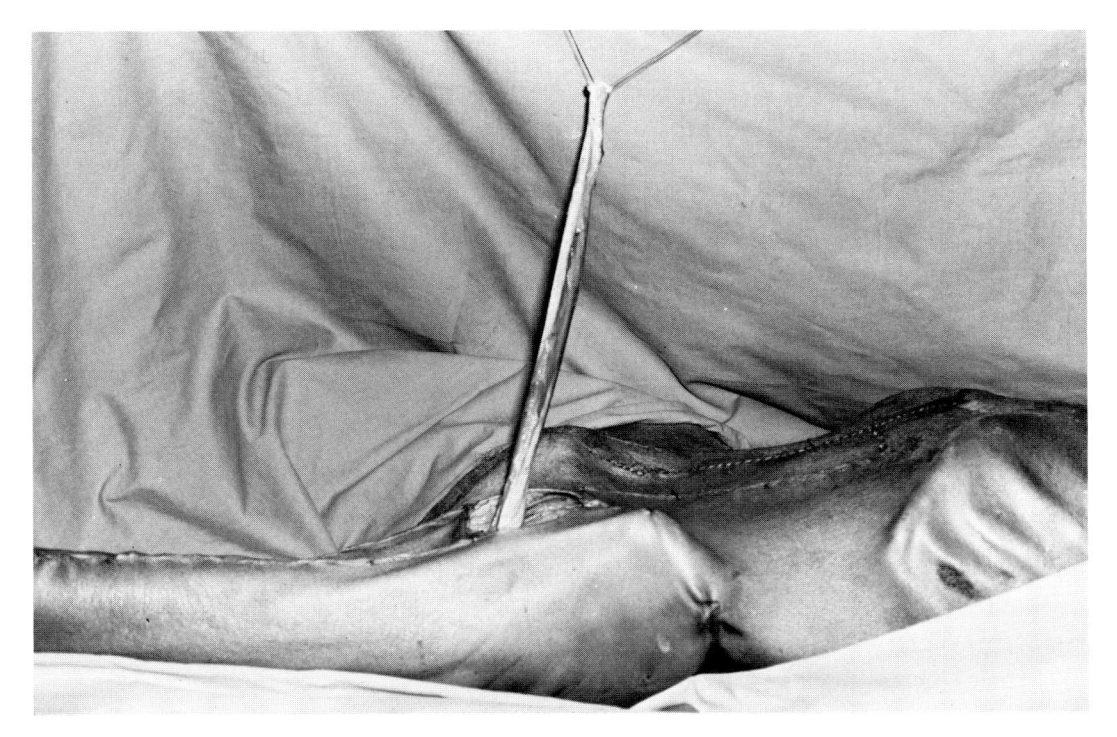

Anterior arc

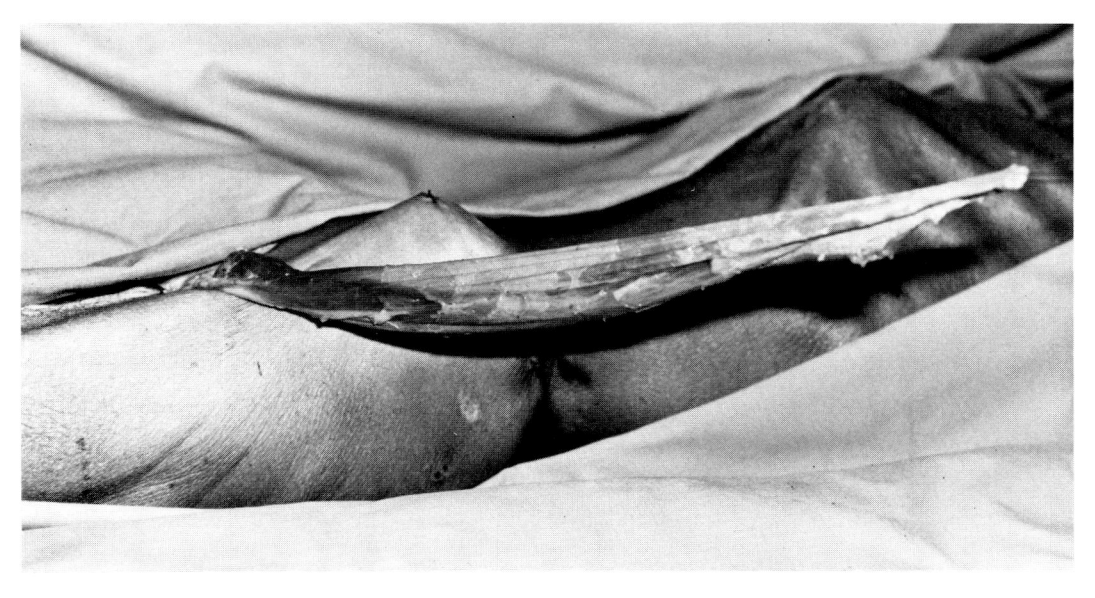

Posterior arc

ANTERIOR THIGH

Rectus femoris

Elevation of flap

The rectus femoris muscle is located by a vertical incision in the anterior mid thigh. The cutaneous territory includes the skin between the sartorius and TFL. The distal third of the cutaneous territory is less reliable in musculocutaneous transposition. The rectus femoris muscle is elevated off the vastus intermedius. When the sartorius muscle is visualized, care is required, as at this level the vascular pedicles enter the proximal belly of the rectus femoris muscle. The donor area will close primarily when the cutaneous territory is small. Larger defects require skin grafts.

Precautions

- Medial extension of cutaneous territory over sartorius muscle may be unreliable.
- Patient may develop loss of full leg extension, especially if other quadriceps muscles are weak.
- Care should be taken during dissection of the vascular pedicles, since the femoral nerve branches to the remaining quadriceps muscles are intimately related to the vascular pedicles of the rectus femoris muscle.

ANTERIOR THIGH

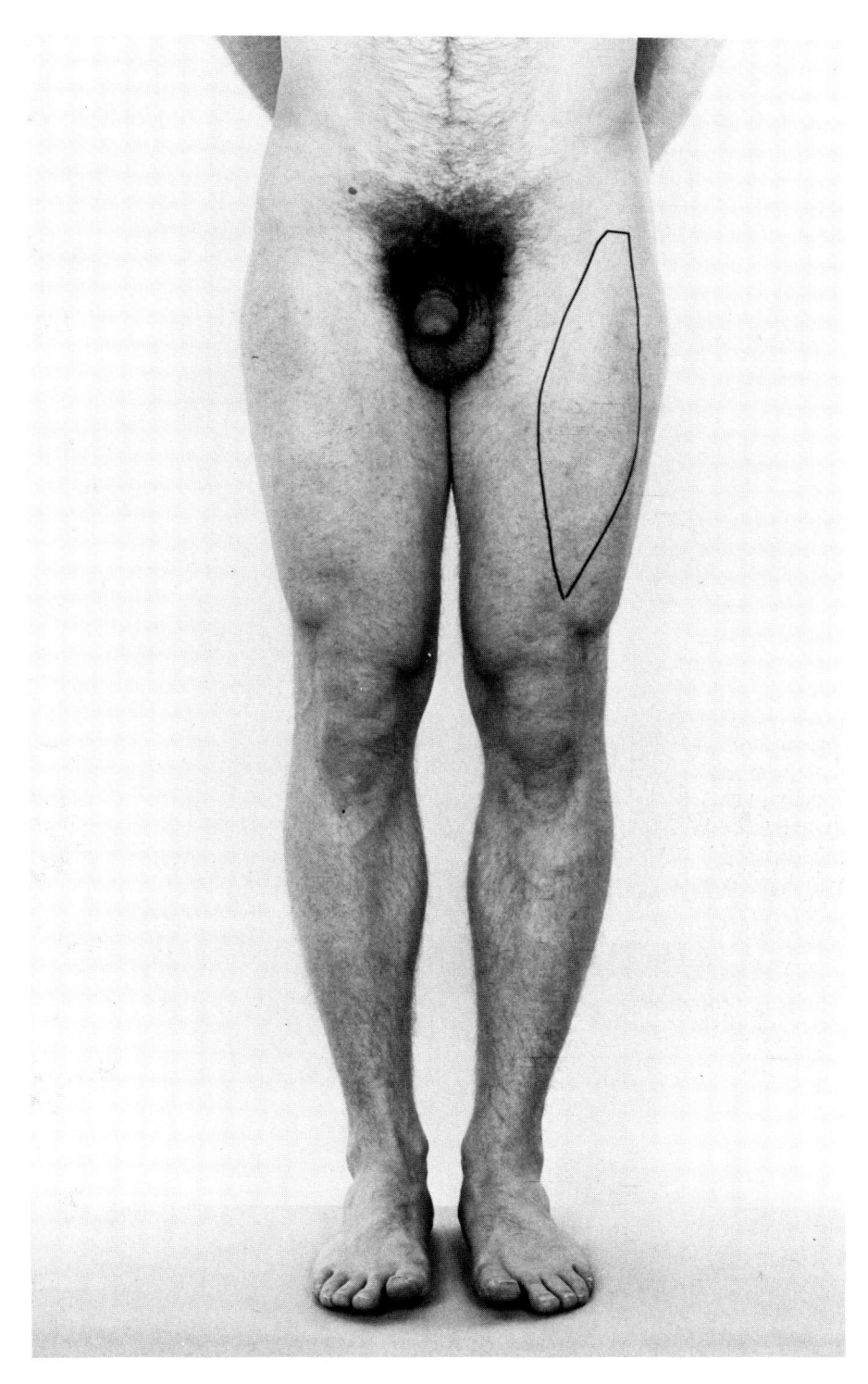

Skin island

ANTERIOR THIGH

Vastus lateralis

MUSCLE FLAP

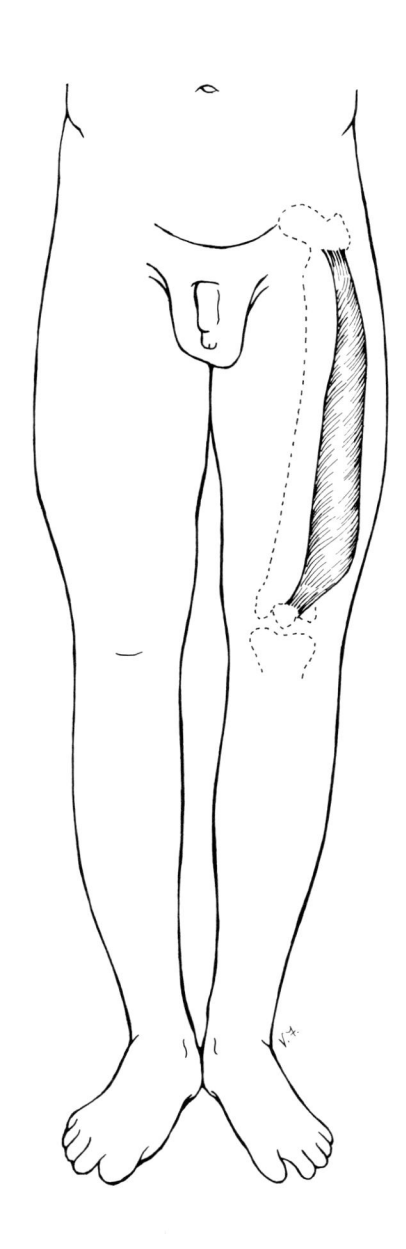

Applications

Reconstruction of:
ABDOMINAL WALL
ACETABULAR FOSSA

Coverage of:
TROCHANTER
ISCHIUM
GROIN
BUTTOCKS

ANTERIOR THIGH

Vastus lateralis

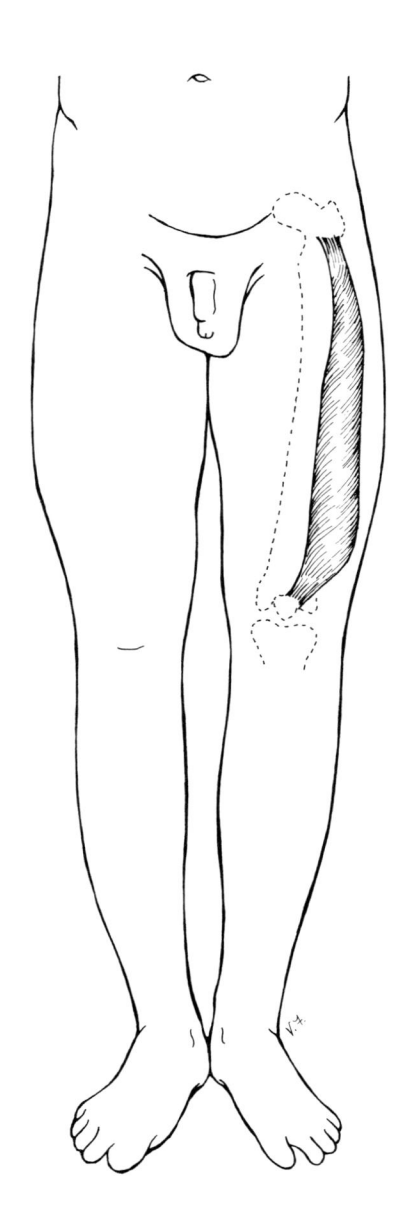

The vastus lateralis is a large, broad muscle located on the anterolateral aspect of the thigh. The muscle lies beneath the TFL and between the vastus intermedius and the biceps femoris muscles.

Origin: Trochanter of femur, gluteal tuberosity, lateral intermuscular system.
Insertion: Patella.

ANTERIOR THIGH

Labels on figure:
- Lateral circumflex femoral branch of profunda artery
- Transverse branch of lateral circumflex femoral artery
- Descending branch of lateral circumflex femoral artery

Blood supply

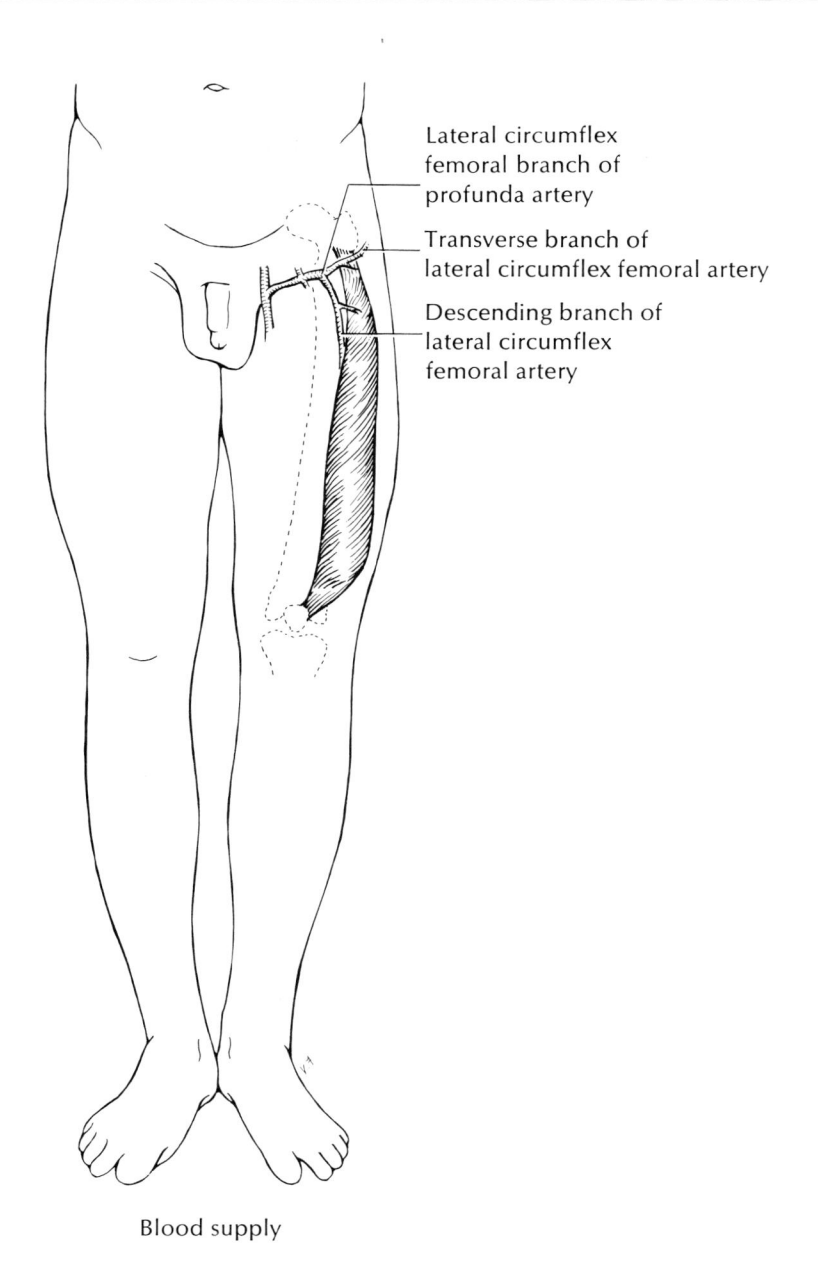

Nerve supply: Muscular branch of femoral nerve. This motor nerve enters the proximal muscle belly at the medial border inferior to the vascular pedicle.

Function: This muscle is expendable. This muscle is a strong leg extensor, but the remaining quadriceps muscles provide synergistic function.

ANTERIOR THIGH

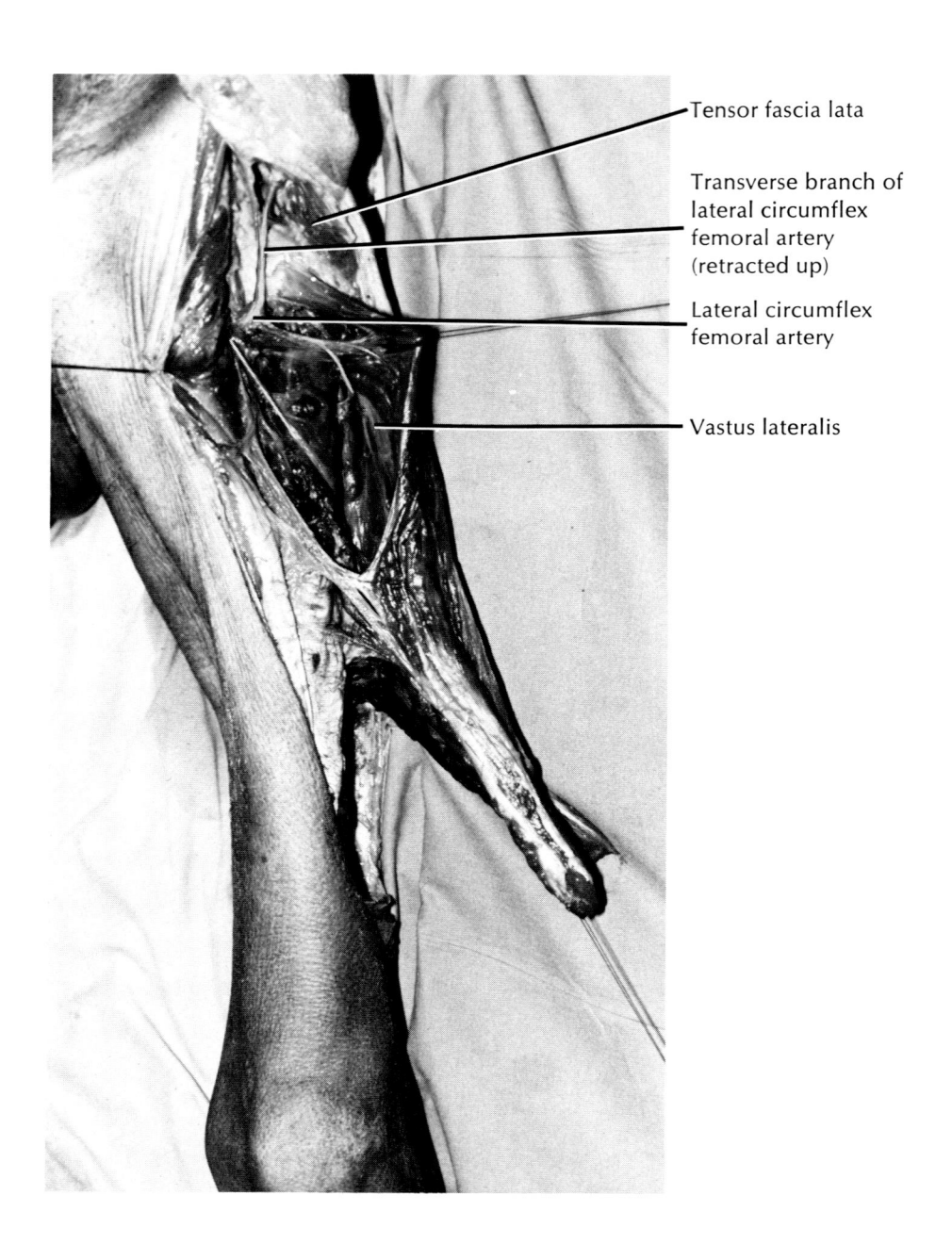

Vastus lateralis

Tensor fascia lata

Transverse branch of lateral circumflex femoral artery (retracted up)

Lateral circumflex femoral artery

Vastus lateralis

ANTERIOR THIGH

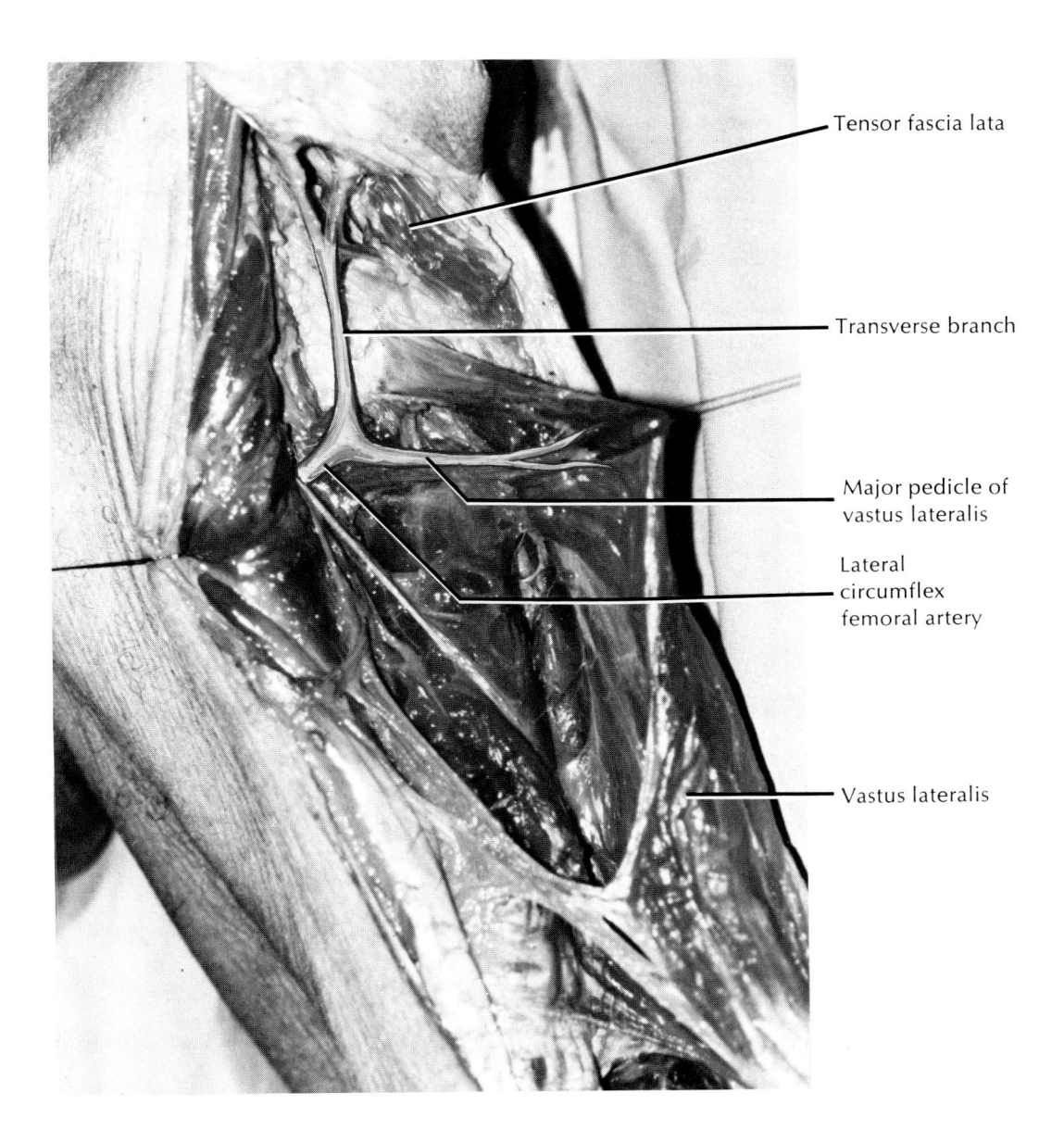

Blood supply

This muscle receives its blood supply from branches of the lateral circumflex femoral artery. After emerging from beneath the rectus femoris muscle, the descending branch of the lateral femoral circumflex artery has several pedicles to the anterior proximal muscle belly. The vascular pedicles may be located approximately 10 cm inferior to the anterior superior iliac crest.

ANTERIOR THIGH

Vastus lateralis

Anterior arc

Posterior arc

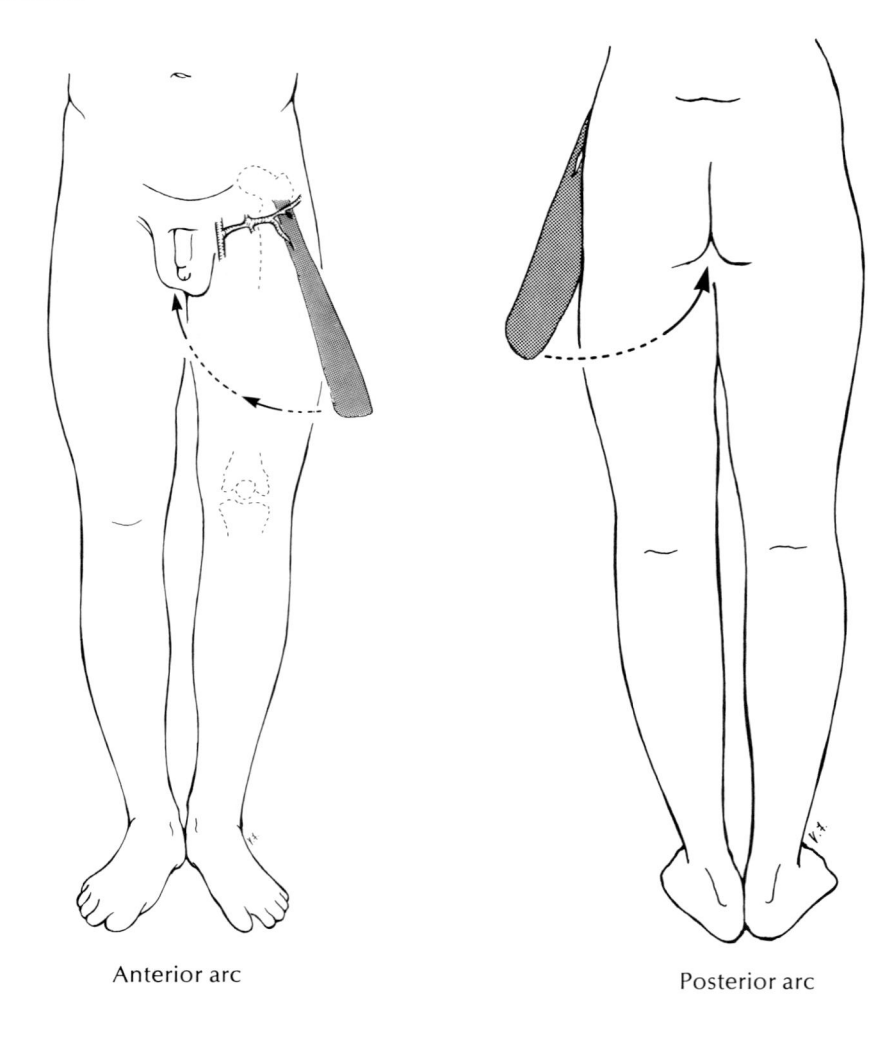

Arc of rotation

ANTERIOR ARC

This large muscle has a wide arc of rotation from the point of entrance of the vascular pedicles approximately 10 cm beneath the anterior superior iliac crest. After incision of the epimysium, the muscle will expand to provide coverage for large defects on the inferior abdominal wall and groin. Since the muscle originates beneath the greater trochanter of the femur, this muscle may be transposed superiorly as an island to fill the acetabular fossa following hip disarticulation.

POSTERIOR ARC

The posterior arc of the vastus lateralis muscle will cover defects of the trochanter, ischium, and buttocks.

ANTERIOR THIGH

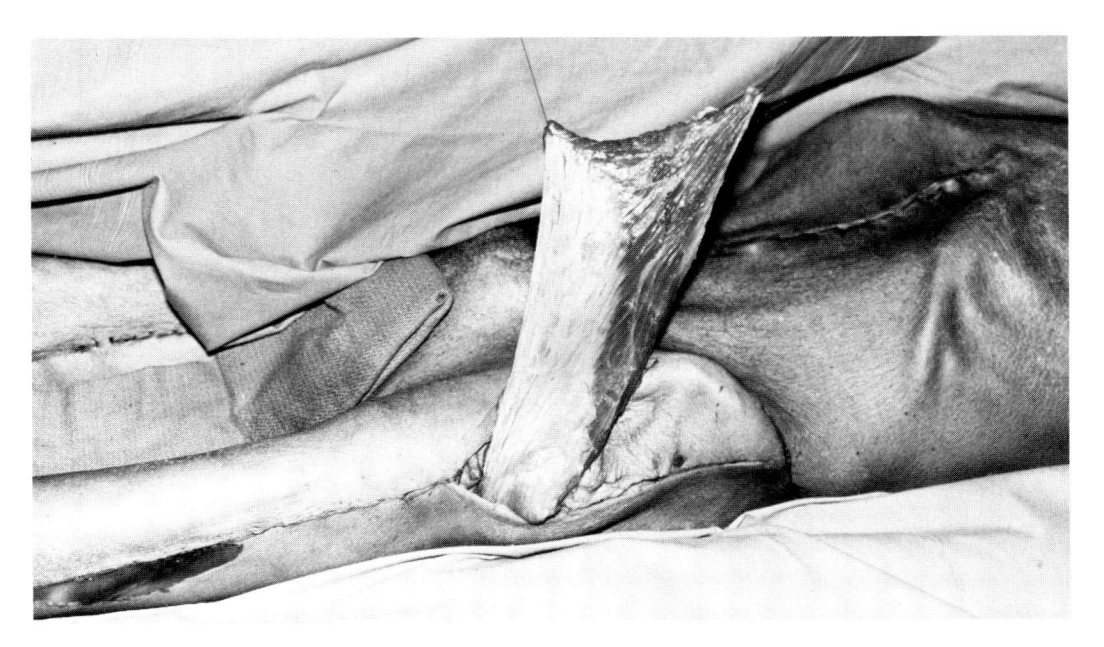

Anterior arc

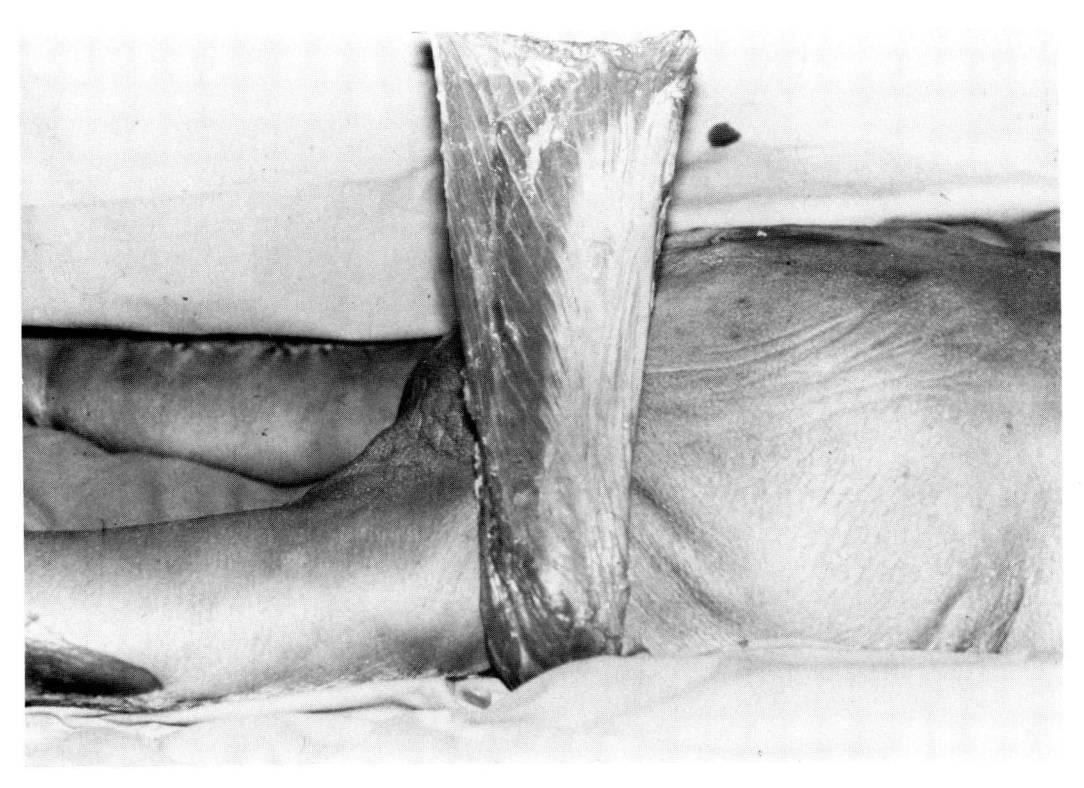

Posterior arc

ANTERIOR THIGH

Vastus lateralis

Elevation of flap

This muscle is located by a lateral incision through the fascia lata. This lateral thigh skin belongs to the TFL cutaneous territory, so a musculocutaneous flap is not possible. At 10 cm inferior to the iliac crest the rectus femoris muscle can be retracted medially to locate the lateral circumflex femoral artery. The vascular pedicles supporting this muscle enter the medial anterior muscle belly. During hip disarticulation the muscle can be elevated as an island and advanced into the acetabular fossa to fill this region and further provide coverage of the lateral thigh in extensive pressure sores. The donor area following vastus lateralis transposition can be closed primarily.

Precautions

- The cutaneous area overlying this muscle is part of the TFL unit.
- This muscle may not be used as a musculocutaneous flap.

ANTERIOR THIGH

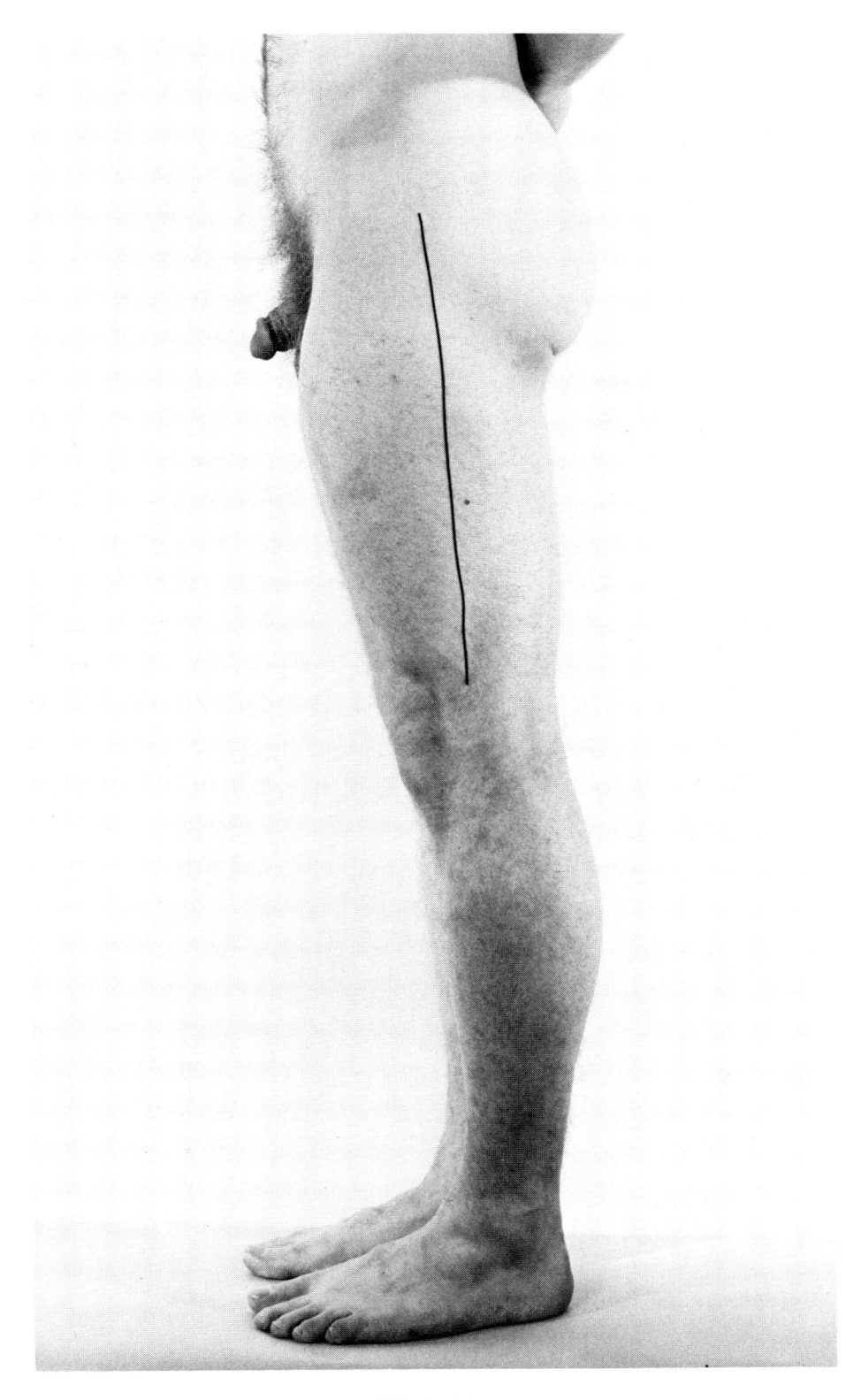

Skin incision

ANTERIOR THIGH

Vastus lateralis

Reconstruction following hip disarticulation for trochanteric pressure sore with septic hip joint

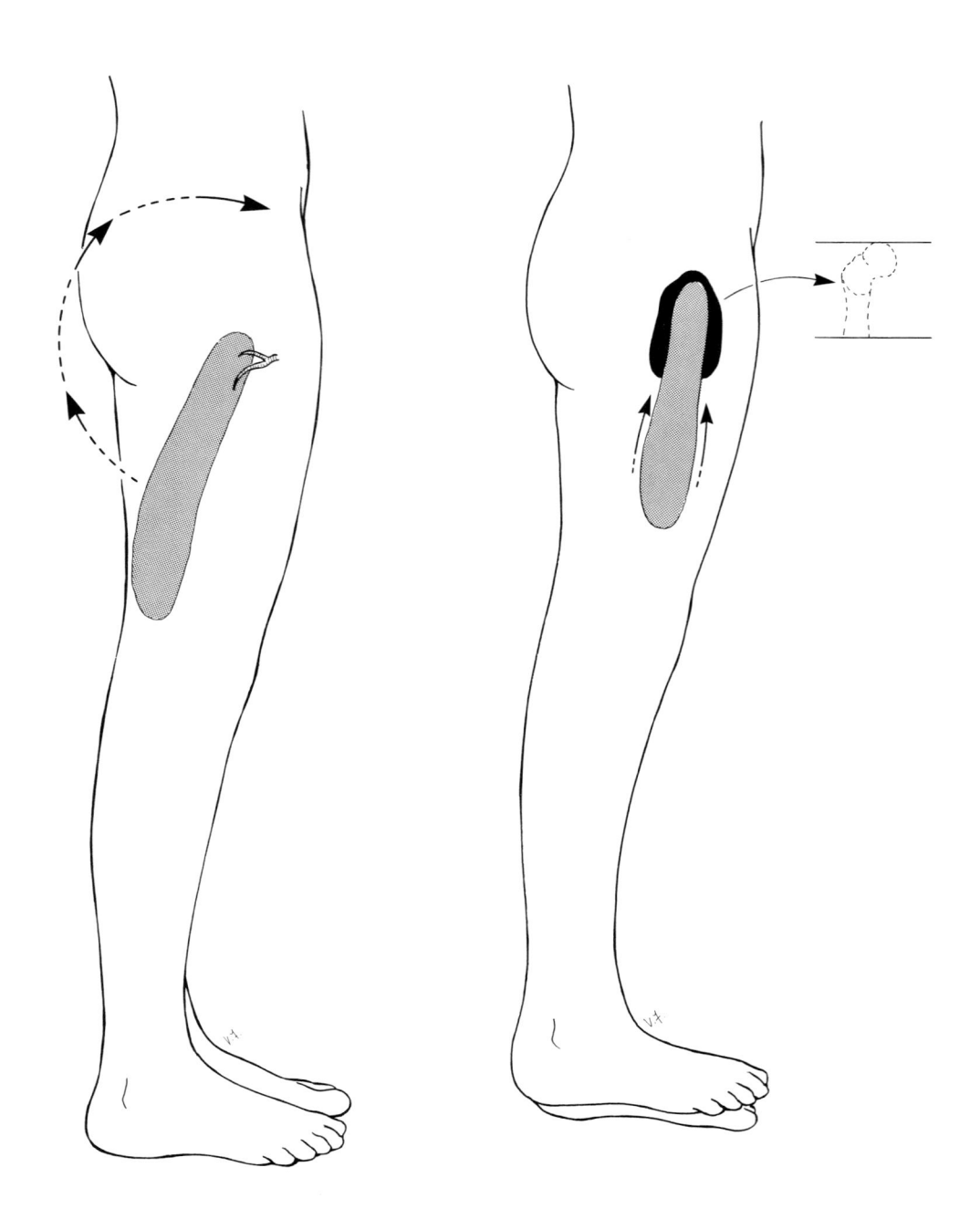

60

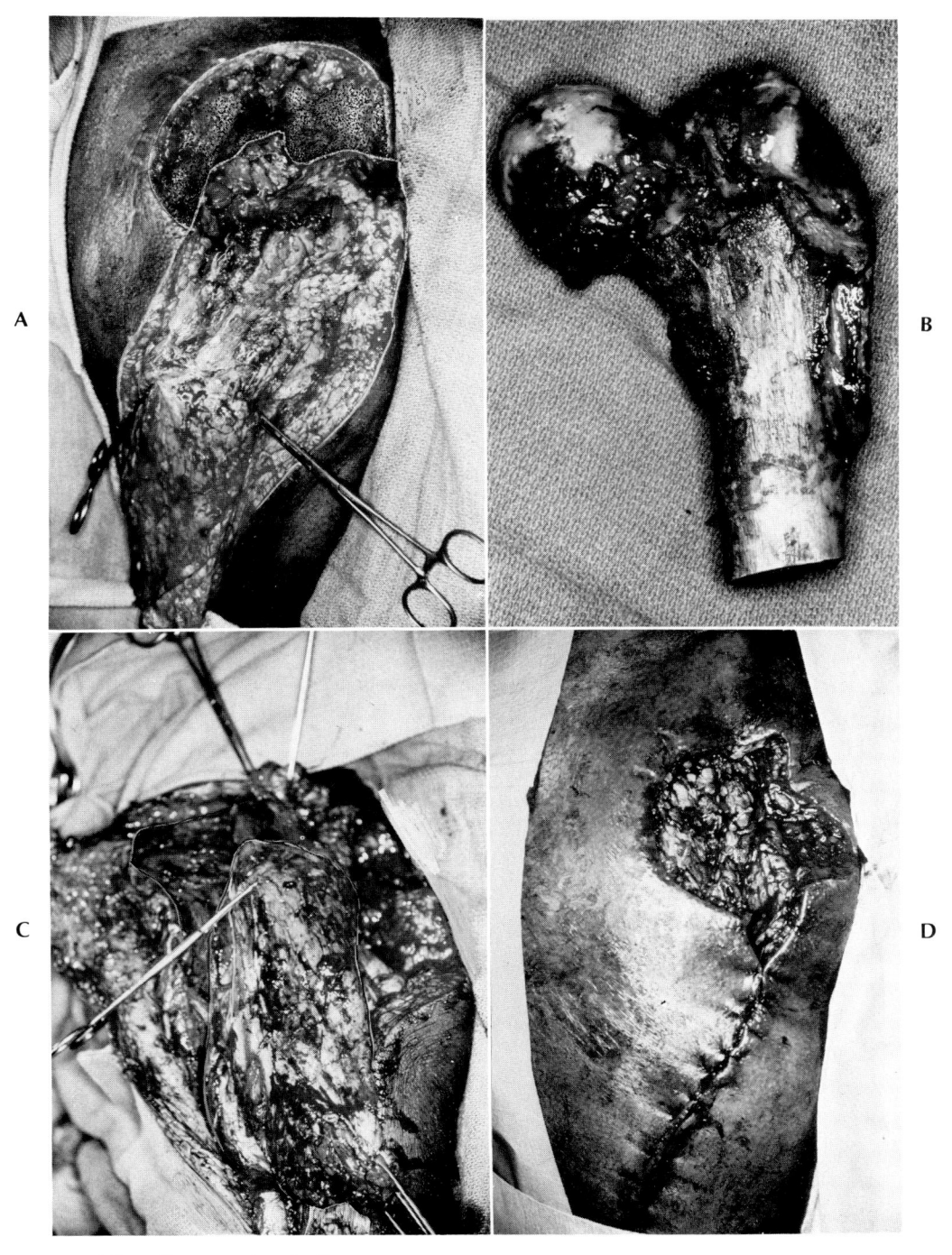

A   Trochanteric defect.
B   Resection of head of femur.
C   Vastus lateralis transposed into acetabulum.
D   Skin incision closed. Exposed muscle grafted.

Tensor fascia lata (TFL)

MUSCLE FLAP
MUSCULOFASCIAL FLAP
MUSCULOCUTANEOUS FLAP
FREE FLAP

Applications

Reconstruction of:
ABDOMINAL WALL
VULVA
INGUINAL HERNIA

Coverage of:
ABDOMEN
GROIN
PERINEUM
TROCHANTER
ISCHIUM
SACRUM

Free flap:
DISTANT COVERAGE
NEUROSENSORY FLAP
FUNCTIONAL MUSCLE
OSSEOUS-MUSCULOCUTANEOUS FLAP

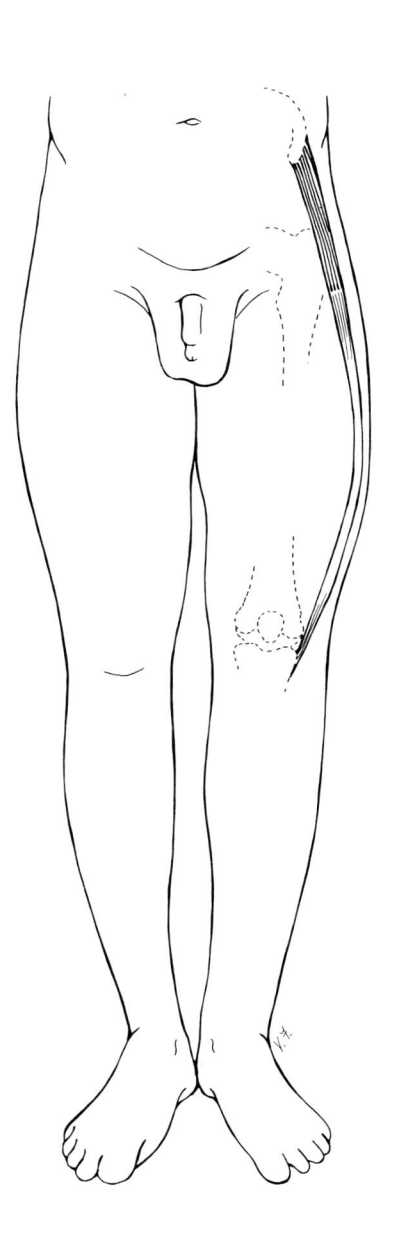

ANTERIOR THIGH

Tensor fascia lata

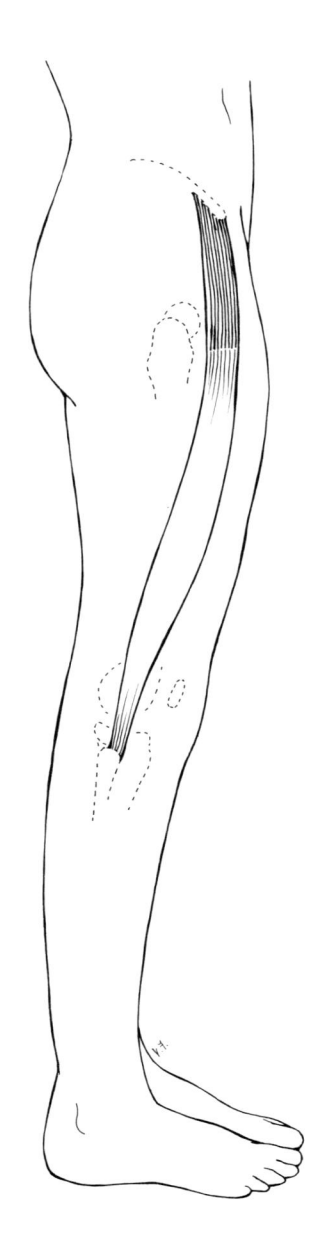

The tensor fascia lata (TFL) is a small, thin, flat muscle on the lateral aspect of the upper thigh. It is lateral to the sartorius and rectus femoris at its origin and lies superficial to the vastus lateralis. This unit is unusual in that this small muscle has a cutaneous territory of skin up to four times greater in size than the muscle.

Origin: Anterior 5 to 8 cm of the outer lip of the iliac crest, lateral to the origin of the sartorius.
Insertion: Iliotibial tract.

ANTERIOR THIGH

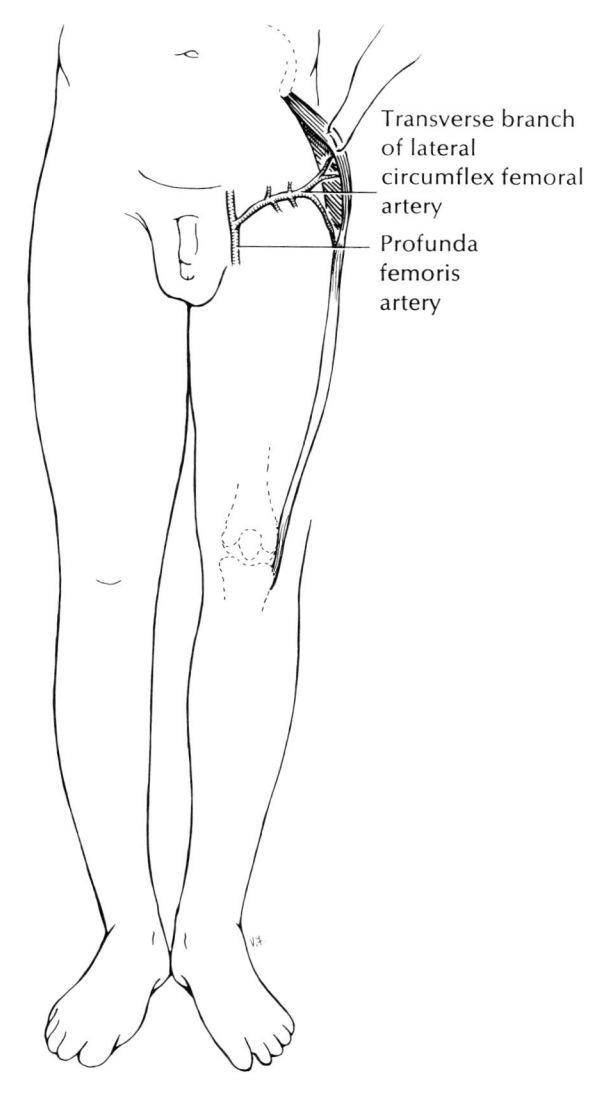

Blood supply

Nerve supply:
  Motor — The superior gluteal nerve is the motor nerve to the muscle. It emerges between the gluteus medius and gluteus maximus muscles to innervate the TFL on its deep superior surface.
  Sensory — The cutaneous branch of T12 innervates the skin over the origin of the muscle from the iliac crest. The lateral femoral cutaneous nerve innervates most of the skin of the anterolateral thigh.
Function: This is an expendable muscle that is an abductor and medial rotator of the thigh.

ANTERIOR THIGH

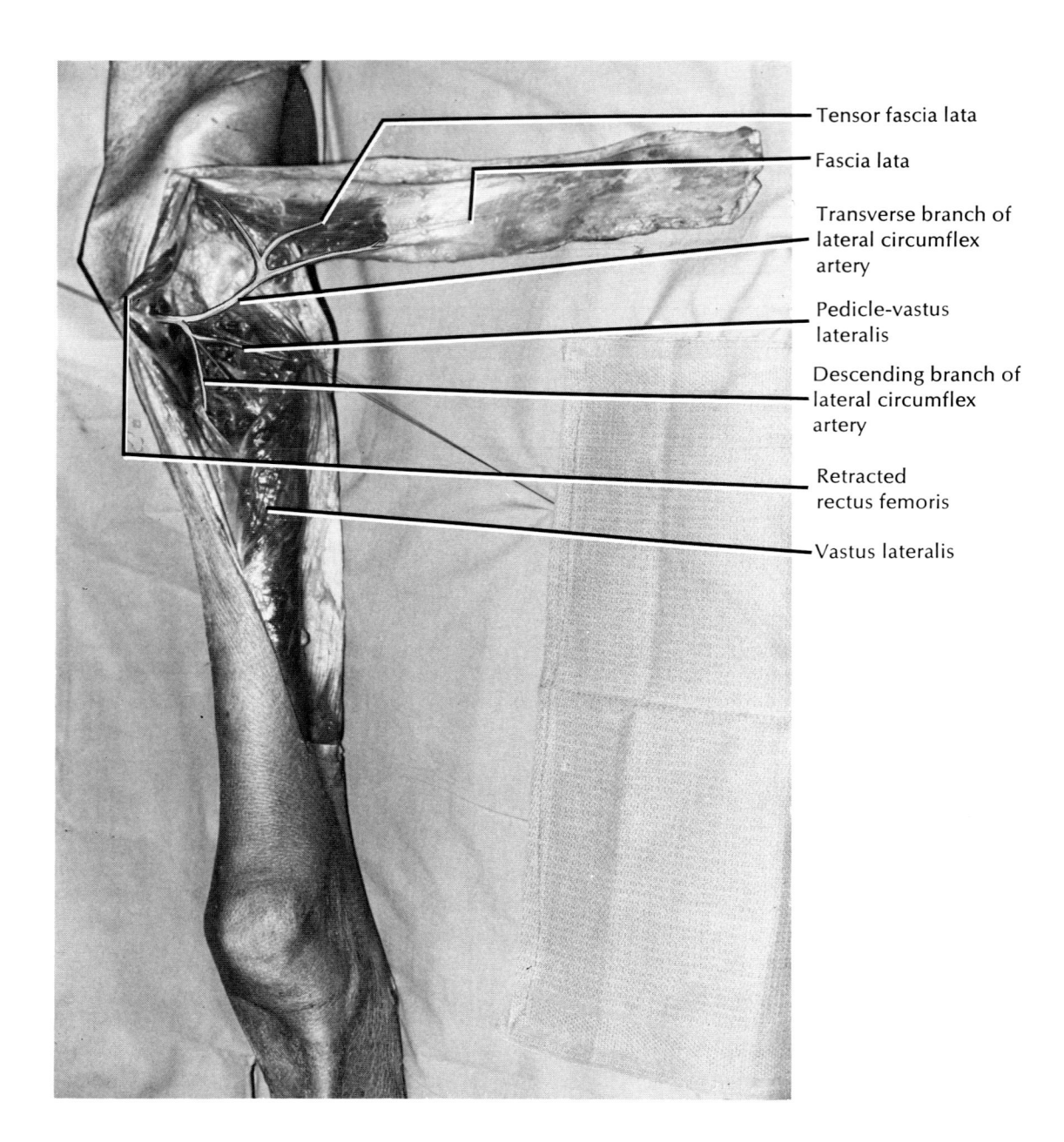

Tensor fascia lata

- Tensor fascia lata
- Fascia lata
- Transverse branch of lateral circumflex artery
- Pedicle-vastus lateralis
- Descending branch of lateral circumflex artery
- Retracted rectus femoris
- Vastus lateralis

ANTERIOR THIGH

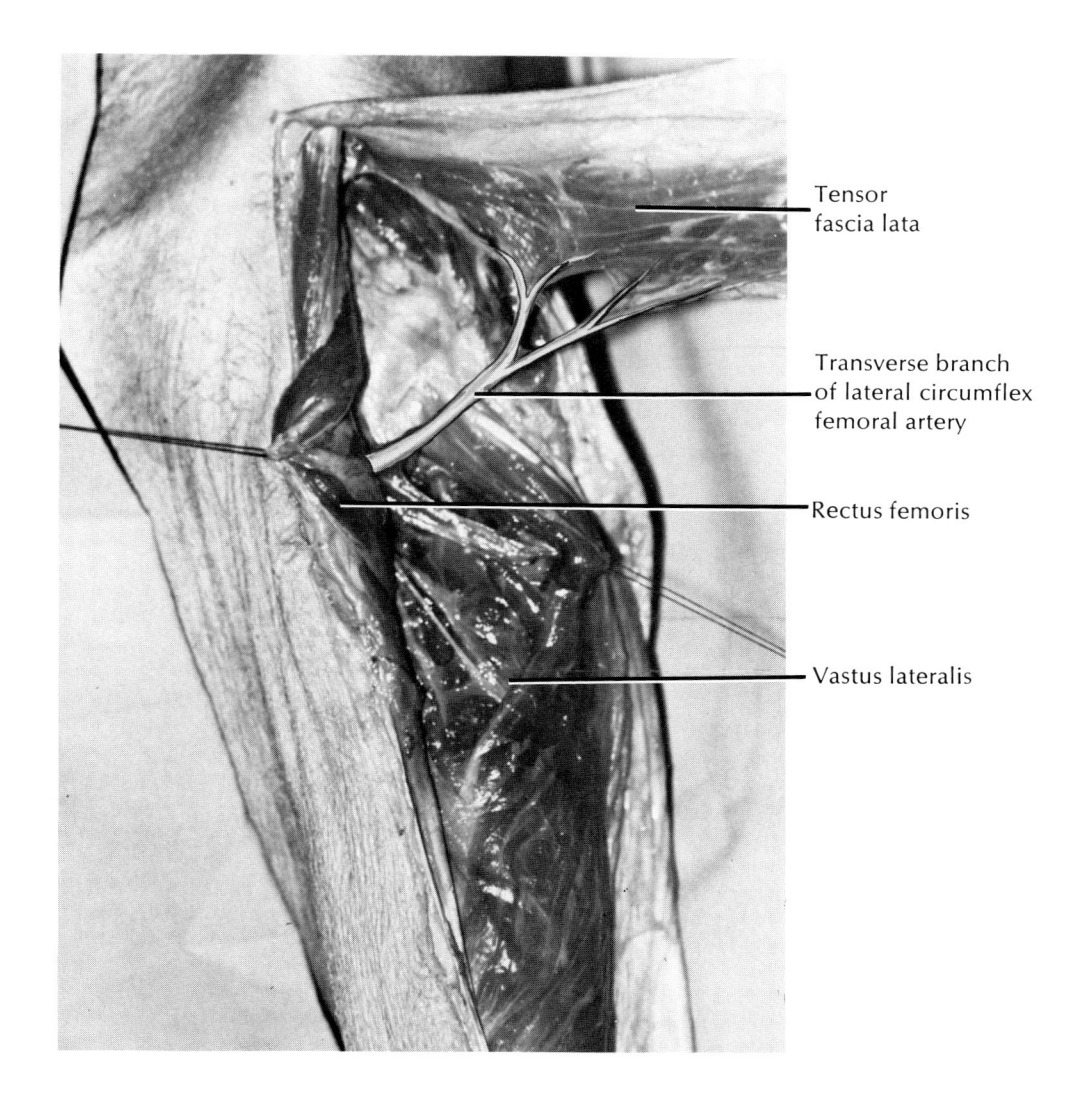

Tensor fascia lata

Transverse branch of lateral circumflex femoral artery

Rectus femoris

Vastus lateralis

Blood supply

The single dominant pedicle enters the medial deep surface of the muscle 8 to 10 cm below the anterior superior iliac spine. The pedicle is based on the lateral circumflex femoral branch of the profunda femoris artery. The vessel emerges deep to the rectus femoris muscle and divides into a descending branch that supplies the vastus lateralis. The transverse branch, which courses laterally, has a small branch to the gluteus minimus. The artery then divides into two or three large terminal branches that comprise the pedicle to the TFL.

ANTERIOR THIGH

Tensor fascia lata

Anterior arc

Posterior arc

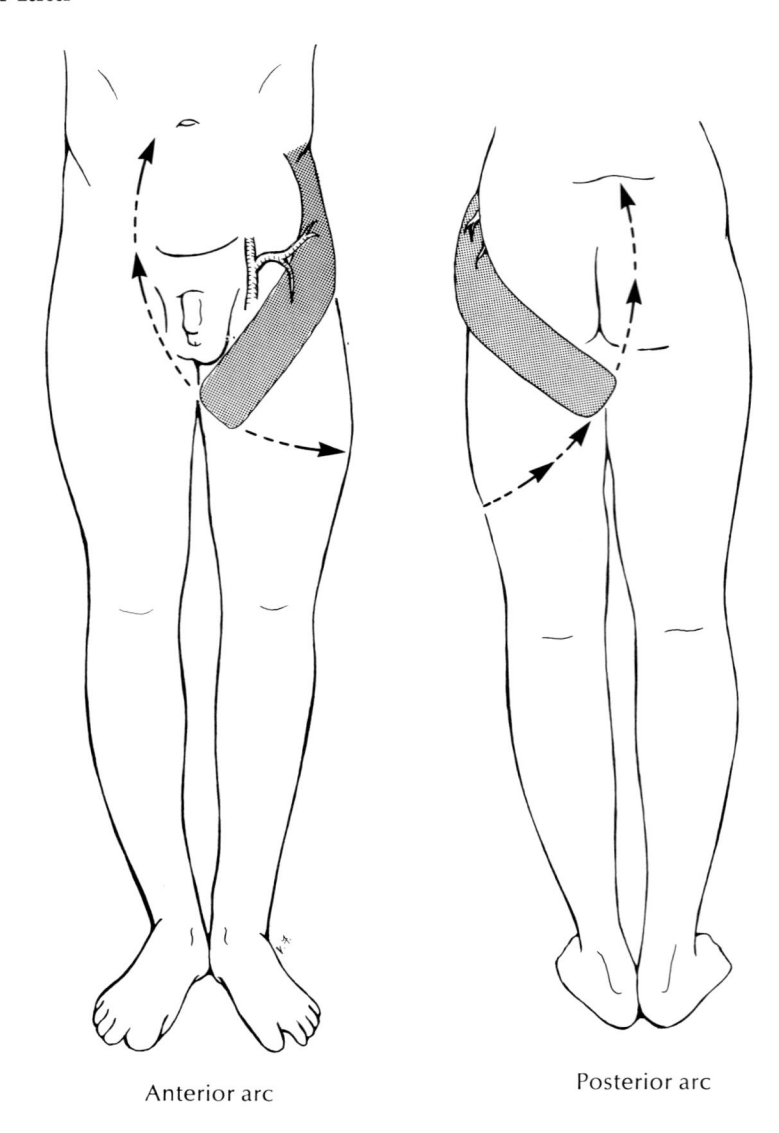

Arc of rotation

Either the muscle or muscle with its overlying skin can be used as a transposition flap to cover trochanteric defects posteriorly or the groin anteriorly. However, the skin of the anterolateral thigh together with the underlying fascia lata can be elevated with the TFL muscle as an extended unit. Based on the dominant pedicle of the TFL, the extended flap has a useful anterior and posterior arc, with the point of rotation 8 to 10 cm below the anterior superior iliac spine.

ANTERIOR THIGH

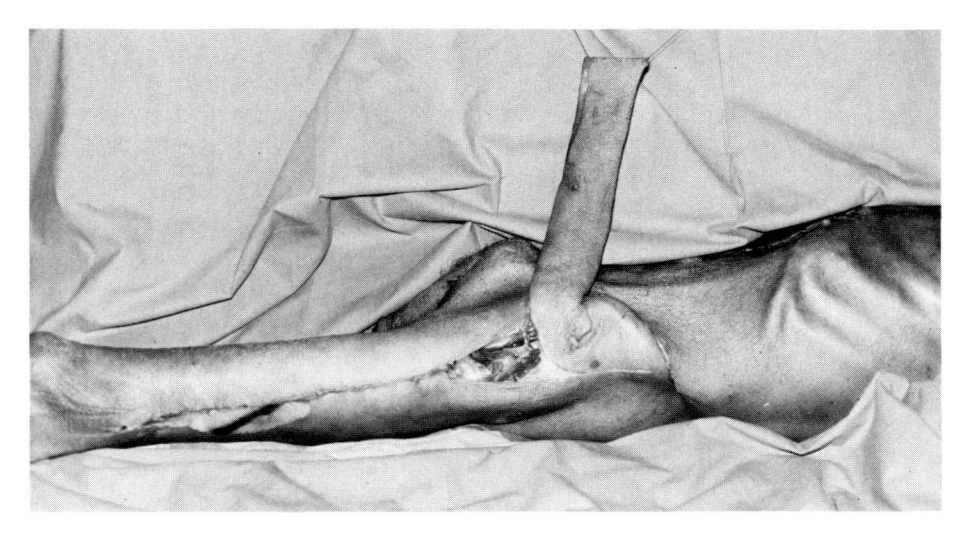

Anterior arc

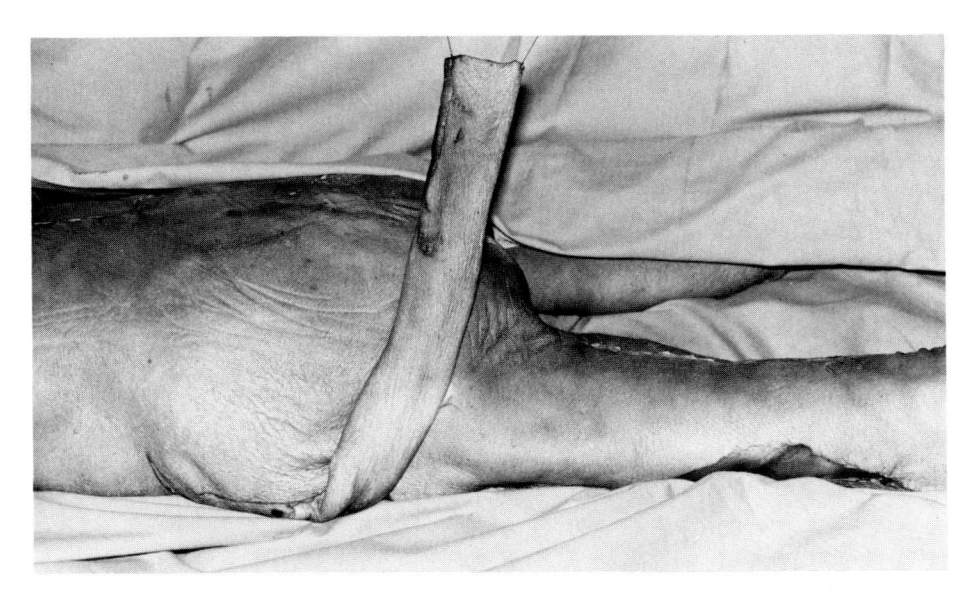

Posterior arc

ANTERIOR ARC

The anterior arc of rotation will cover the groin, perineum, and abdominal wall.

POSTERIOR ARC

The posterior arc of rotation will cover the trochanter, ischium, anal region, and sacrum.

ANTERIOR THIGH

Tensor fascia lata

Anterior, superior iliac spine

Fascia lata covering tensor fascia lata

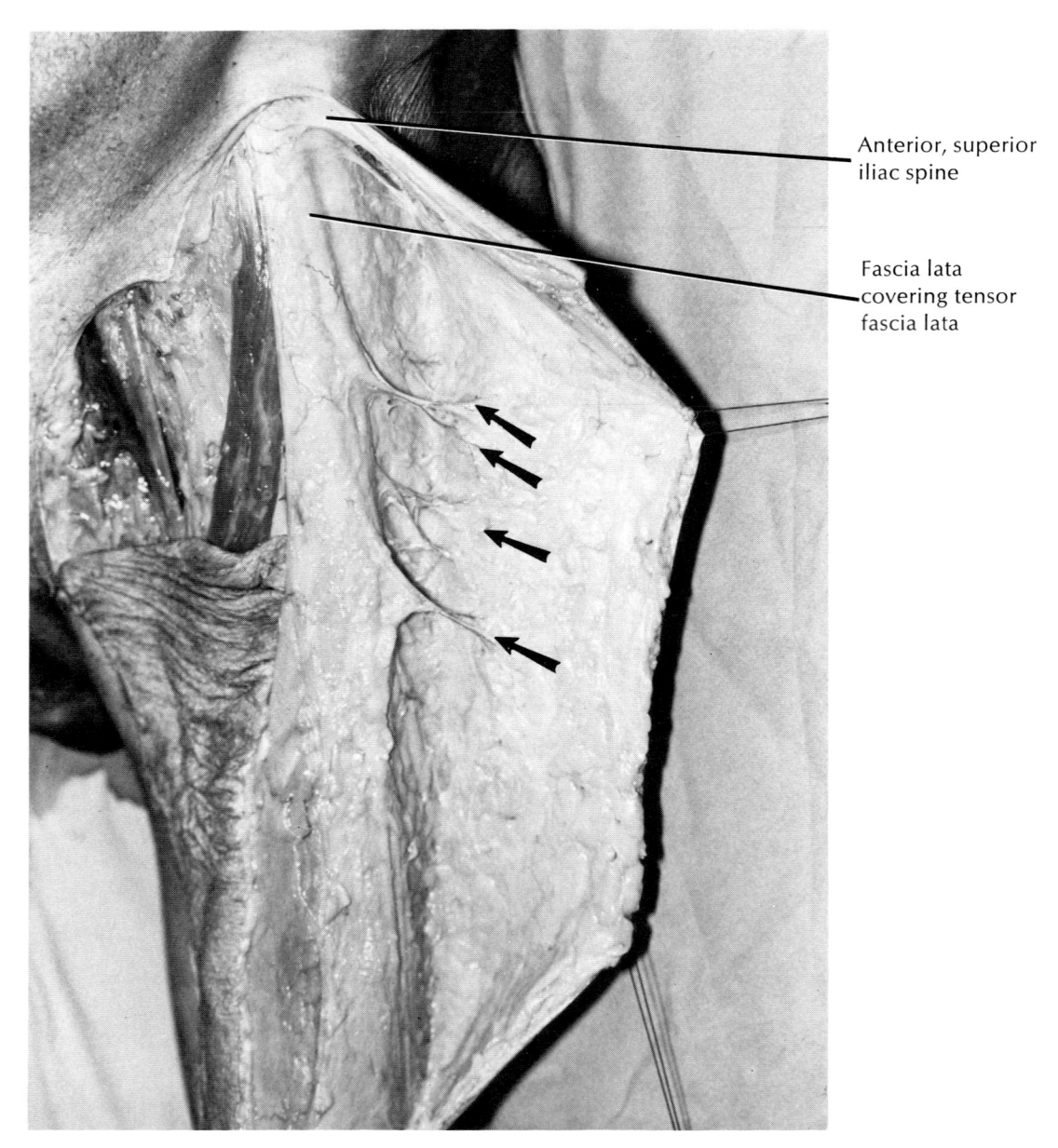

Axially directed musculocutaneous perforators are noted by arrows

MUSCULOCUTANEOUS PERFORATORS

There are several large perforators from the muscle into the overlying skin. The perforators supply most of the skin of the anterolateral thigh and are proximally located. The inferior muscle perforators extend as axial vessels distally along the fascia lata and are the basis of the extended TFL flap.

ANTERIOR THIGH

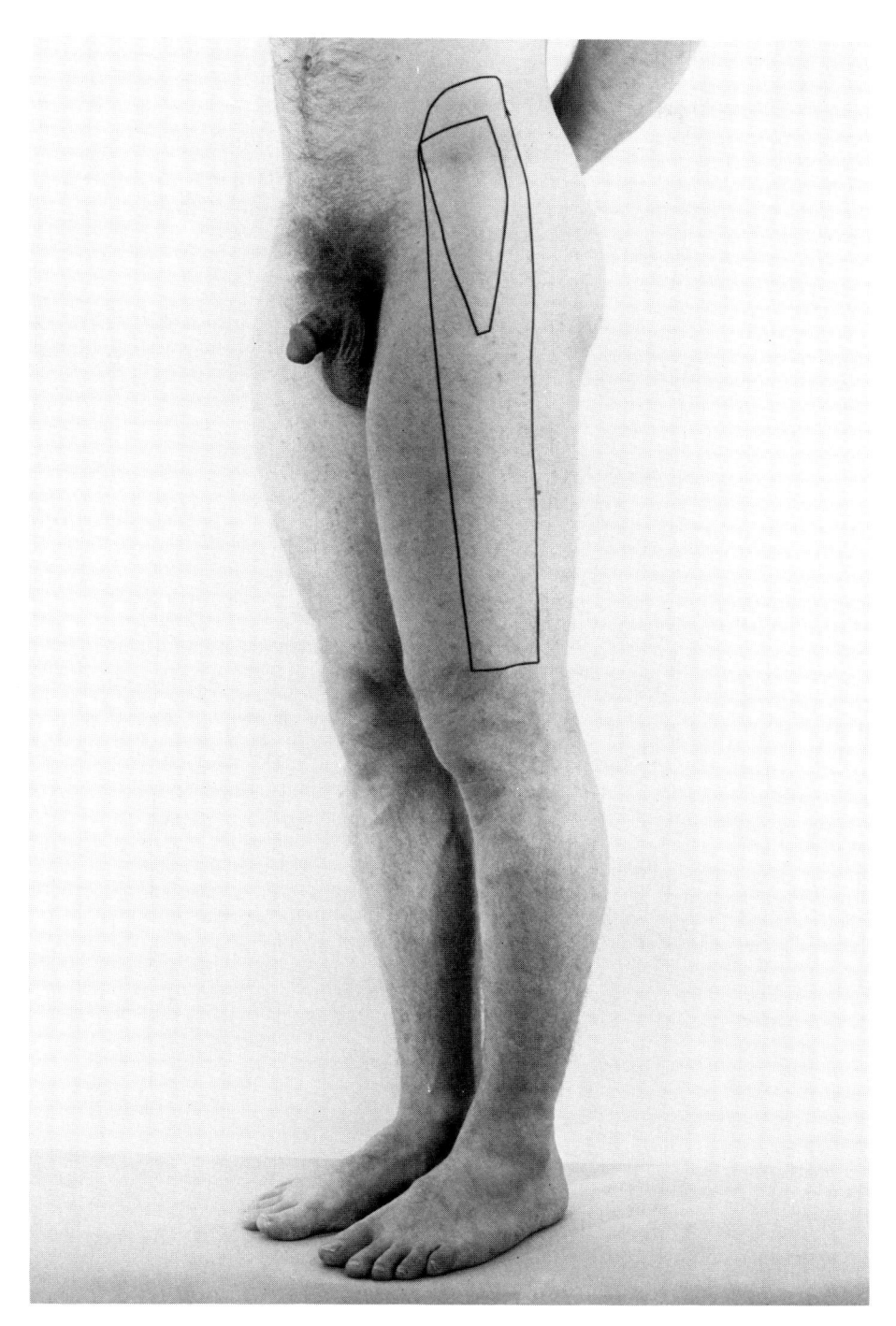

Muscle margins and cutaneous territory

ANTERIOR THIGH

Tensor fascia lata

Elevation of flap

A line drawn from the anterior superior iliac spine to the lateral condyle of the tibia marks the anterior border of the unit. This musculocutaneous unit may be designed to include skin and fascia lata extending to 5 to 8 cm of the knee. In width, the greater trochanter marks the posterior boundary. If necessary, the territory can be safely extended anteriorly over the rectus femoris muscle.

The flap is elevated from distal to proximal. The fascia lata is elevated with the overlying skin and is sutured to the skin temporarily during flap elevation to protect the perforating vessels. Flap elevation reveals the vastus lateralis.

The vascular pedicle is visualized on the deep medial aspect of the muscle 8 to 10 cm below the anterior superior iliac spine by medial retraction of the rectus femoris muscle. For free flap transfer the entire muscle and, if desired, iliac crest bone can be included. However, if a thinner flap is desired, the muscle can be safely divided 5 cm distal to its origin between the upper and middle branches of the vascular pedicle. This also minimizes the secondary defect by eliminating the depression beneath the iliac crest resulting from removal of all of the TFL muscle.

The donor defect in most instances may be closed directly, but if a wide flap is elevated, skin grafting over the vastus lateralis is necessary.

ANTERIOR THIGH

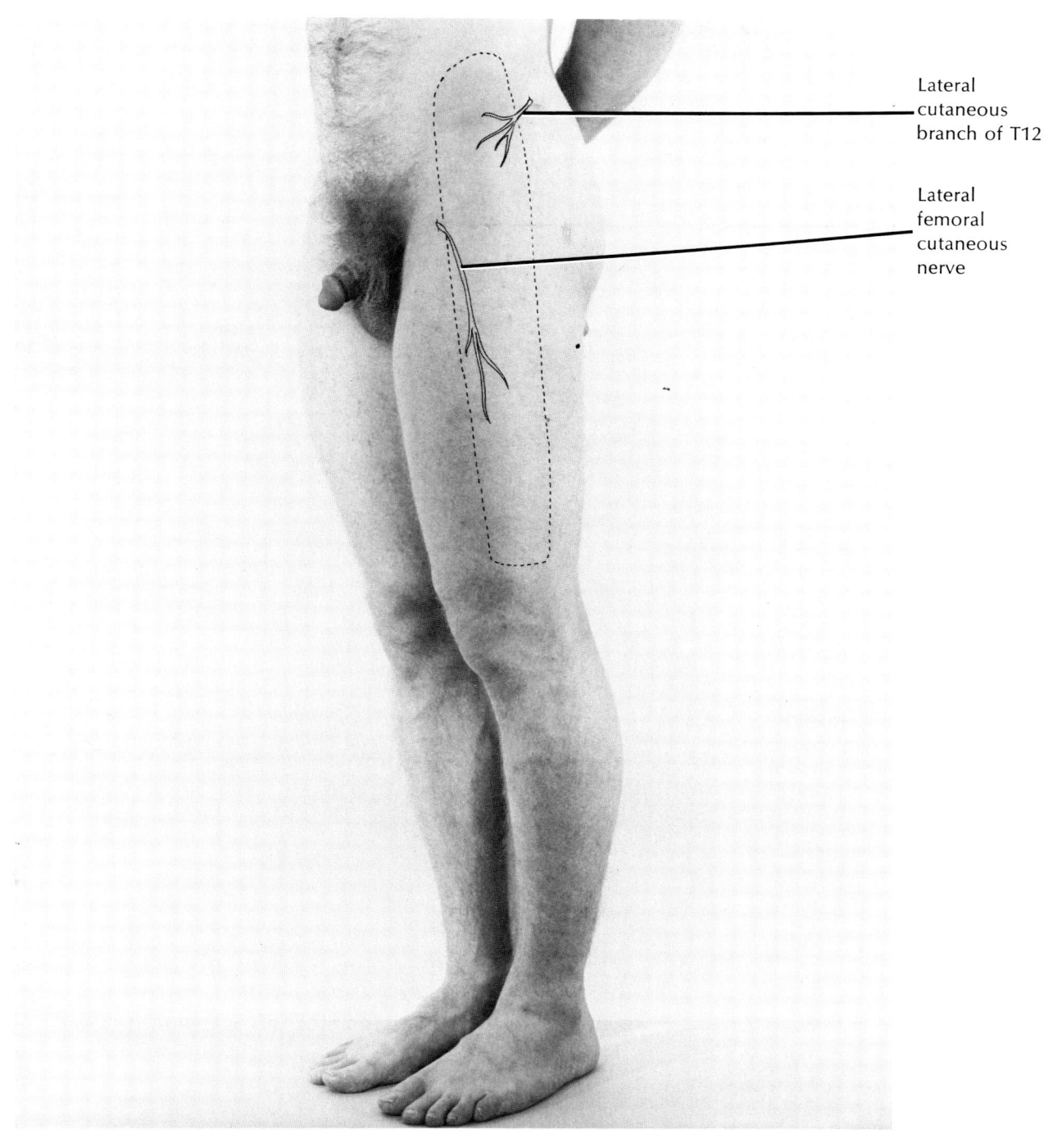

Lateral cutaneous branch of T12

Lateral femoral cutaneous nerve

Sensory distribution

ANTERIOR THIGH

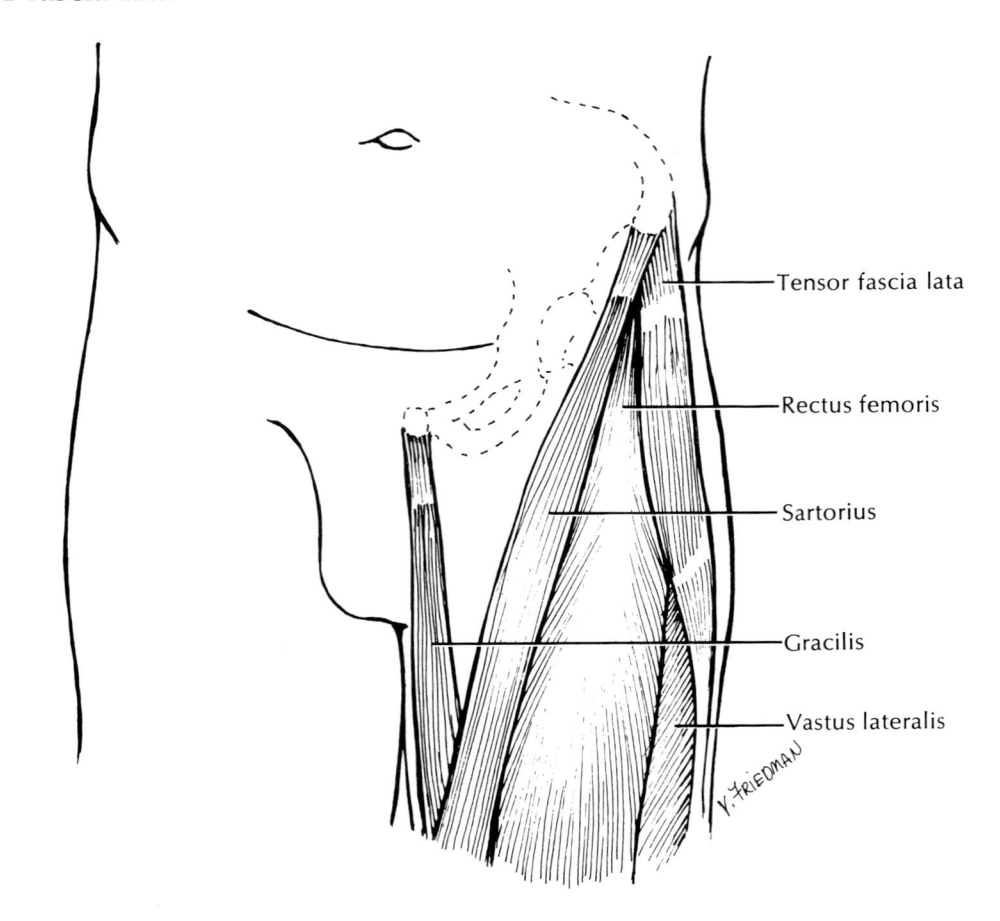

Tensor fascia lata

MUSCULOFASCIAL FLAP

The TFL muscle and the fascia lata without overlying skin can be elevated as a musculofascial unit. This unit is useful for reconstruction of recurrent groin hernias and abdominal wall hernia where skin coverage is adequate but a strong fascial sheet is needed.

Precautions

- In isolating the unit as a free flap transfer, the small TFL muscle can be located at the iliac spine in close relationship to the medially located sartorius. Care should be taken posteriorly where the muscle is firmly adherent to the gluteus minimus muscle, which can be easily elevated with this flap.
- For free flap transfer, the lateral circumflex femoral artery can be dissected proximally to gain length, but care should be taken deep to the rectus femoris muscle where the vessel is intimately related to the muscular branches of the femoral nerve.

ANTERIOR THIGH

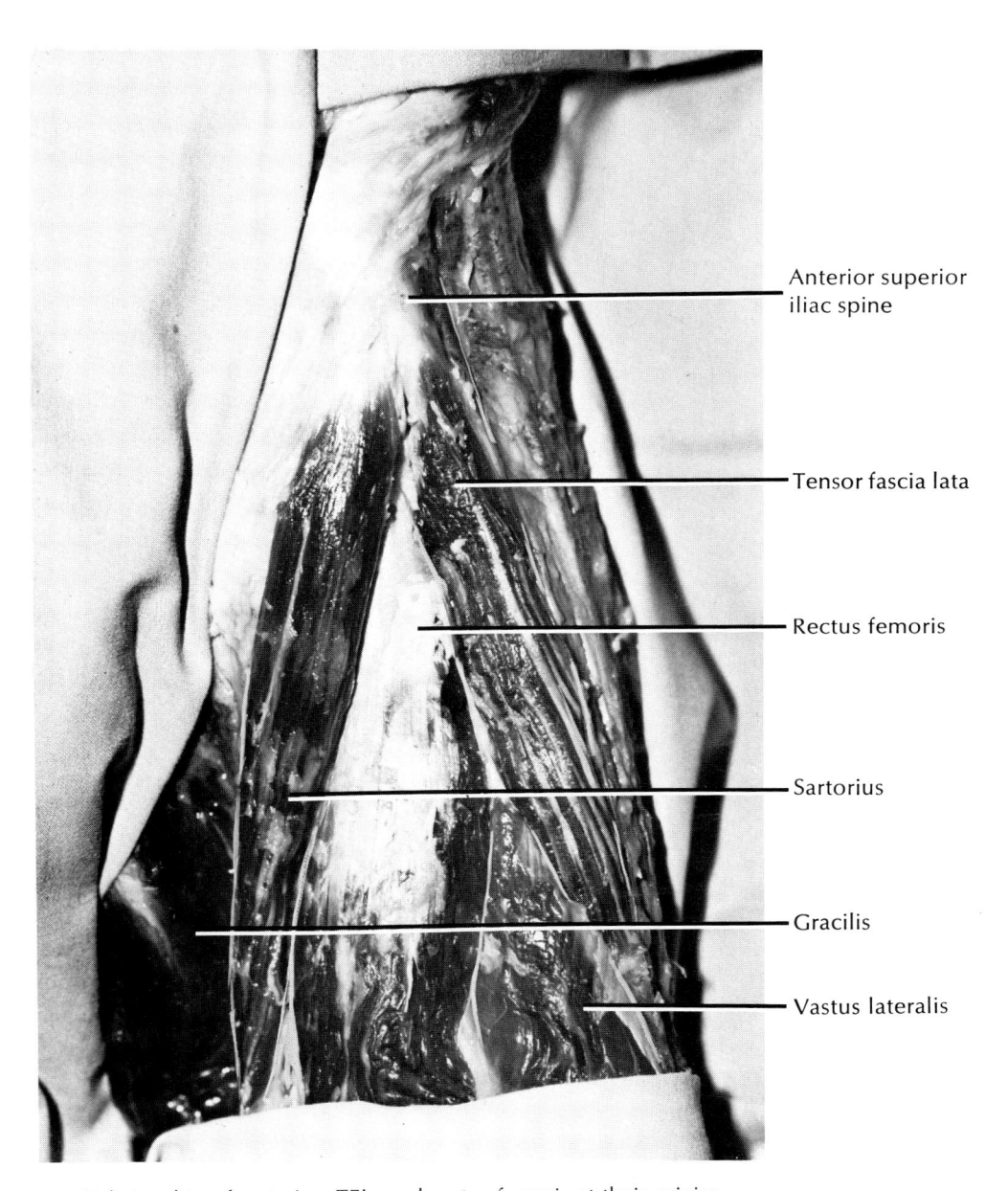

- Anterior superior iliac spine
- Tensor fascia lata
- Rectus femoris
- Sartorius
- Gracilis
- Vastus lateralis

Relationship of sartorius, TFL, and rectus femoris at their origins

ANTERIOR THIGH

Tensor fascia lata

Musculocutaneous flap for abdominal wall reconstruction

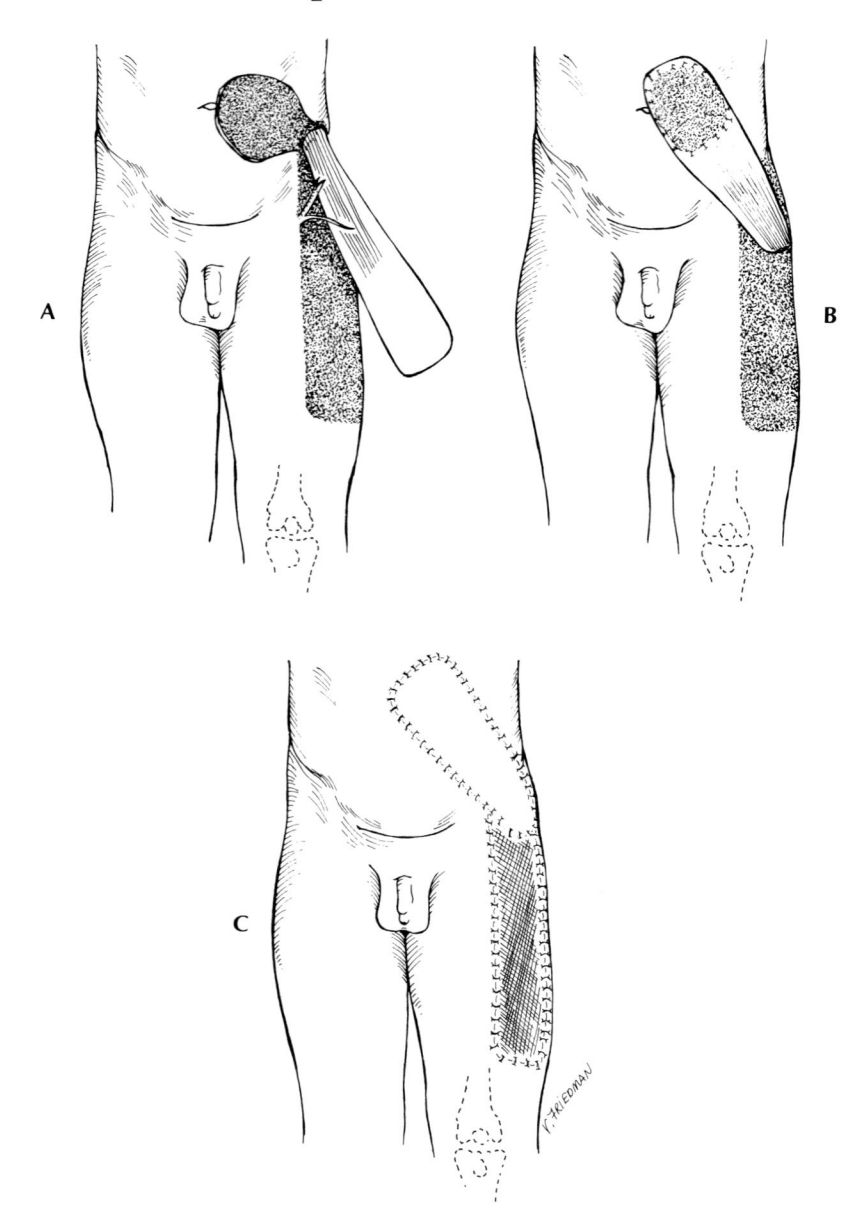

A Flap elevation.

B Fascia lata sutured into abdominal wall.

C Secondary defect will close primarily for small islands. Larger islands may need skin grafting of donor site.

ANTERIOR THIGH

Anterior arc transposition for groin reconstruction

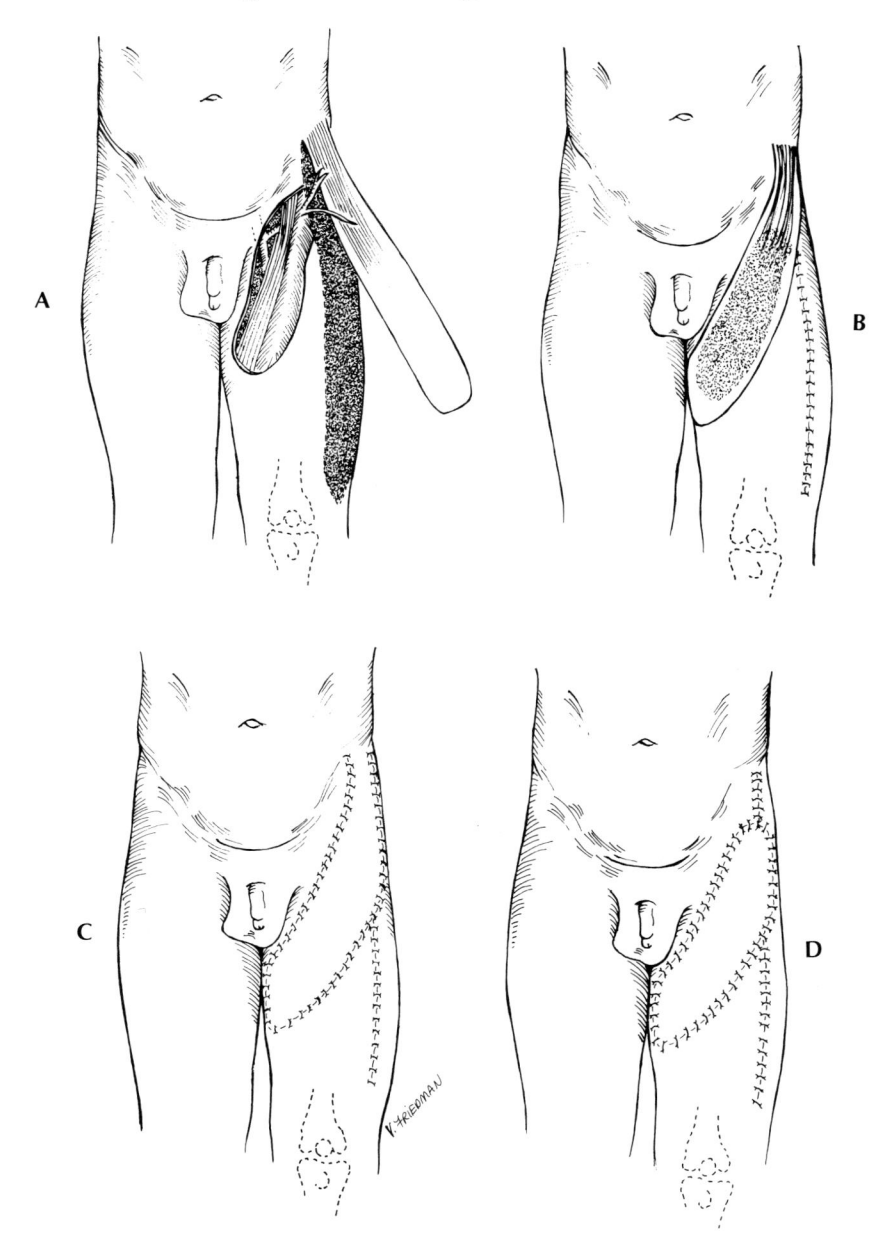

A Flap elevated.

B Transposed.

C Secondary defect is usually closed directly.

D In order to avoid "dog ears" over iliac crest, flap can be safely made an island.

ANTERIOR THIGH

Tensor fascia lata

Anterior arc transposition for groin coverage

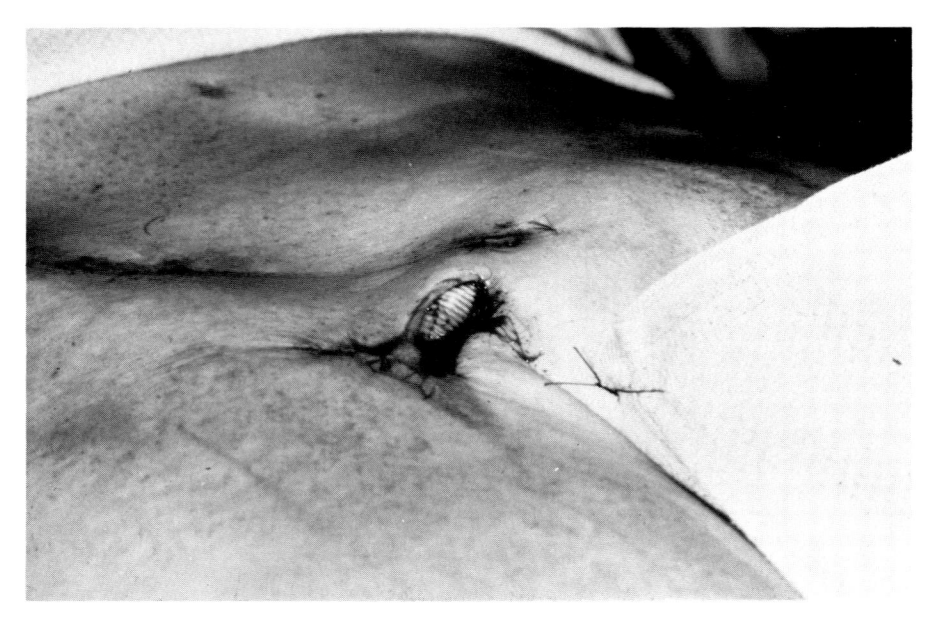

Exposed vascular femero-femoral bypass graft in groin.

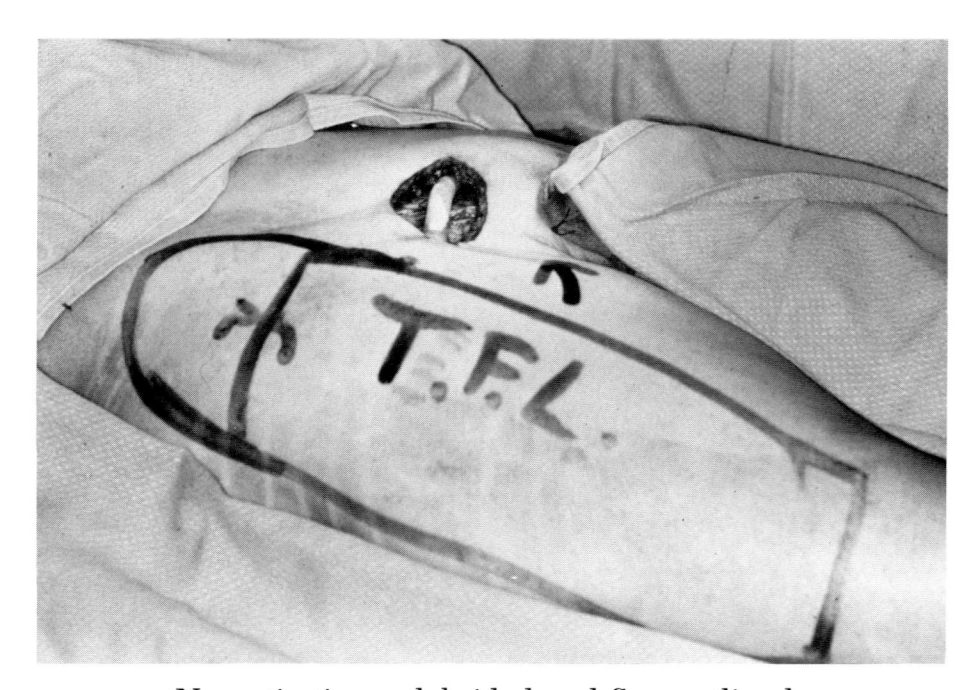

Necrotic tissue debrided and flap outlined.

ANTERIOR THIGH

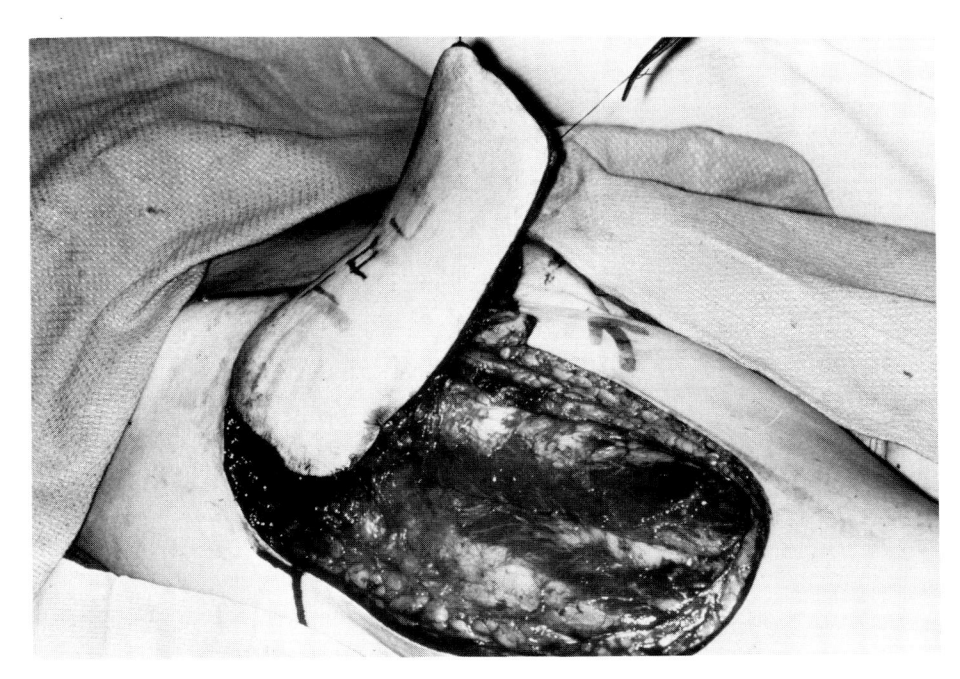

Island flap anterior transposition.

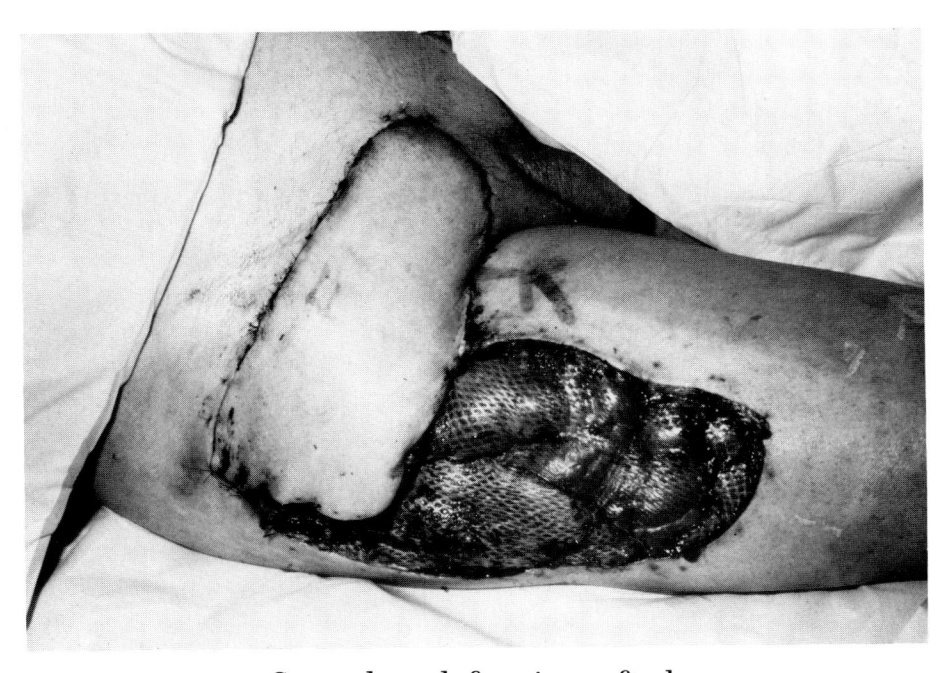

Secondary defect is grafted.

ANTERIOR THIGH

Tensor fascia lata

Ischial and trochanteric pressure sore coverage (posterior arc)

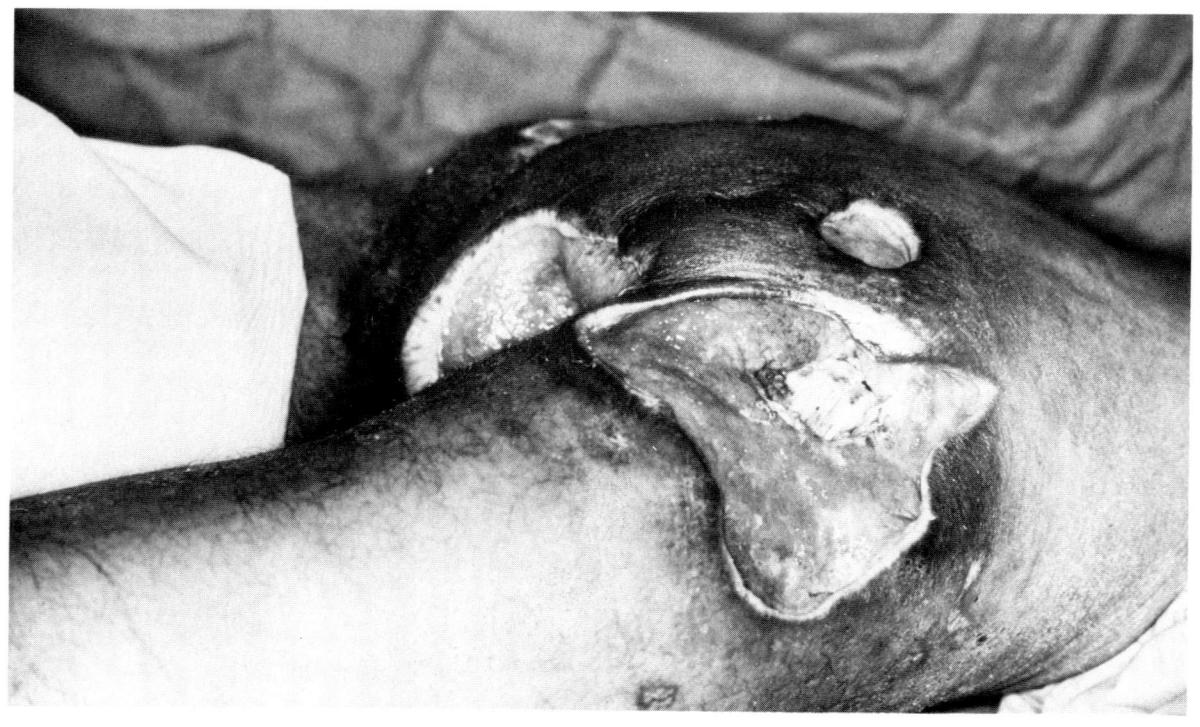

Large ischial and trochanteric pressure sores.

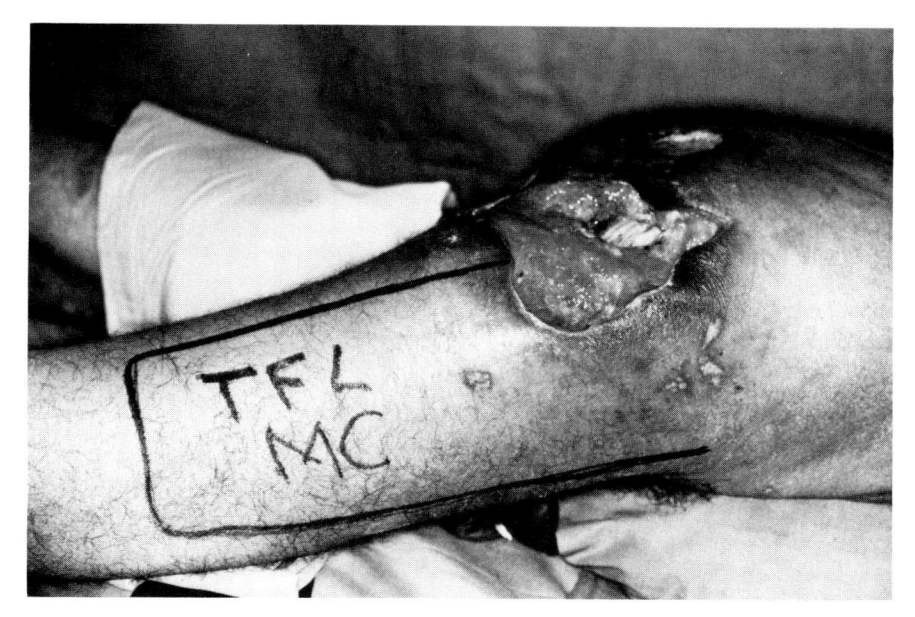

Flap outlined: extended TFL musculocutaneous flap.

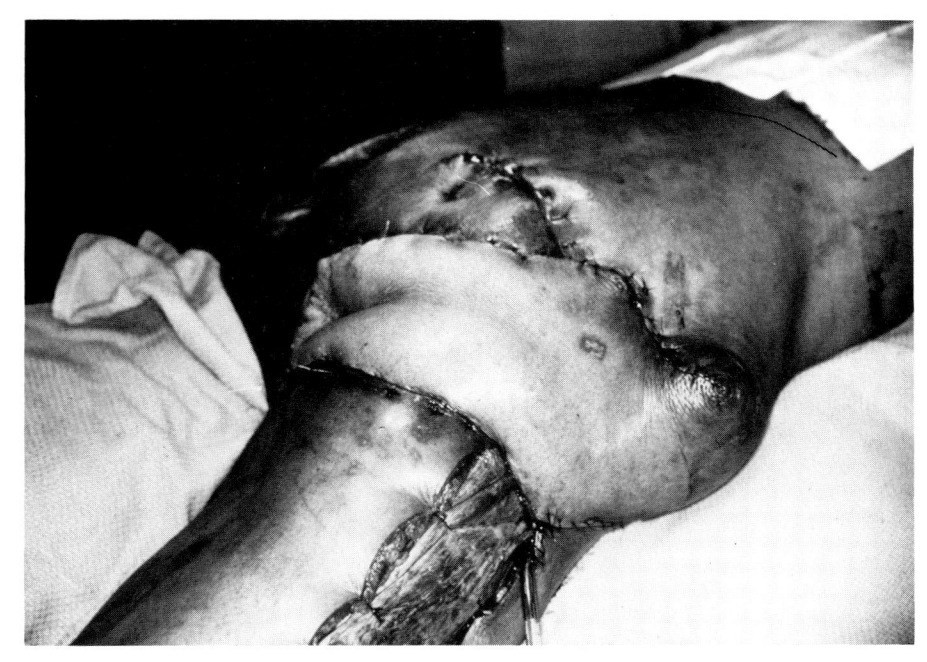

Flap transposed to cover ischium and trochanter; secondary defect grafted.

ANTERIOR THIGH

Tensor fascia lata

Trochanteric pressure sore coverage

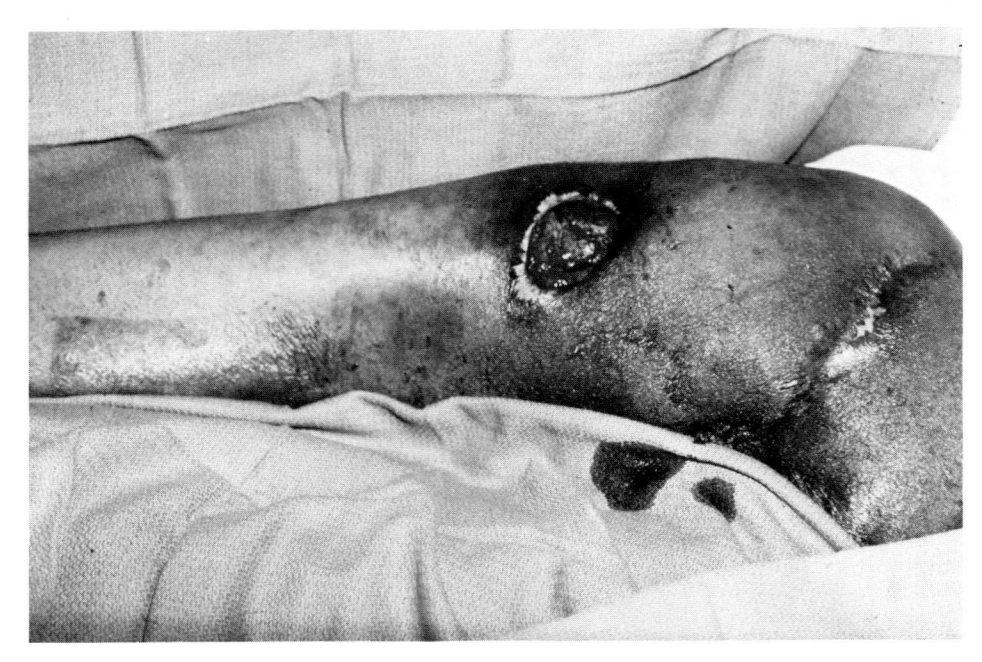

Trochanteric pressure sore.

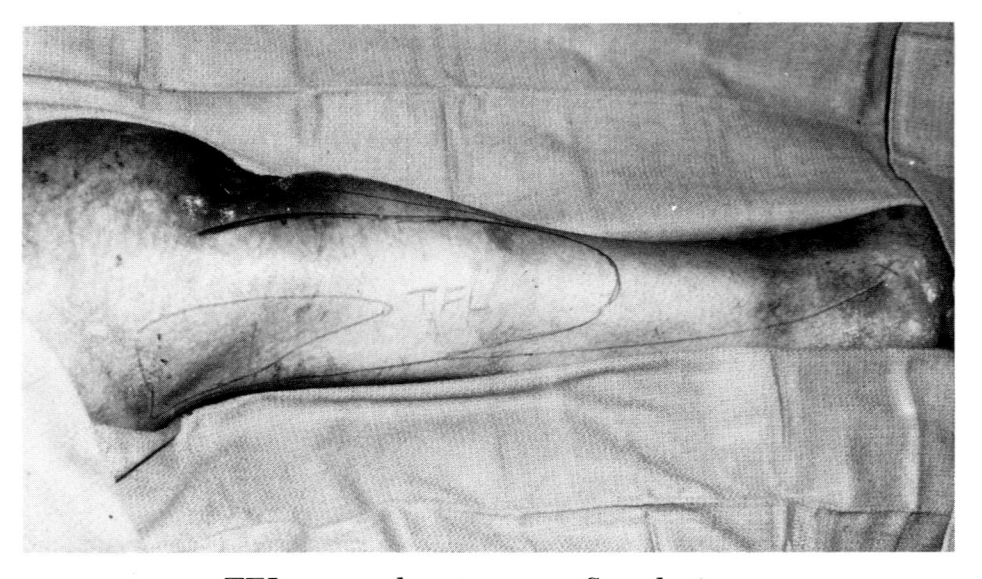

TFL musculocutaneous flap design.

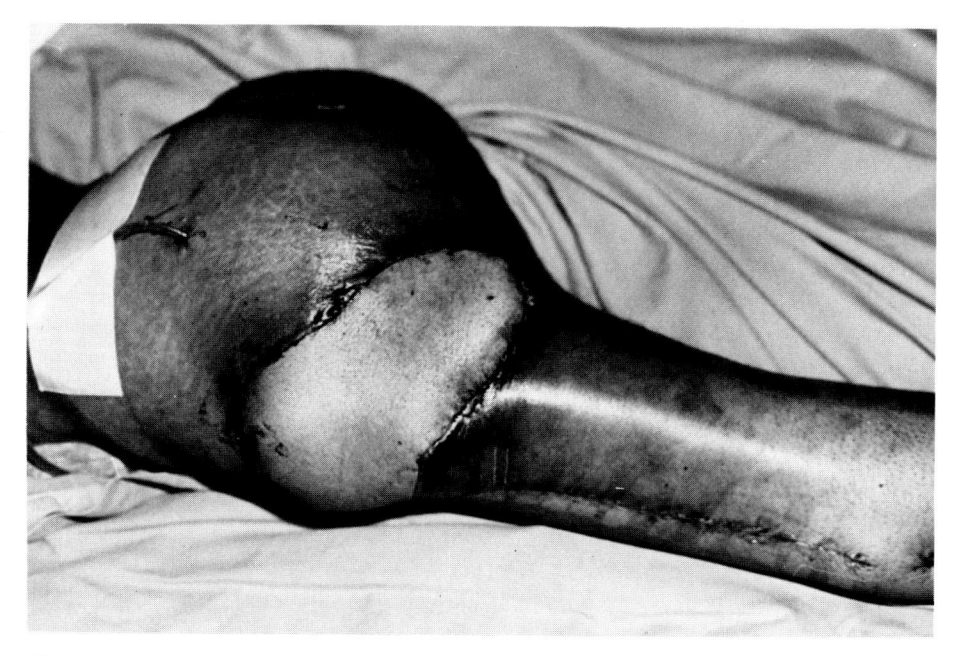

Trochanteric defect closed with TFL island musculocutaneous flap (posterior arc). Secondary defect is closed directly.

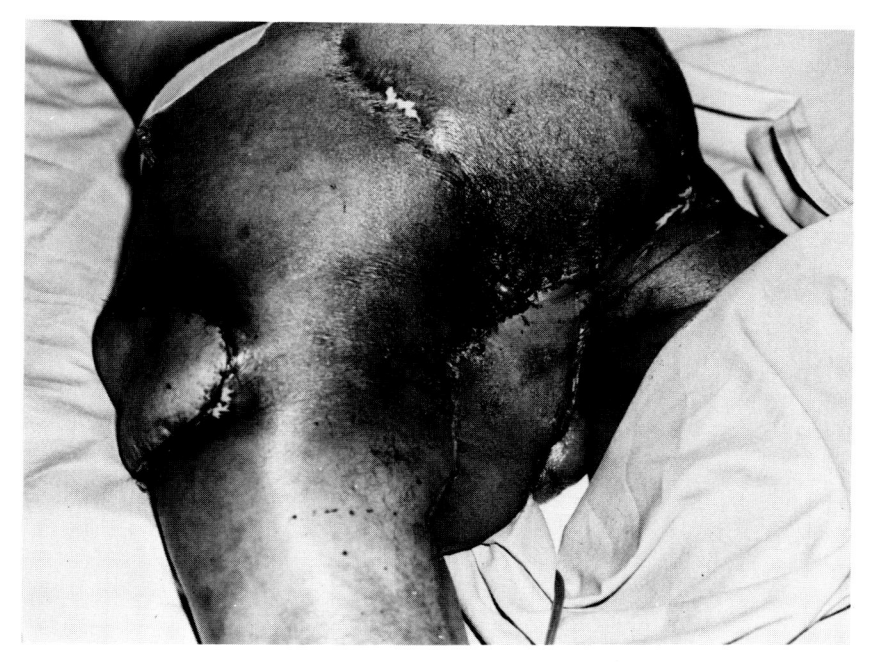

TFL flap for closure of trochanteric area. Note gracilis musculocutaneous flap closure of ischial defect.

ANTERIOR THIGH

Tensor fascia lata

Tubed pedicle flap

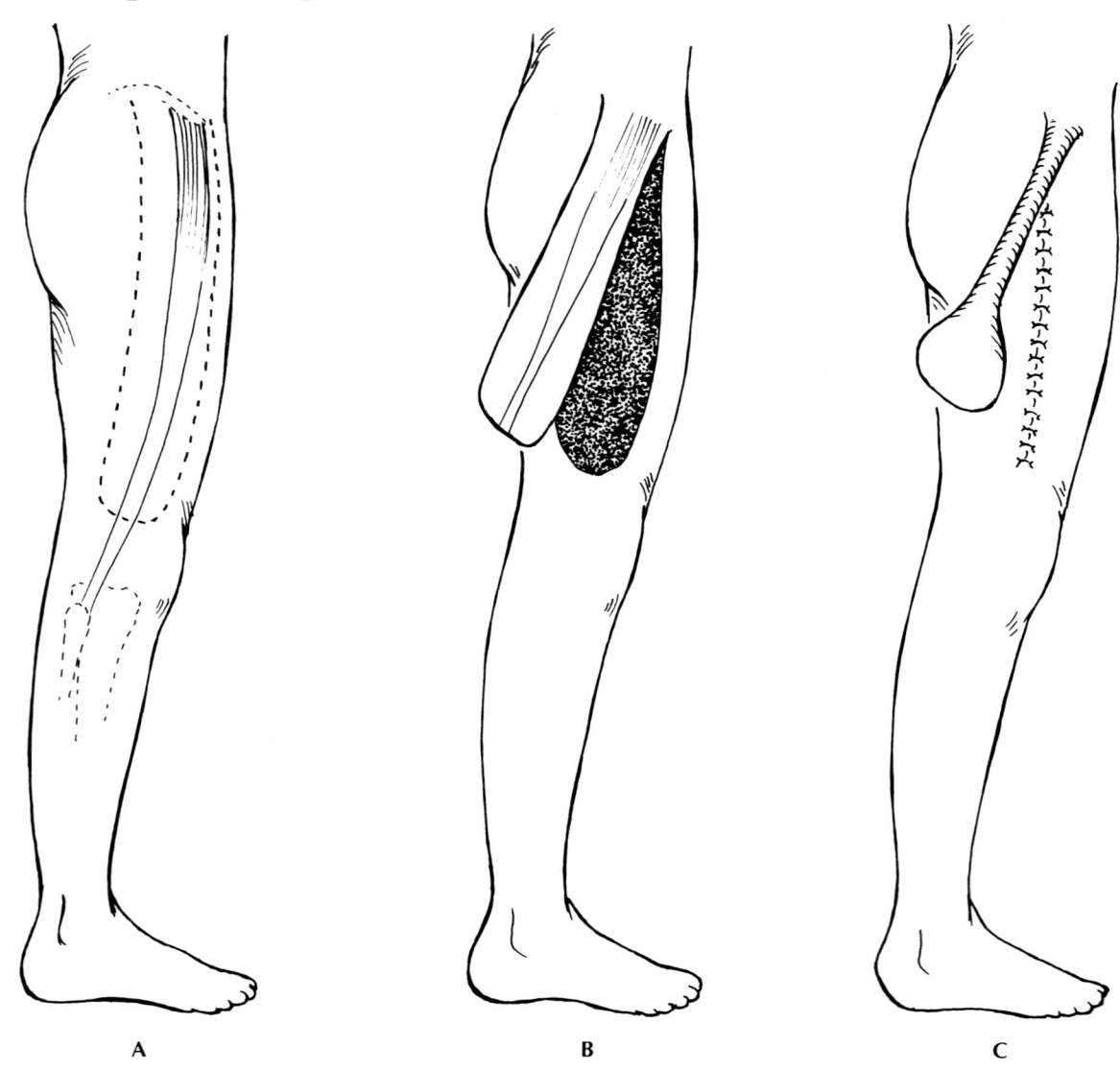

A   Flap outline.

B   Flap elevated.

C   Flap tubed and secondary defect closed directly.

ANTERIOR THIGH

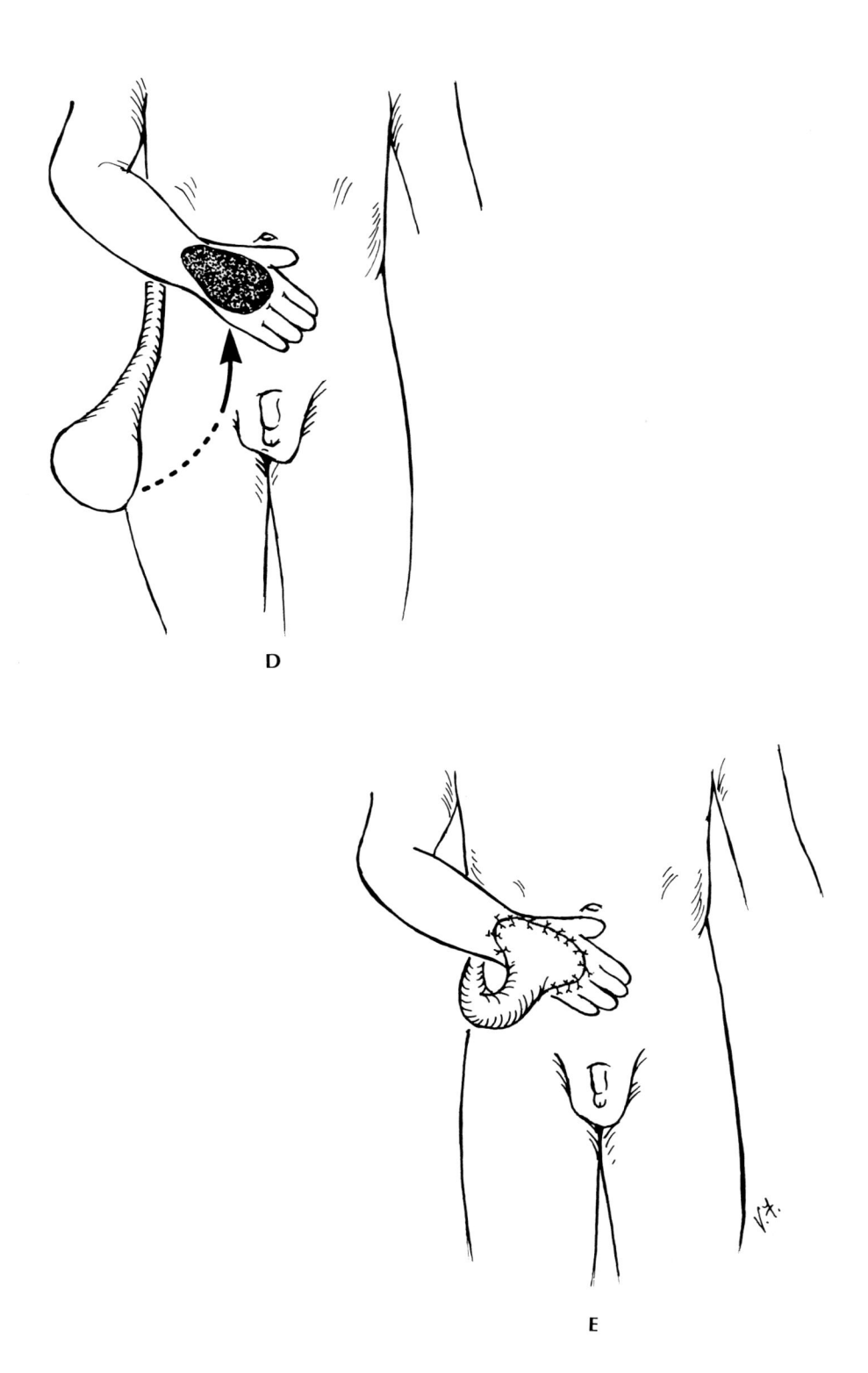

D and E Flap transposed to cover defect on dorsum of hand.

85

SECTION TWO
POSTERIOR THIGH

Gluteus maximus

Biceps femoris

Semitendinosus

Semimembranosus

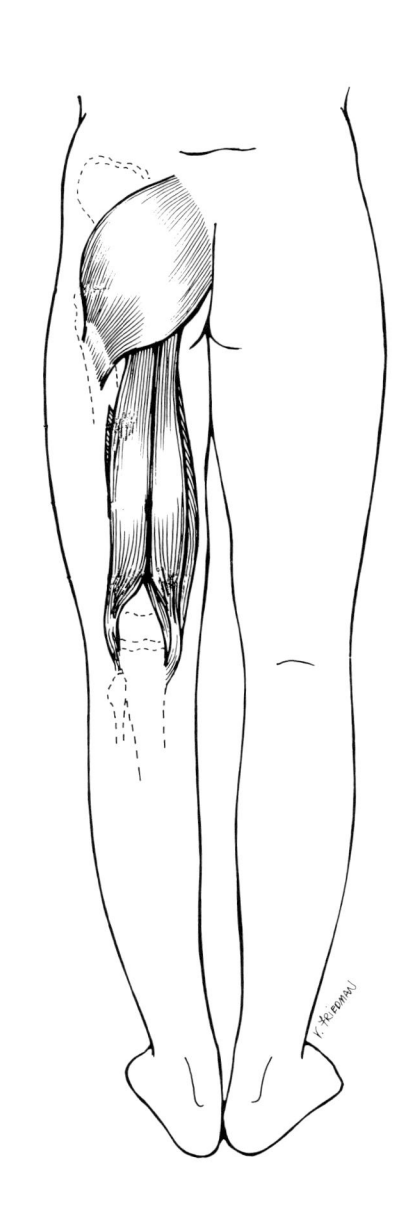

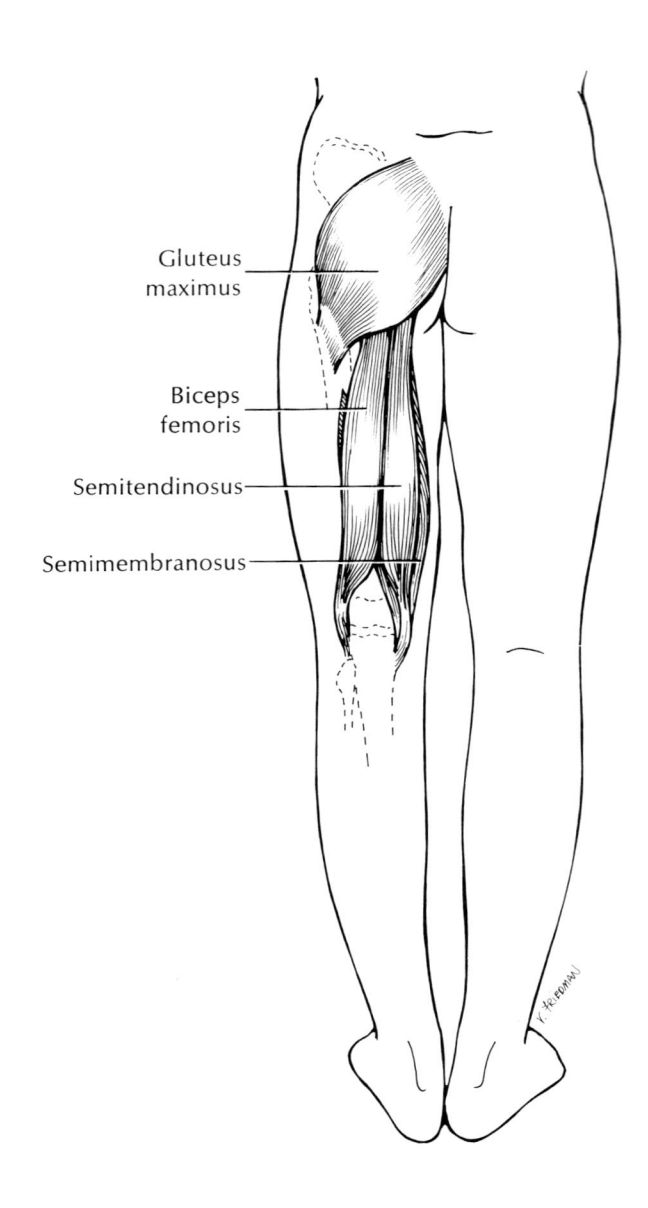

POSTERIOR THIGH

POSTERIOR THIGH

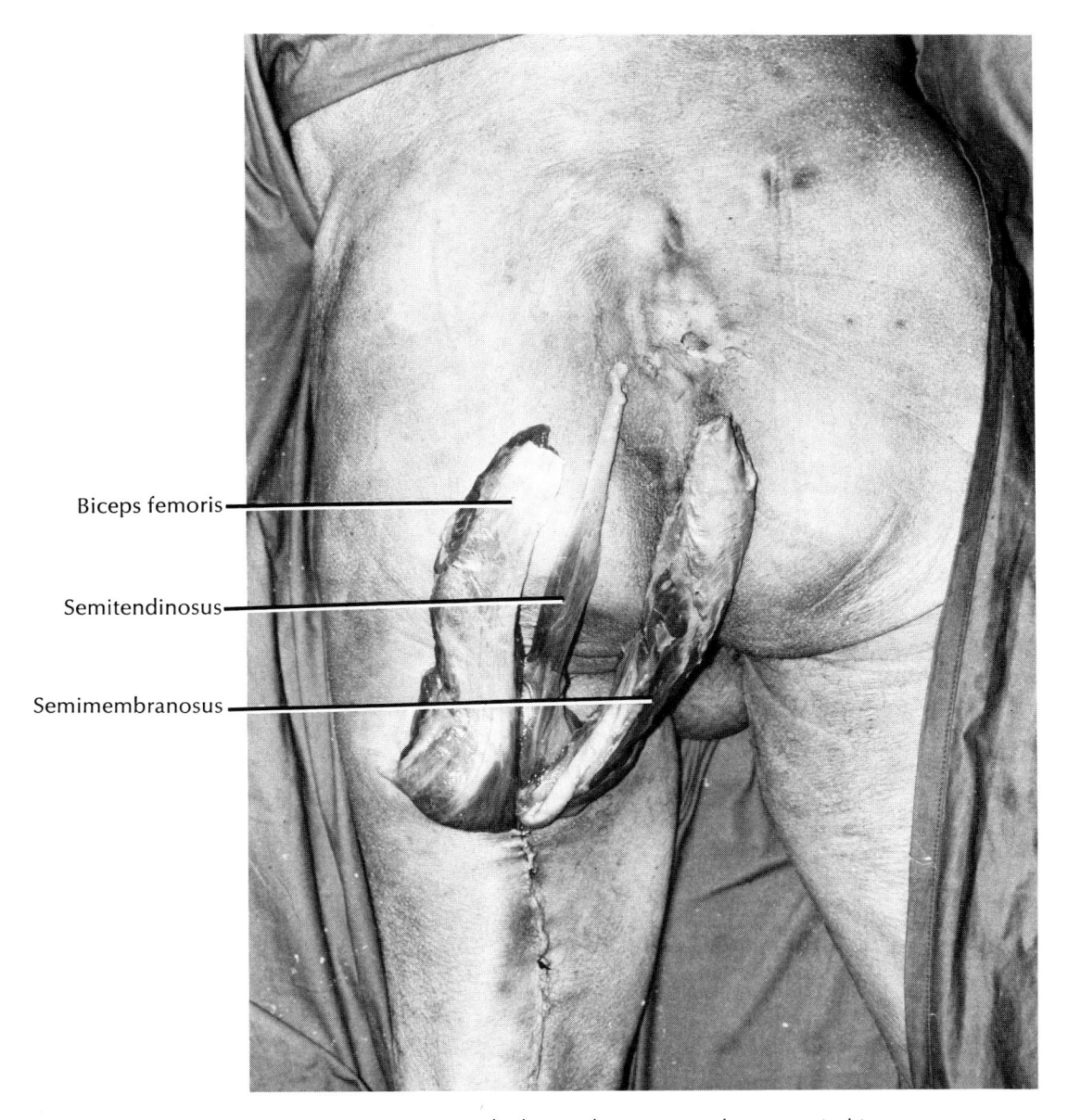

Posterior thigh muscles transposed to cover ischium

The posterior thigh muscles are most useful as muscle and musculocutaneous flaps for coverage of the ischial and sacral areas.

POSTERIOR THIGH

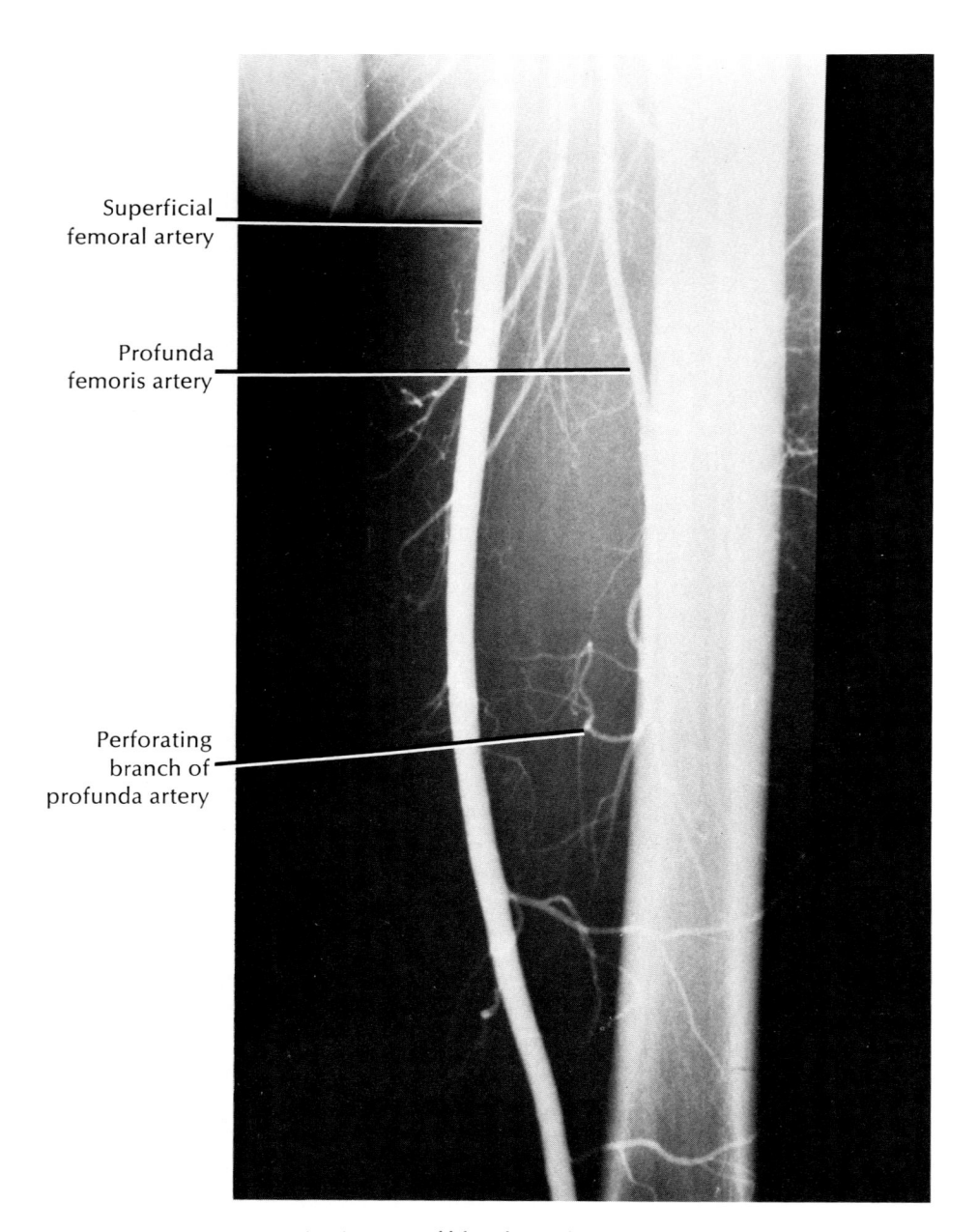

Angiogram of blood supply to posterior thigh muscles

(Labels: Superficial femoral artery; Profunda femoris artery; Perforating branch of profunda artery)

Blood supply

The blood supply to the posterior thigh (hamstring) muscles is based on the perforating branches of the profunda femoris and direct branches of the superficial femoral artery. The gluteus maximus is supplied by the superior and inferior gluteal branches of the hypogastric artery.

… POSTERIOR THIGH

Gluteus maximus

MUSCLE FLAP
MUSCULOCUTANEOUS FLAP

Applications

Coverage of:
SACRUM
ISCHIUM

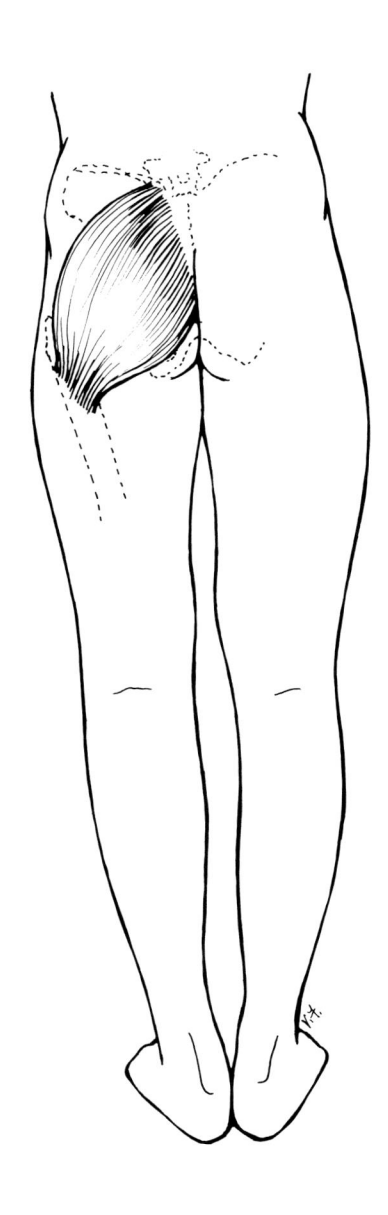

POSTERIOR THIGH

Gluteus maximus

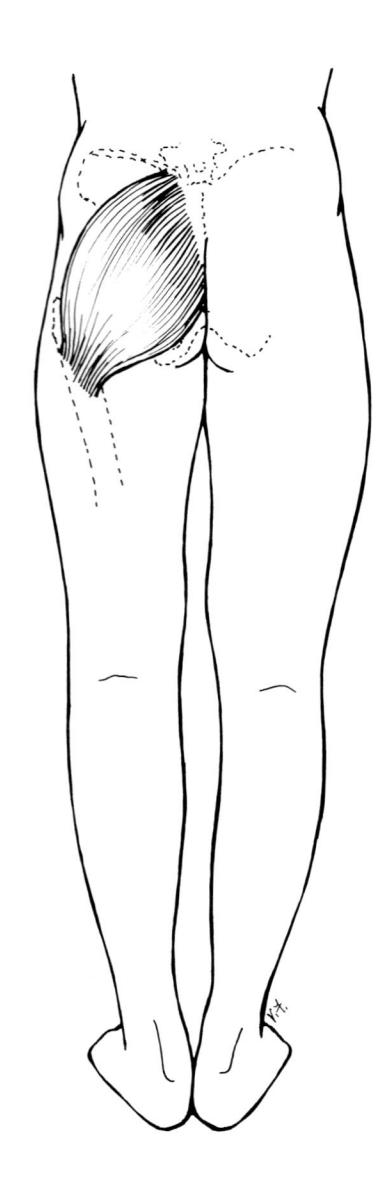

The gluteus maximus is a thick, broad muscle, which is the most superficial of the gluteal muscle group.

Origin: Gluteal line of ilium and sacrum.
Insertion: The greater tuberosity of the femur and the iliotibial tract.

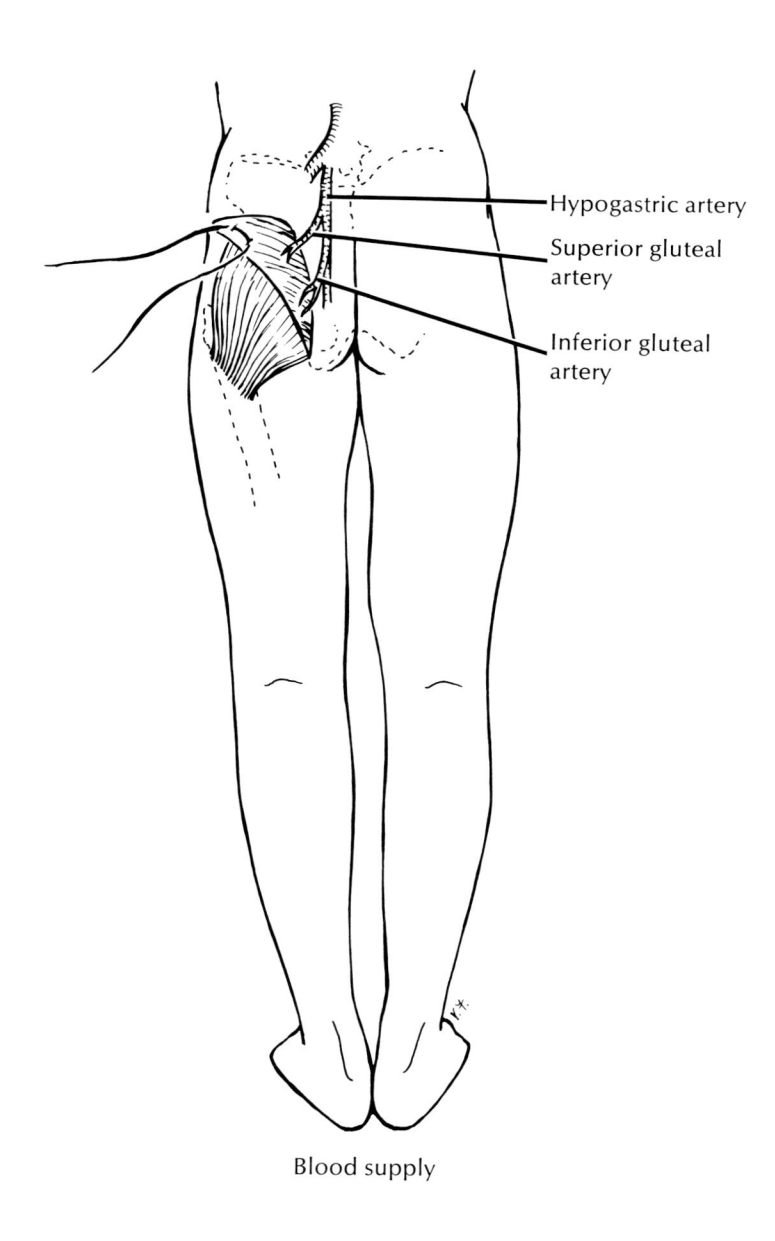

POSTERIOR THIGH

Blood supply

- Hypogastric artery
- Superior gluteal artery
- Inferior gluteal artery

Nerve supply:
  Motor—The inferior gluteal nerve innervates the muscle on its deep surface.
Function: As a strong extensor and lateral thigh rotator, the gluteus maximus is not an expendable muscle. However, either the superior or inferior half of the muscle can be used without significant functional loss in the ambulatory patient. In the paraplegic the entire muscle is expendable and may be used as a transposition flap.

POSTERIOR THIGH

Gluteus maximus

Blood supply

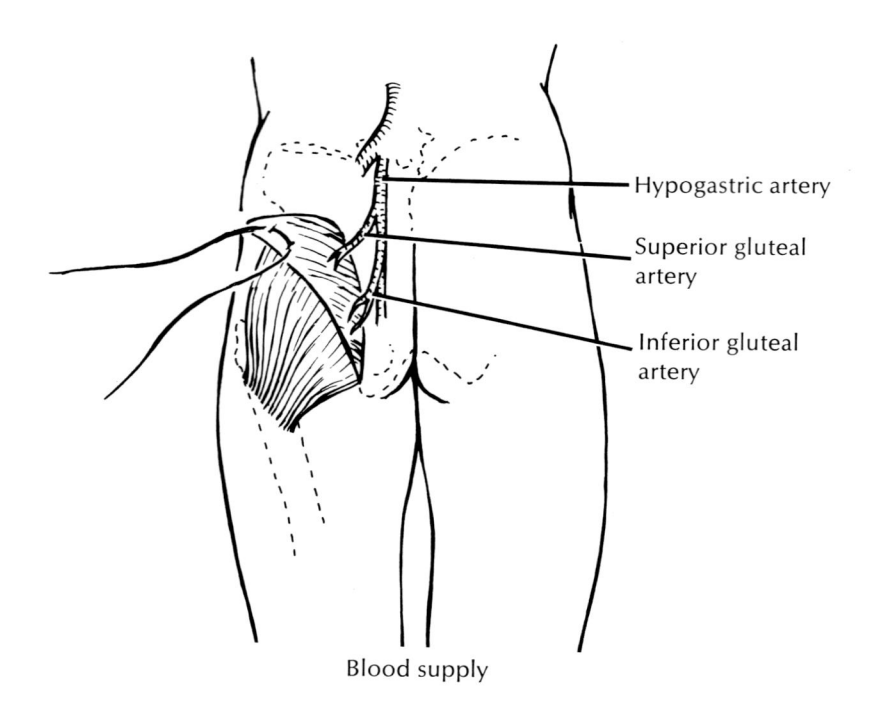

Blood supply

The superior gluteal artery supplies the upper part of the muscle and the inferior gluteal artery the lower half. Based on these two separate upper and lower pedicles, the muscle can be divided into upper and lower halves. The superior and inferior gluteal arteries are separate branches of the hypogastric artery. The superior gluteal artery is a large vessel that courses posteriorly out of the pelvis above the piriform muscle. It then supplies a branch to the upper half of the gluteus maximus muscle. The artery then courses laterally to supply the gluteus medius and gluteus minimus muscles. The inferior gluteal artery enters the gluteal region through the sciatic foramen in close relationship to the sciatic nerve. This vessel supplies the lower half of the gluteus maximus muscle. Both vascular pedicles are medially located.

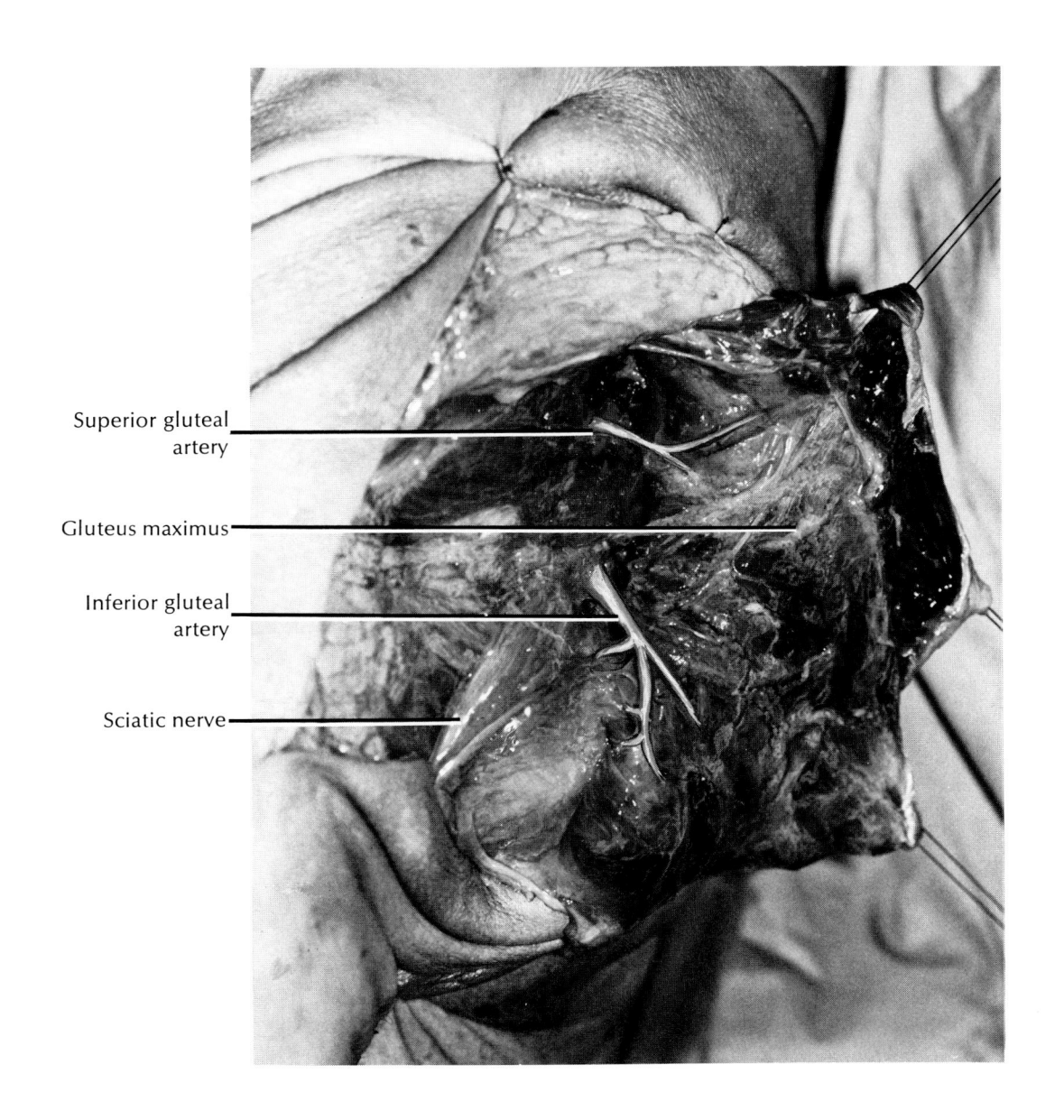

POSTERIOR THIGH

- Superior gluteal artery
- Gluteus maximus
- Inferior gluteal artery
- Sciatic nerve

POSTERIOR THIGH

Gluteus maximus

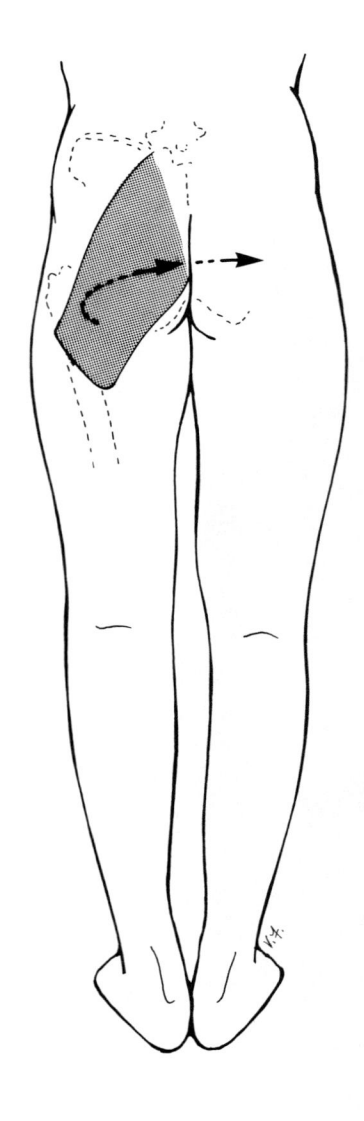

Total muscle transposition for sacral cover.

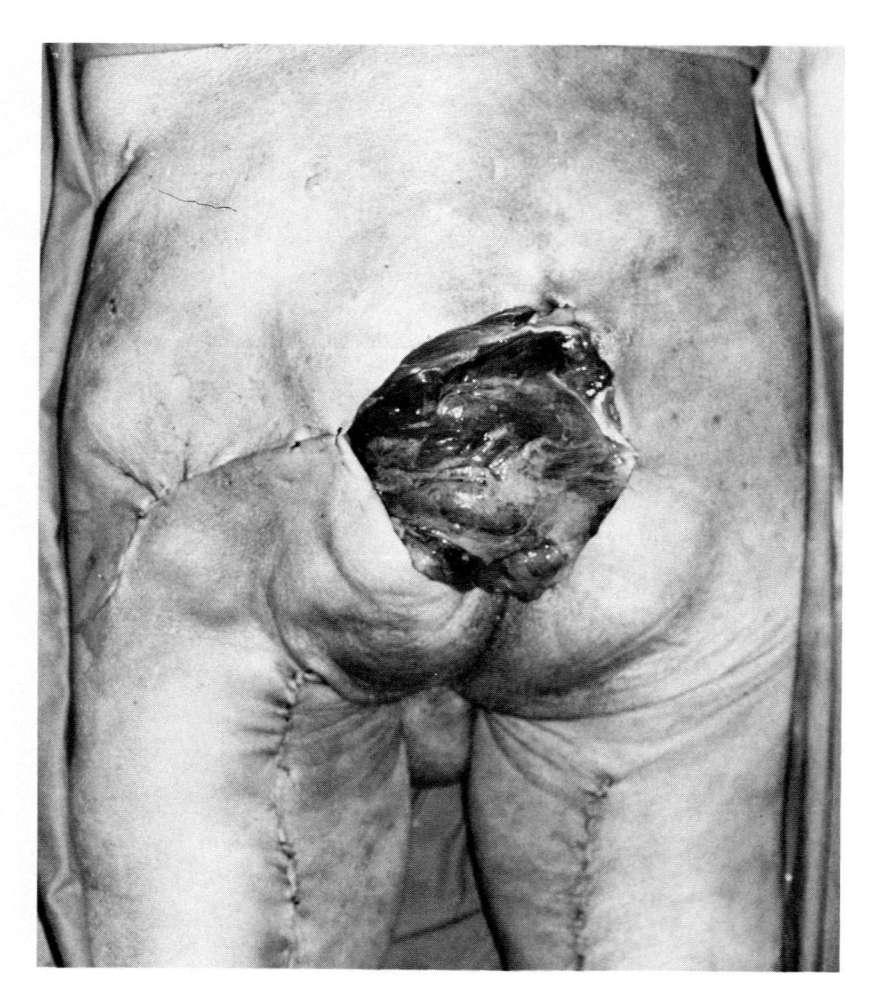

Arc of rotation

Based on the medially located vascular pedicles, the muscle with the entire overlying buttock skin or with a selected island of skin can be transposed superiorly or inferiorly. Inferior transposition will cover the ischial area. Either half or the entire muscle without skin can be folded on itself to cover the sacrum.

POSTERIOR THIGH

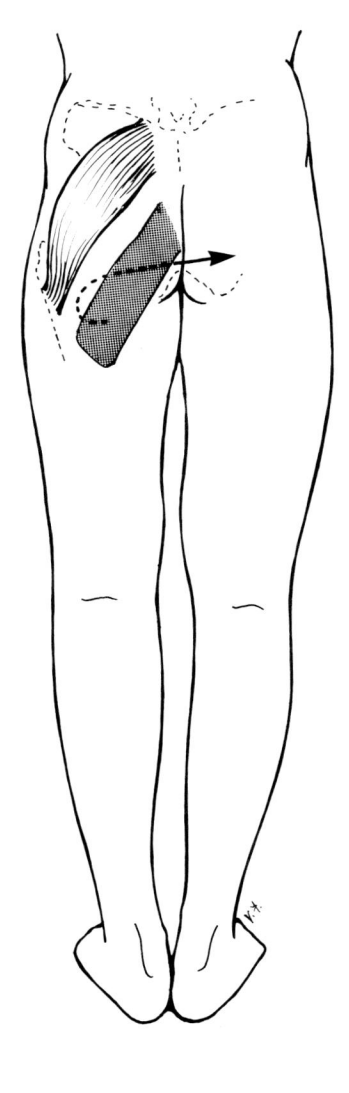

Half muscle transposition for sacral cover.

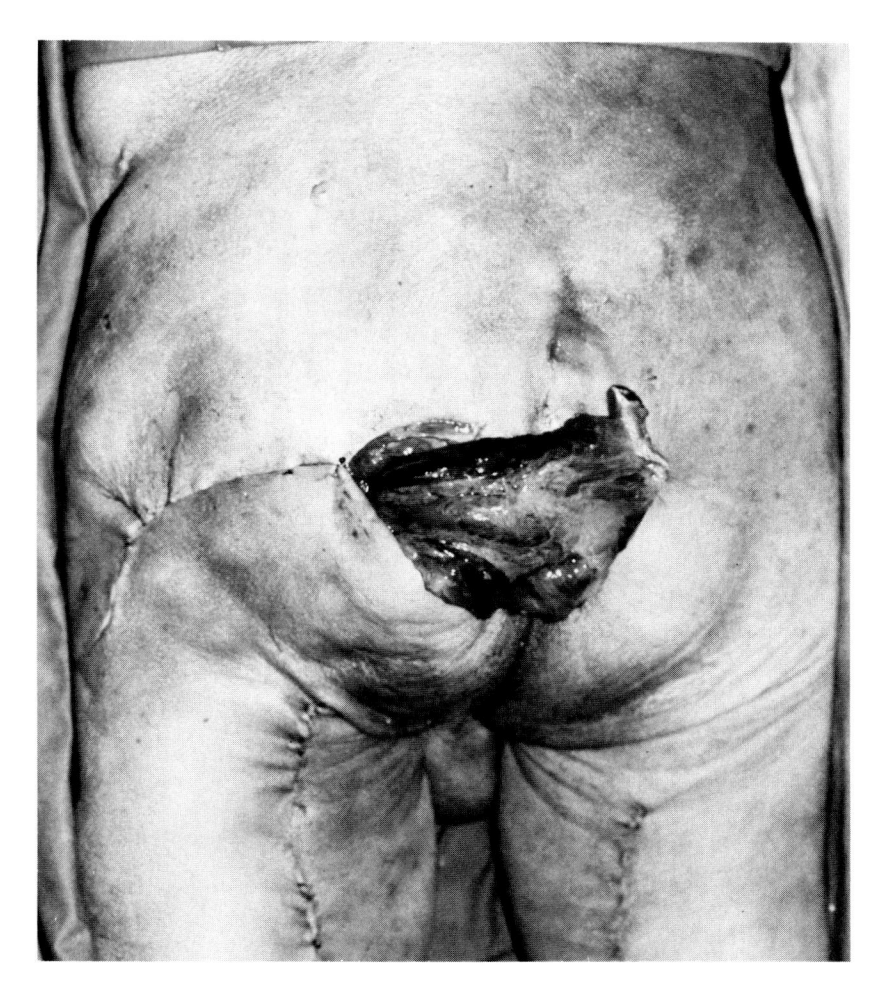

97

POSTERIOR THIGH

Gluteus maximus

Elevation of flap

The gluteus maximus muscle is identified either through an oblique incision on the buttock or a curvilinear incision extending over the iliac crest. The latter incision is often necessary for paraplegics who have previously undergone rotational buttock flaps. For musculocutaneous flaps the cutaneous territory may include the entire buttock skin or a selected island. In the ambulatory patient only an inferior or superior skin island should be incorporated with the underlying half of the muscle. An entire gluteus maximus muscle transposition in an ambulatory patient may result in hip instability. After the superior and inferior borders of the muscle are isolated, the muscle is bluntly elevated from the underlying gluteus medius muscle through the trochanteric bursae. Either the entire insertion or the appropriate half of the insertion is divided. The muscle or appropriate half of muscle is then elevated in a medial direction. Approximately 5 cm from the sacral edge the main trunks of the superior and inferior gluteal arteries can be visualized anterior to the muscle and preserved. The sciatic nerve is closely related to the inferior gluteal artery. The muscle or musculocutaneous unit is then transposed for reconstructive purposes. The donor area can be closed primarily with suction drains placed to collapse the large cavity in the former muscle bed.

Precautions

- Transposition of the entire gluteus maximus muscle in the ambulatory patient may cause a functional disability.
- The lower half of the inferior gluteus maximus muscle has close proximity to the sciatic nerve.

POSTERIOR THIGH

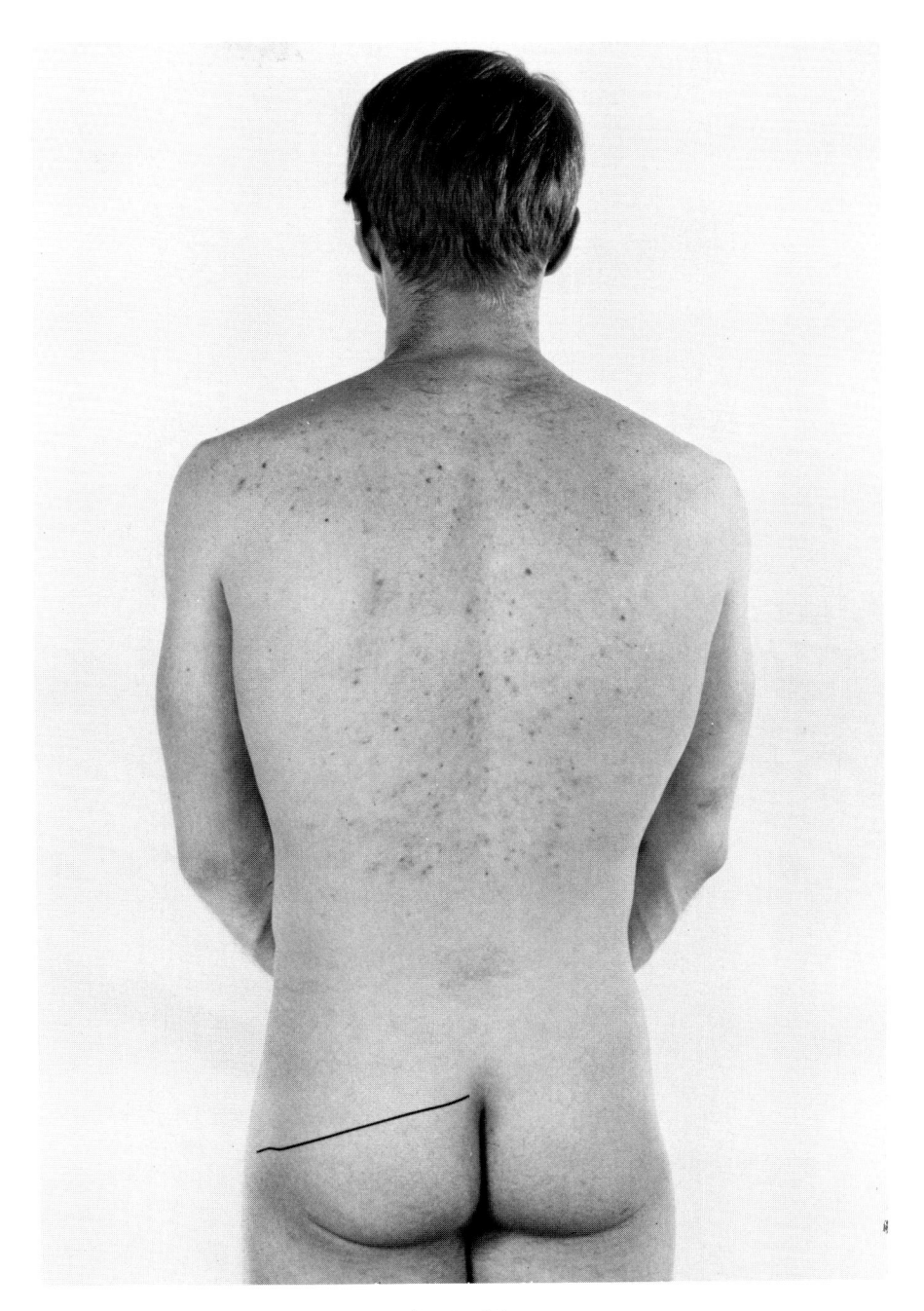

Skin incision

POSTERIOR THIGH

Gluteus maximus

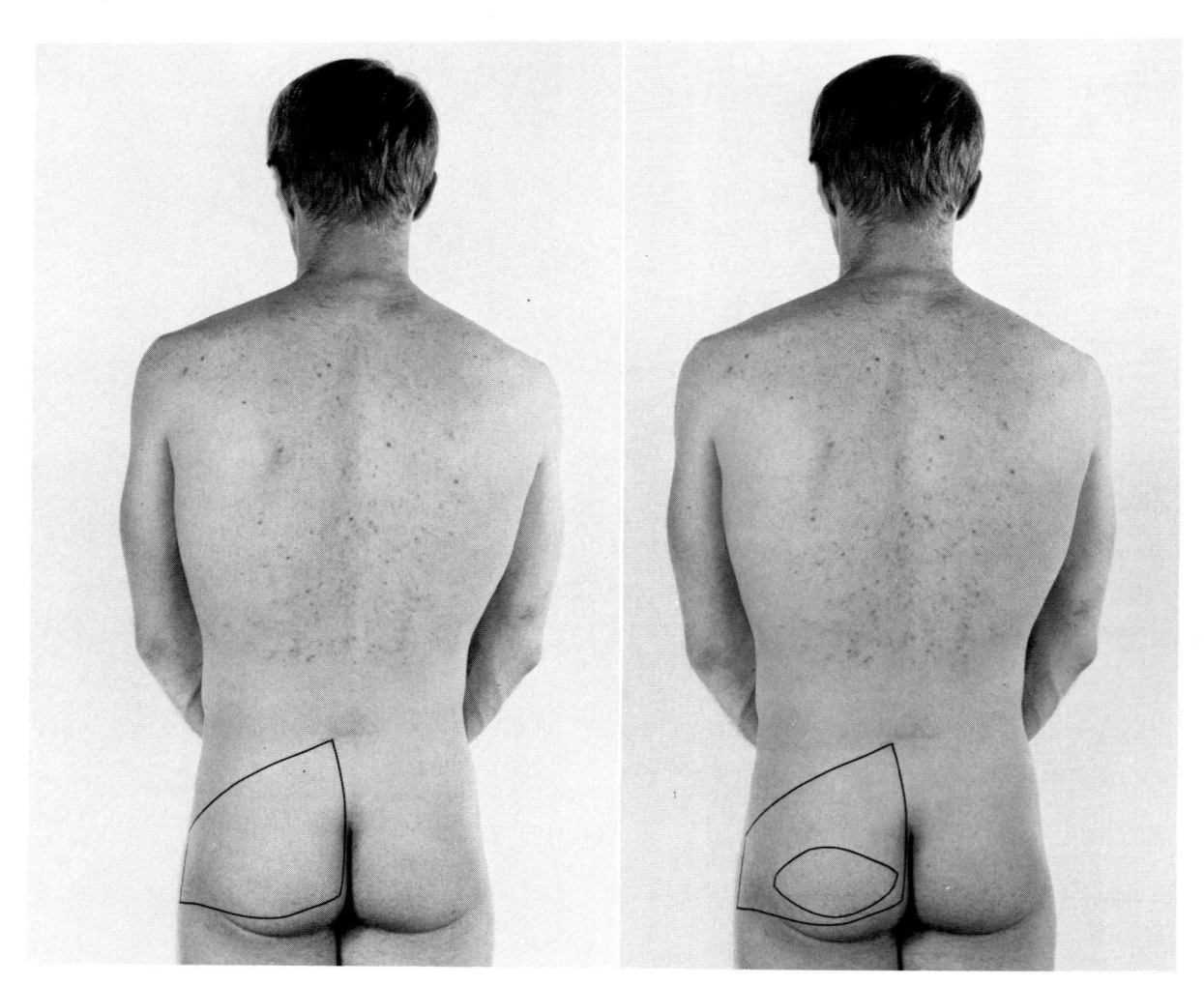

Skin markings outline muscle margins

Skin markings outline margins of muscle and skin island

POSTERIOR THIGH

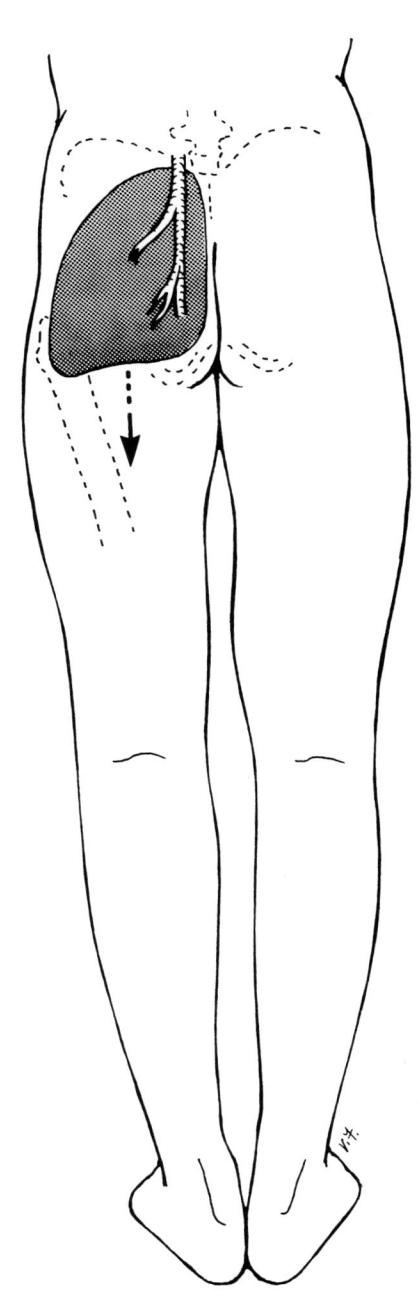

Inferior transposition
for coverage of ischium

POSTERIOR THIGH

Gluteus maximus

Musculocutaneous island flap for coverage of ischial pressure sore

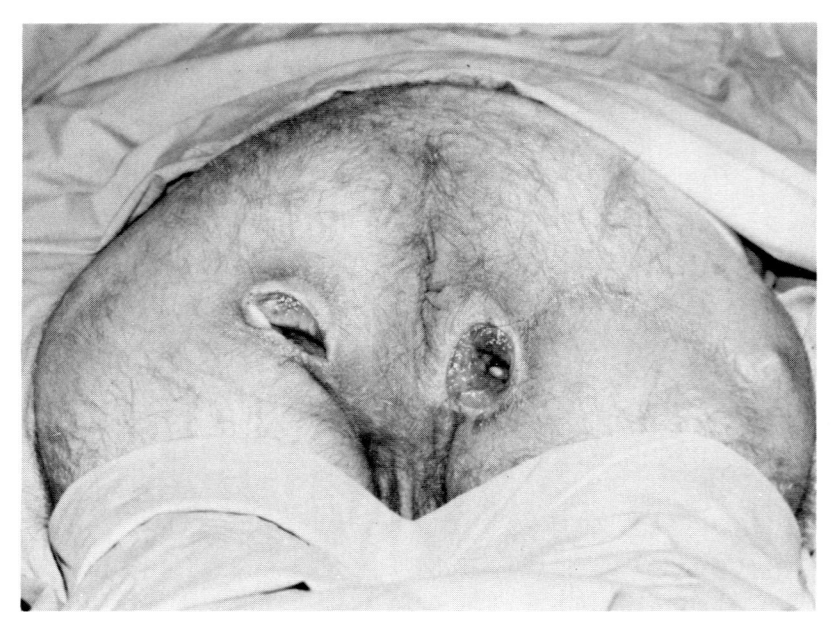

Paraplegic patient with bilateral ischial pressure sores.

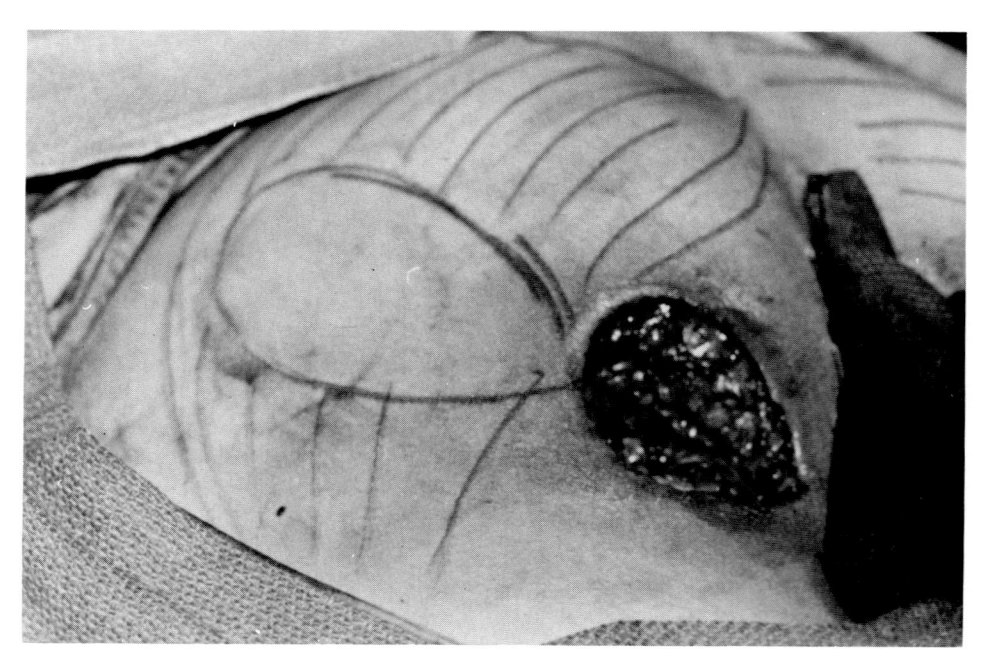

Sore excised, ischiectomy performed, and skin island over gluteus maximus is outlined.

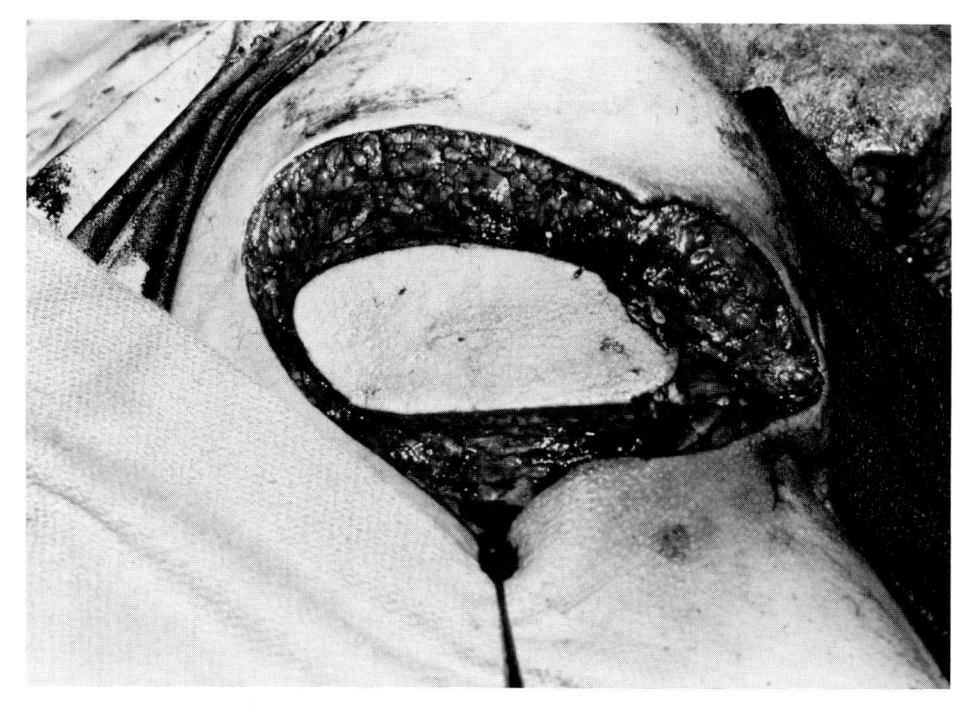

Muscle and skin island elevated and transposed inferiorly.

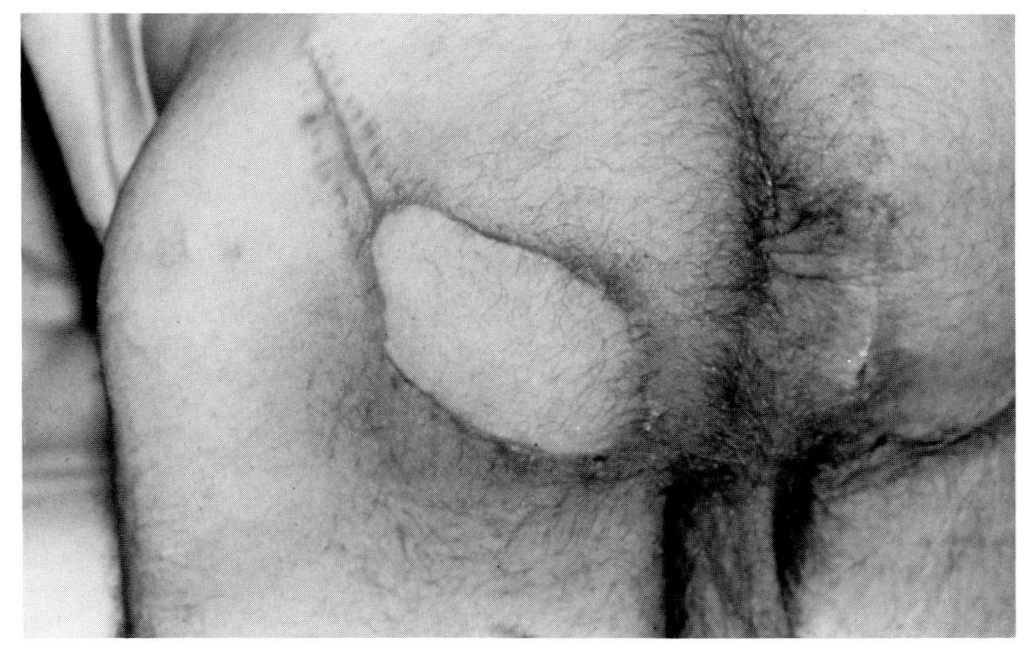

Healed flap; contralateral ischial sore was closed with rotation gluteus maximus musculocutaneous flap.

Biceps femoris

MUSCLE FLAP
MUSCULOCUTANEOUS FLAP

Applications

Coverage of:
ISCHIUM
PERINEUM
BUTTOCKS
TROCHANTER

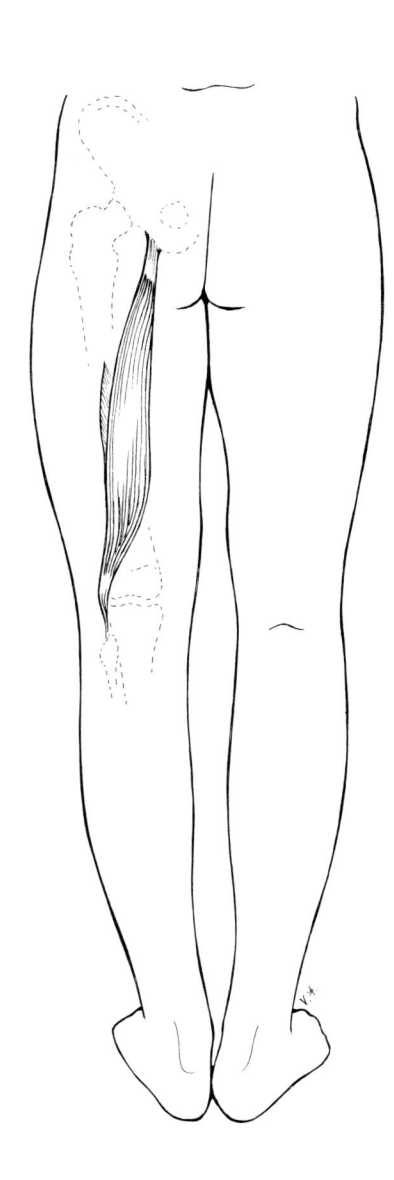

POSTERIOR THIGH

Biceps femoris

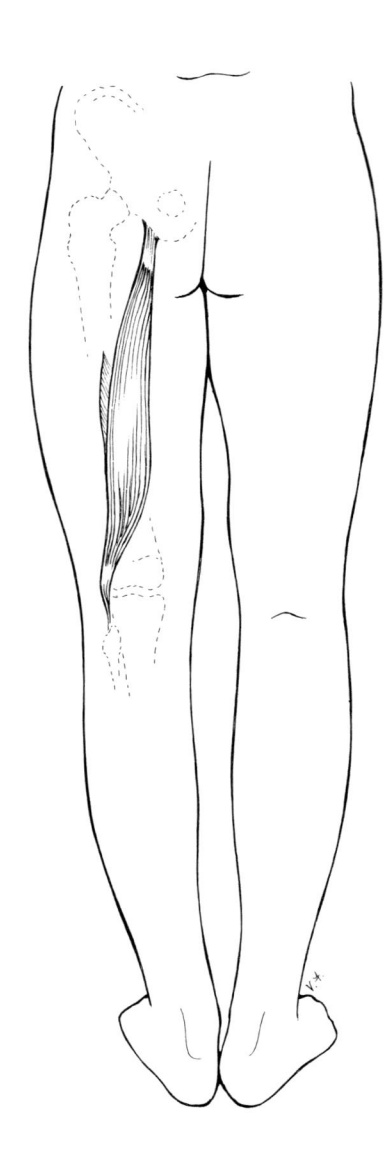

The biceps femoris is the most lateral and largest of the posterior thigh (hamstring) muscle group. It has a tendinous origin and insertion, but a large fusiform muscle belly. It lies deep to the gluteus maximus at its origin and is lateral to the semitendinosus throughout its length. As it crosses from medial to lateral, it covers the sciatic nerve, and at its insertion is closely related to the common peroneal nerve.

Origin: Long head—ischial tuberosity; short head—linea aspera of femur.
Insertion: Head of fibula.

POSTERIOR THIGH

Profunda femoris artery

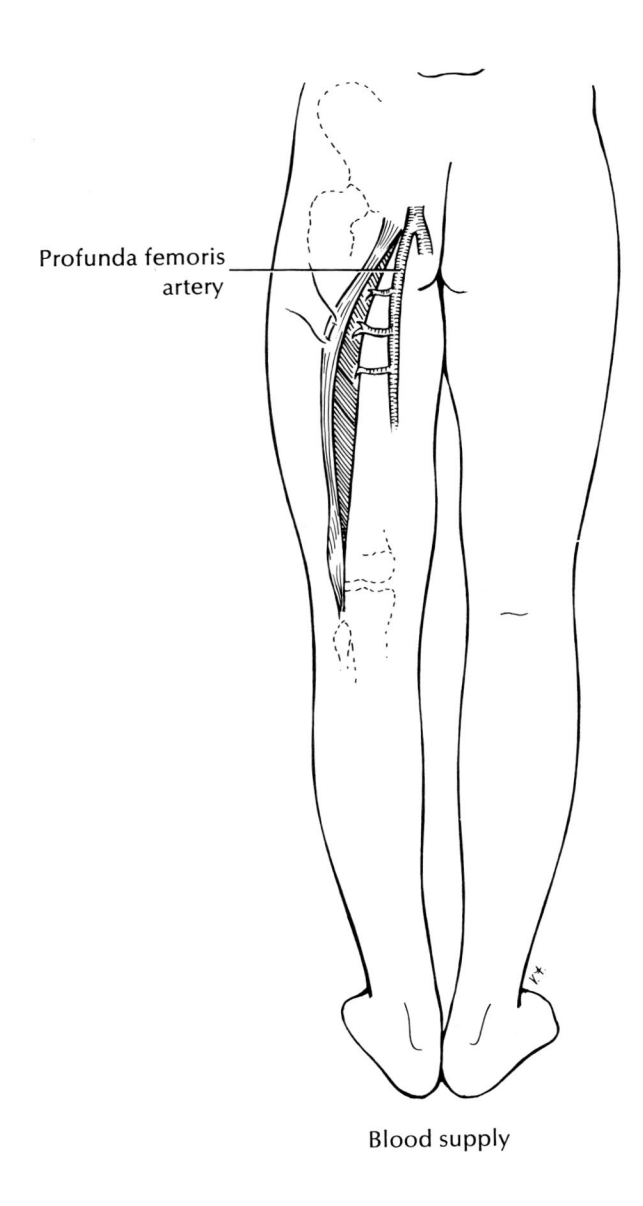

Blood supply

Nerve supply:
    Motor— Branches of sciatic nerve. The motor branches enter the proximal muscle on its deep surface.
    Sensory— Posterior cutaneous nerve of the thigh.
Function: This is not an expendable muscle, since it is a powerful flexor of the leg.

POSTERIOR THIGH

Biceps femoris

Profunda femoris artery

Blood supply

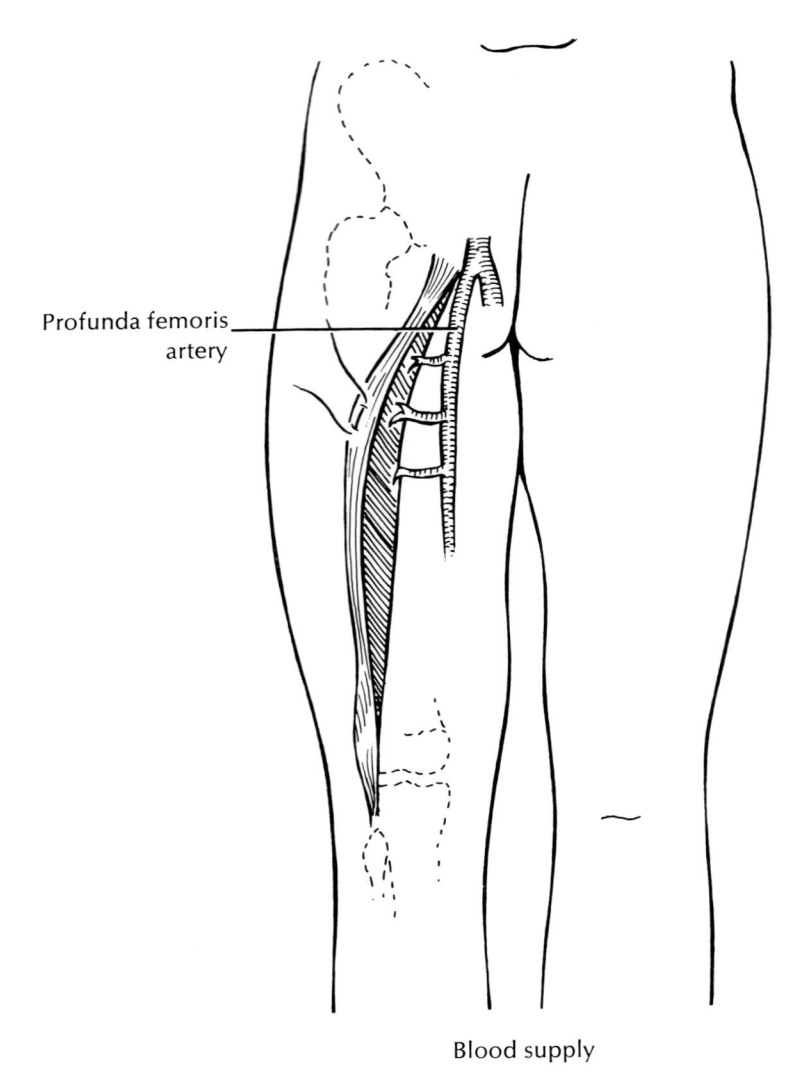

Blood supply

The blood supply to the biceps femoris is based on the perforating branches of the profunda femoris that course through the adductor magnus. The adductor magnus is deep to the biceps femoris. The vessels enter the muscle on its anterior surface. There are usually two or three branches to the long head and two to the short head. The most proximal vessels must be preserved to ensure muscle survival. There are small, perforating musculocutaneous vessels from the biceps femoris into its overlying cutaneous territory.

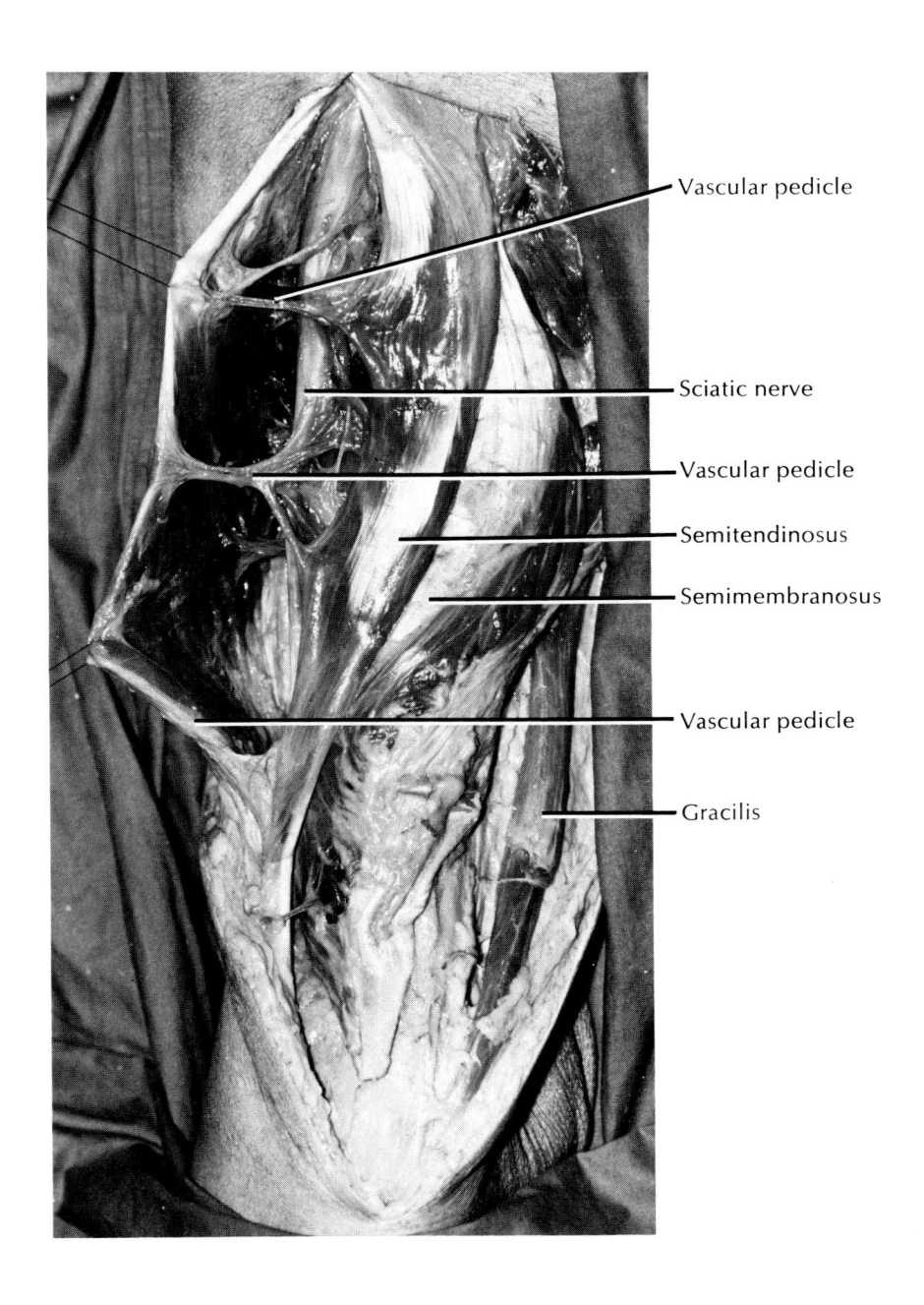

POSTERIOR THIGH

- Vascular pedicle
- Sciatic nerve
- Vascular pedicle
- Semitendinosus
- Semimembranosus
- Vascular pedicle
- Gracilis

POSTERIOR THIGH

Biceps femoris

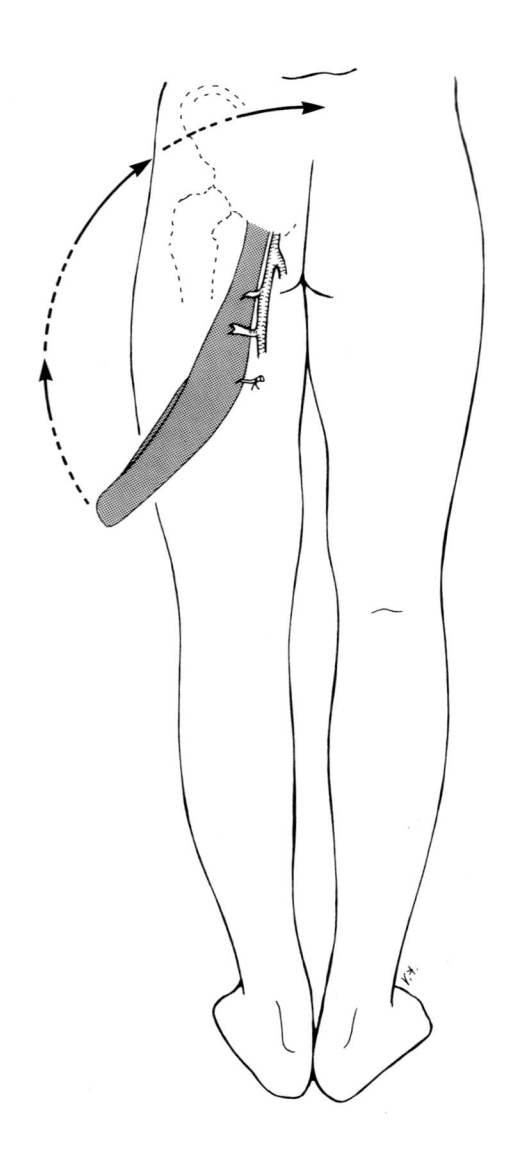

Arc of rotation

This large fusiform muscle has a wide arc of rotation. Based on the most proximal vascular pedicle 5 to 8 cm below the ischial tuberosity, the biceps femoris will easily reach and cover the ischial and perineal areas. It will also cover the trochanteric area.

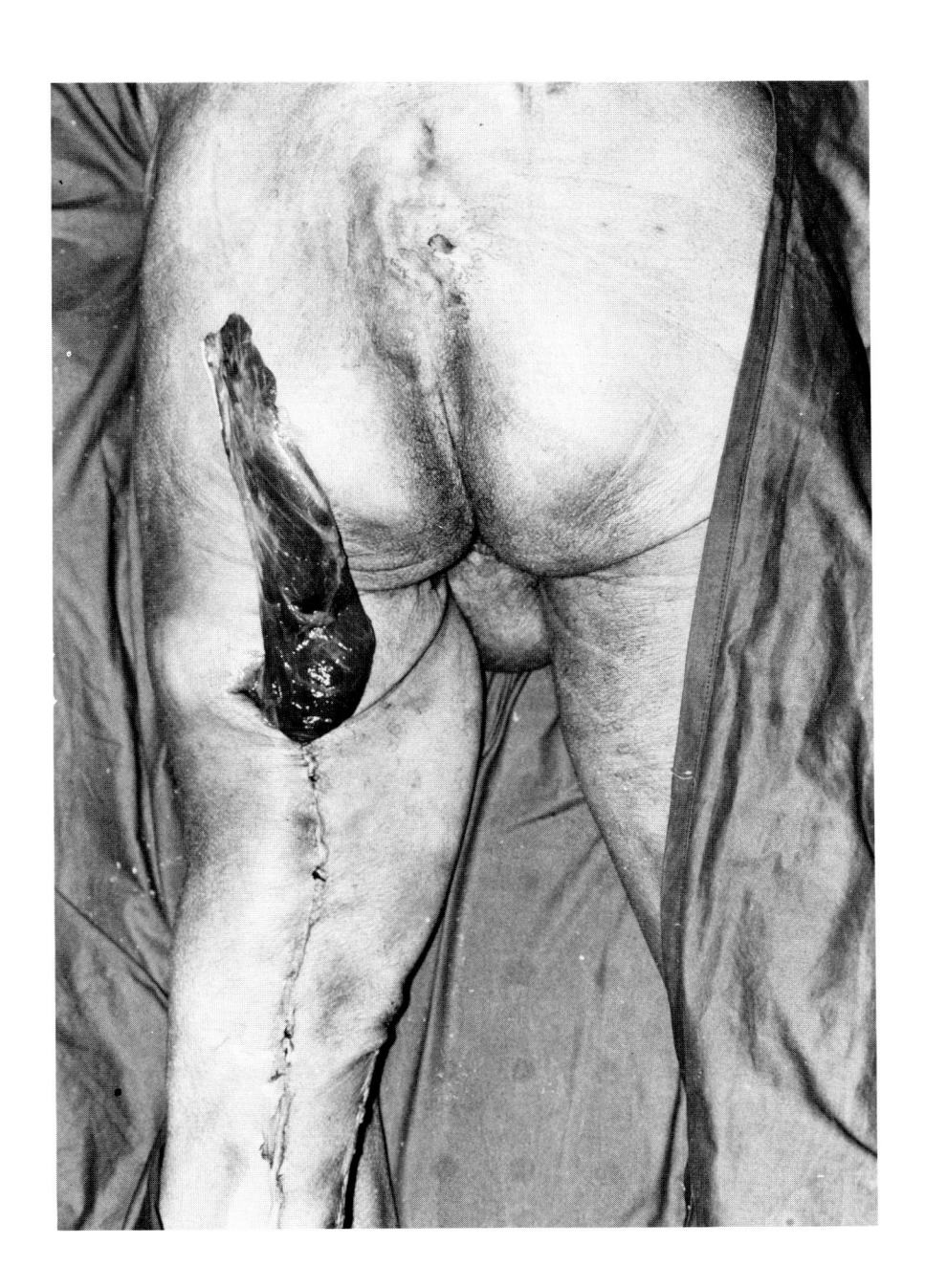

POSTERIOR THIGH

POSTERIOR THIGH

Biceps femoris

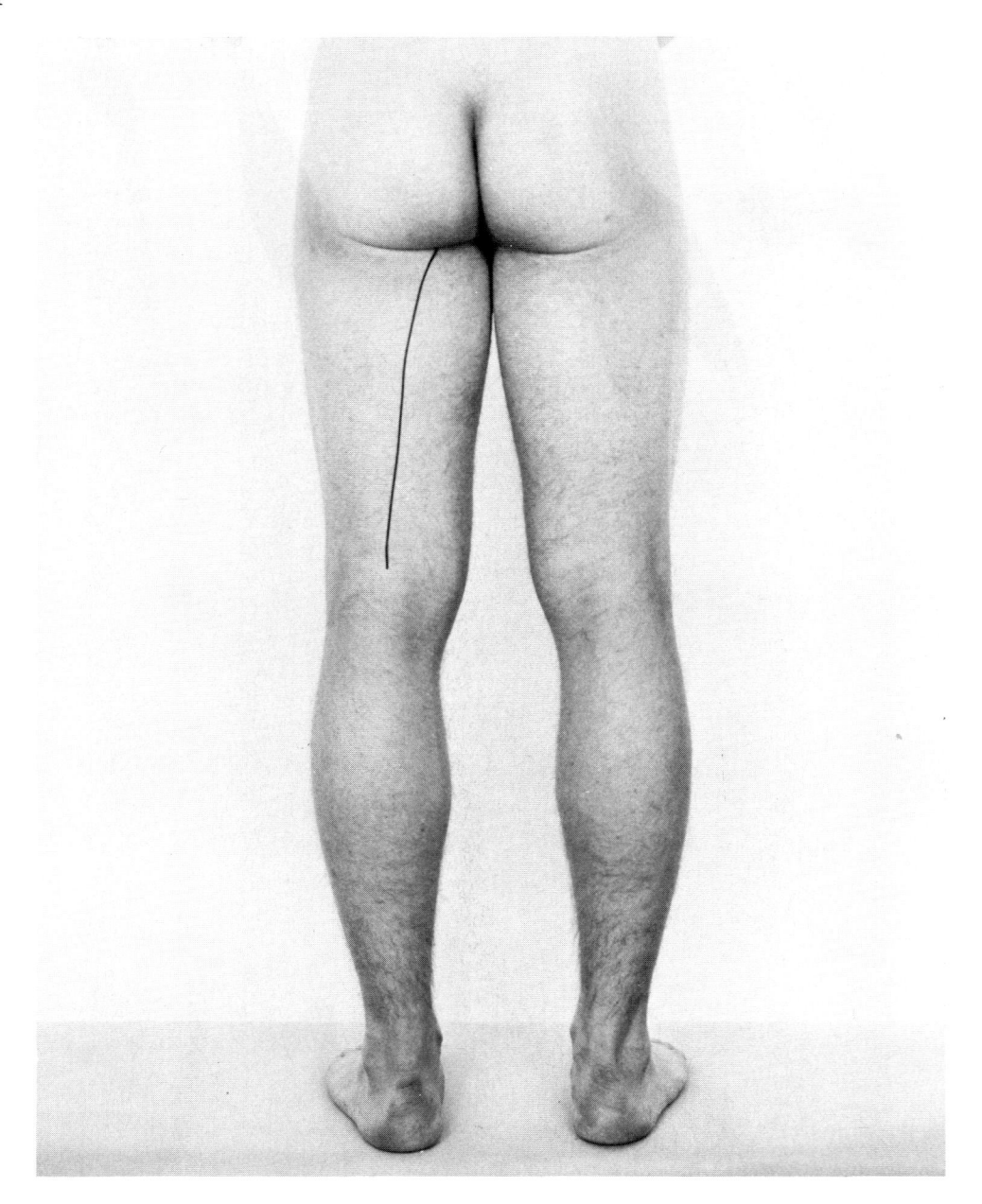

Skin incision

Elevation of flap

A vertical incision along the middle of the posterior thigh will expose the biceps femoris. The tendinous insertion into the head of the fibula is divided. At this point it is intimately related to the medially located common peroneal nerve. The muscle is then elevated by dissecting the short head from the linea aspera. Here the muscle lies

POSTERIOR THIGH

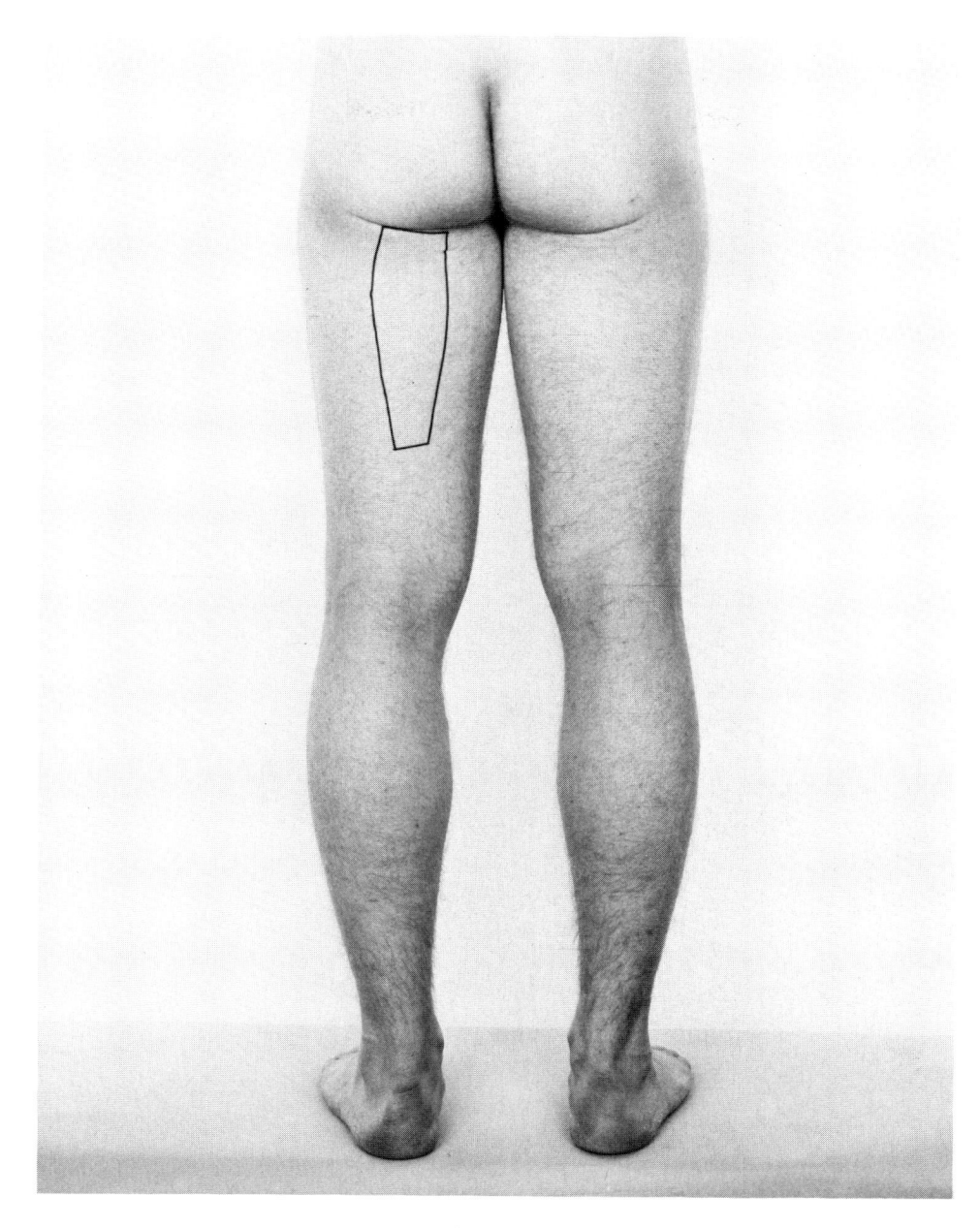

Skin island

superficial to the sciatic nerve. The pedicle to the short head is ligated and the muscle elevated to within 10 cm of its origin. All pedicles proximal to this level must be preserved to ensure muscle survival.

A small island of skin over the proximal half of the muscle can be incorporated with the muscle as a musculocutaneous transposition flap. The secondary defect can be closed directly.

POSTERIOR THIGH

Biceps femoris

Biceps femoris

Perforating musculocutaneous vessel

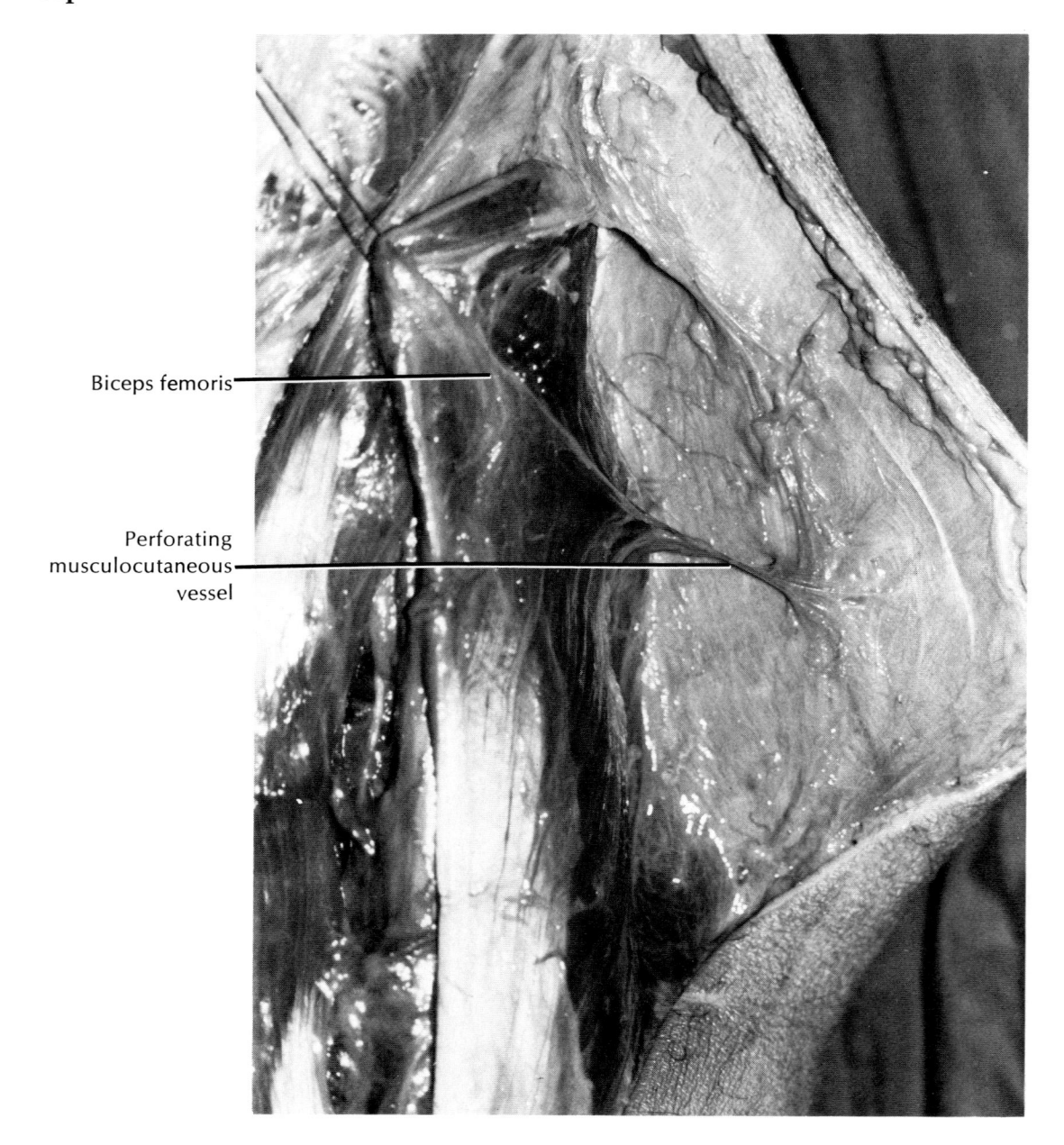

Precautions

- In an ambulatory patient the use of this muscle may lead to disability.
- During flap elevation, care should be taken to avoid injury to the sciatic nerve, especially the common peroneal branch, which is closely related to the muscle.

Semitendinosus

MUSCLE FLAP
MUSCULOCUTANEOUS FLAP

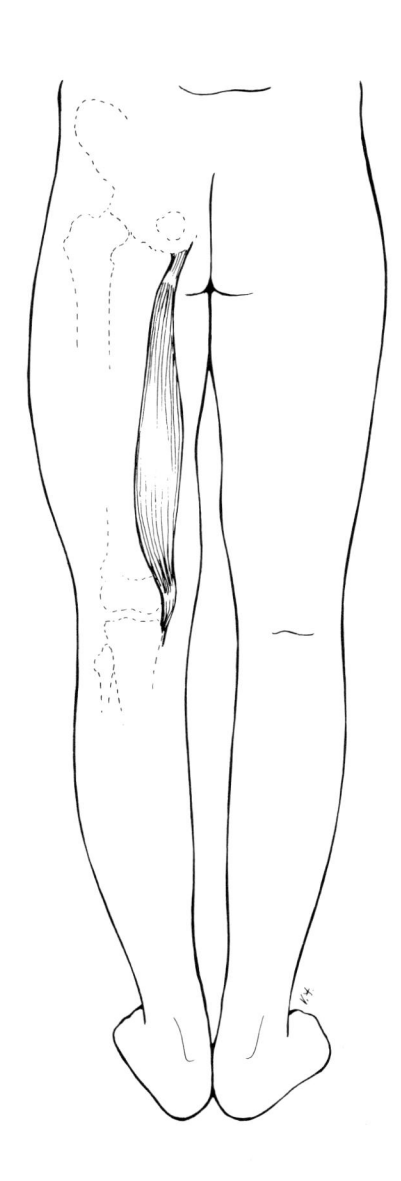

Applications

Coverage of:
ISCHIUM
BUTTOCKS
PERINEUM

POSTERIOR THIGH

Semitendinosus

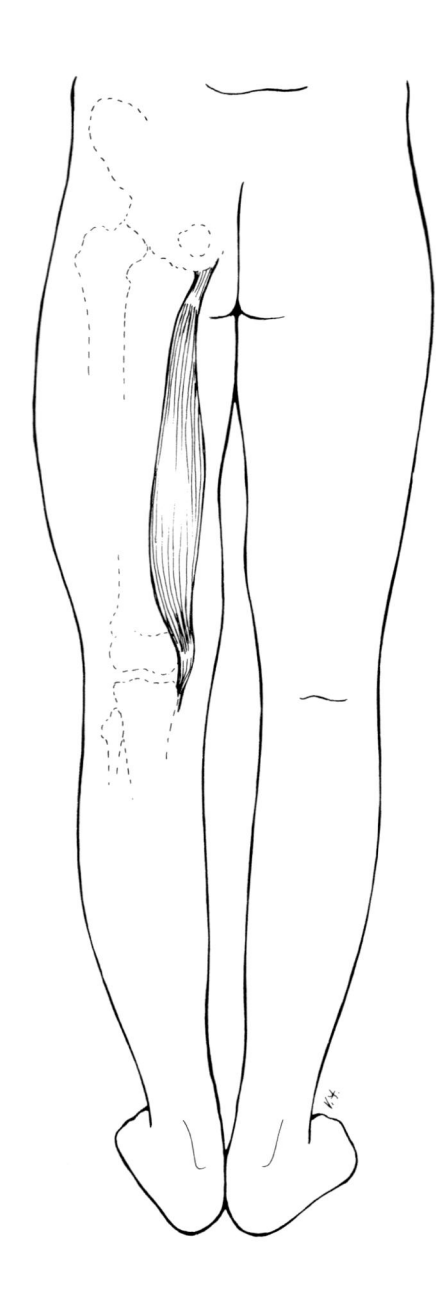

The long, thin semitendinosus muscle is a superficial muscle that lies between the semimembranosus medially and the biceps femoris laterally. The origin and insertion are in close relationship with the tendons of the semimembranosus.

Origin: Ischial tuberosity.
Insertion: Medial condyle of tibia.

POSTERIOR THIGH

Profunda femoris artery

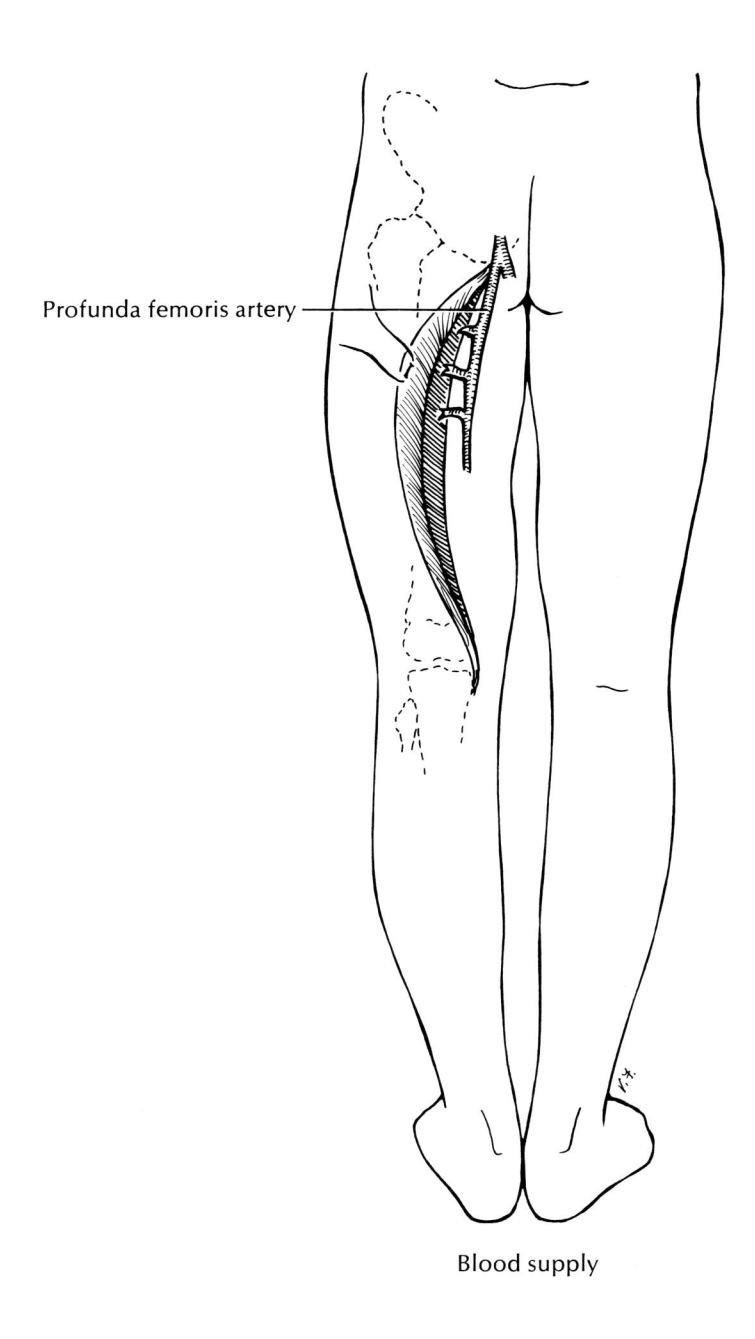

Blood supply

Nerve supply:
    Motor — Branch of sciatic nerve. The motor branches enter the proximal muscle on its deep surface.

Function: This is an expendable muscle. Leg flexion and thigh extension and medial rotation will be preserved by the remaining posterior thigh (hamstring) muscles.

POSTERIOR THIGH

Semitendinosus

Profunda femoris artery

Blood supply

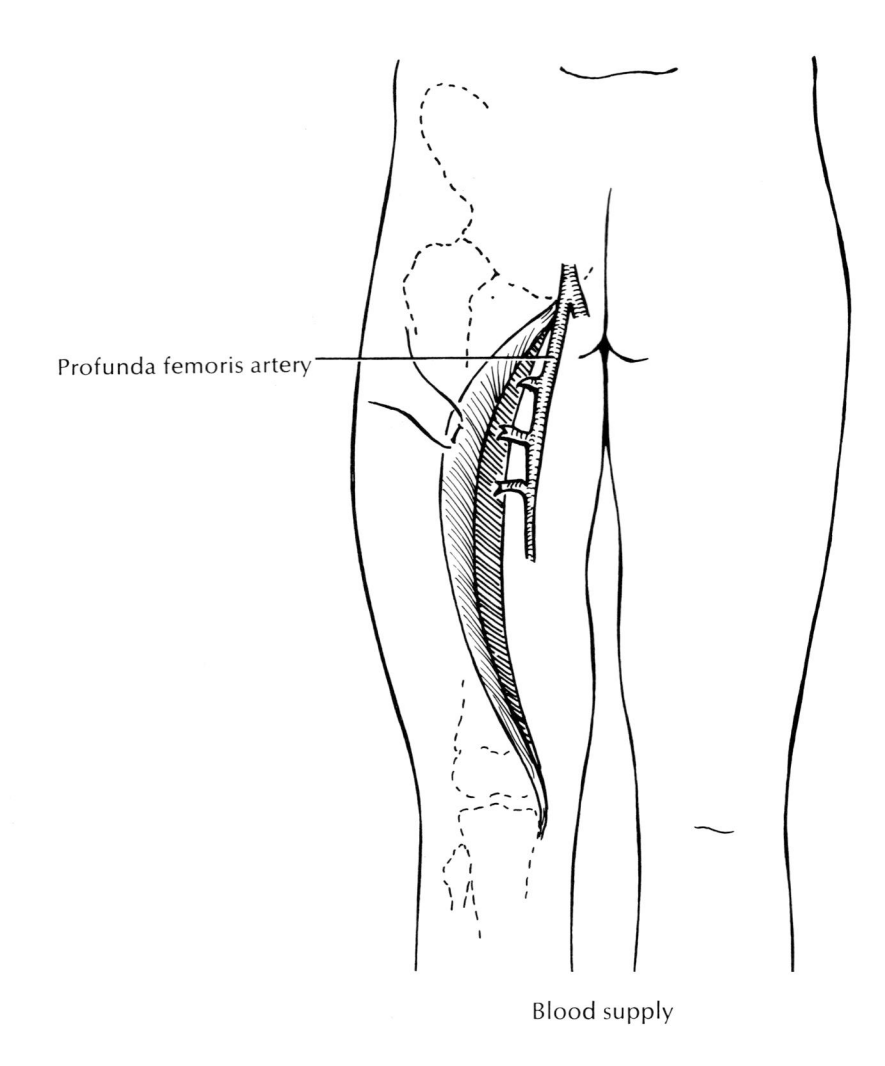

Blood supply

The semitendinosus muscle receives its blood supply from perforating branches of the profunda femoris. The vessels pass through the adductor magnus into the proximal posterior aspect of the semitendinosus muscle. The most proximal perforating vessels must be preserved to ensure muscle survival following transposition.

POSTERIOR THIGH

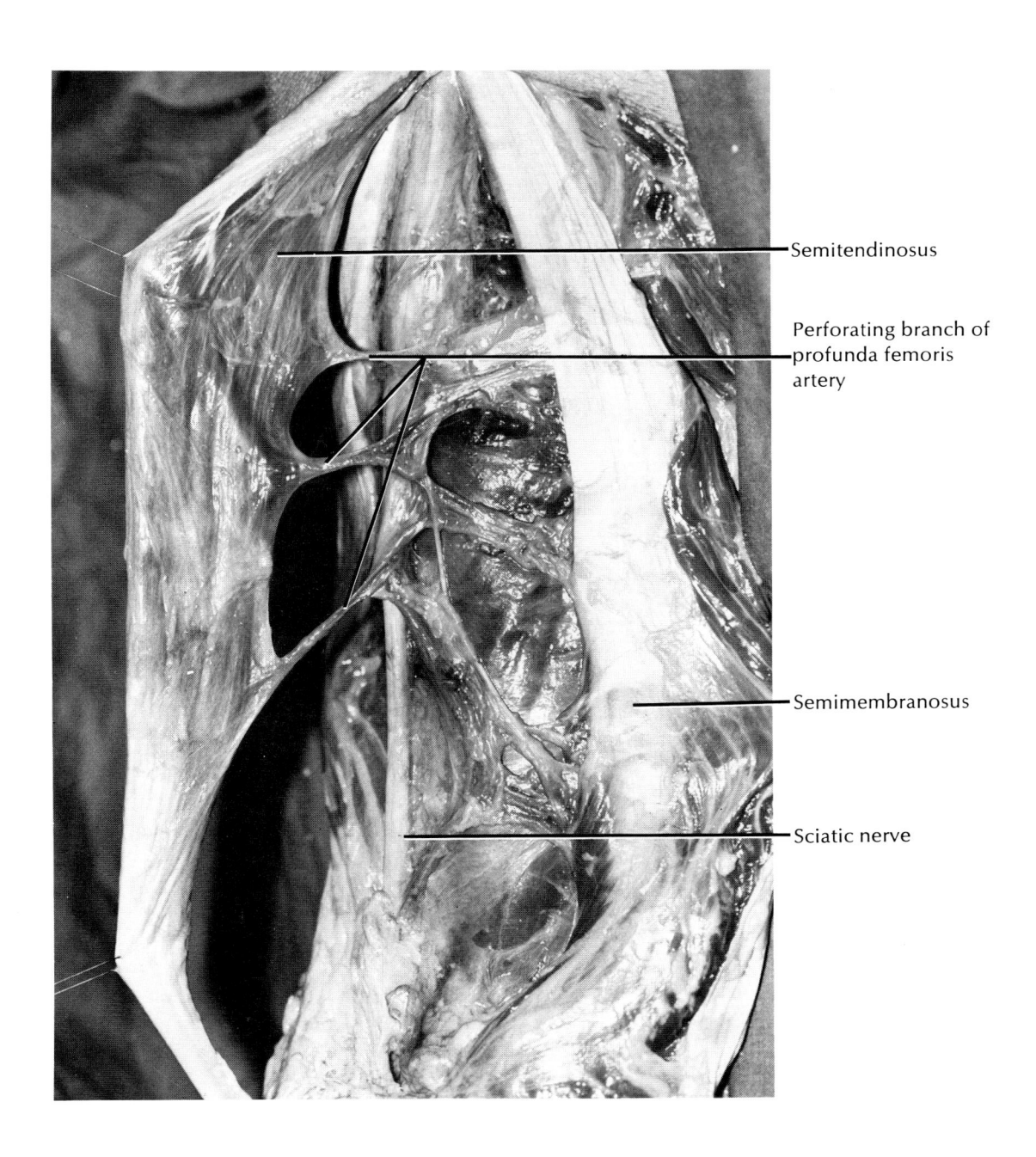

Semitendinosus

Perforating branch of profunda femoris artery

Semimembranosus

Sciatic nerve

POSTERIOR THIGH

Semitendinosus

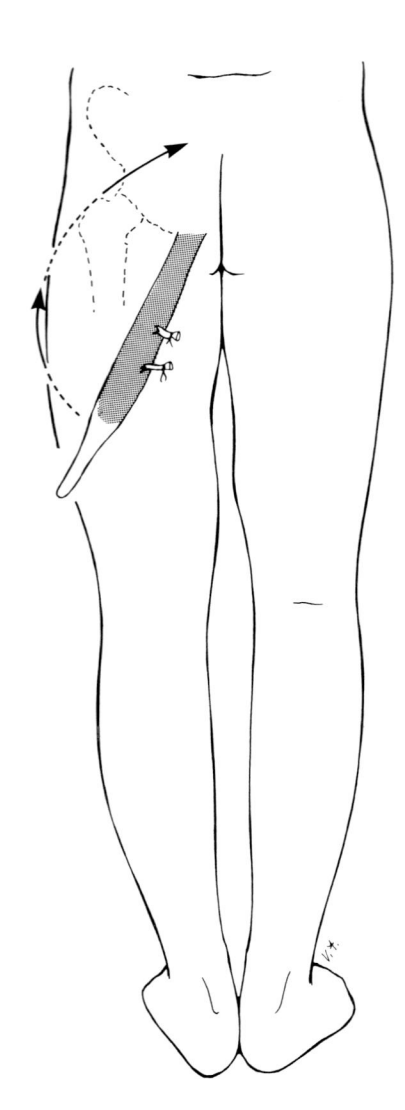

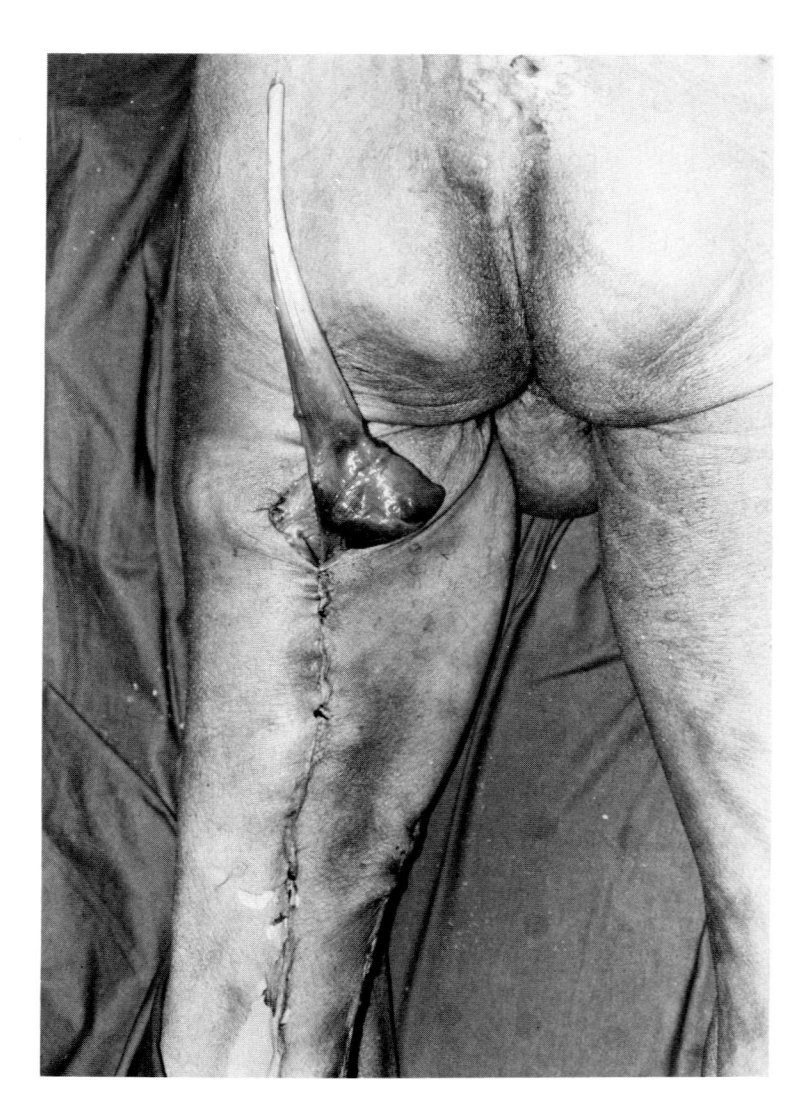

Arc of rotation

This long and thin muscle belly has a wide arc of rotation from the point of entrance of its vascular pedicles 10 cm inferior to the ischial tuberosity. The muscle, although narrow and tendinous, will partly cover the ischium, inferior buttock, and perineum.

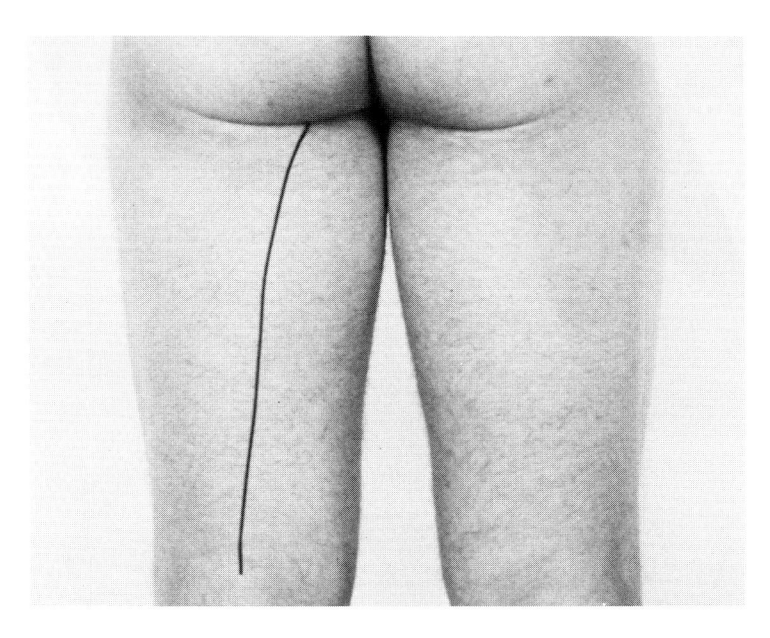

Skin incision

Elevation of flap

The semitendinosus muscle may be located by a medial vertical thigh incision. It has a narrow cutaneous territory that must be located between the semimembranosus and biceps femoris muscles. With elevation of both semitendinosus and semimembranosus muscles, a larger and more useful cutaneous territory could be elevated. After identification of the muscle near its insertion behind the semimembranosus, the muscle is elevated from the adductor magnus muscle posteriorly. Perforating vessels from the profunda femoris artery can be divided until the level of the proximal muscle approximately 10 cm inferior to the ischial tubercle. At this level the two or three vascular pedicles must be preserved to ensure muscle or musculocutaneous survival. After transposition of this muscle, the donor site can be closed primarily.

Precautions

- This is a small, narrow muscle and is best used in combination with the semimembranosus muscle for coverage of ischial pressure sores.
- Its usefulness as a musculocutaneous flap is limited by its narrow size.
- The sciatic nerve must be protected during elevation of the muscle flap.

POSTERIOR THIGH

Semimembranosus

MUSCLE FLAP
MUSCULOCUTANEOUS FLAP

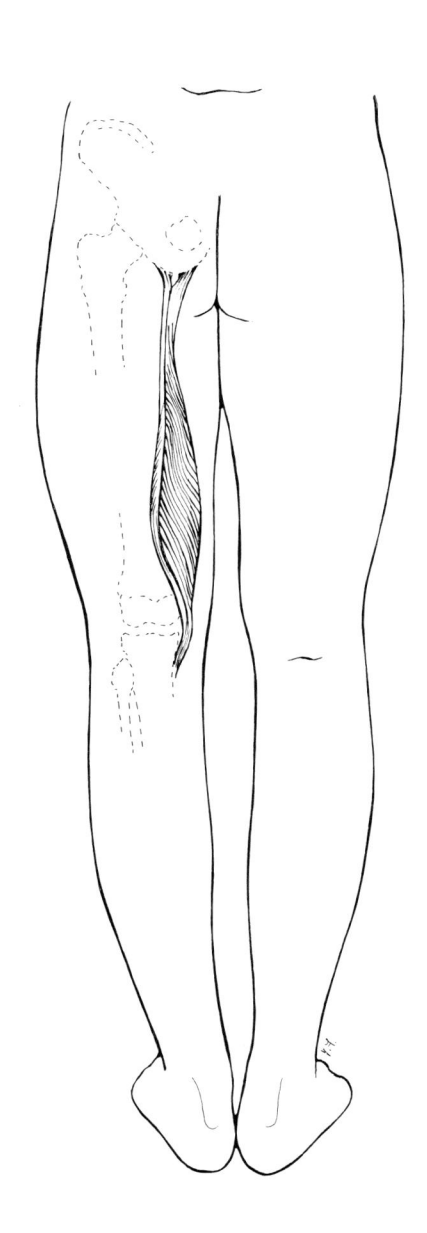

Applications

Coverage of:
TROCHANTER
ISCHIUM
SACRUM
PERINEUM

POSTERIOR THIGH

Semimembranosus

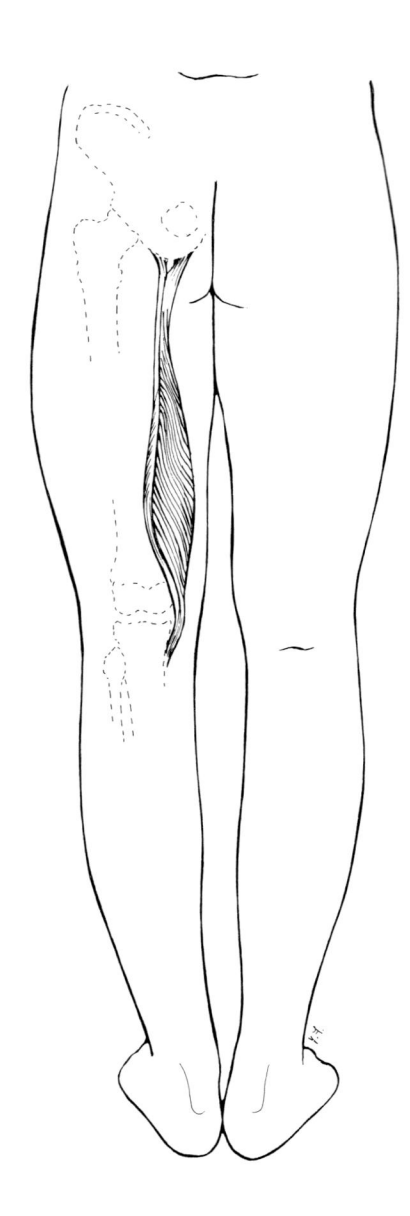

The long, thin semimembranosus muscle is the most medial member of the posterior thigh (hamstring) group. This superficial muscle lies between the semitendinosus muscle laterally and the gracilis and adductor magnus muscles medially. The origin and insertion are in close relationship with the tendon of the semitendinosus.

Origin: Ischial tuberosity.
Insertion: Medial condyle of tibia.

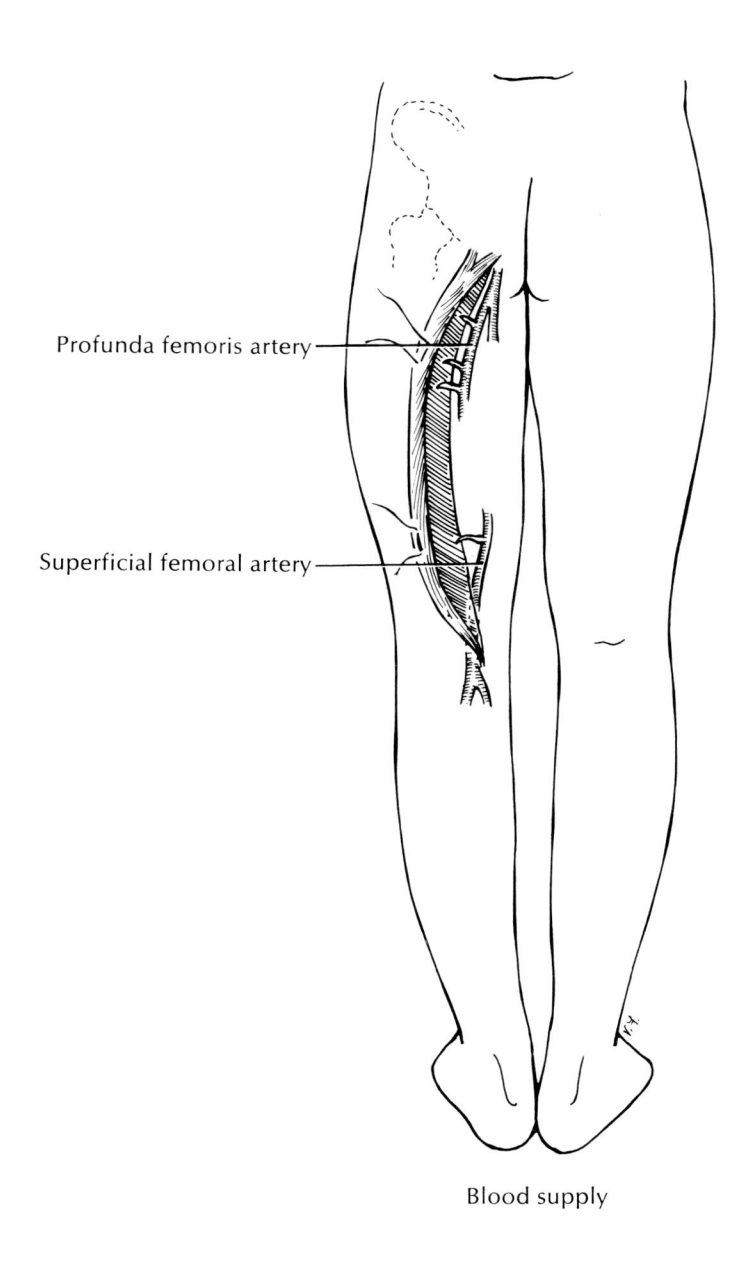

POSTERIOR THIGH

- Profunda femoris artery
- Superficial femoral artery

Blood supply

Nerve supply:
    Motor—Branch of sciatic nerve. The motor branches enter the proximal muscle on its deep surface.

Function: This is an expendable muscle. Leg flexion, thigh extension, and medial rotation will be preserved by the remaining hamstring muscles.

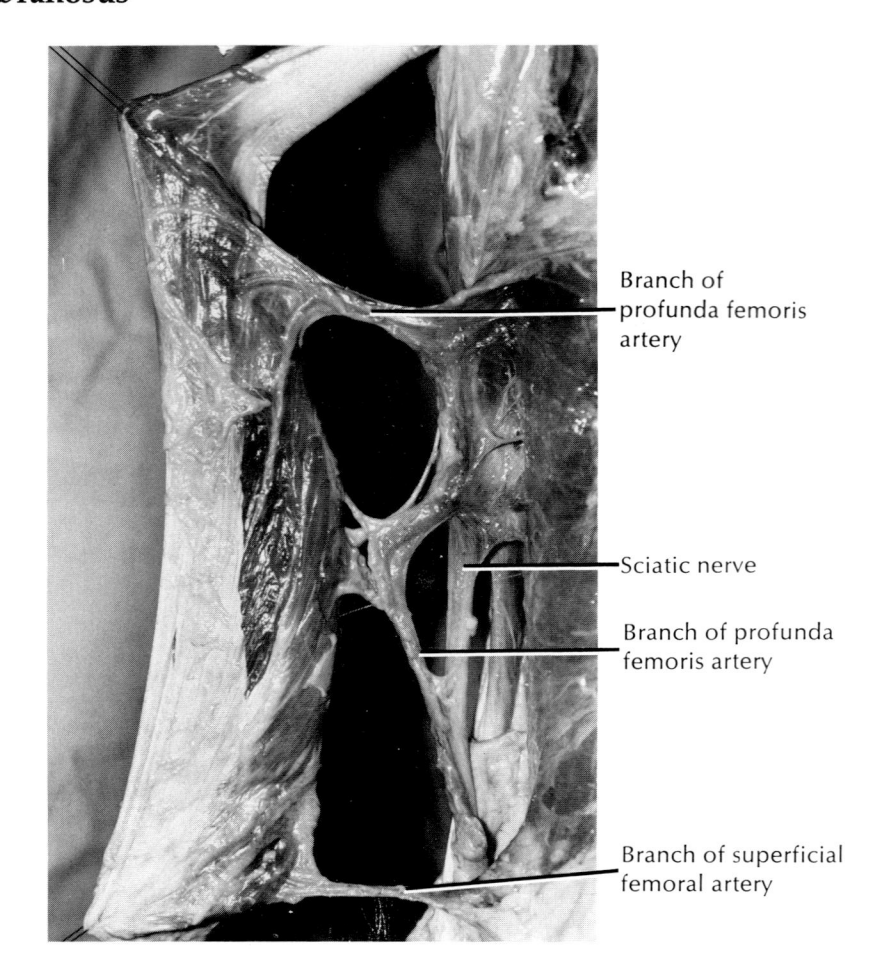

POSTERIOR THIGH

Semimembranosus

- Branch of profunda femoris artery
- Sciatic nerve
- Branch of profunda femoris artery
- Branch of superficial femoral artery

Blood supply

The semimembranosus muscle has a dual blood supply from the profunda femoris artery proximally and the superficial femoral artery distally. Either system will support the muscle, allowing a superior or inferior arc of rotation. The proximal pedicles are perforating branches of the profunda femoris artery. The vessels pass through the adductor magnus into the proximal posterior aspect of the semimembranosus muscle. Perforators proximal to 10 cm inferior to the ischial tuberosity should be preserved when superior muscle transposition is planned. The inferior pedicle of the semimembranosus is a branch of the superficial femoral artery, entering the medial muscle belly at the junction of the middle and distal thirds. This pedicle must be preserved when an inferior muscle transposition is planned.

POSTERIOR THIGH

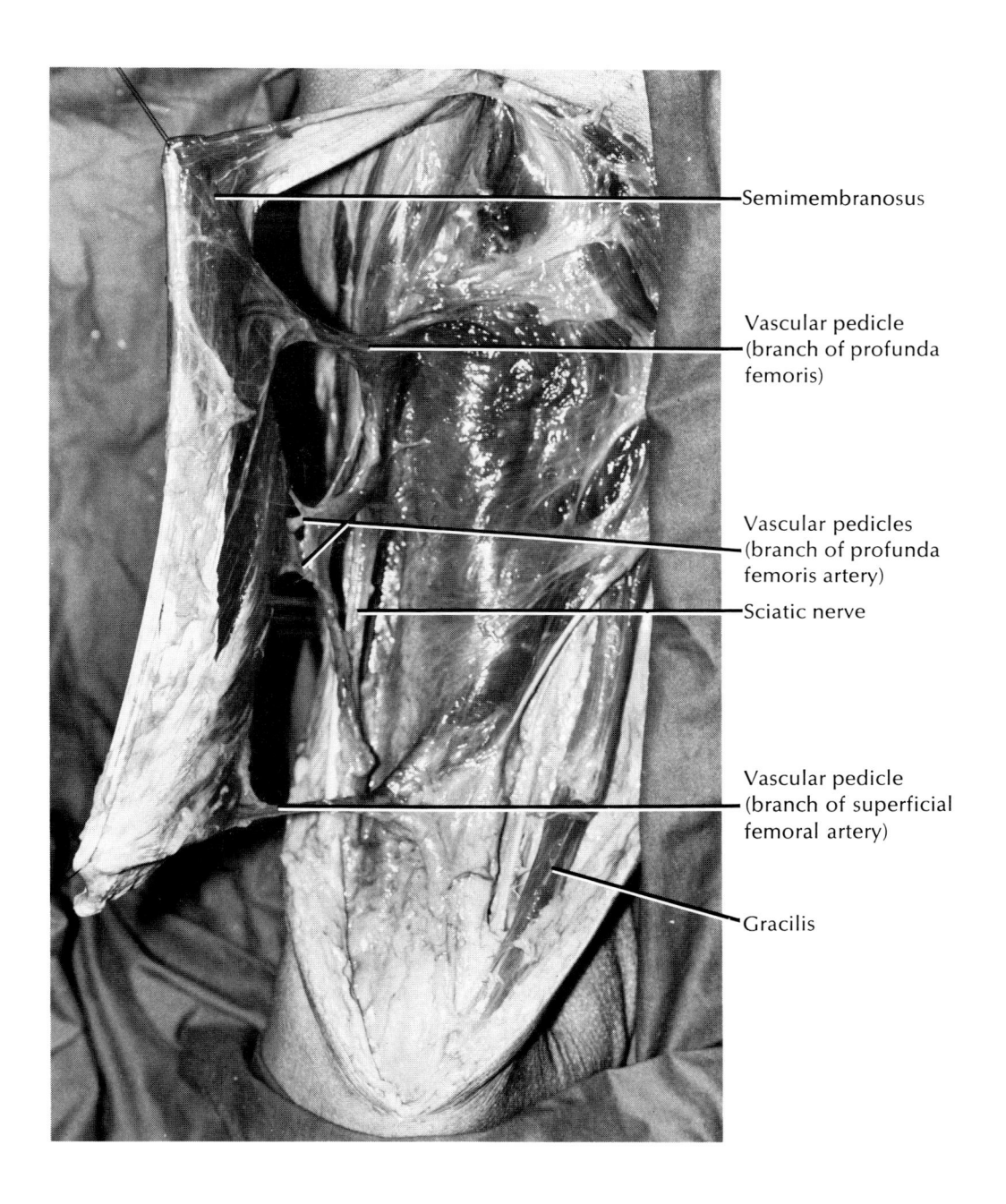

- Semimembranosus
- Vascular pedicle (branch of profunda femoris)
- Vascular pedicles (branch of profunda femoris artery)
- Sciatic nerve
- Vascular pedicle (branch of superficial femoral artery)
- Gracilis

POSTERIOR THIGH

Semimembranosus

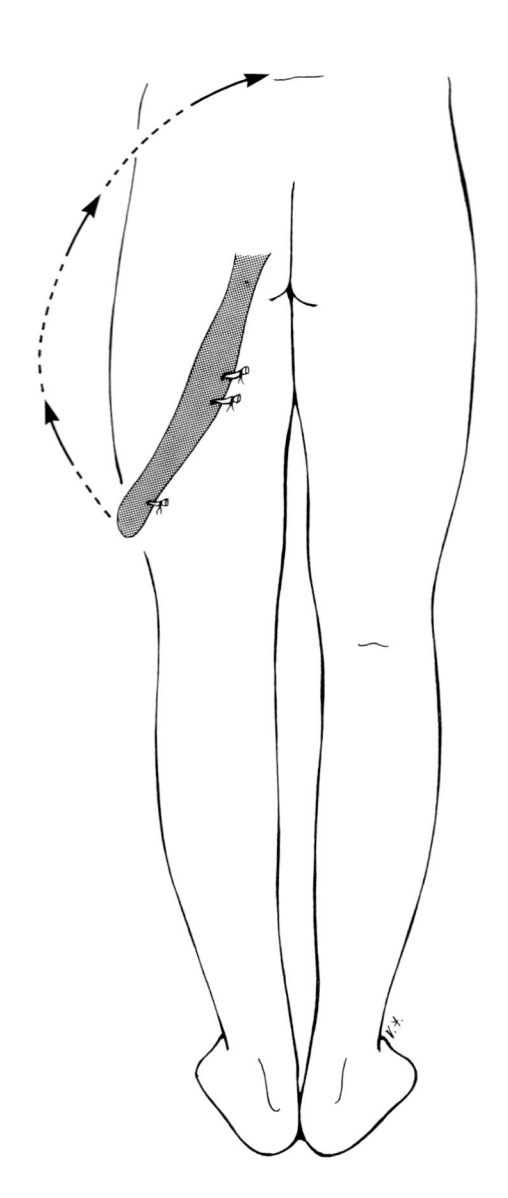

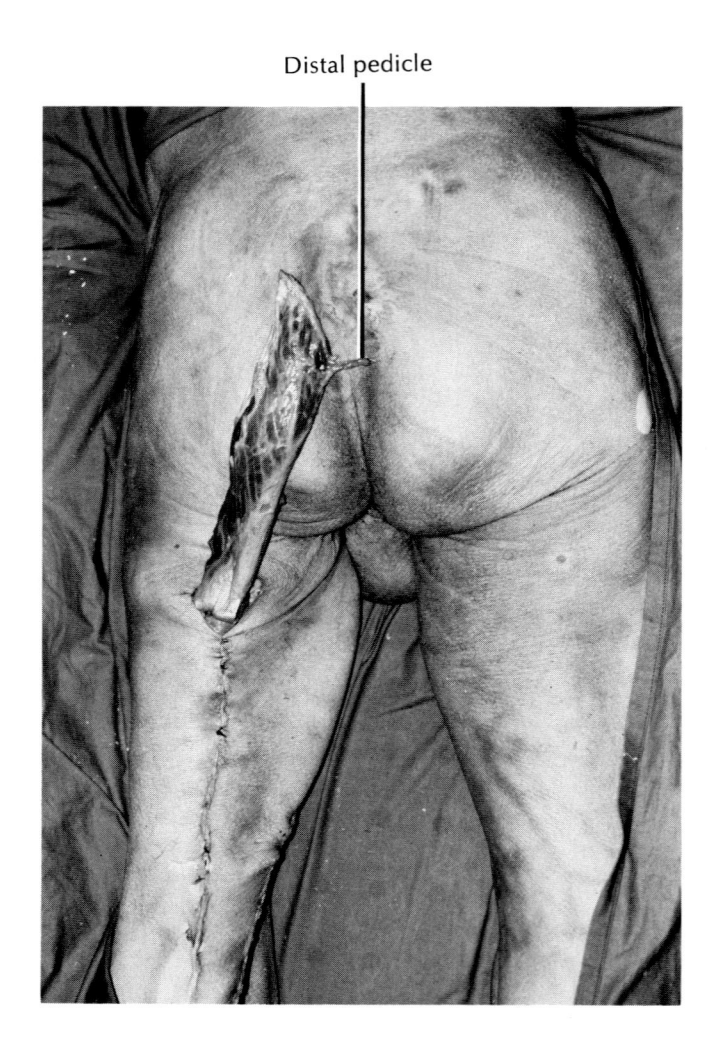

Distal pedicle

Arc of rotation
SUPERIOR ARC

This long and thin muscle belly has a wide arc of rotation from the point of entrance of its vascular pedicles 10 cm inferior to the ischial tubercle. The muscle will cover the trochanter, ischium, inferior sacrum, and perineum.

POSTERIOR THIGH

Inferior vascular pedicle

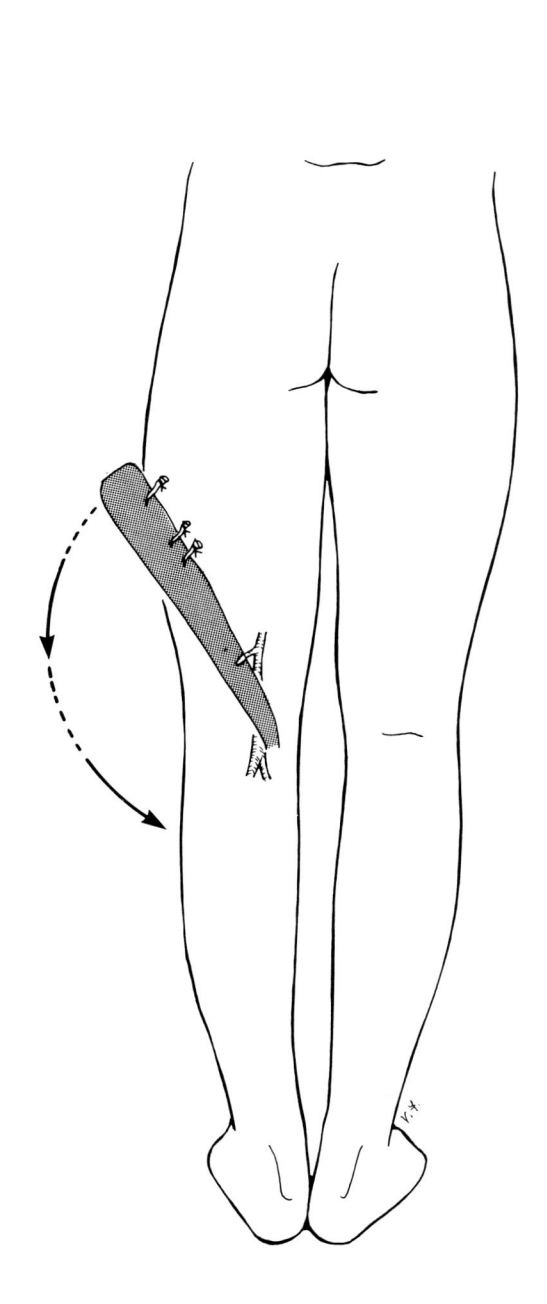

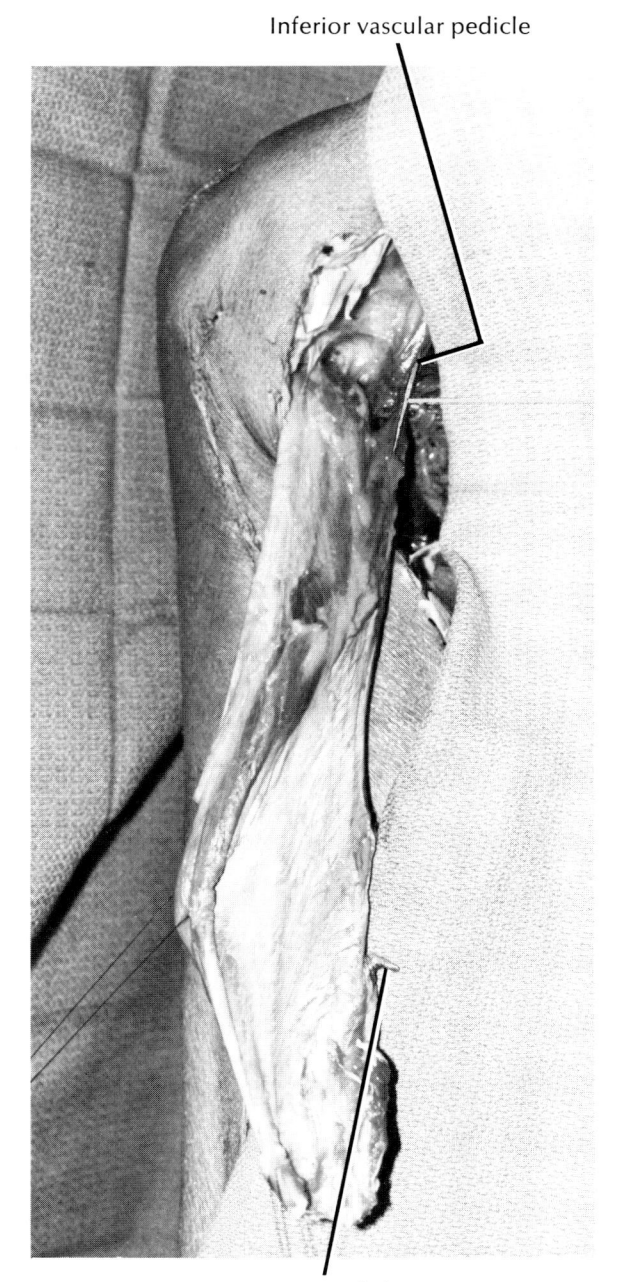

Superior pedicle

INFERIOR ARC

Although less reliable, the muscle can be transposed inferiorly based on the inferior pedicles from the superficial femoral artery. This muscle will then cover the posterior knee and inferior thigh.

POSTERIOR THIGH

Semimembranosus

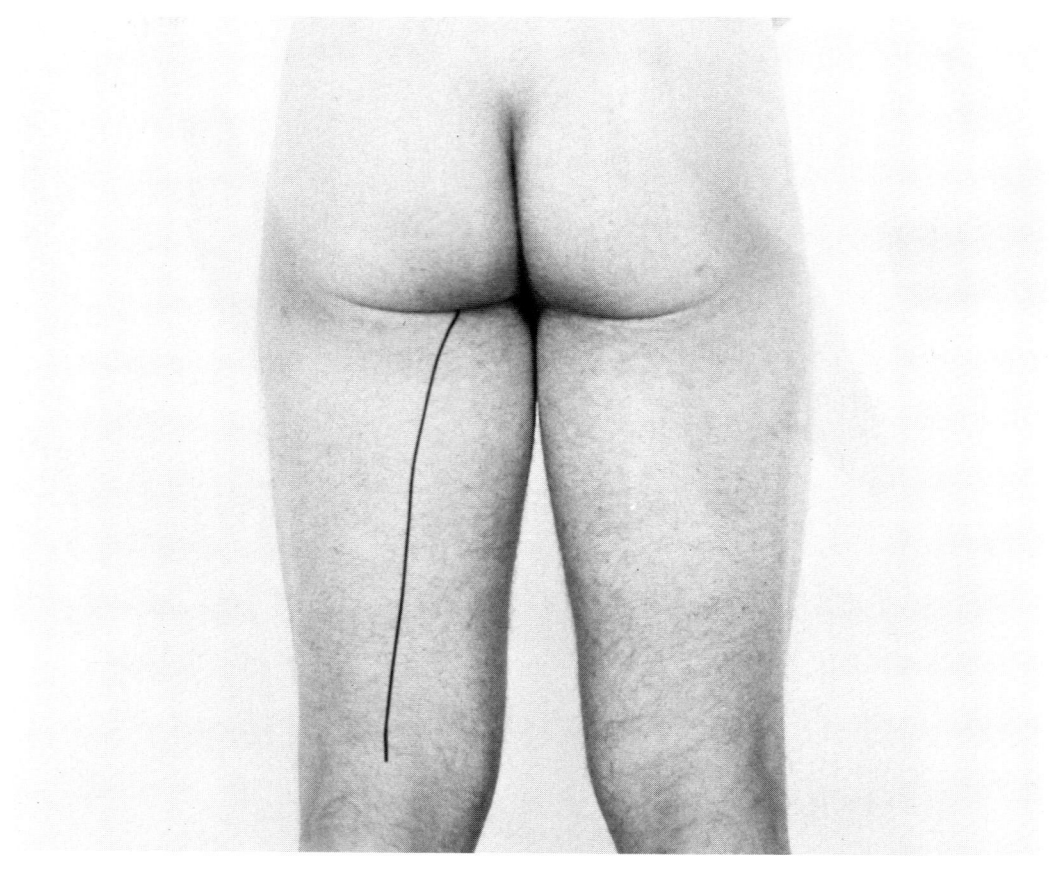

Skin incision

Elevation of flap

The semimembranosus muscle may be located by a medial vertical thigh incision. The tendon of insertion has a close relationship with the gracilis tendon medially and semitendinosus tendon laterally. After identification of the muscle, it is elevated from the adductor magnus muscle. Distally the pedicle from the superficial femoral artery is ligated and divided. The pedicles to the muscle are not divided beyond a point approximately 10 cm inferior to the ischial tubercle. This point of rotation allows safe muscle transposition. The donor region can be closed primarily.

The cutaneous territory overlying the muscle between the gracilis muscle and semitendinosus is small but may be incorporated with the muscle for reconstructive purposes. The donor area can be closed primarily.

Precautions

- The sciatic nerve has a close proximity to the proximal vascular pedicles to the semimembranosus muscle.
- The inferior muscle transposition based on the inferior pedicle may not be reliable.

SECTION THREE
MEDIAL LEG

Gastrocnemius

Soleus

Flexor digitorum longus

Flexor hallucis longus

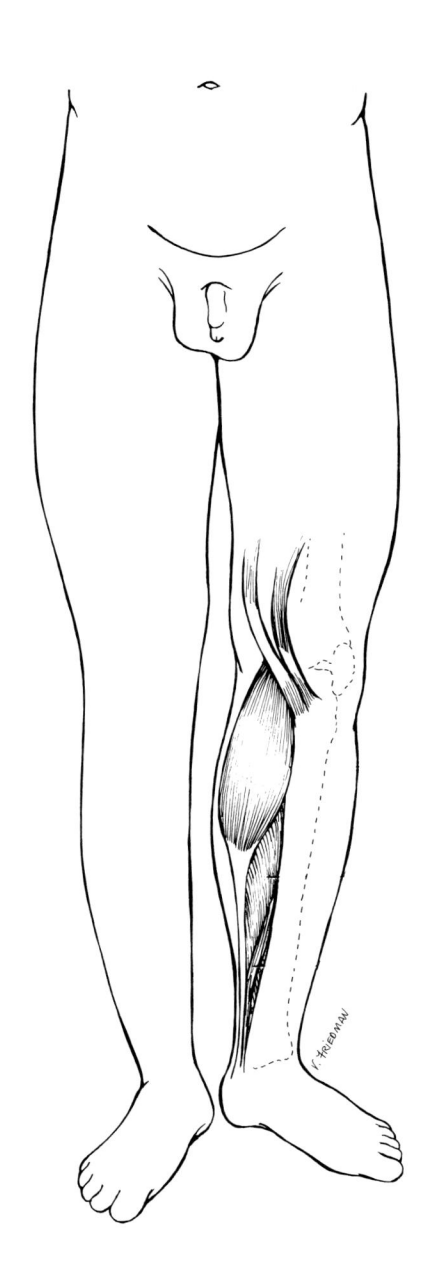

MEDIAL LEG

Gastrocnemius

Soleus

Flexor digitorum longus

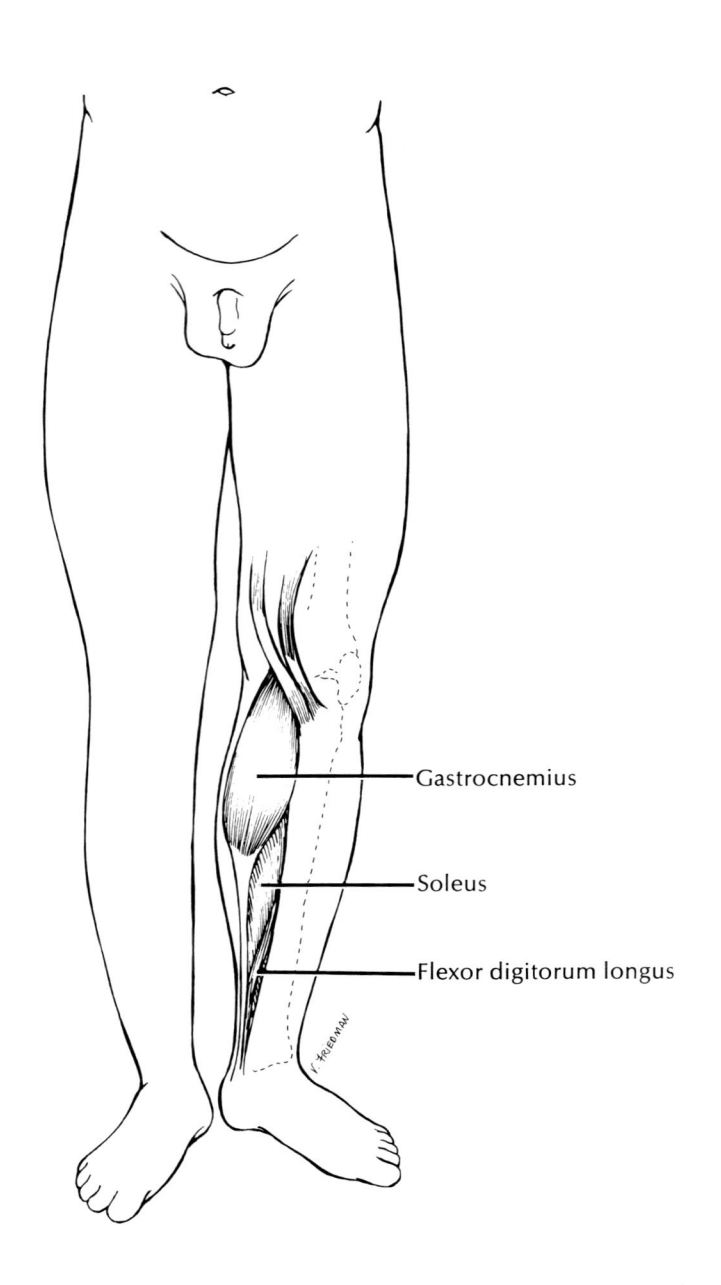

134

MEDIAL LEG

- Gastrocnemius
- Soleus
- Flexor digitorum longus

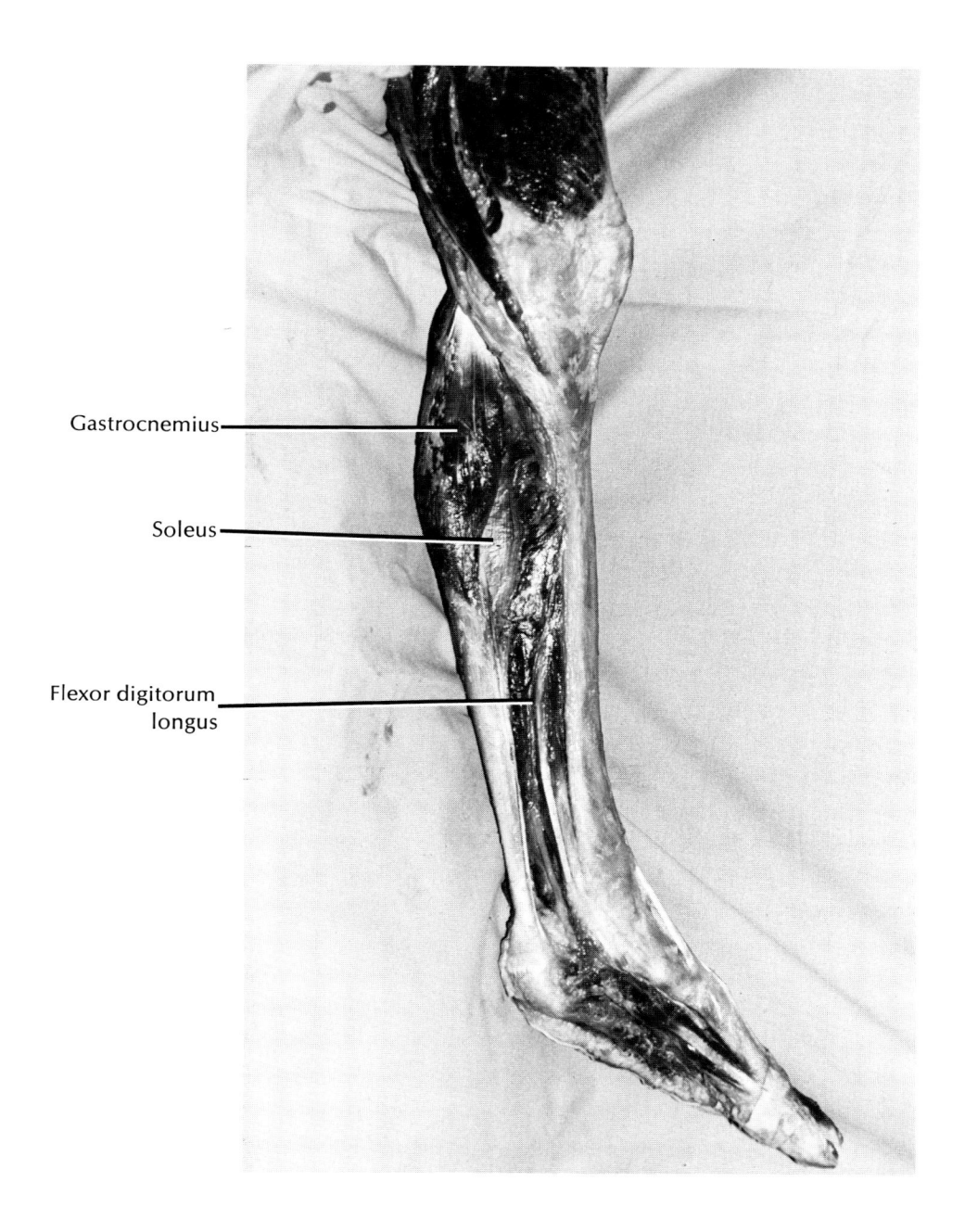

Although the medial leg muscles are located posteriorly in the leg, they are referred to as the medial group because they are most easily approached through an incision medial to the tibia.

MEDIAL LEG

- Gastrocnemius
- Soleus
- Flexor digitorum longus
- Flexor hallucis longus

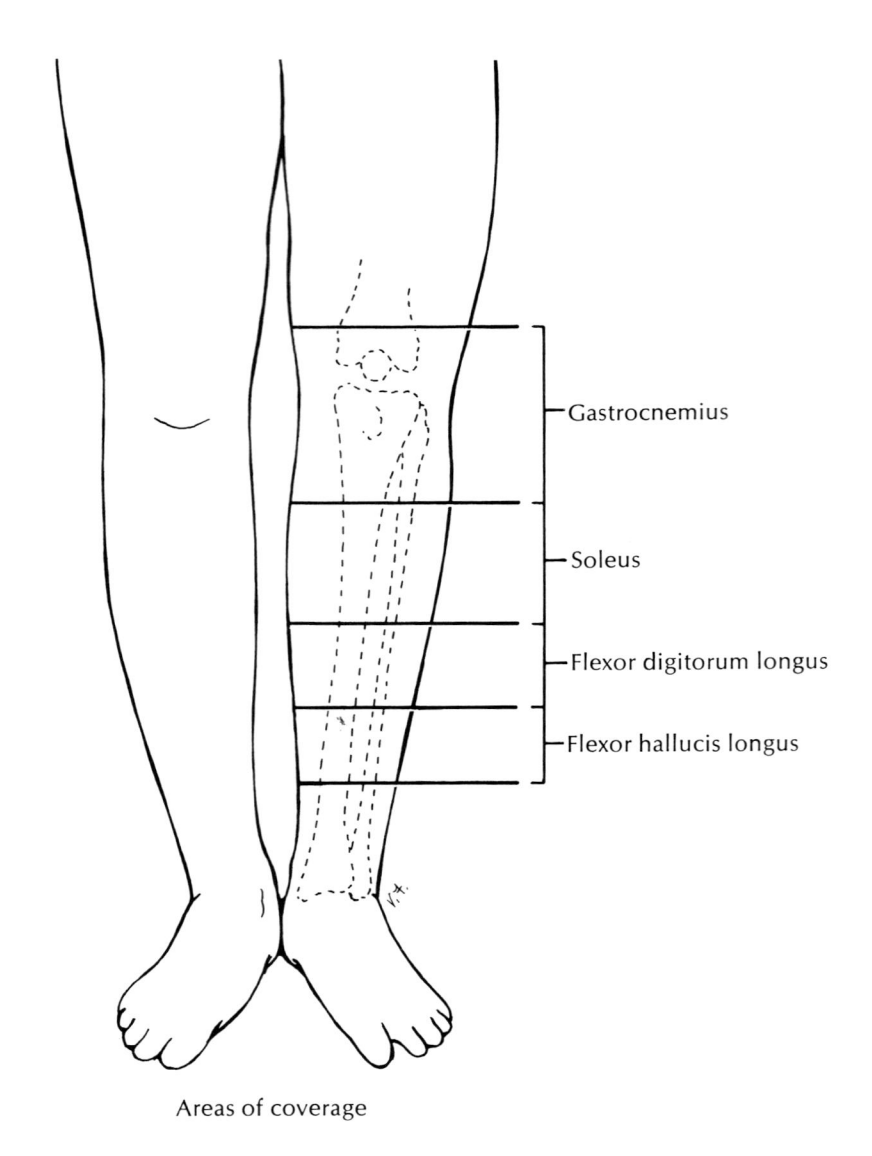

Areas of coverage

This is a most useful and frequently used group of muscles for coverage of the knee and upper three-fourths of the tibia. Based on the proximal dominant vascular pedicles, the muscles are safely transposed laterally to cover the tibia.

The gastrocnemius and soleus are excellent and commonly used muscle flaps that will cover large defects. However, the flexor digitorum longus and flexor hallucis longus are small muscles and are best used with the soleus to extend the lower area of tibial coverage.

MEDIAL LEG

Gastrocnemius

Soleus

Flexor digitorum longus

Flexor hallucis longus

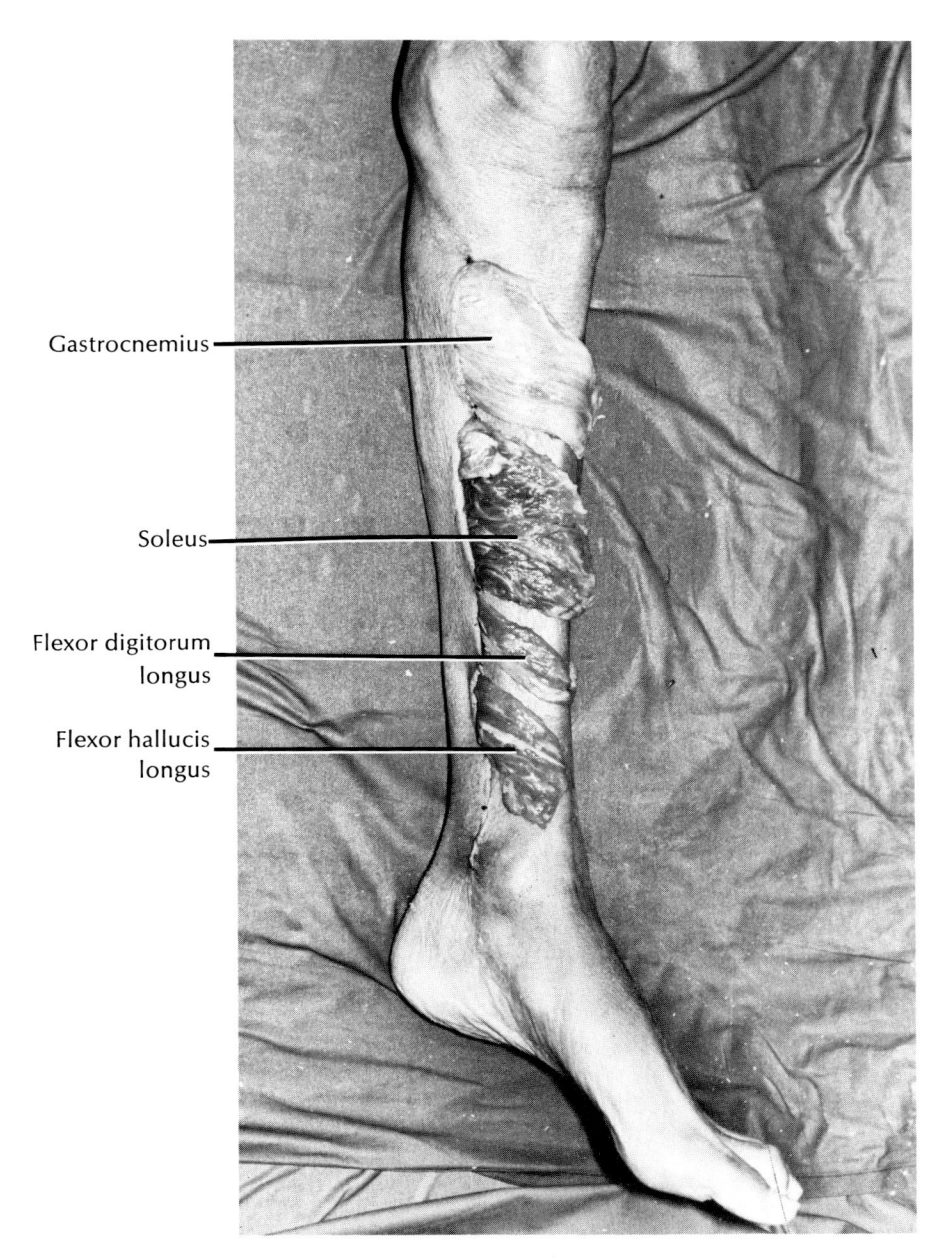

Arcs of rotation

MEDIAL LEG

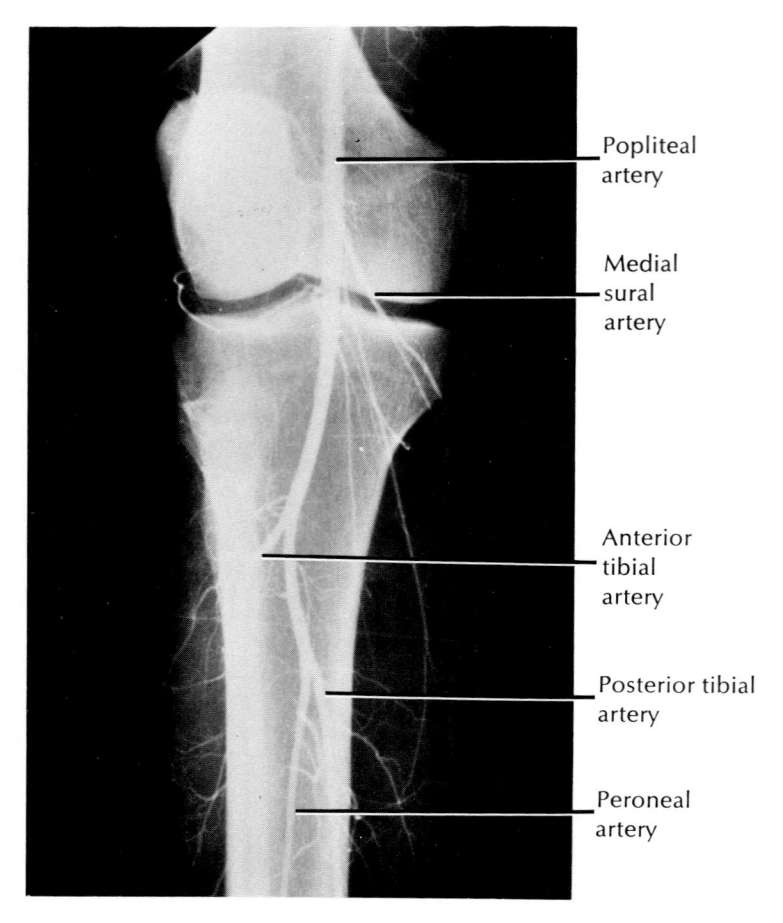

Angiogram of major vessels of leg

- Popliteal artery
- Medial sural artery
- Anterior tibial artery
- Posterior tibial artery
- Peroneal artery

Blood supply

The popliteal, posterior tibial, and peroneal arteries supply the medial leg muscles. The popliteal artery through the sural branches supplies the gastrocnemius muscle. The tibial origin of the soleus is supplied by the posterior tibial artery, and the fibular origin is supplied by the peroneal artery. The flexor digitorum longus has direct branches from the posterior tibial artery. The flexor hallucis longus has direct branches from the peroneal artery.

The proximal location of the dominant vascular pedicles in this muscle group permits safe muscle flap transposition.

MEDIAL LEG

- Popliteal artery
- Sural artery
- Soleus
- Anterior tibial artery
- Posterior tibial artery
- Peroneal artery
- Flexor hallucis longus
- Flexor digitorum longus

Blood supply

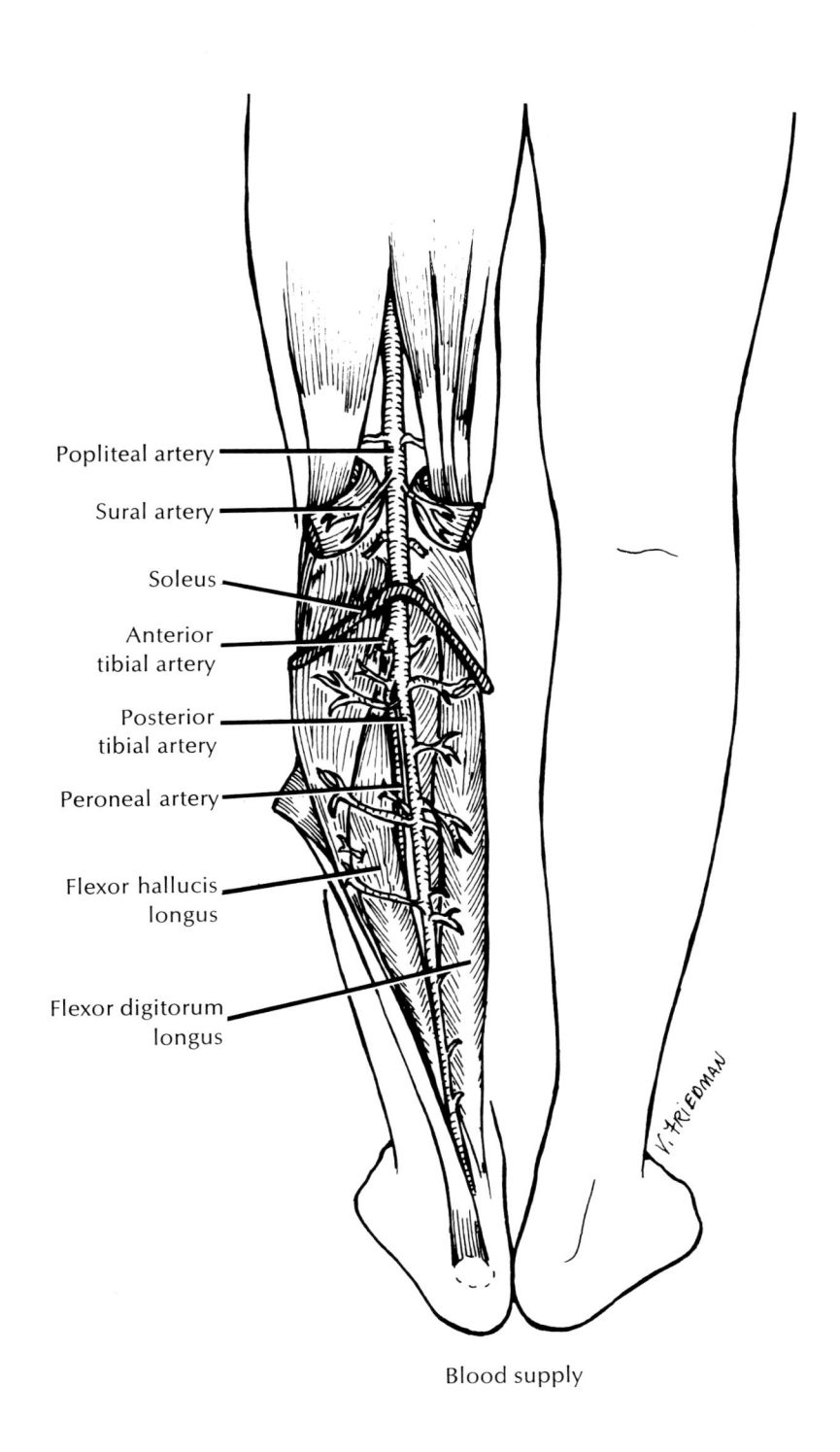

Gastrocnemius

MUSCLE FLAP
MUSCULOCUTANEOUS FLAP
FREE FLAP

Applications

Coverage of:

KNEE
UPPER THIRD OF TIBIA

Cross leg flap

Free flap:

DISTANT COVERAGE

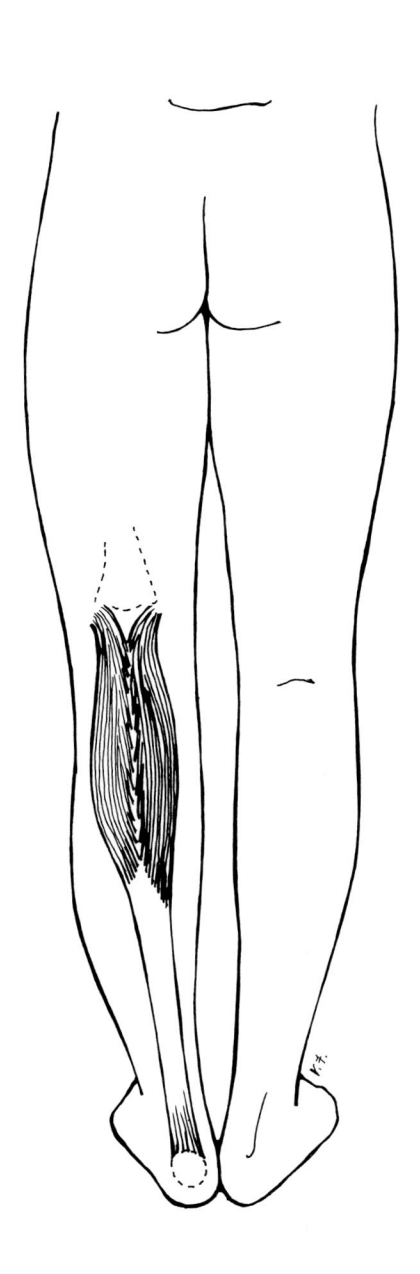

MEDIAL LEG

Gastrocnemius

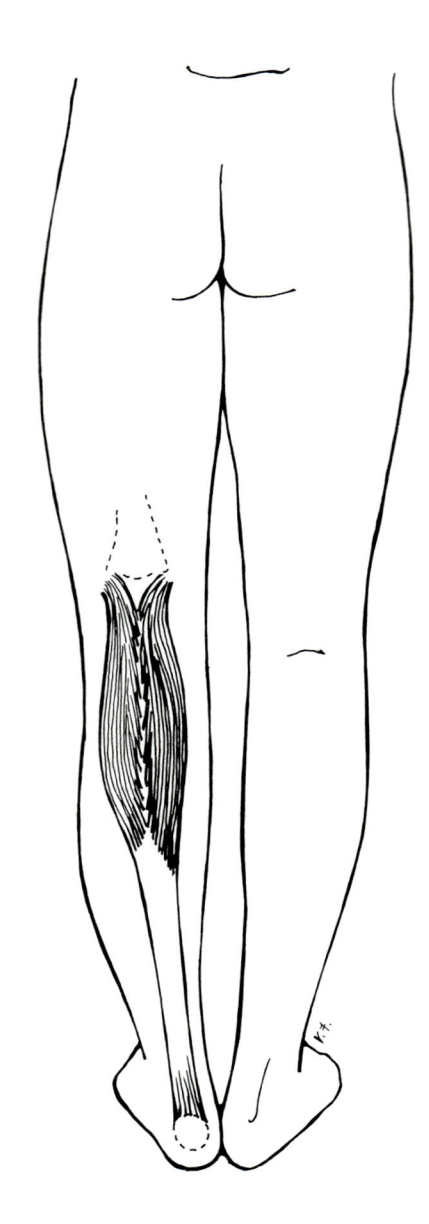

The gastrocnemius is the most superficial and largest calf muscle. It has a lateral and medial head and lies superficial to the plantaris, popliteus, and soleus muscles. The medial head is larger and extends a greater distance inferiorly. The two heads unite and join the tendon of the soleus to form the Achilles tendon.

> **Origin:** Medial head—medial condyle of femur; lateral head—lateral condyle of femur.
> **Insertion:** Calcaneus through the Achilles tendon.

Superficial femoral artery

Sural arteries

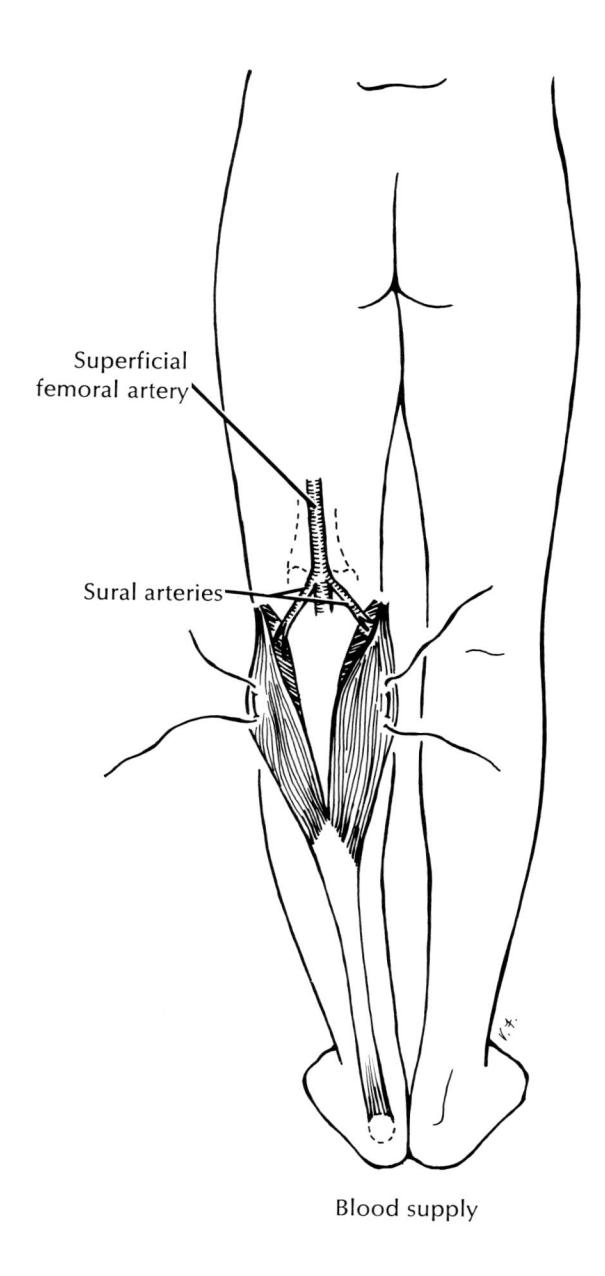

Blood supply

Nerve supply:
  Motor—Branches of tibial nerve. The paired motor nerves enter the proximal heads of the gastrocnemius muscle in the popliteal fossa adjacent to the vascular pedicles.
Function: Plantar flexion of the foot. With the soleus intact, one head of the gastrocnemius can be used without creating any significant functional disability.

MEDIAL LEG

Gastrocnemius

Superficial femoral artery

Sural arteries

Blood supply

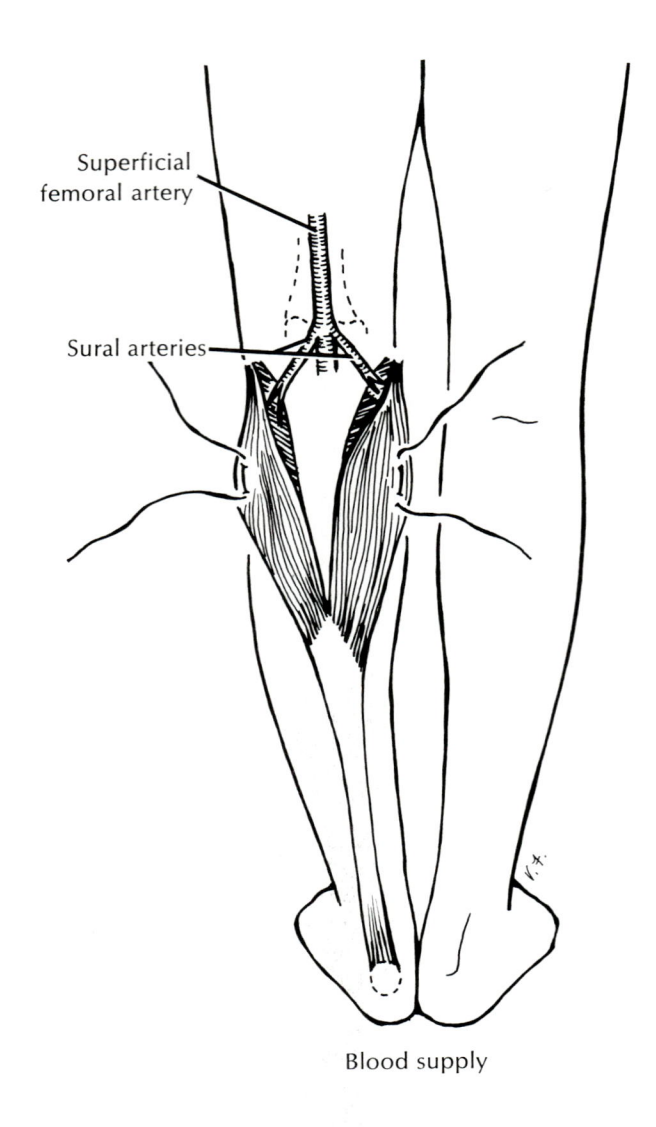

Blood supply

Each head of the gastrocnemius has a dominant vascular pedicle, the sural branches of the popliteal artery. These are large vessels entering each head proximally close to the origin of the muscle at the level of the femoral condyles. The vessels divide in the muscle and run longitudinally parallel to the muscle fibers. Each head has an independent vascular unit. Musculocutaneous perforators through the muscular portion of the gastrocnemius supply the overlying skin and part of the skin lying over the Achilles tendon. There are no perforators through the Achilles tendon into the overlying skin.

MEDIAL LEG

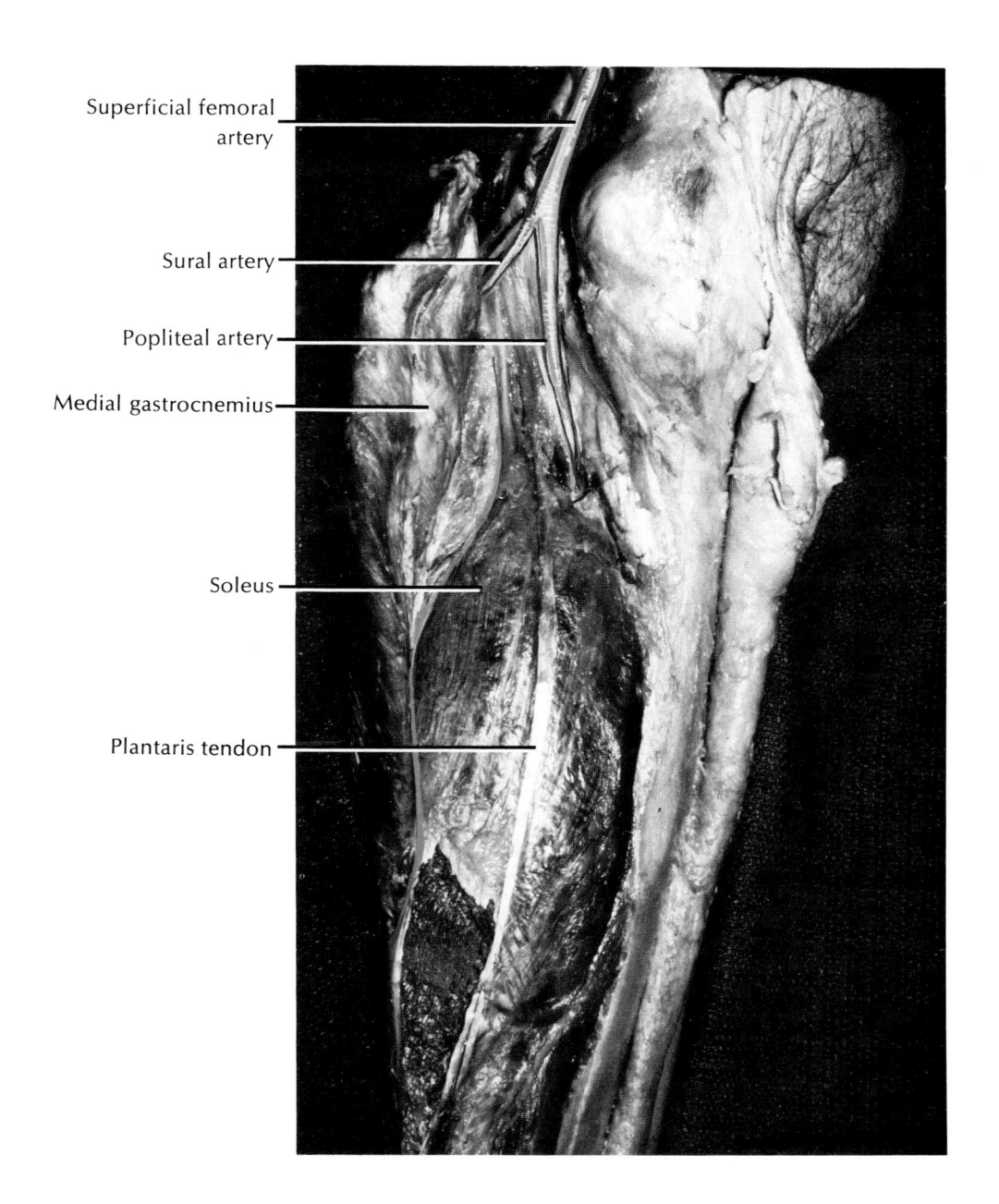

- Superficial femoral artery
- Sural artery
- Popliteal artery
- Medial gastrocnemius
- Soleus
- Plantaris tendon

MEDIAL LEG

Gastrocnemius

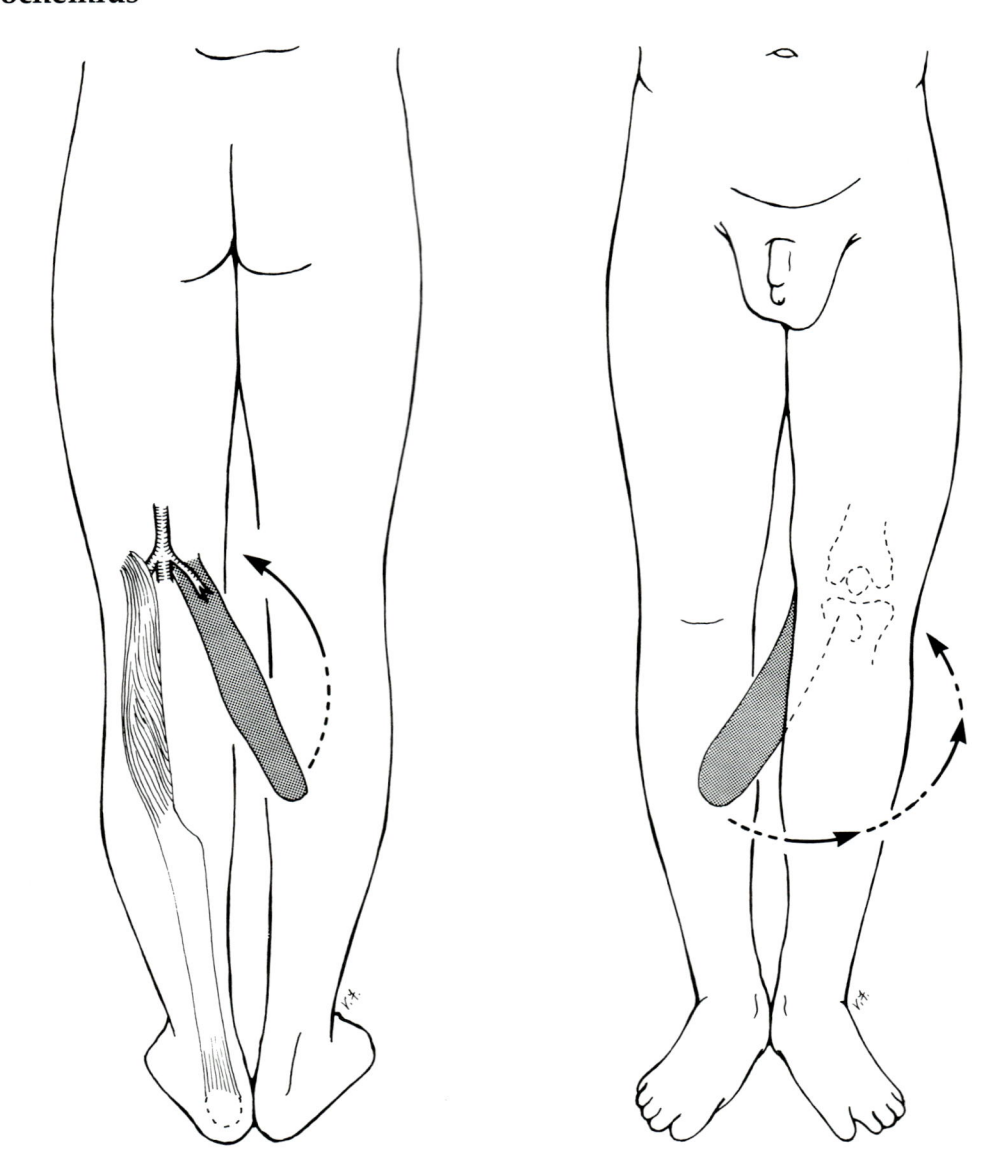

Medial gastrocnemius

Arc of rotation (medial head)

Based on the proximal dominant pedicle, which is 4 or 5 cm above the popliteal crease, the medial gastrocnemius head may be transposed as a muscle or musculocutaneous unit to cover the knee or the upper third of the leg. A slightly greater arc of rotation can be obtained by division of the muscle origin and rotation of the muscle as either a muscle or musculocutaneous island unit.

MEDIAL LEG

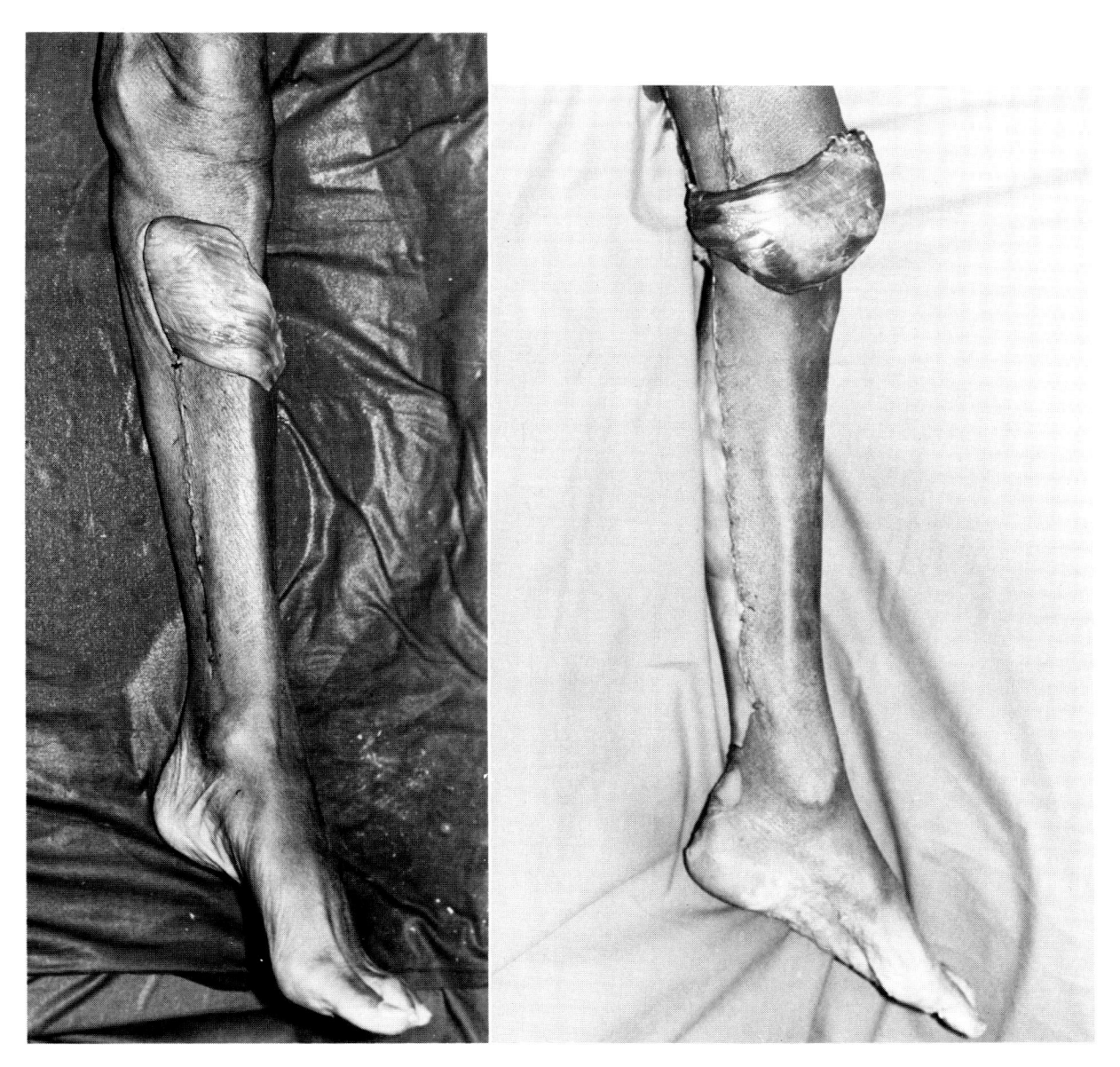

Medial gastrocnemius

MEDIAL LEG

Gastrocnemius

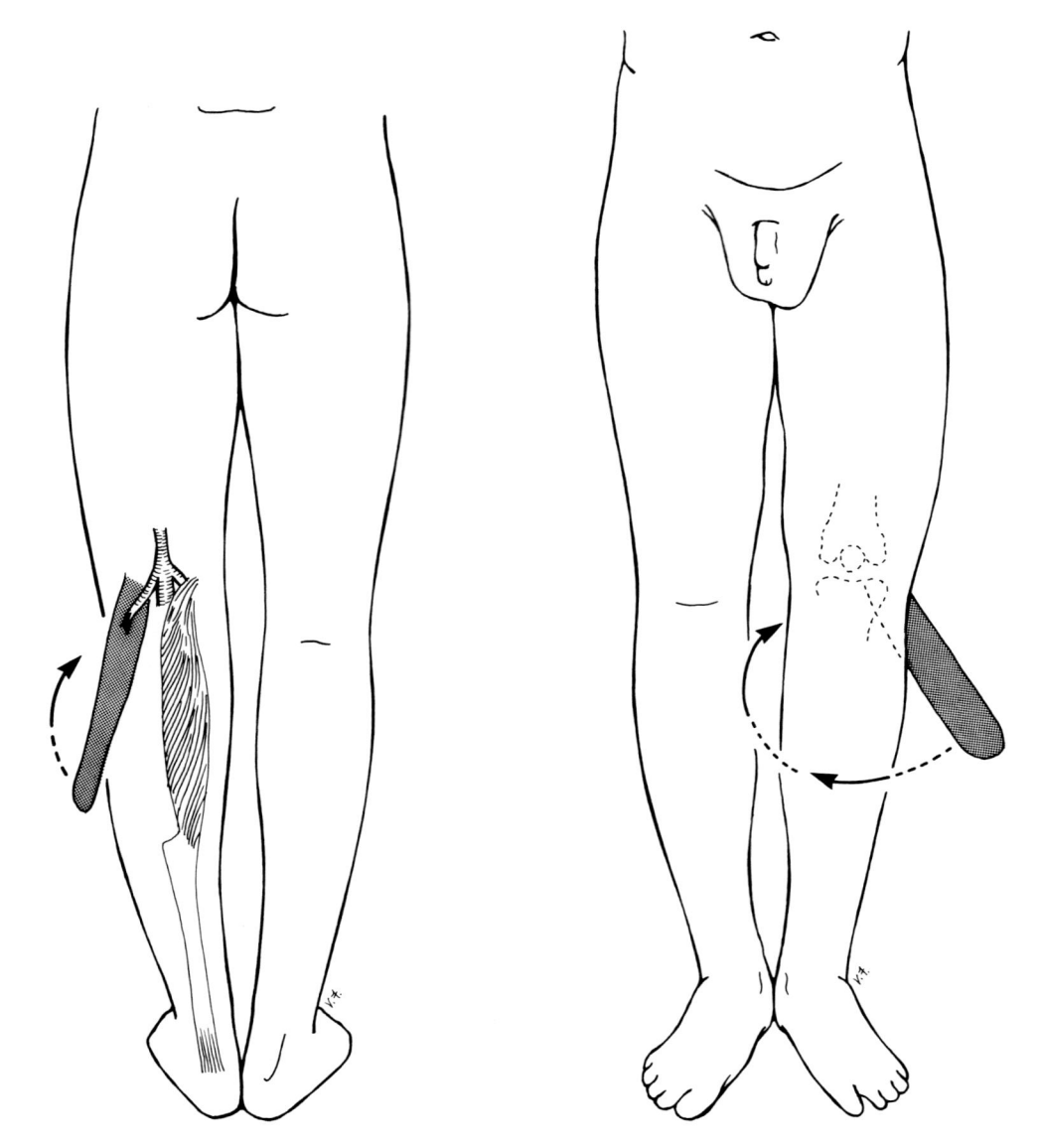

Lateral gastrocnemius

Arc of rotation (lateral head)

Based on the proximal dominant pedicle, which is 4 or 5 cm above the popliteal crease, the lateral gastrocnemius head may be transposed as a muscle or musculocutaneous unit to cover the knee or the upper third of the leg. A slightly greater arc of rotation can be obtained by division of the muscle origin and rotation of the muscle as either a muscle or musculocutaneous island unit.

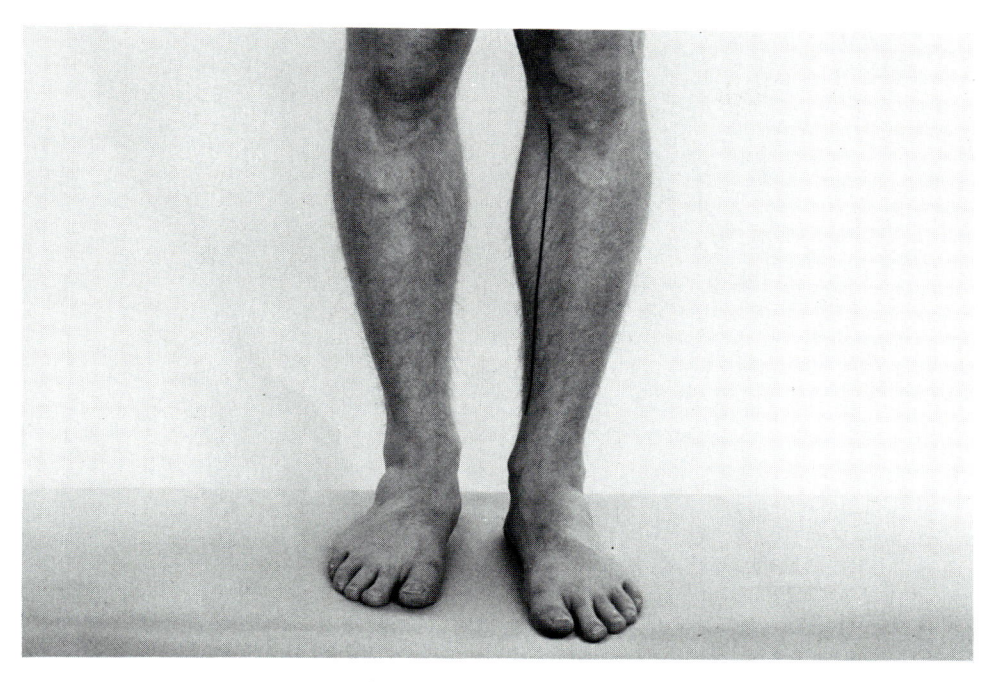

Medial gastrocnemius skin incision

Elevation of flap
MEDIAL HEAD

An incision is made along the medial border of the tibia, and the gastrocnemius is easily identified and separated from the soleus muscle. The plantaris tendon is easily viewed beneath the gastrocnemius medial muscle belly. The muscle fibers of the gastrocnemius are separated from the Achilles tendon. The raphe between the medial and lateral fibers is identified and divided from distal to proximal. The muscle is then elevated up to the level of the tibial condyles, where it generally has an adequate arc of rotation for anterior tibial coverage. If a greater arc of rotation is desirable, this unit can be elevated as a muscle or musculocutaneous island flap. During the more proximal dissection the popliteal artery and tibial nerve must be exposed within the popliteal fossa. However, it is rarely necessary to dissect the muscle more proximally than the inferior aspect of the femoral condyles. At this level the muscle has a wide, safe arc of rotation.

LATERAL HEAD

The lateral head is elevated in a similar maneuver through a lateral vertical incision. In elevating the lateral head, extreme caution is necessary proximally where the peroneal nerve lies superficial to the gastrocnemius and deep to the tendon of the biceps femoris muscle.

MEDIAL LEG

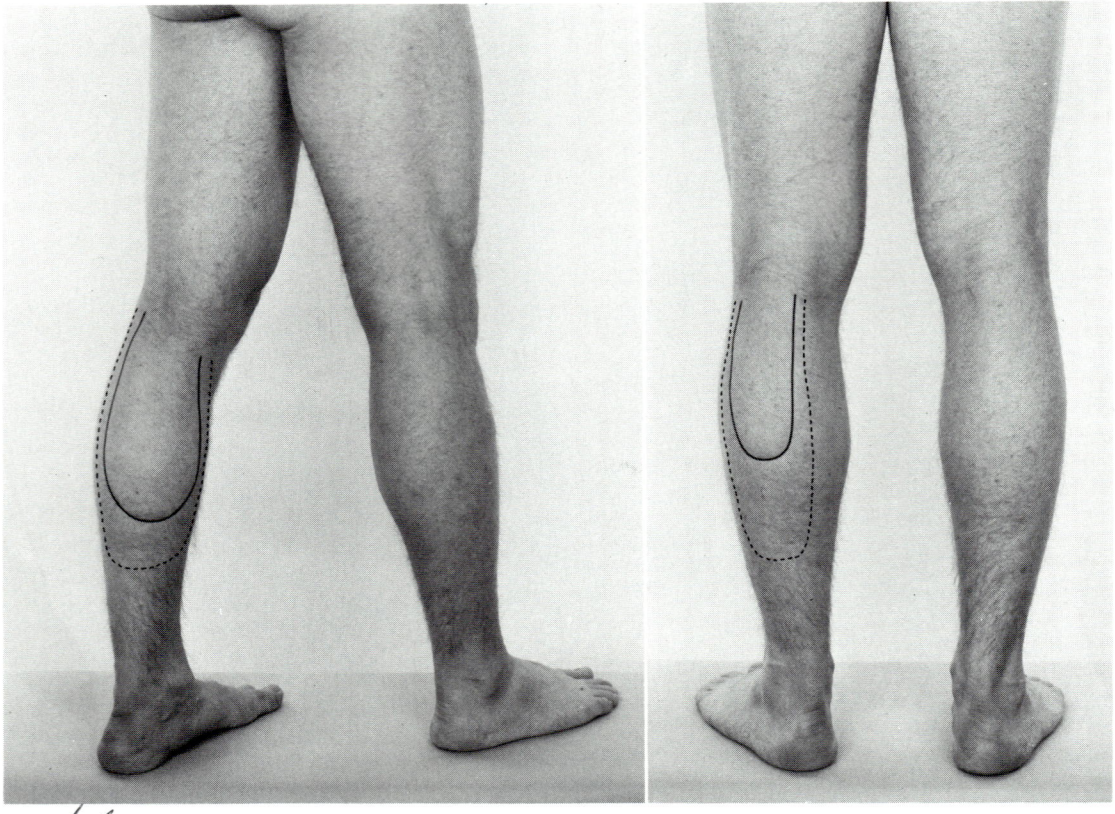

Gastrocnemius

Solid line outlines muscle margin   Dotted line outlines skin territory

MUSCULOCUTANEOUS UNIT

The overlying skin of the medial or lateral gastrocnemius muscle can be elevated as an island with the underlying muscle. An extension of the skin over the Achilles tendon for several centimeters can be safely elevated with the muscle. The donor defect for the musculocutaneous flap will require skin grafting.

Precautions

- The peroneal nerve, located between the origin of the lateral head of the gastrocnemius muscle and insertion of the biceps femoris, is at risk during proximal dissection for the lateral gastrocnemius head.
- Proximal elevation of both heads, but especially the medial head, exposes the popliteal artery and tibial nerve within the popliteal fossa. (It should be noted that it is rarely necessary to dissect this far proximally on this muscle so as to expose or endanger these vessels and nerves.)

MEDIAL LEG

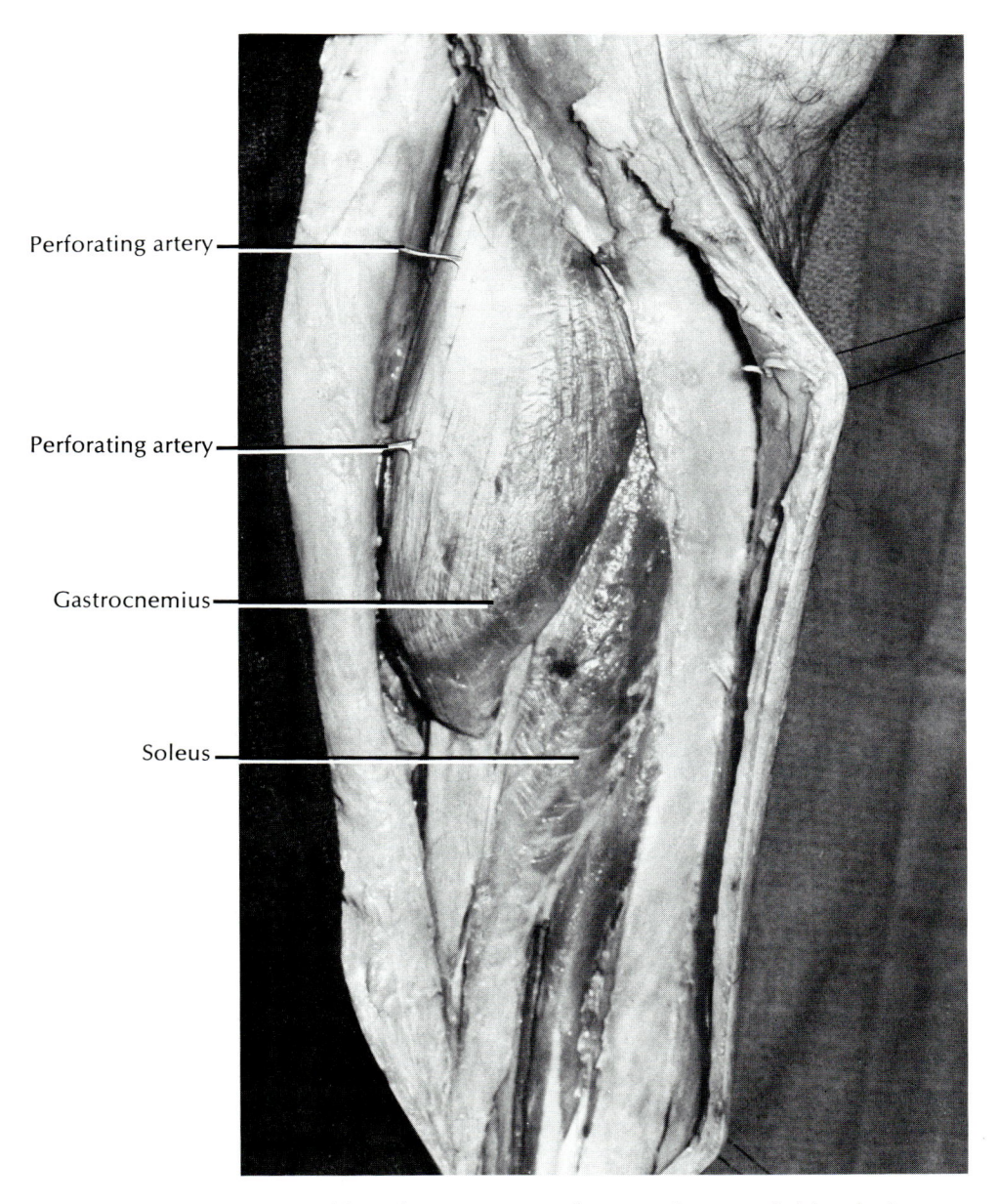

Labels on figure:
- Perforating artery
- Perforating artery
- Gastrocnemius
- Soleus

Musculocutaneous perforators from medial head of gastrocnemius into overlying skin

- The sural nerve runs in the lateral gastrocnemius skin territory and should be preserved if possible.
- The saphenous vein runs in the anterior portion of the medial gastrocnemius skin territory and should be preserved if possible.

MEDIAL LEG

Gastrocnemius

Medial musculocutaneous flap

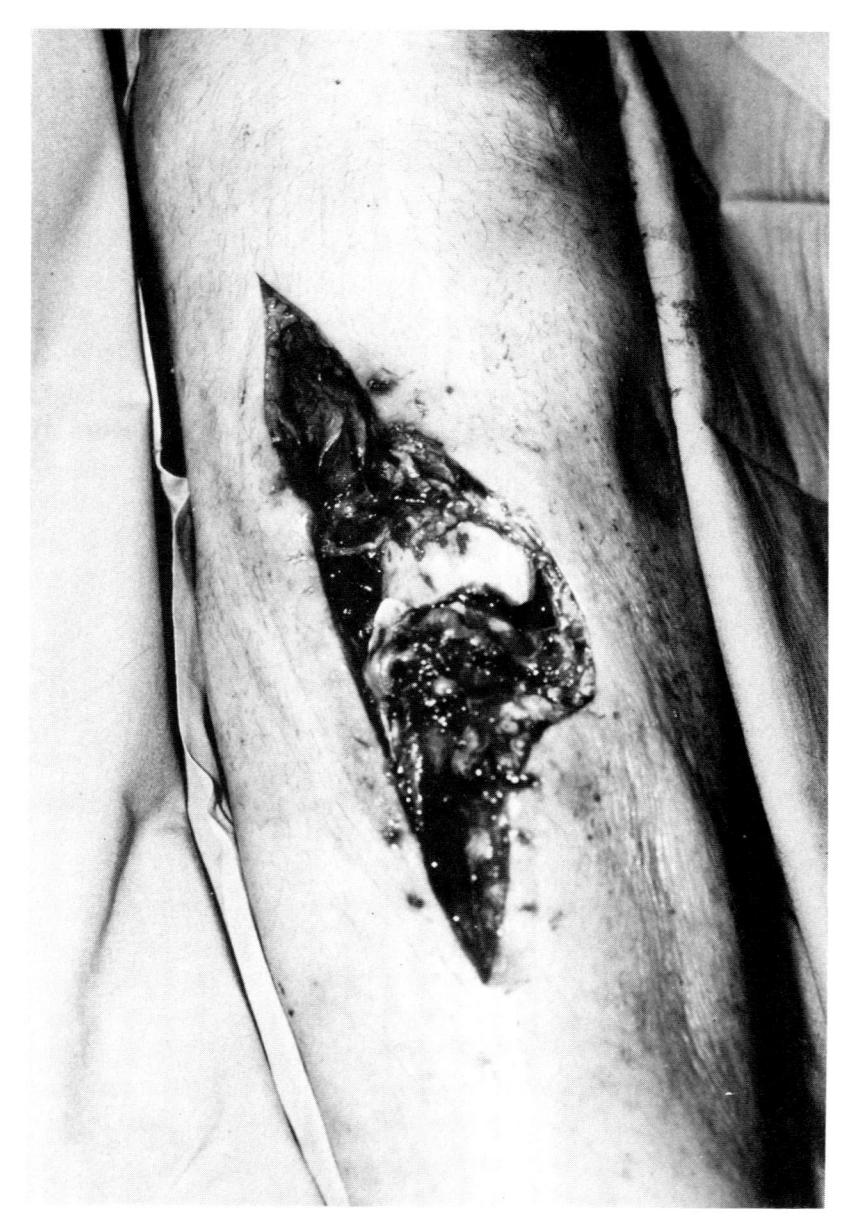

Open comminuted tibial fracture with loss of soft tissue cover.

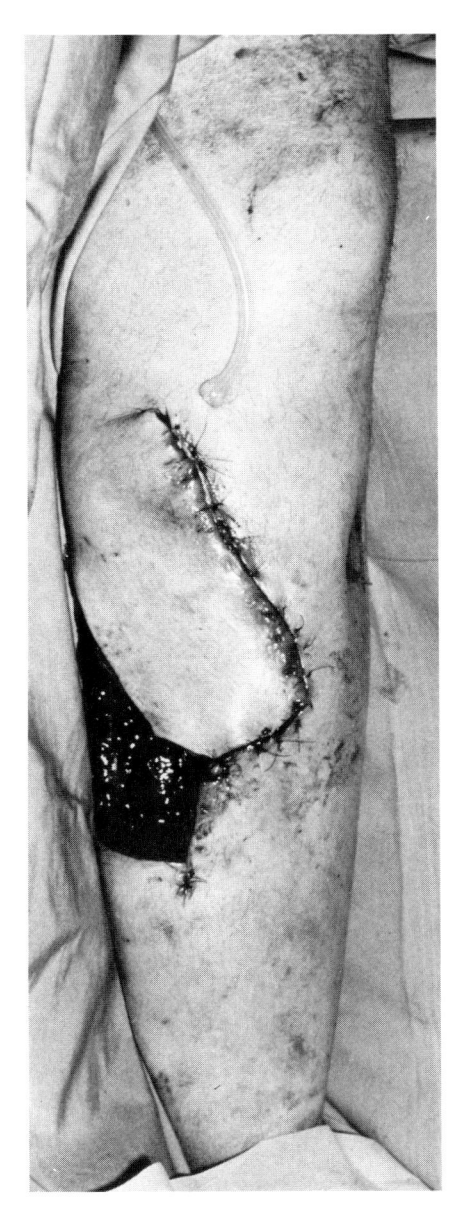

Medial gastrocnemius musculocutaneous flap transposed over fracture.

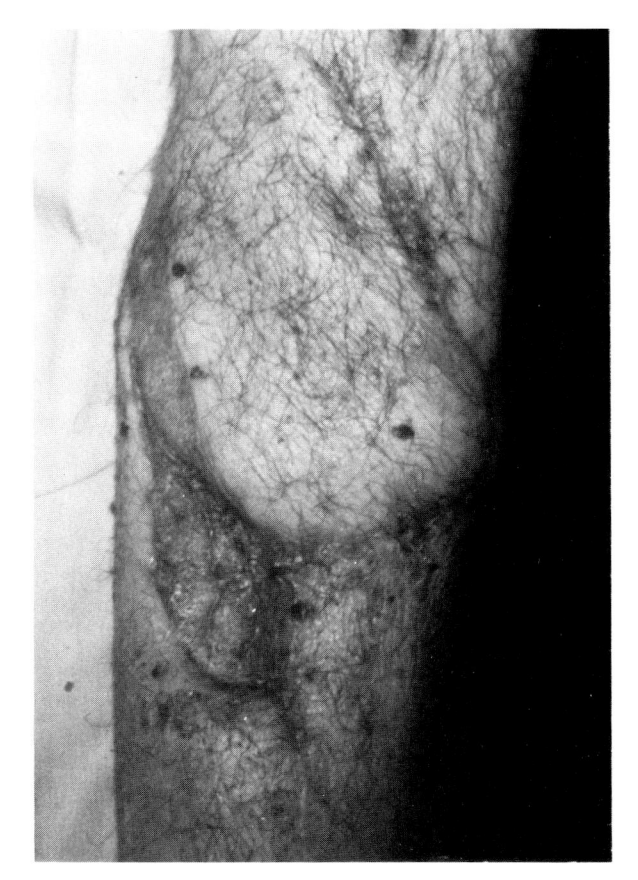

Long-term result, bone healed. Single-stage closure of defect.

MEDIAL LEG

Gastrocnemius

Lateral musculocutaneous flap

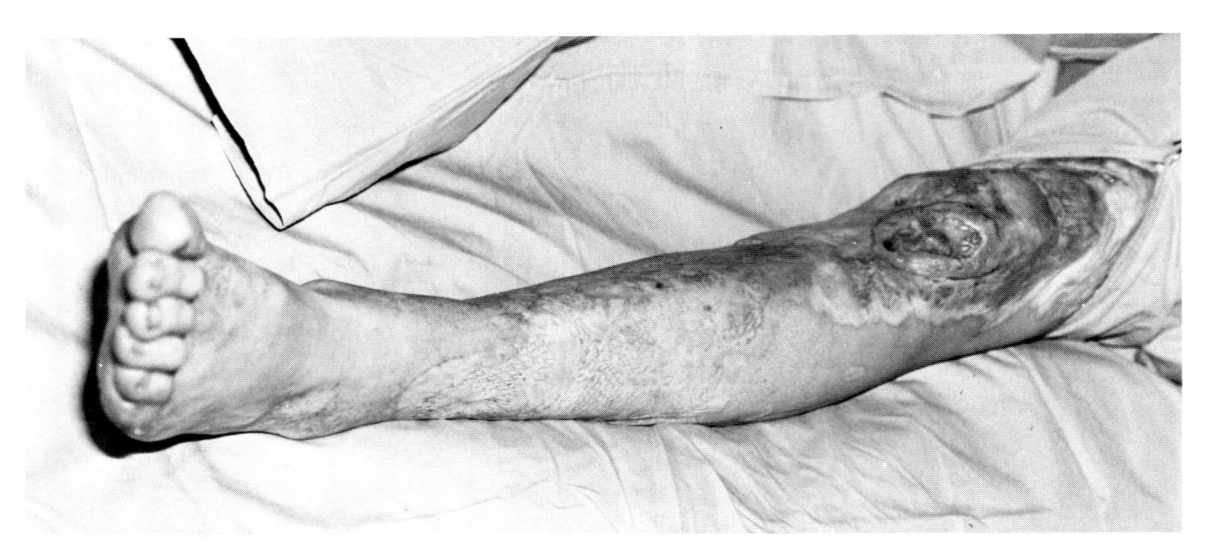

Unstable burn scar over knee with exposed joint.

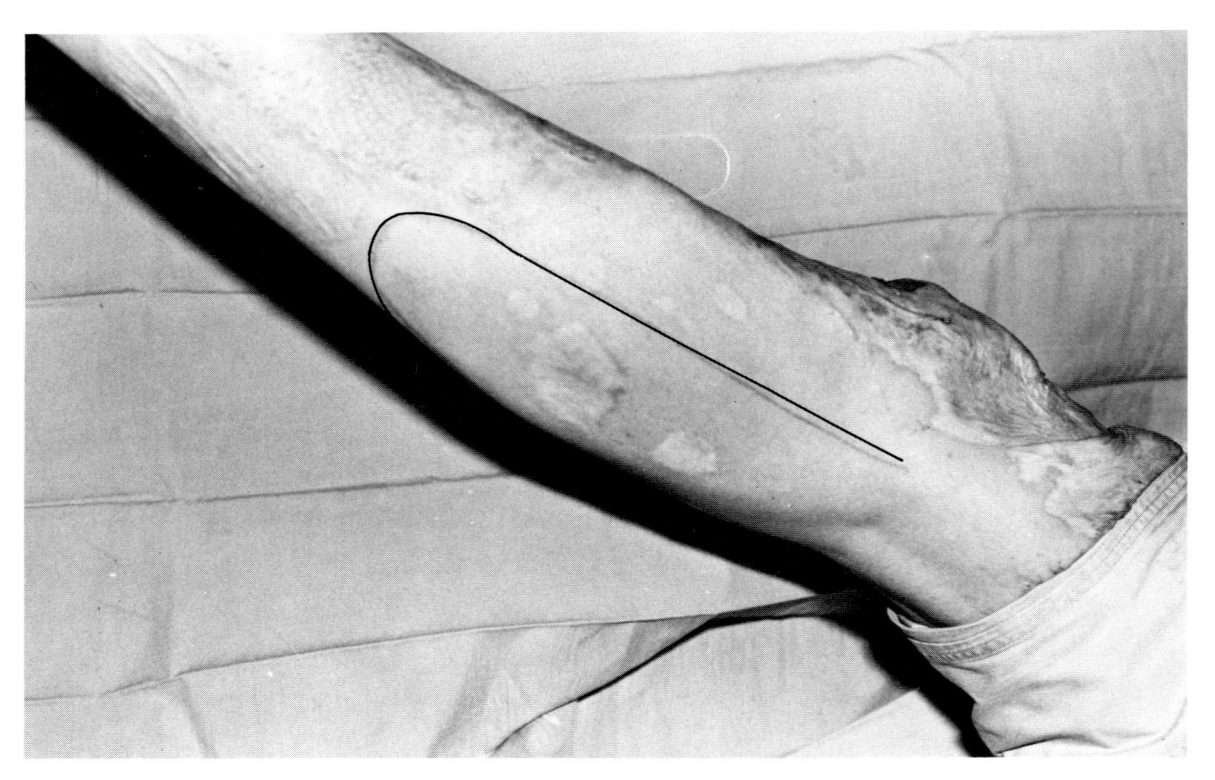

Lateral gastrocnemius musculocutaneous flap outlined.

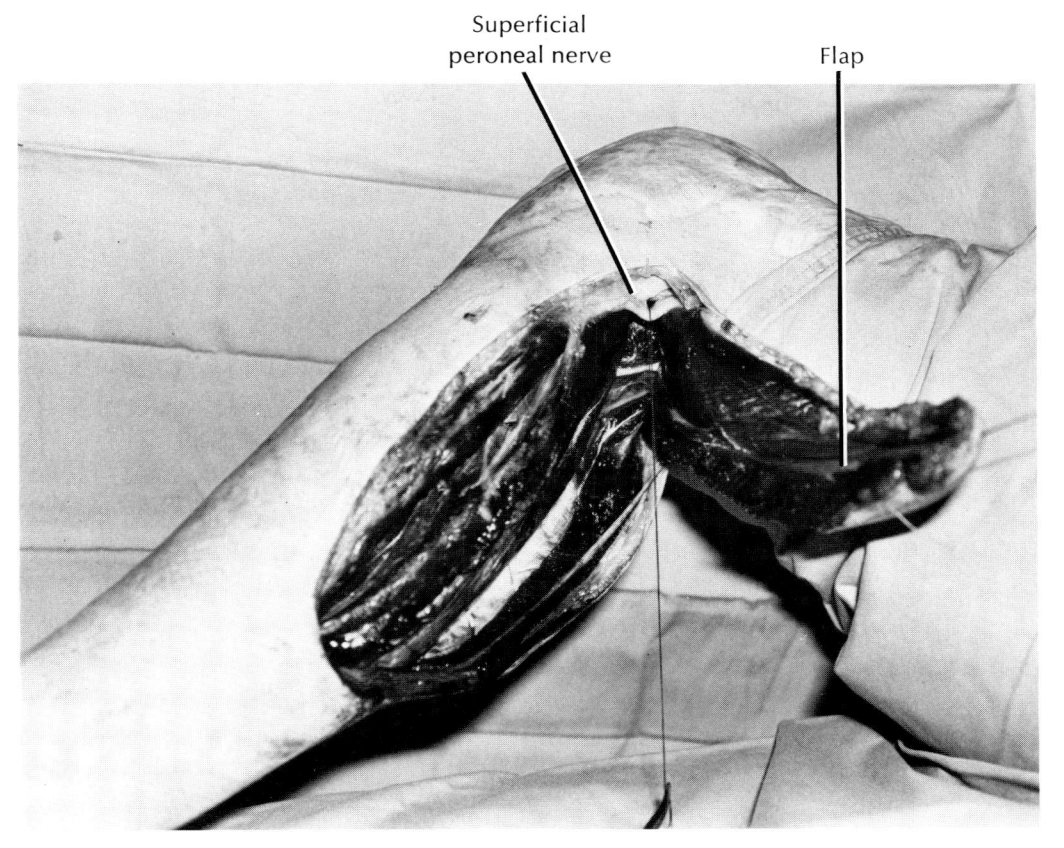

Flap elevated. Note superficial peroneal nerve.

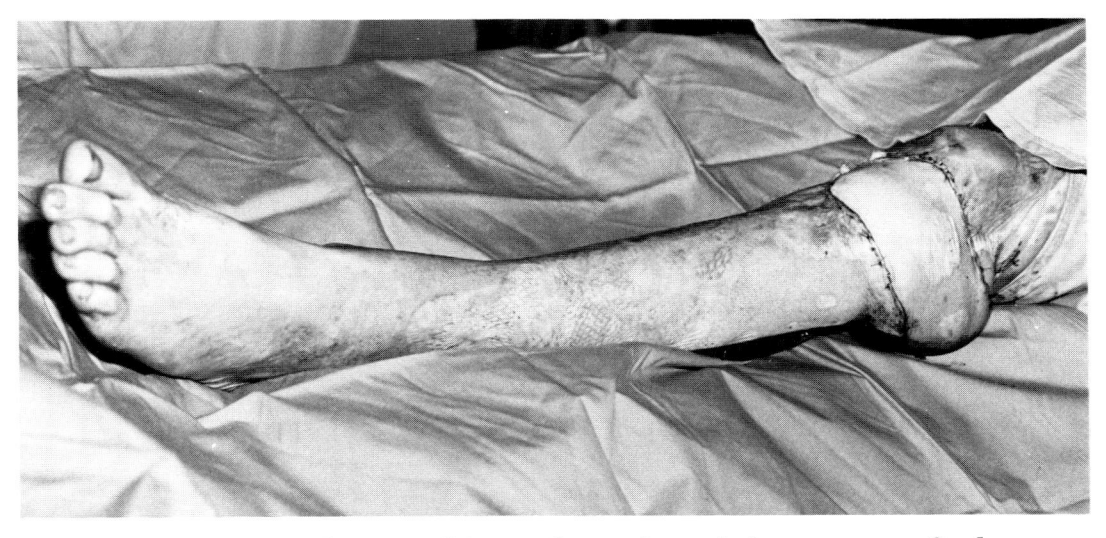

Single-stage closure of joint. Secondary defect was grafted.

MEDIAL LEG

Soleus

MUSCLE FLAP

Application

  Coverage of:

    MEDIAL THIRD OF LEG

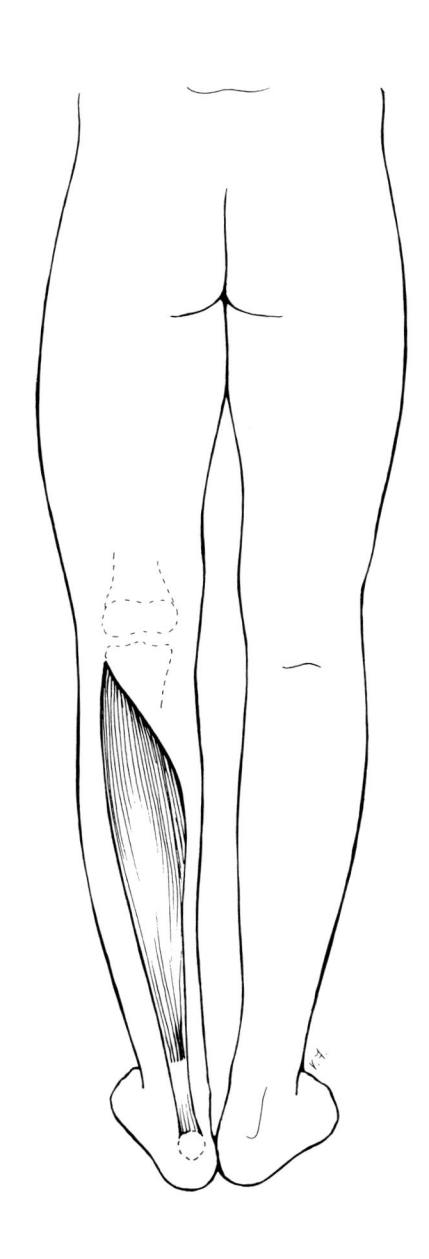

MEDIAL LEG

Soleus

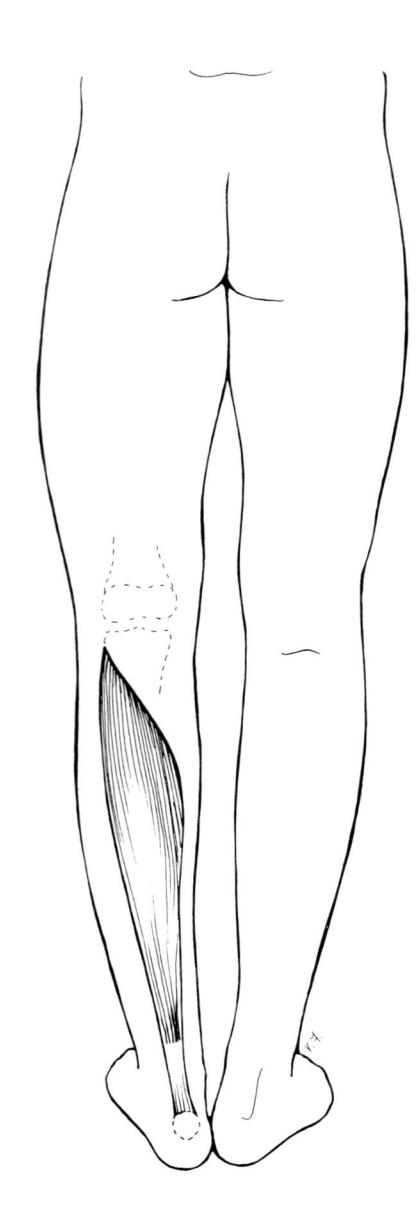

The soleus is a large, broad, flat muscle lying immediately deep to the gastrocnemius muscle. The muscle fibers converge into an aponeurosis that joins the tendon of the gastrocnemius to form the Achilles tendon.

Origin: Fibular—posterior aspect of head and upper third of fibula; tibial—popliteal line on posterior aspect of tibia.
Insertion: Calcaneus through the Achilles tendon.

MEDIAL LEG

Superficial femoral artery
Pedicle off peroneal artery
Pedicle off posterior tibial artery
Peroneal artery
Minor pedicles off posterior tibial artery

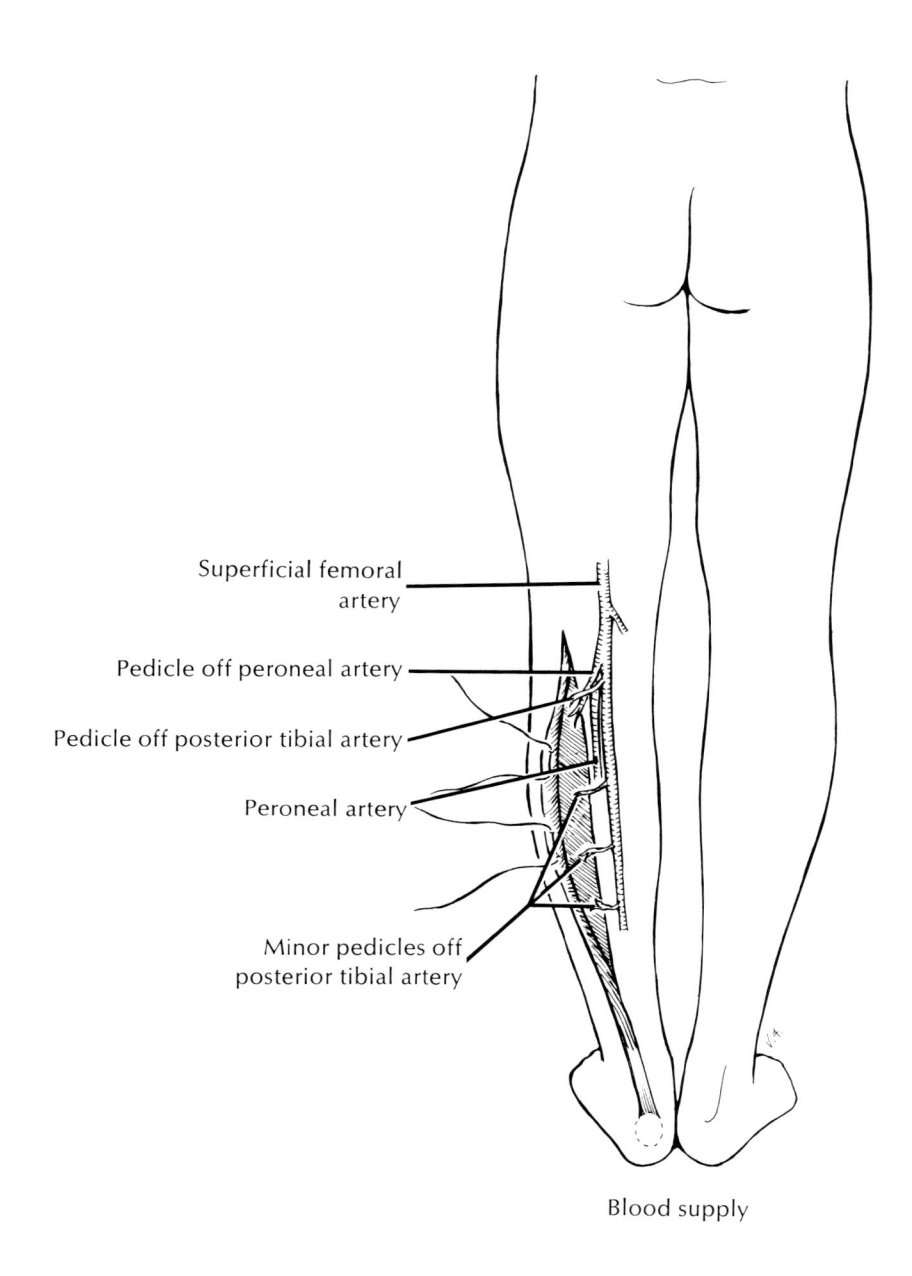

Blood supply

Nerve supply:
> **Motor**—Branches of tibial nerve. The posterior tibial nerve lies deep to the soleus muscle along the entire muscle belly. The motor branches enter the proximal deep muscle belly.

Function: Plantar flexion of the foot. If the gastrocnemius is intact, the transposition of this muscle will not cause any significant disability in an ambulatory patient.

MEDIAL LEG

Soleus

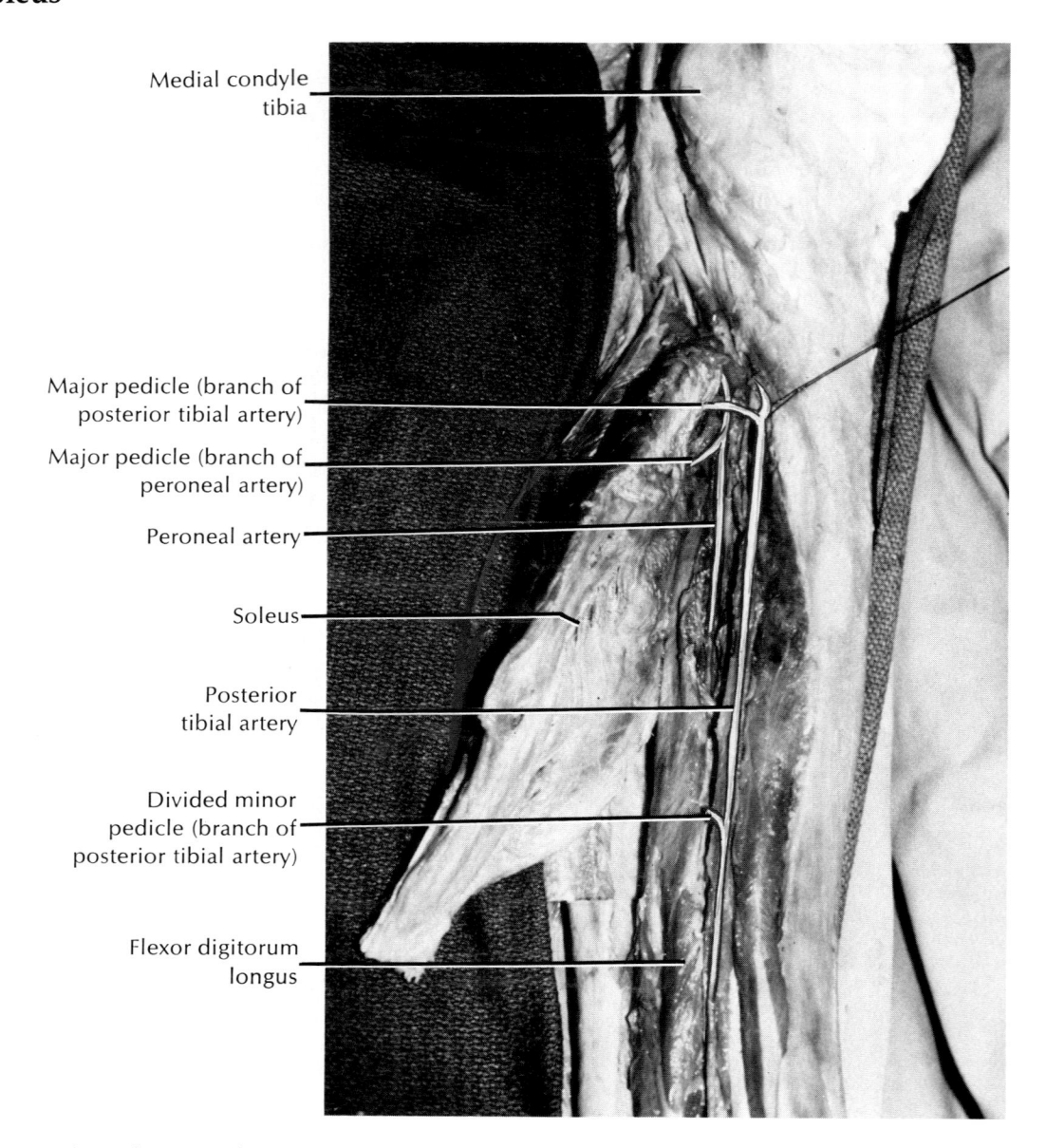

- Medial condyle tibia
- Major pedicle (branch of posterior tibial artery)
- Major pedicle (branch of peroneal artery)
- Peroneal artery
- Soleus
- Posterior tibial artery
- Divided minor pedicle (branch of posterior tibial artery)
- Flexor digitorum longus

Blood supply

The fibular origin has a dominant proximal pedicle based on the peroneal artery, and the tibial portion has a proximal dominant pedicle and three or more distal pedicles that are branches from the posterior tibial artery.

Deep to the upper free border of the soleus muscle between the tibial and fibular origins, the popliteal artery divides into the peroneal and posterior tibial vessels. The tibial nerve follows the posterior tibial vessels deep to the soleus muscle.

Posterior tibial artery

Minor pedicle

Soleus

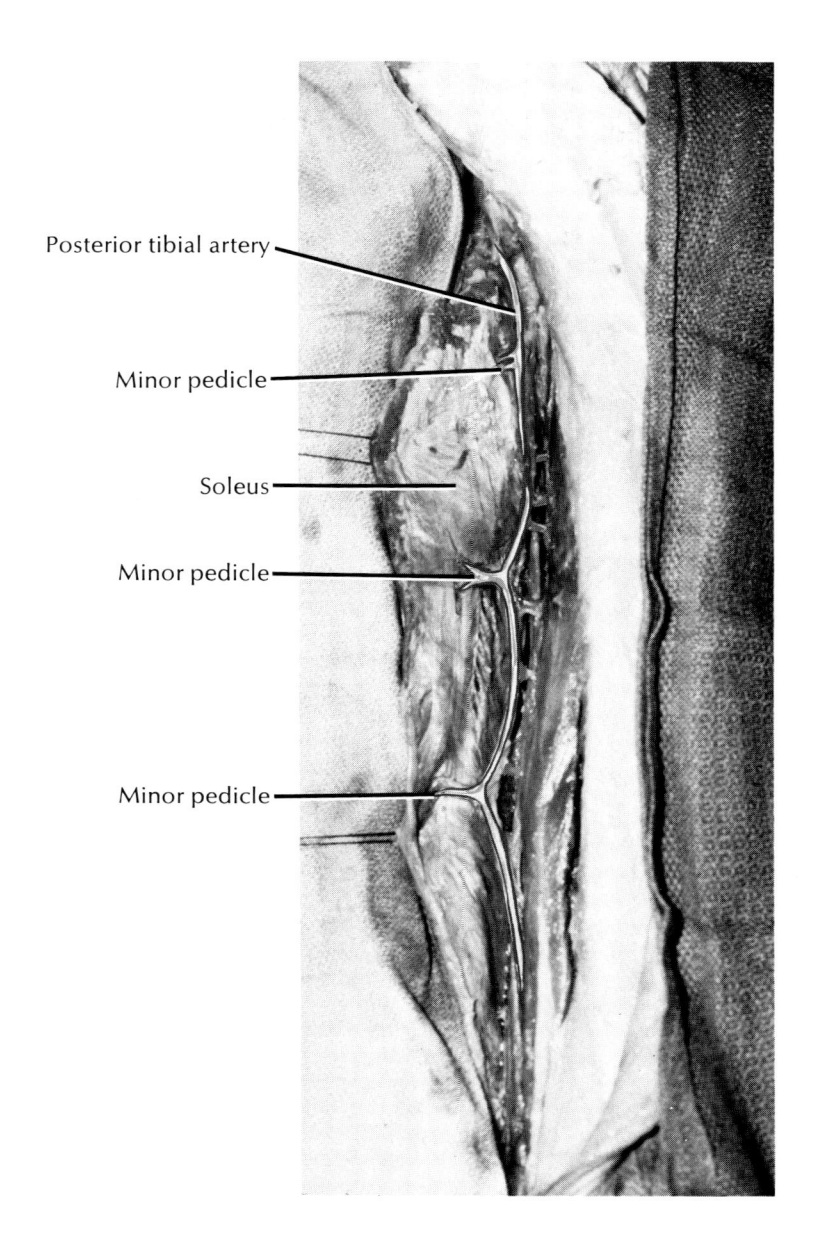

Minor pedicle

Minor pedicle

The peroneal artery supplies the fibular origin, then runs deep to the soleus and deep to or through the flexor hallucis longus to the peroneal muscles. The posterior tibial artery runs deep to the soleus behind the tibialis posterior and lateral to the flexor digitorum longus. The branches to the soleus enter the muscle on its deep surface.

The branch from the peroneal artery and the proximal pedicle from the posterior tibial artery will support the entire muscle for transposition.

MEDIAL LEG

Soleus

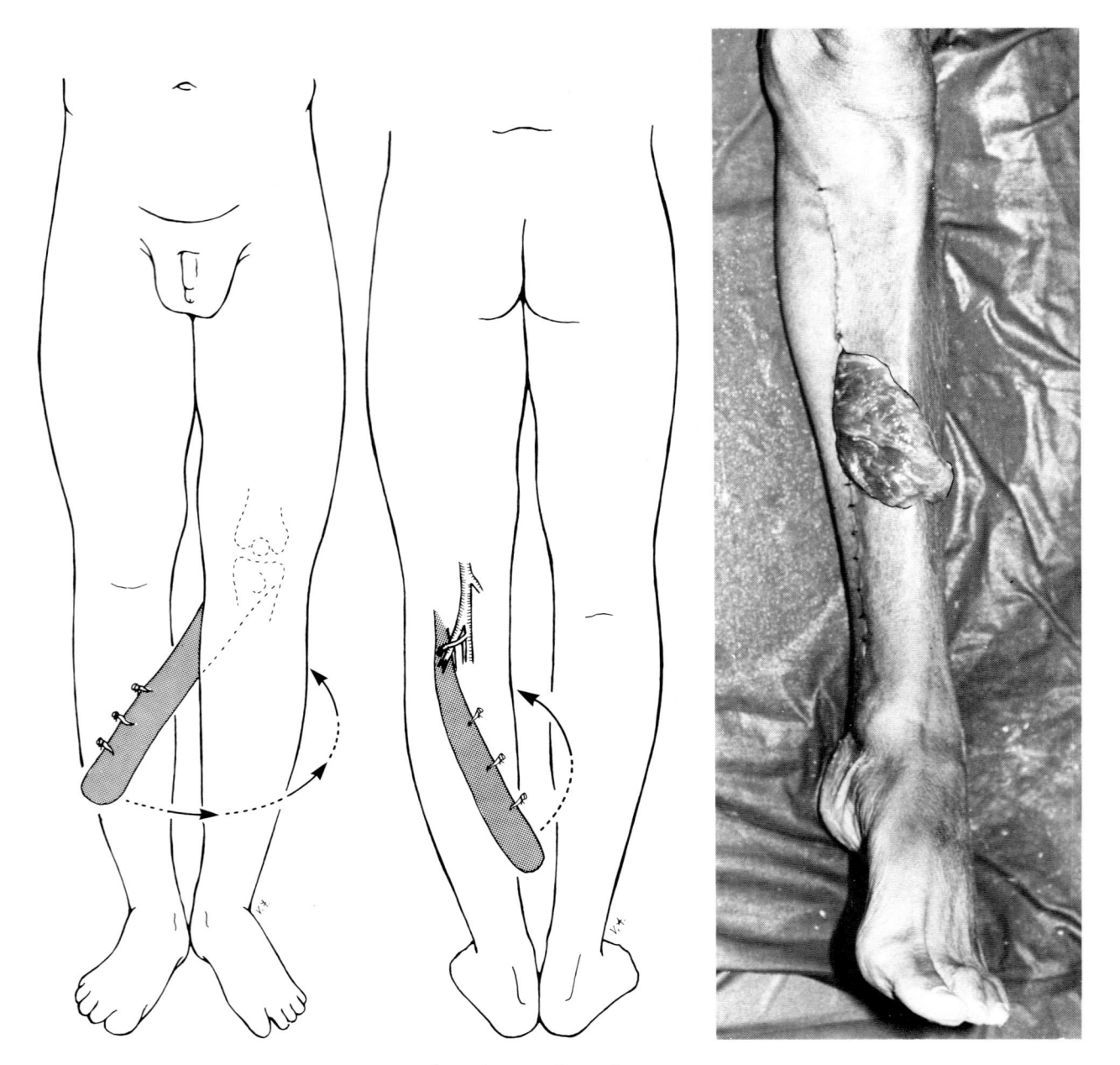

Superior arc (lateral transposition)

Arc of rotation

With the distal pedicles from the posterior tibial artery ligated, and based on the proximal pedicles from the posterior tibial and peroneal arteries, the muscle can be transposed medially or laterally to cover the middle third of the leg. This area of coverage anteriorly is located

MEDIAL LEG

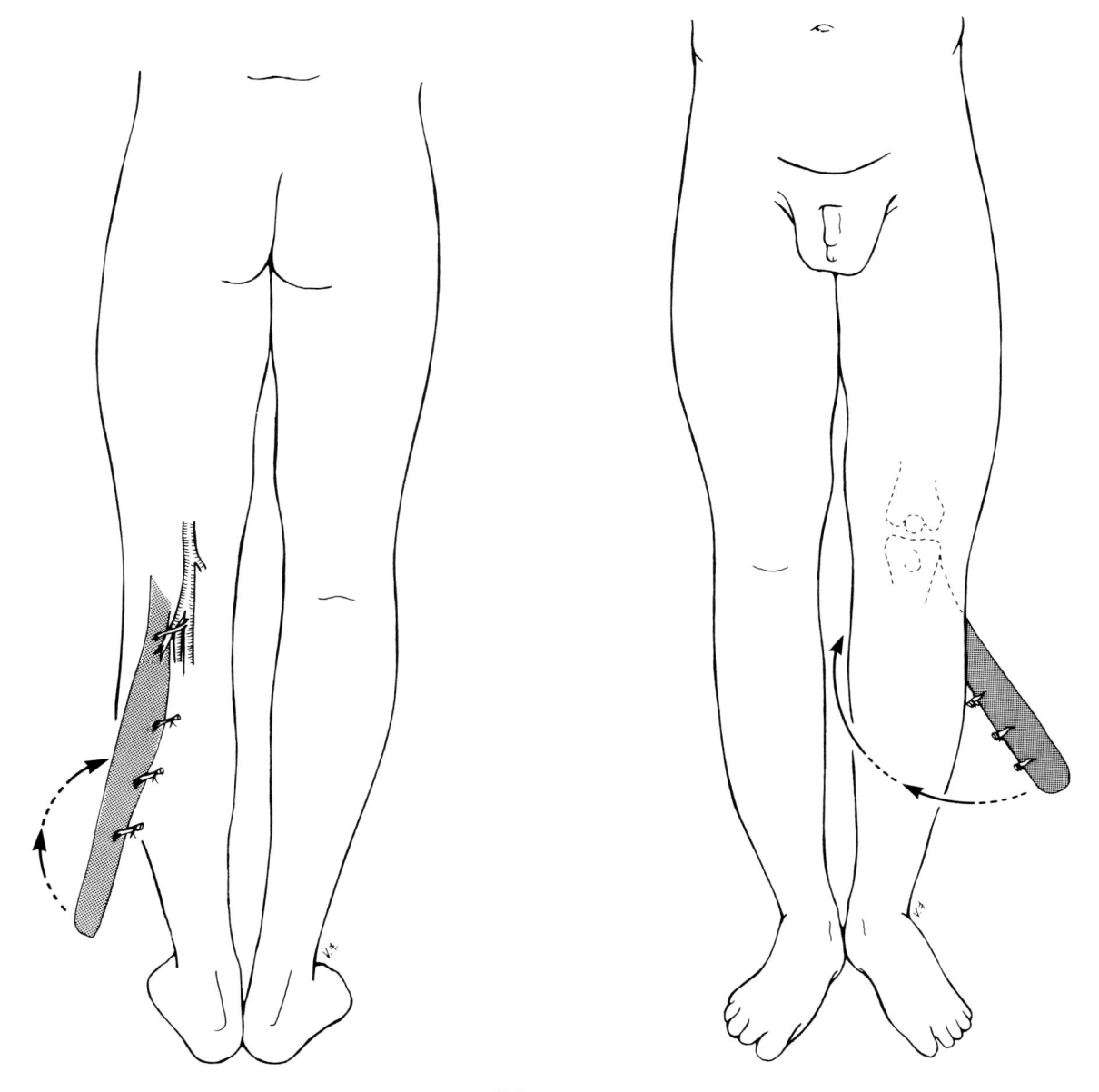

Medial transposition

between the arc of rotation for the gastrocnemius superiorly and the flexor digitorum longus inferiorly. The point of rotation is about 10 to 12 cm below the knee. By incising the epimysium, the muscle can be expanded to cover a larger area.

MEDIAL LEG

Soleus

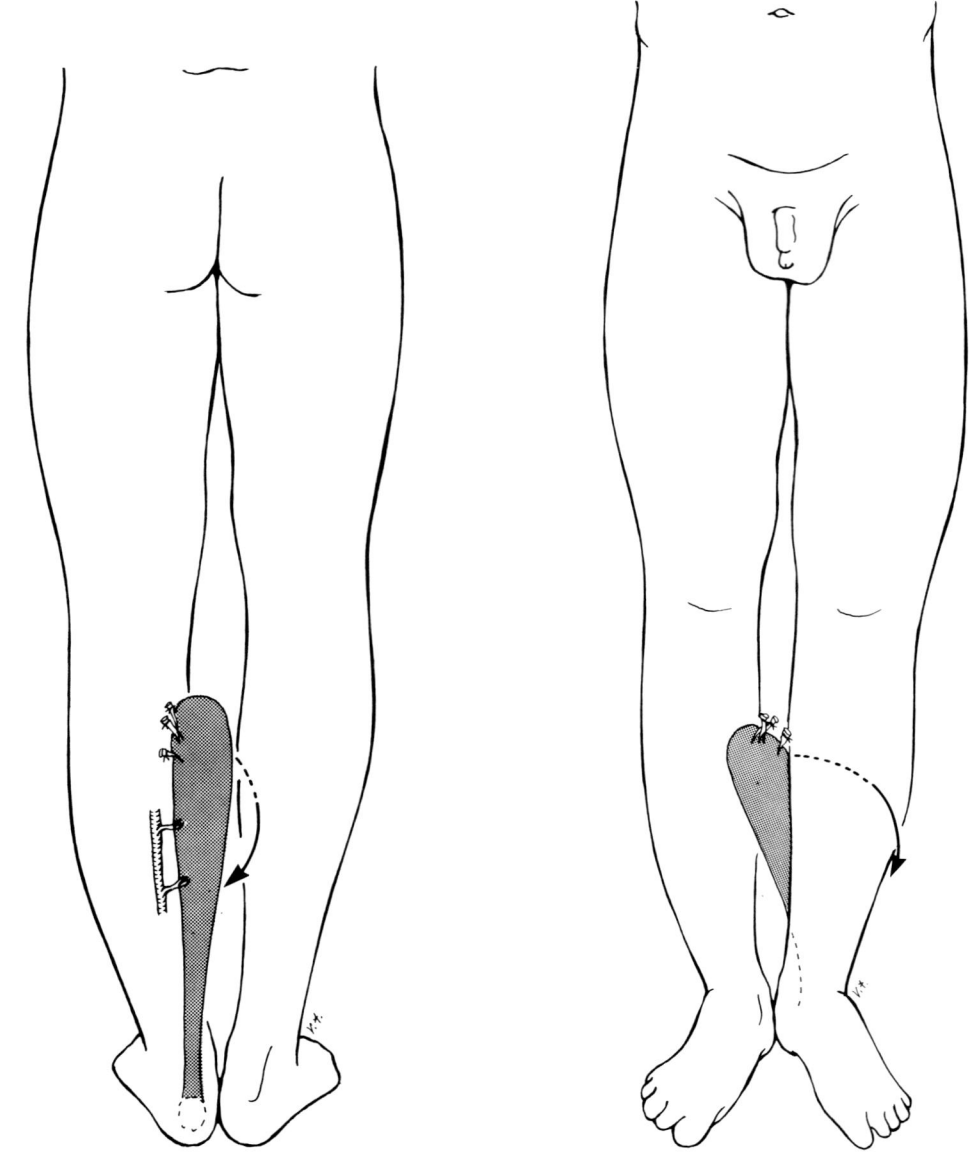

Inferior arc

DISTALLY BASED SOLEUS

The soleus muscle can be rotated inferiorly based on the two or three pedicles from the posterior tibial artery entering the inferior muscle. However, this transposition is rarely indicated, since the dissection in the proximal soleus muscle is difficult and these smaller inferior pedicles are less predictable in location and size.

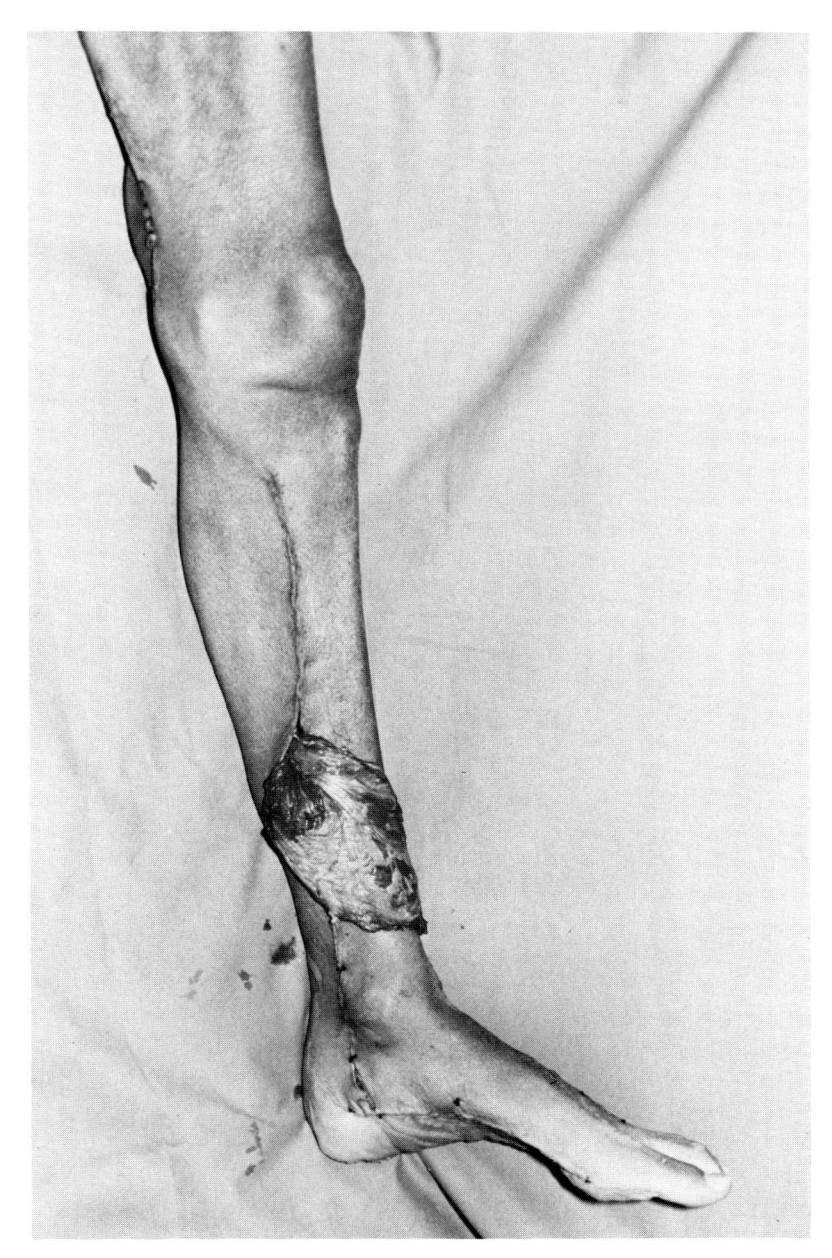

Inferior arc

MEDIAL LEG

Soleus

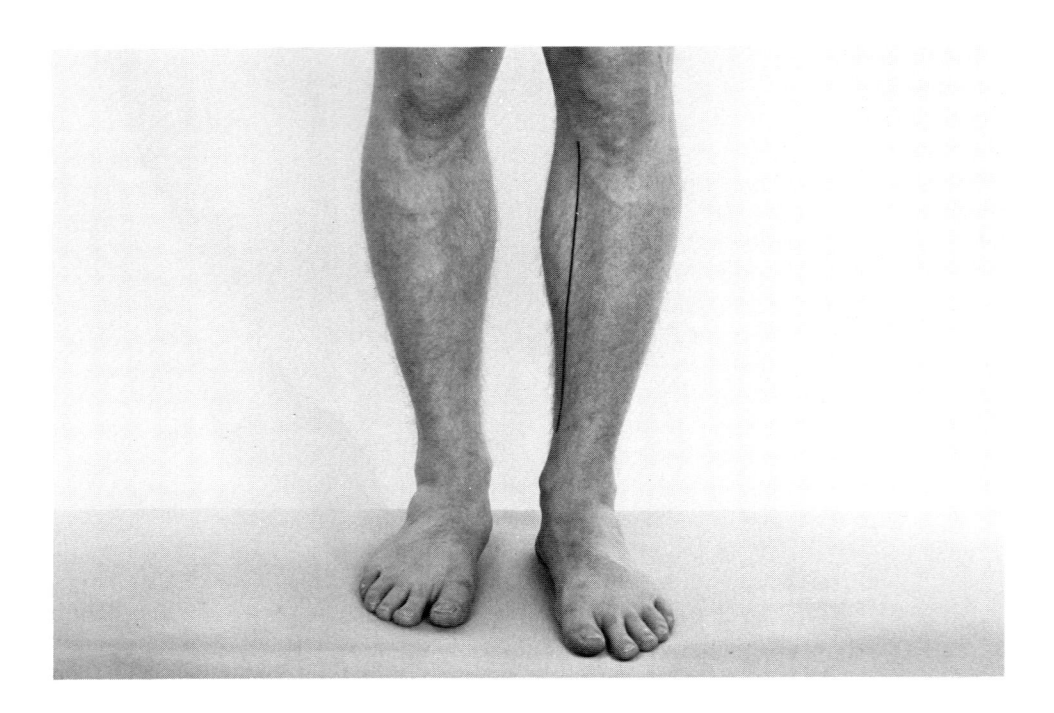

Elevation of flap
MEDIAL ARC (MEDIAL EXPOSURE AND LATERAL TRANSPOSITION)

An incision in the lower half of the leg medial to the tibia is made. The muscle is identified immediately deep to the Achilles tendon. To avoid confusion with the gastrocnemius, dissection is continued proximally to identify the plantaris tendon and gastrocnemius, which lie superficial to the soleus. From medial to lateral, the flexor digitorum longus, posterior tibial artery, tibial nerve, tibialis posterior muscle, and the flexor hallucis longus muscle lie deep to the soleus muscle. The soleus muscle fibers are separated from the deep surface of the Achilles tendon, and the muscle is retracted outward. The distal pedicles from the posterior tibial artery are ligated, and the muscle is transposed over the defect to be covered. This is the standard elevation and transposition of this flap.

LATERAL ARC (LATERAL EXPOSURE AND MEDIAL TRANSPOSITION)

Under certain circumstances it may be necessary to expose the soleus laterally and transpose it from lateral to medial for coverage of the middle third of the leg. This may be useful if the lateral gastrocnemius muscle is not available. When transposition is performed through a vertical incision lateral to the fibula, the soleus muscle is identified beneath the gastrocnemius and elevated in a manner similar to that described for the medial exposure. In order to accomplish this transposition, it may be necessary to divide the proximal pedicle from the posterior tibial artery. Further length may be obtained, if necessary, for coverage of a large tibial defect by excision of the fibula to allow a better anterior arc of rotation for the medial transposition of the soleus muscle.

The donor defect in both medial arc and lateral arc transposition may be closed primarily.

Precautions

- The posterior tibial artery and the tibial nerve lie deep to the soleus and may be at risk during flap elevation.
- Distally based inferior transposition of the muscle is not entirely reliable.

MEDIAL LEG

Soleus

Medial arc — coverage of exposed tibia
(medial exposure and lateral transposition)

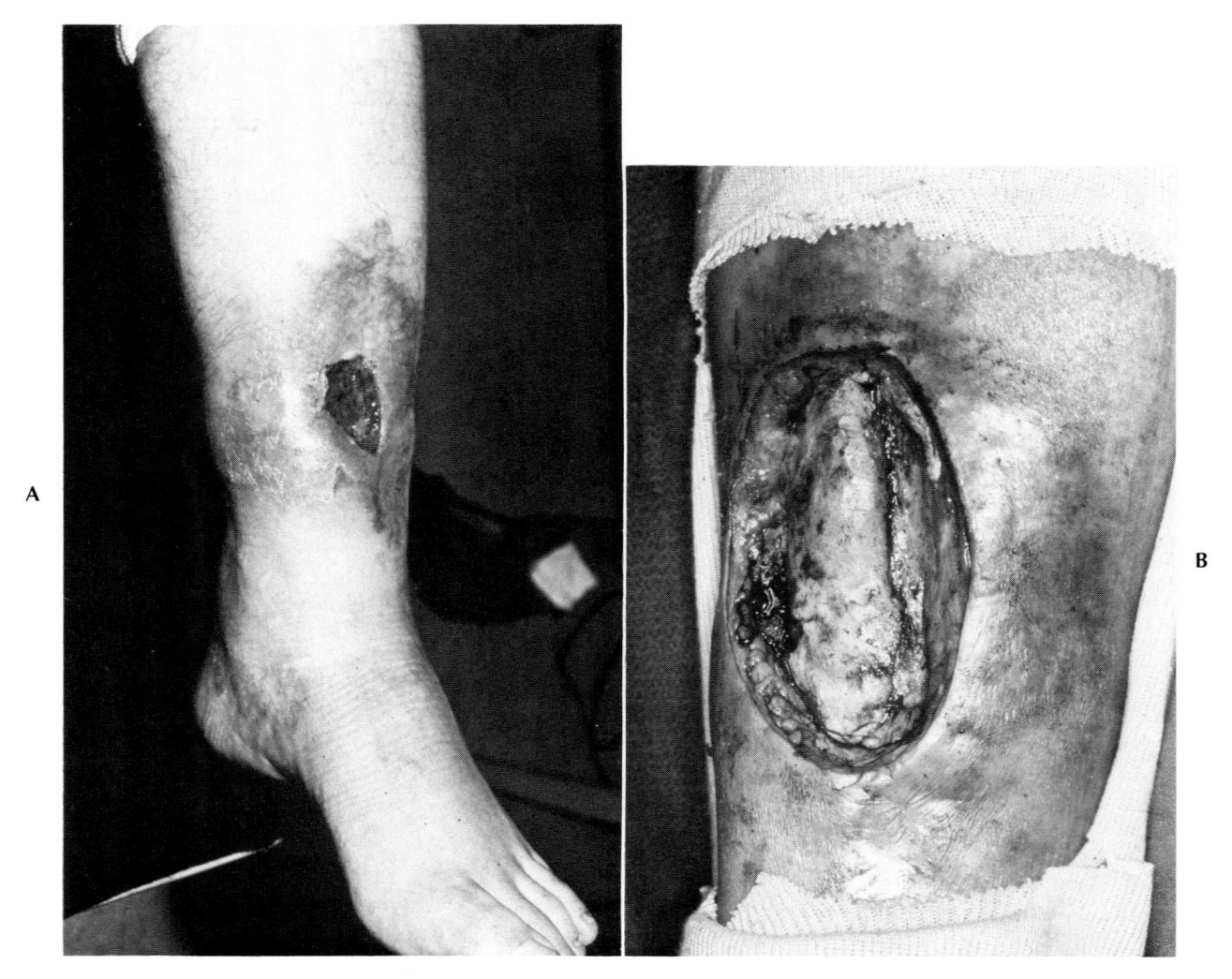

A Traumatic defect. Left leg with exposed tibia.
B Wound debrided.

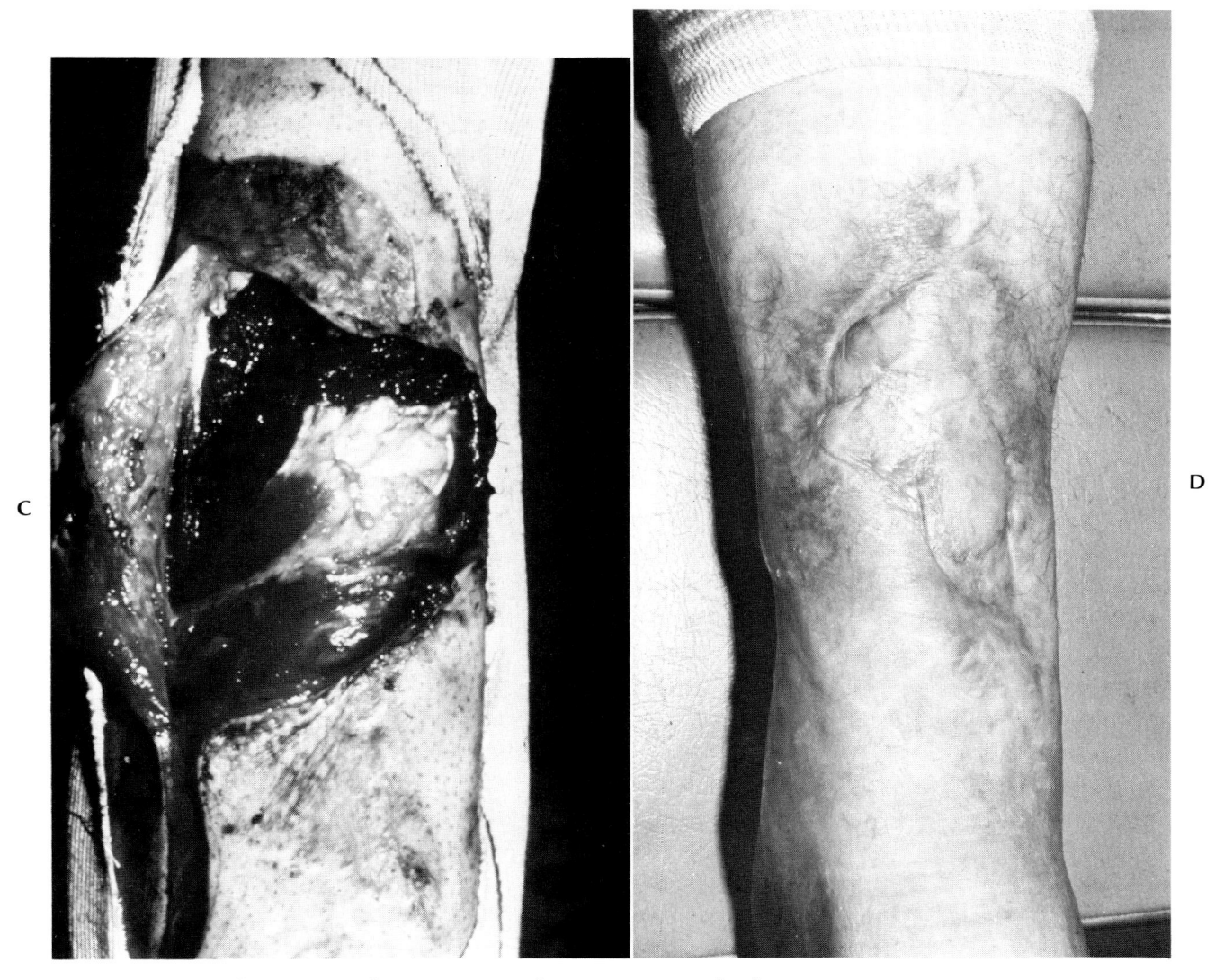

C   Soleus muscle transposed over exposed tibia.

D   Final result. Skin graft over soleus muscle. Single-stage coverage of exposed tibia.

MEDIAL LEG

Soleus

Coverage of exposed tibia

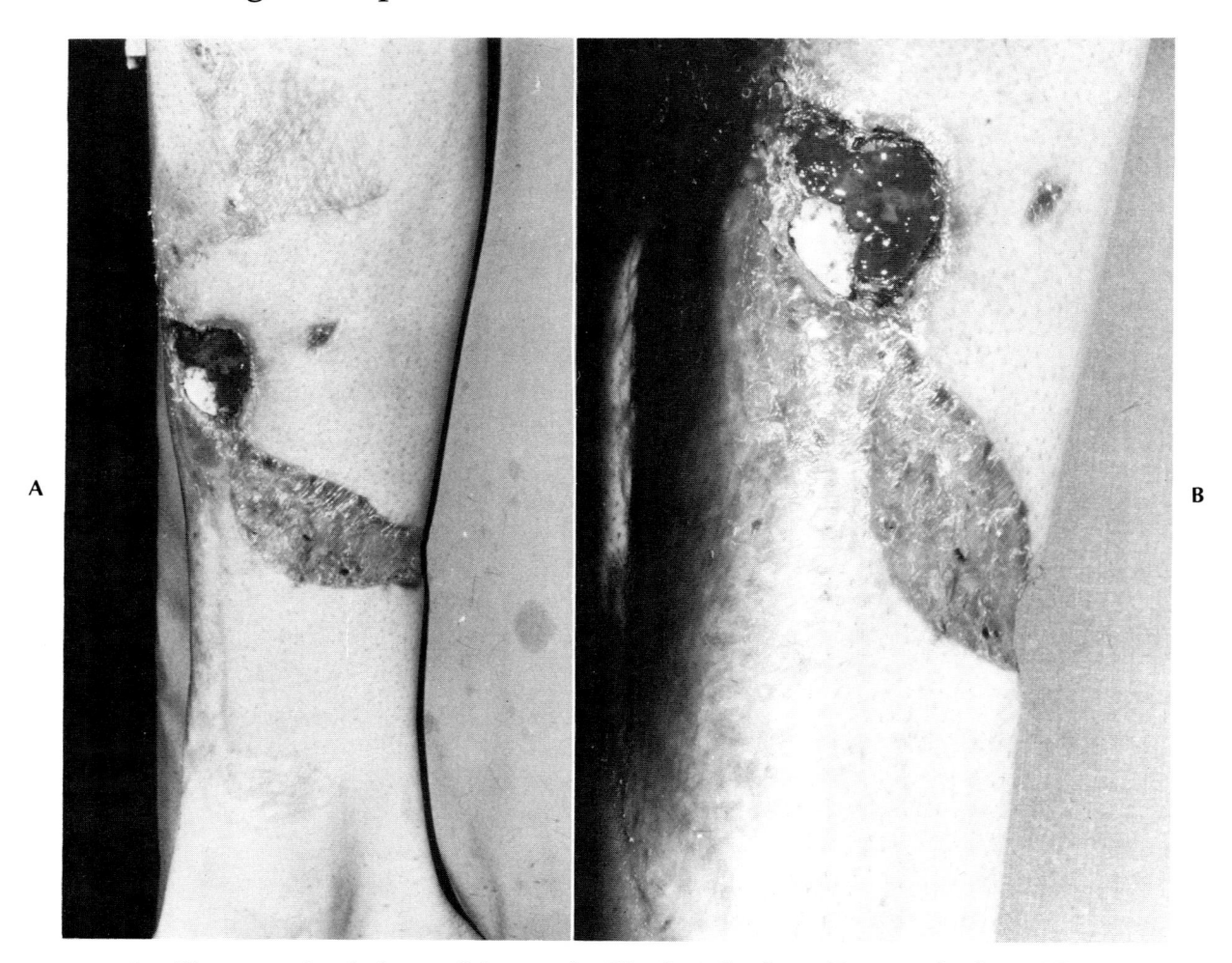

A Traumatic defect of lower half of right leg. Exposed tibia. Note failed skin transposition flap.

B Close-up of exposed tibia and granulation tissue.

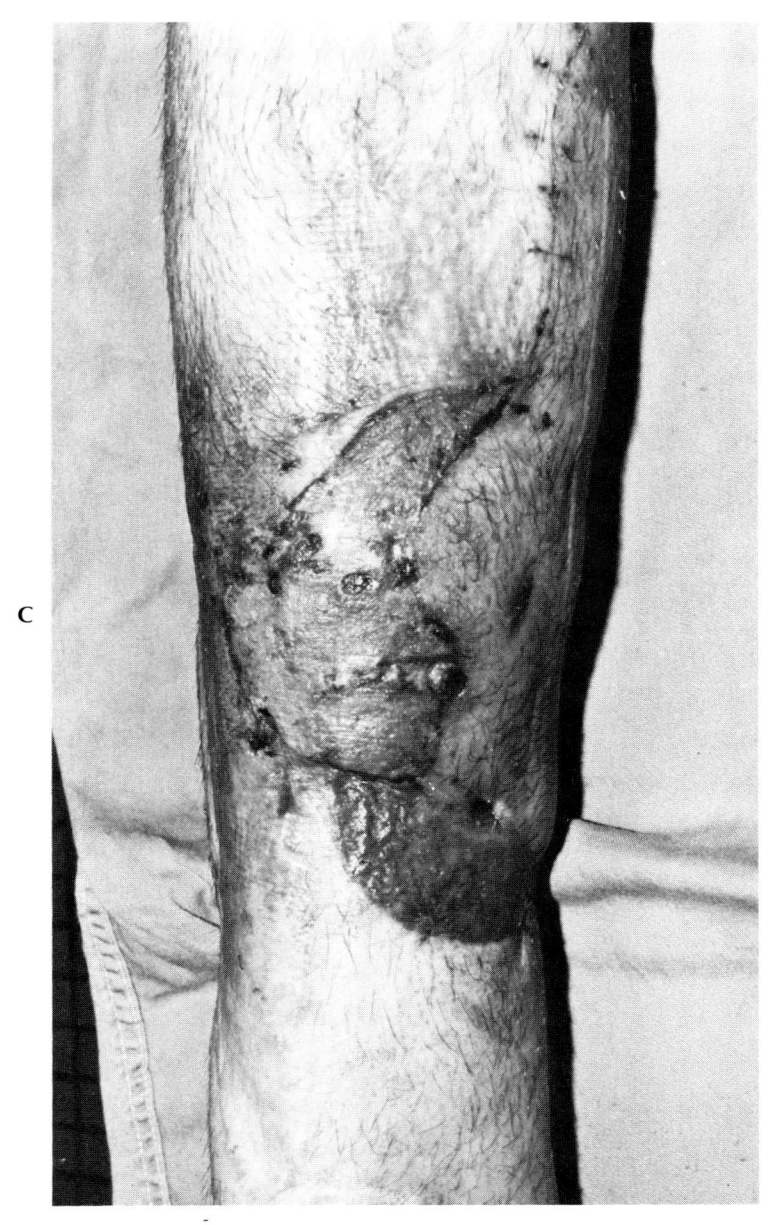

C Final result. Muscle transposition and split-skin coverage of tibia in a single stage.

MEDIAL LEG

Soleus

Coverage of exposed vascular graft

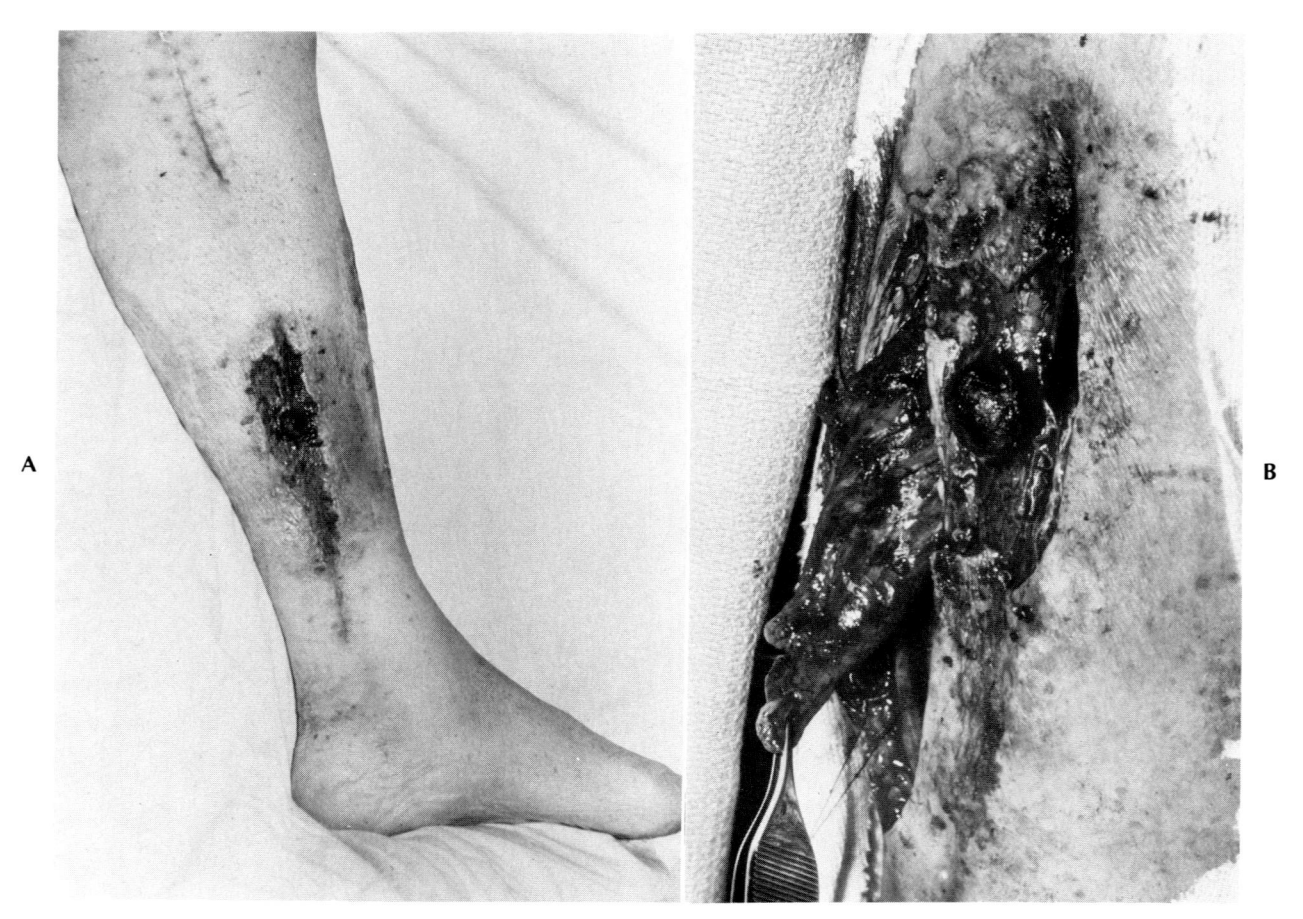

A Exposed vein graft in left leg following vascular bypass surgery.

B Close-up of exposed vein-graft aneurysm. Note elevated soleus muscle.

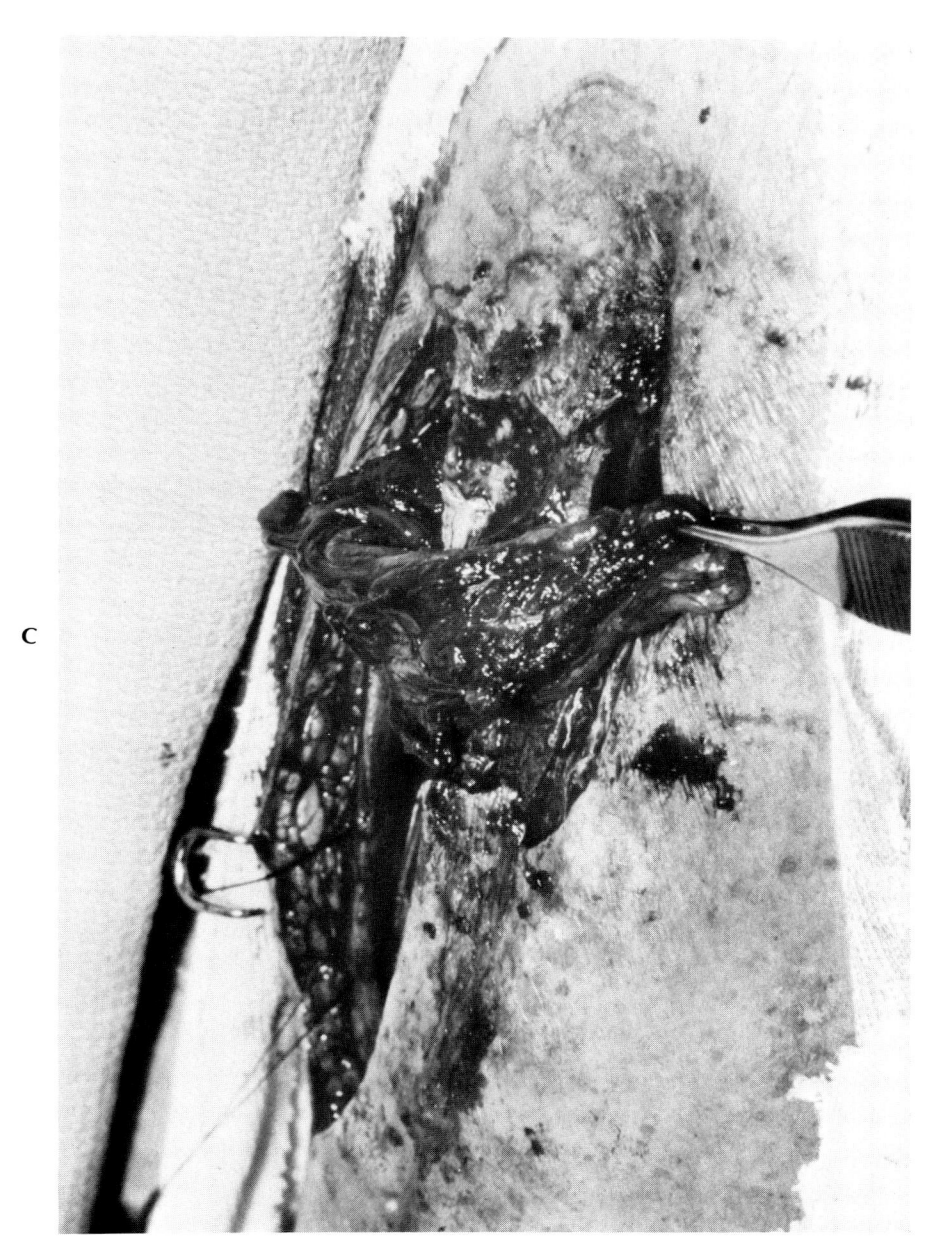

C Vein-graft aneurysm was resected and soleus muscle transposed to cover exposed vascular suture line.

MEDIAL LEG

Soleus

Coverage of exposed Achilles tendon

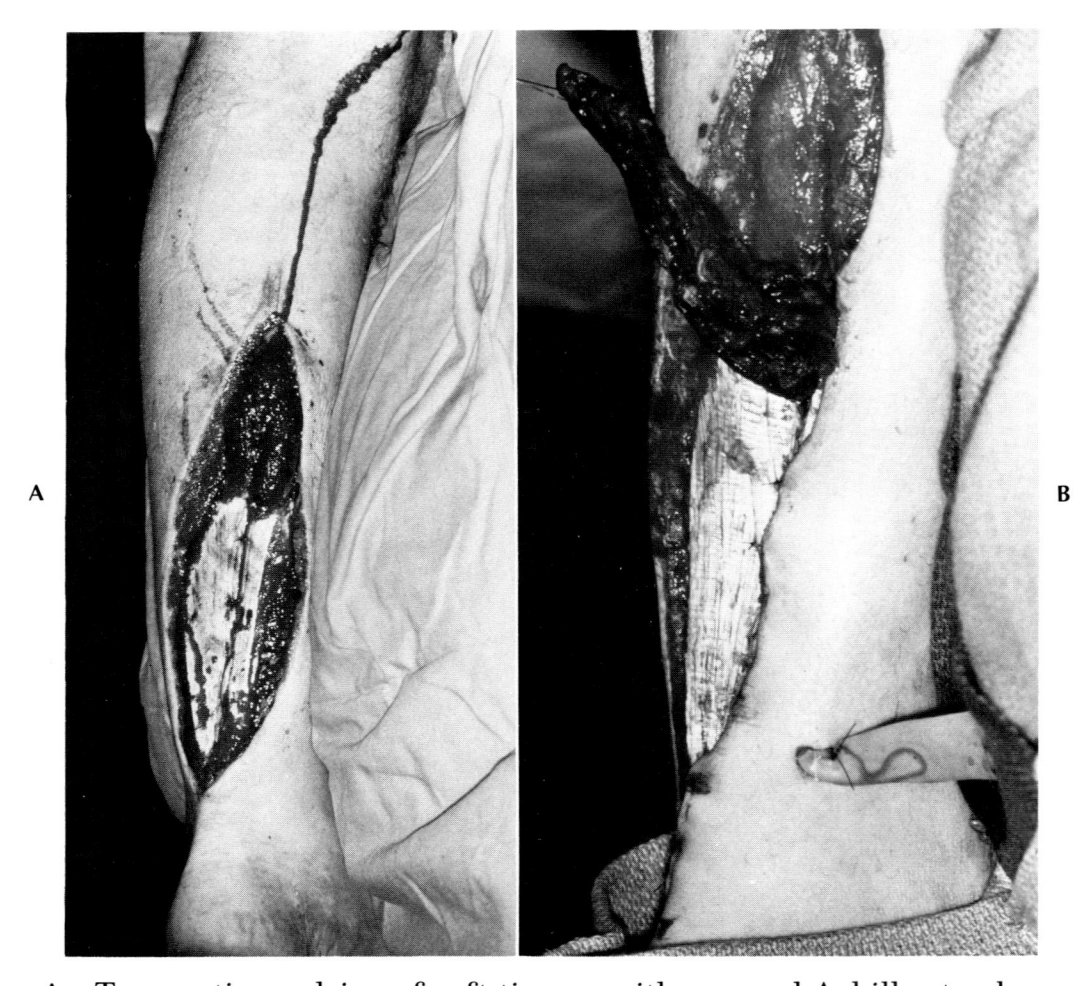

A   Traumatic avulsion of soft tissues with exposed Achilles tendon.
B   Soleus muscle flap elevated.

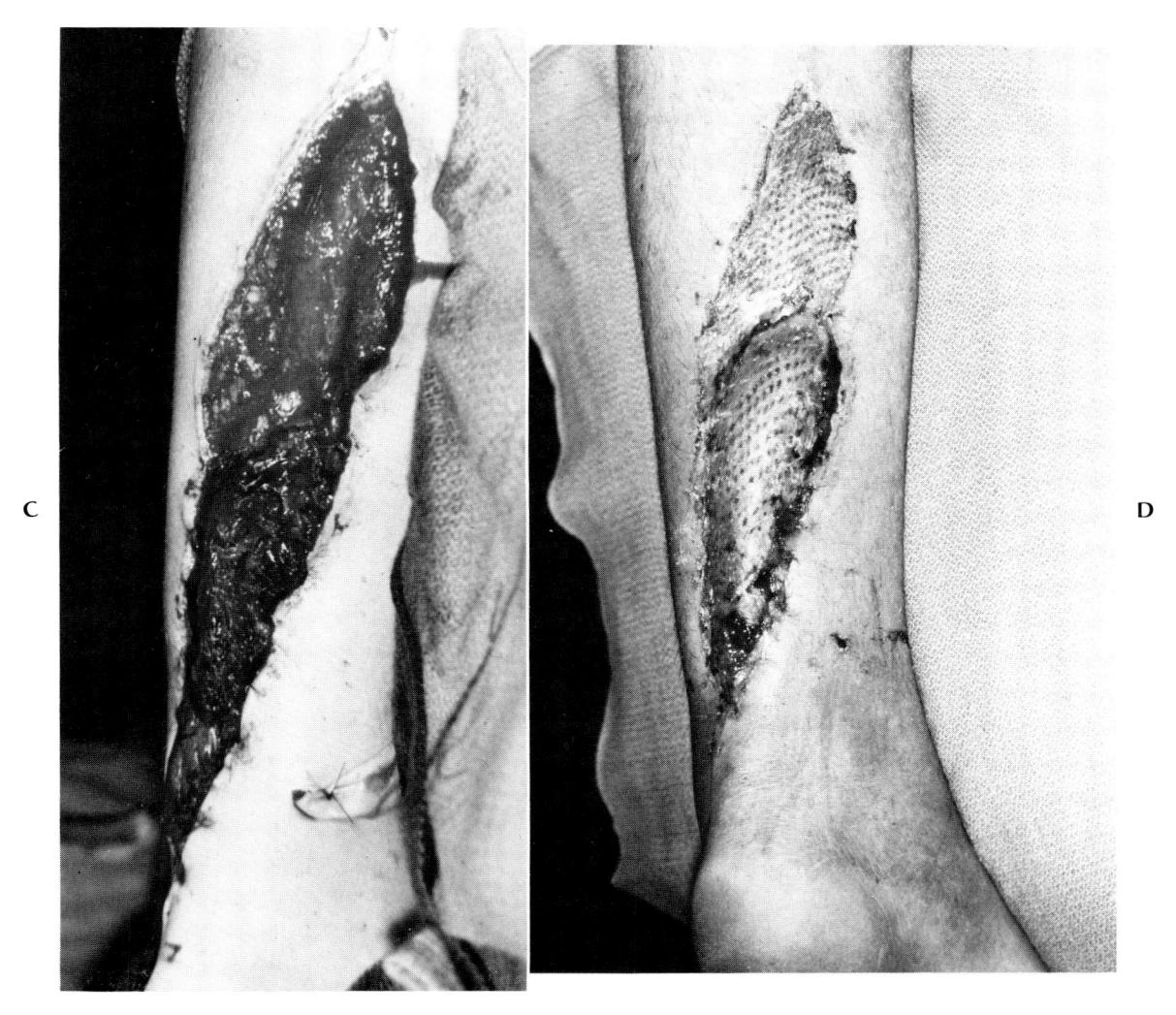

C Muscle transposed posteriorly to cover exposed tendon.

D Skin graft over muscle. Single-stage coverage of exposed Achilles tendon. (B and D reproduced with permission from Vasconez, L. O., Schneider, W. J., and Jurkiewicz, M. J.: Pressure sores. In Ravitch, M. M., et al., editors: Current problems in surgery, April 1977. Copyright © 1977 by Year Book Medical Publishers, Inc., Chicago.)

MEDIAL LEG

Soleus

Coverage of exposed tibia (lateral exposure and medial transposition)

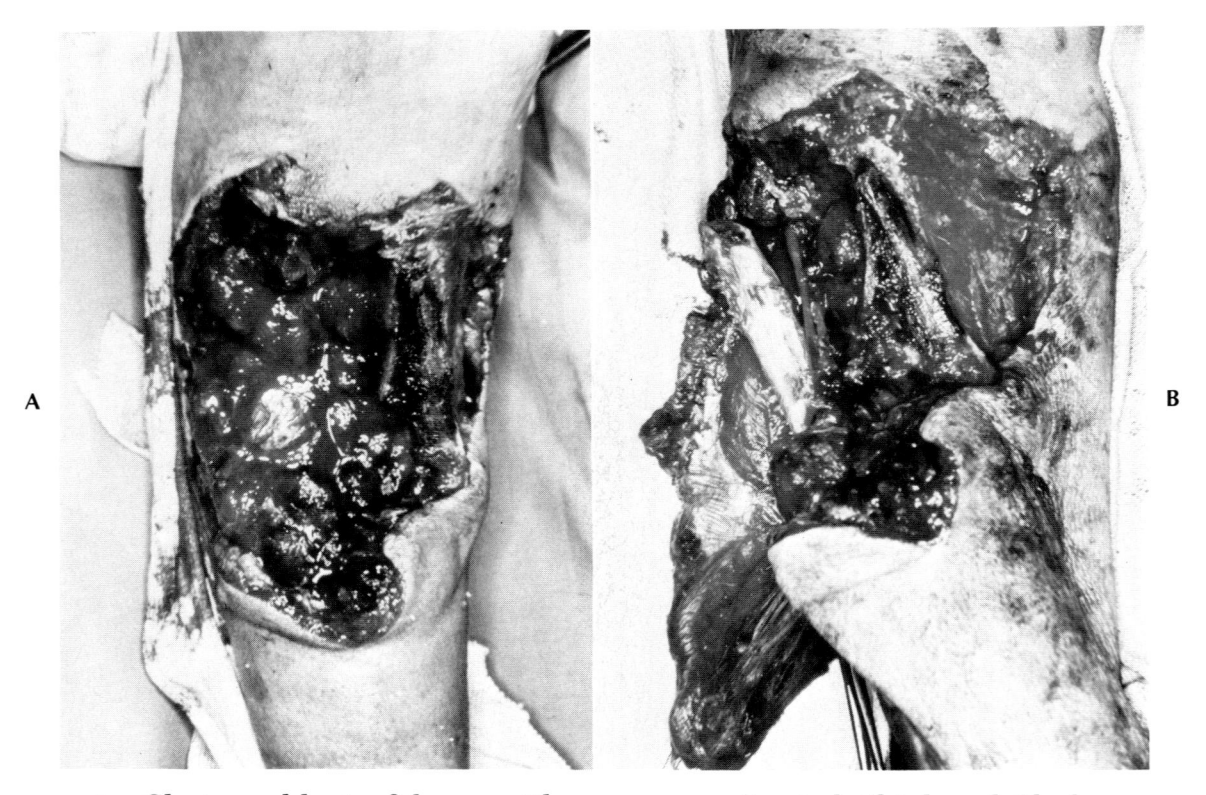

A Shotgun blast of knee with open comminuted tibial and fibular fractures.

B Medial gastrocnemius and medial portion of soleus were injured; therefore a lateral-to-medial soleus flap was selected.

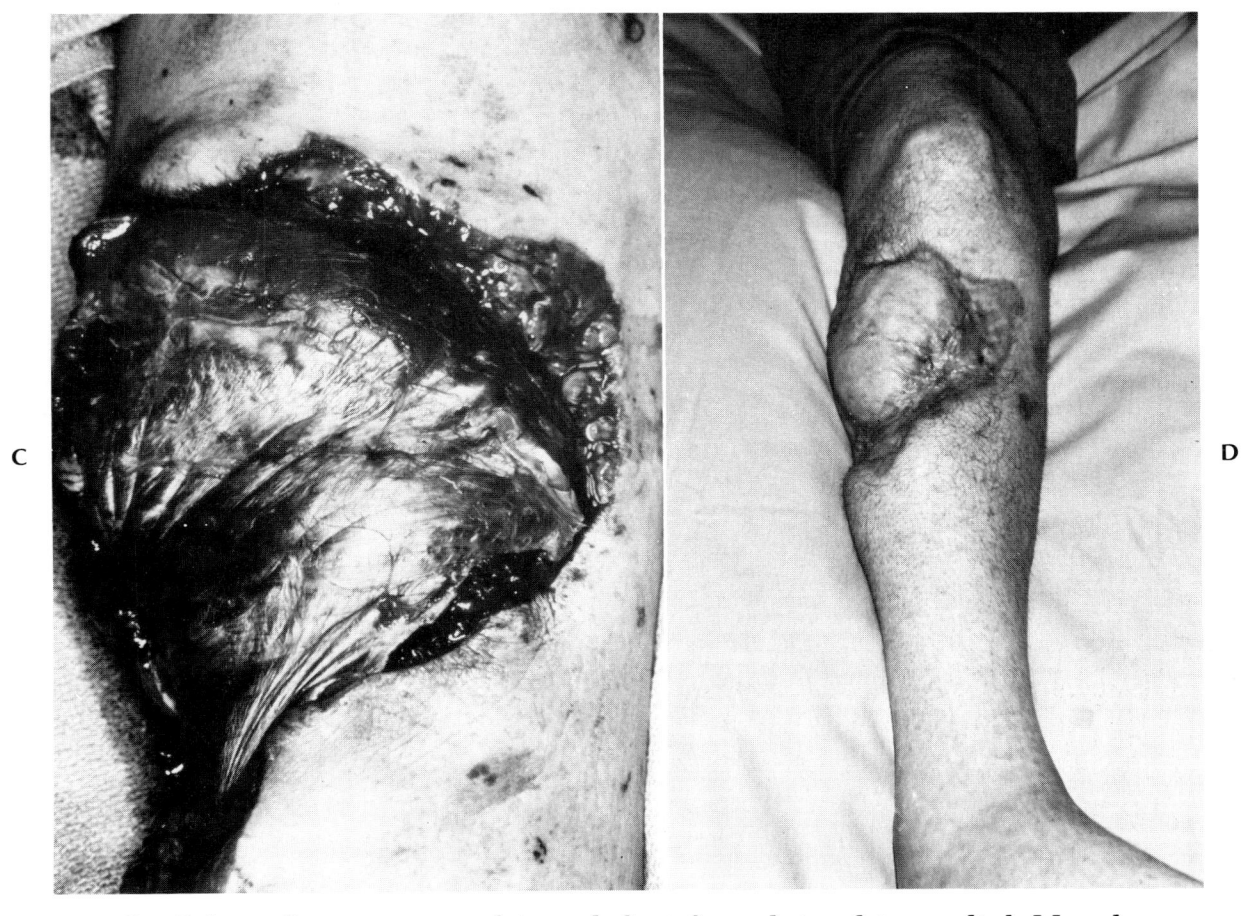

C Soleus flap transposed into defect from lateral to medial. Muscle was grafted.

D Final result. Single-stage closure of wound with healed fracture.

Flexor digitorum longus

MUSCLE FLAP

Application

   Coverage of:

   LOWER THIRD OF LEG

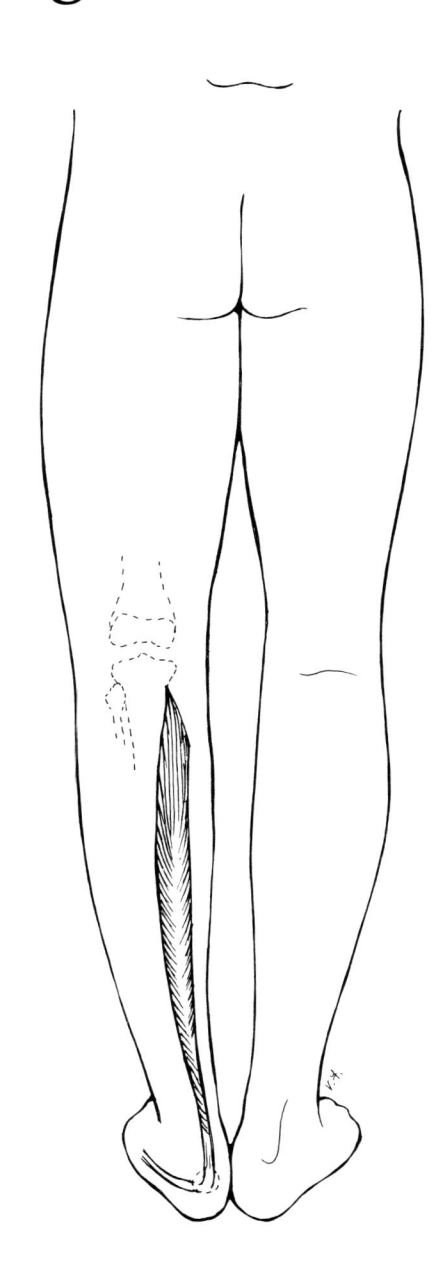

MEDIAL LEG

Flexor digitorum longus

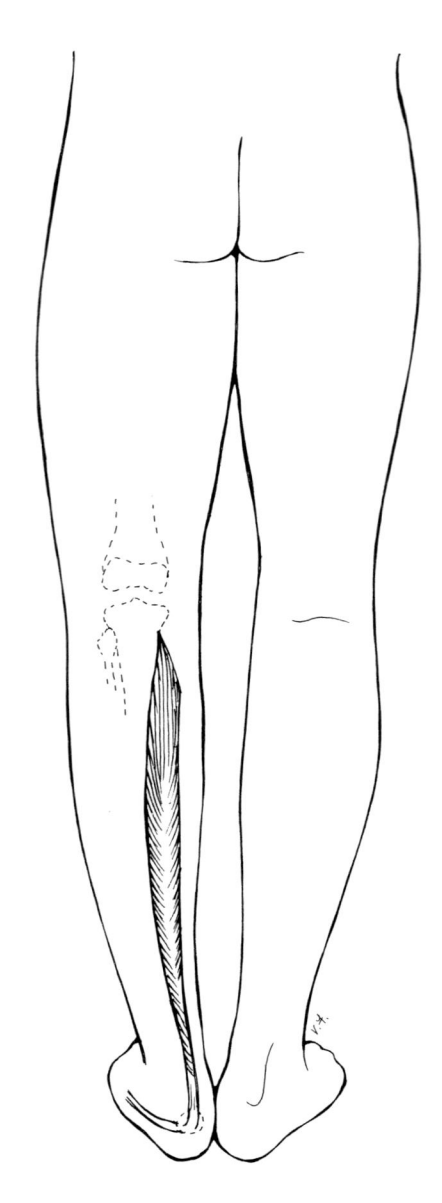

The flexor digitorum longus is located deep to the soleus on the tibial side of the leg. It is a thin, flat muscle that becomes tendinous at the level of the ankle.

> **Origin:** Posterior surface of body of tibia.
> **Insertion:** Base of distal phalanges of the second, third, fourth, and fifth toes.
> **Nerve supply:**
> > **Motor** — Branch of the tibial nerve. The motor branches enter the muscle proximally on its deep surface.

Blood supply

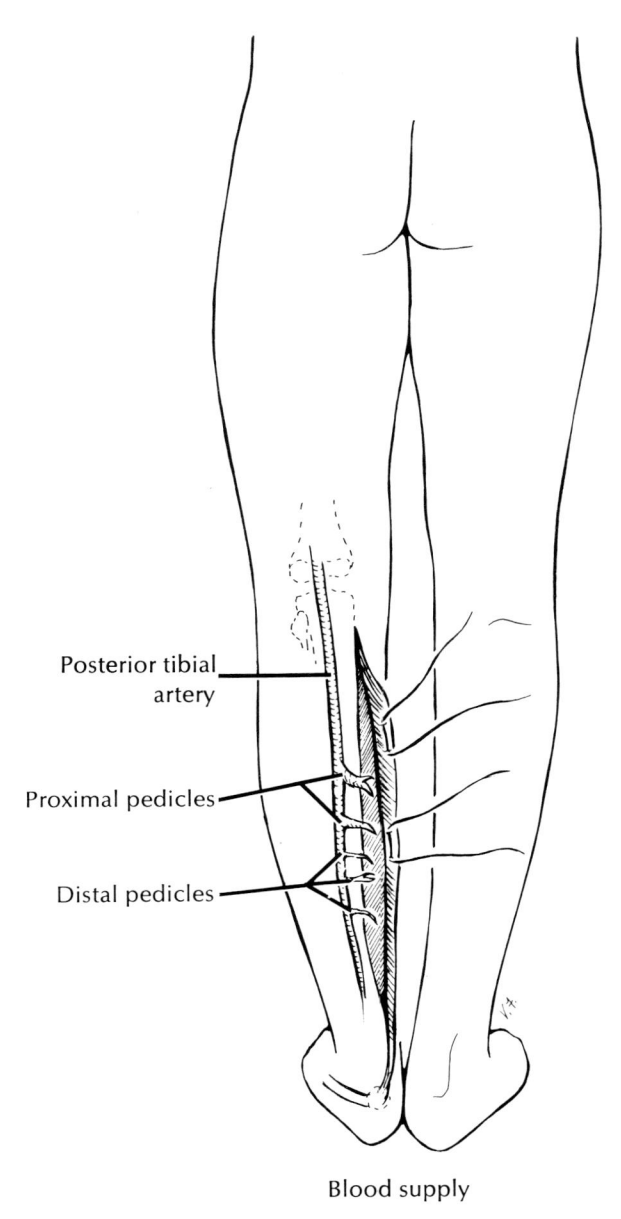

Function: Flexion of the terminal phalanx of the second, third, fourth, and fifth toes. This muscle is expendable provided the flexor digitorum brevis muscle is intact, but flexion of the terminal phalanges of the second, third, fourth, and fifth toes is lost.

Our anatomic studies have shown that the tendon will remain vascularized if the muscular fibers of the muscle are separated from the distal tendon and transposed, thus sparing the tendon. From our clinical experience it appears that muscle and tendon function can be preserved by merely separating 10 to 12 cm of muscle from the tendon for transposition.

MEDIAL LEG

Flexor digitorum longus

Blood supply

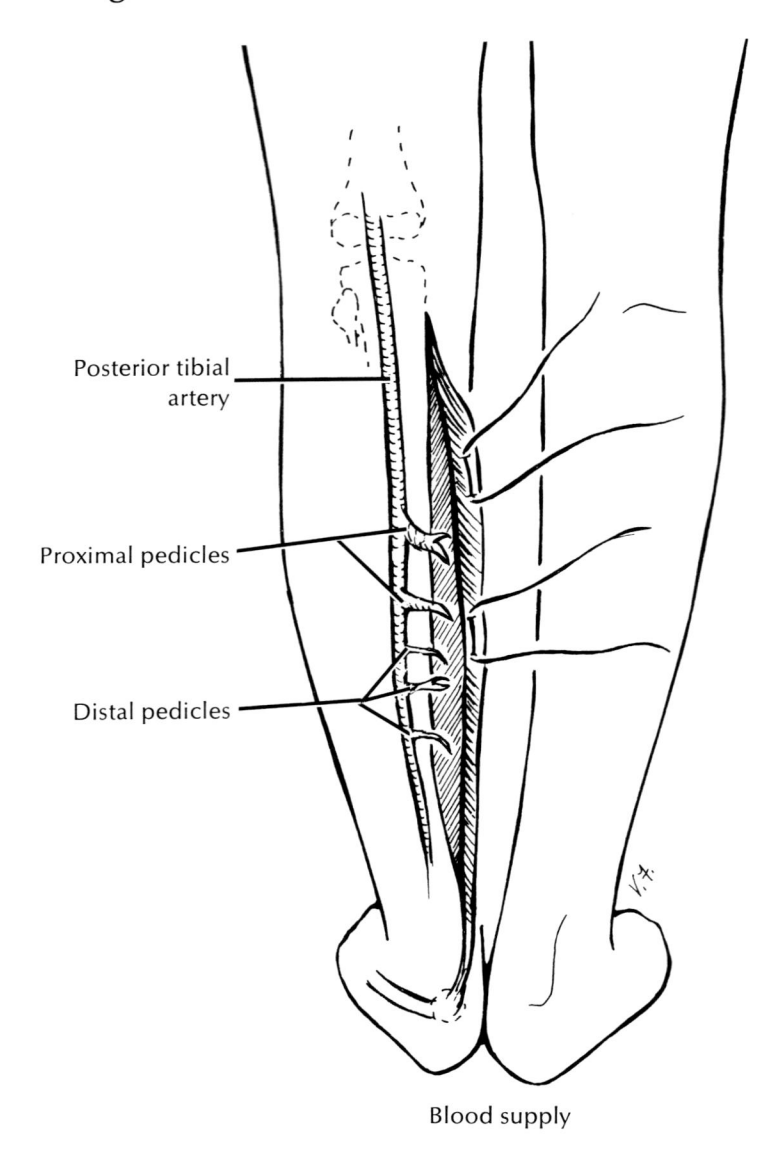

Blood supply

The blood supply is segmental through the posterior tibial artery. The artery runs along the lateral border of the muscle and sends three or more direct branches into the muscle on its deep lateral surface. Two or three proximal pedicles must be maintained in order to allow safe transposition of this muscle.

MEDIAL LEG

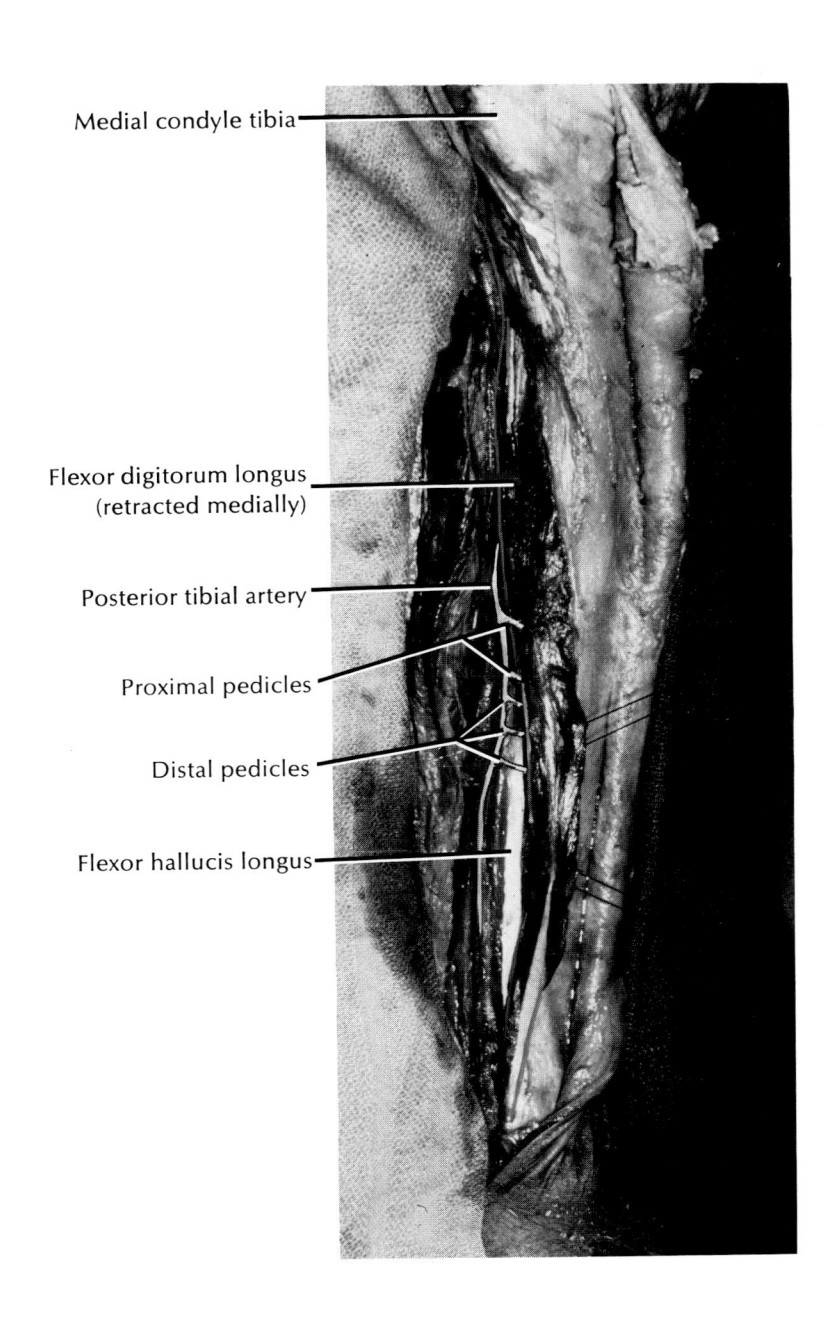

- Medial condyle tibia
- Flexor digitorum longus (retracted medially)
- Posterior tibial artery
- Proximal pedicles
- Distal pedicles
- Flexor hallucis longus

MEDIAL LEG

Flexor digitorum longus

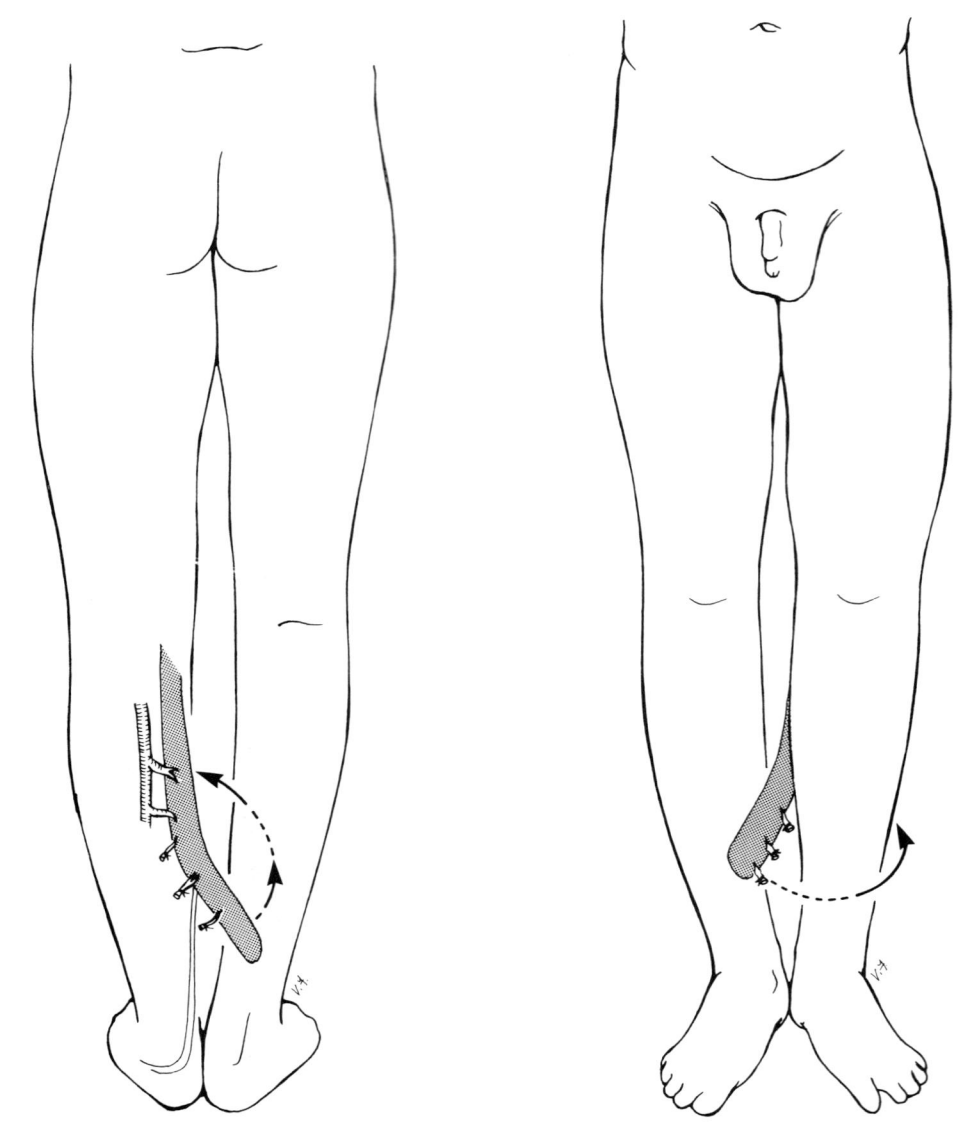

Anterior arc

Arc of rotation

The distal one or two vascular pedicles are divided, and the distal 15 to 20 cm of the muscle fibers are separated from the tendon and transposed laterally. This is a thin, flat muscle that will only cover a small defect. The area covered is below the arc of the soleus and above the arc of the flexor hallucis longus in the proximal distal third of the tibia. This small flap has limited usefulness alone, but is best used with the soleus to extend the lower reach of the soleus in muscle transposition.

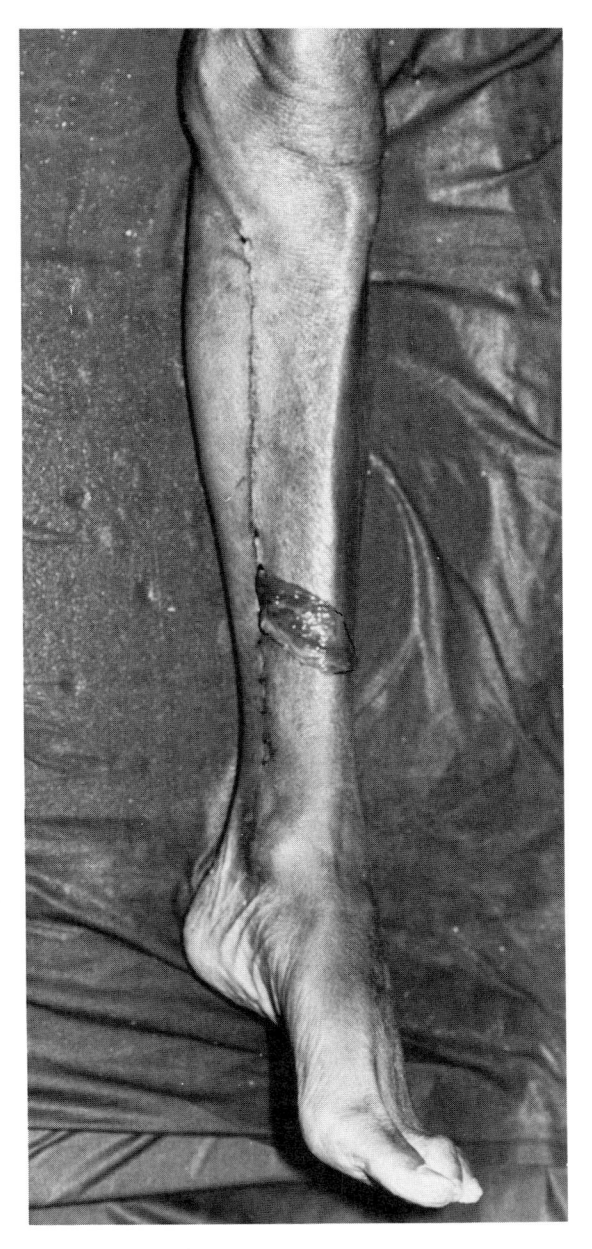

Anterior transposition (tendon intact)

It is possible that the flexor digitorum longus, based on the posterior tibial artery, could be transferred with its motor nerve as a functional free muscle flap. In order to accomplish this transfer a difficult dissection would have to be undertaken and the posterior tibial artery sacrificed. When there are suitable alternatives available, this type of free transfer may not be justified.

MEDIAL LEG

Flexor digitorum longus

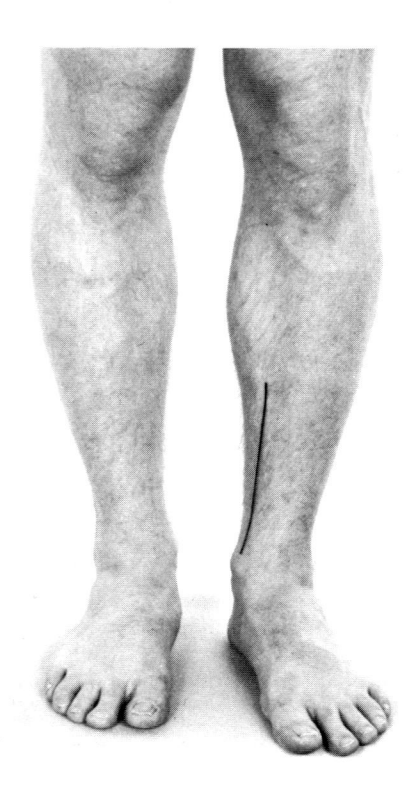

Skin incision

Elevation of flap

A medial incision similar to that for the soleus is made. As this muscle is usually used in conjunction with the soleus, the soleus is first identified, elevated, and transposed. If coverage is still needed below the transposition of the soleus, then this flap is elevated. The flexor digitorum longus muscle lies deep to the soleus on the tibial side and medial to the posterior tibial artery and tibial nerve. The muscle fibers are separated from the tendon for a distance of 15 to 20 cm, the distal one or two vascular pedicles are ligated, and the tendon is preserved. The muscle is then transposed from medial to lateral over the exposed tibia.

Precautions

- A small, thin muscle with limited applications alone, but useful to augment the lower limit of the arc of the soleus.
- The posterior tibial artery and tibial nerve are located immediately lateral to this muscle.

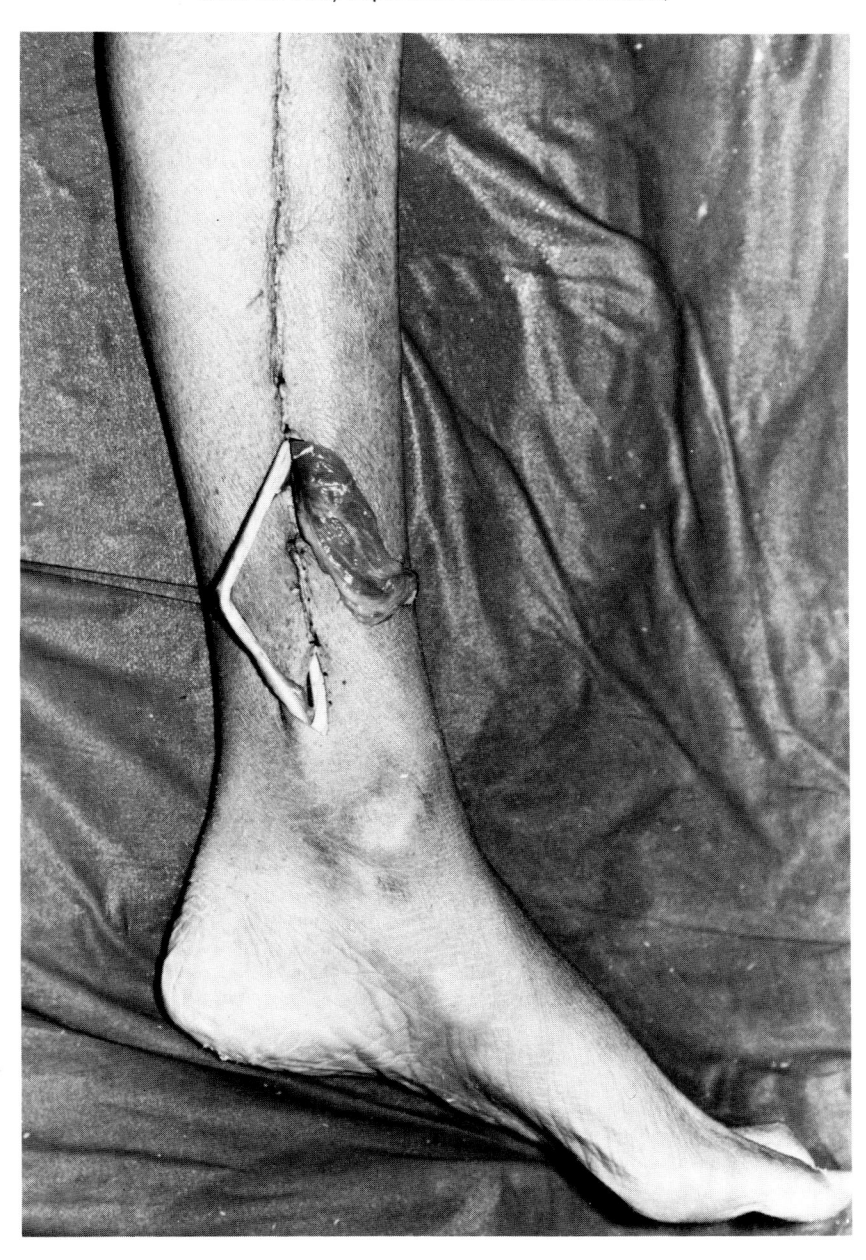

Function preserving elevation of muscle
(muscle belly separated from intact tendon)

Flexor hallucis longus

MUSCLE FLAP

Application

Coverage of:

LOWER THIRD OF LEG

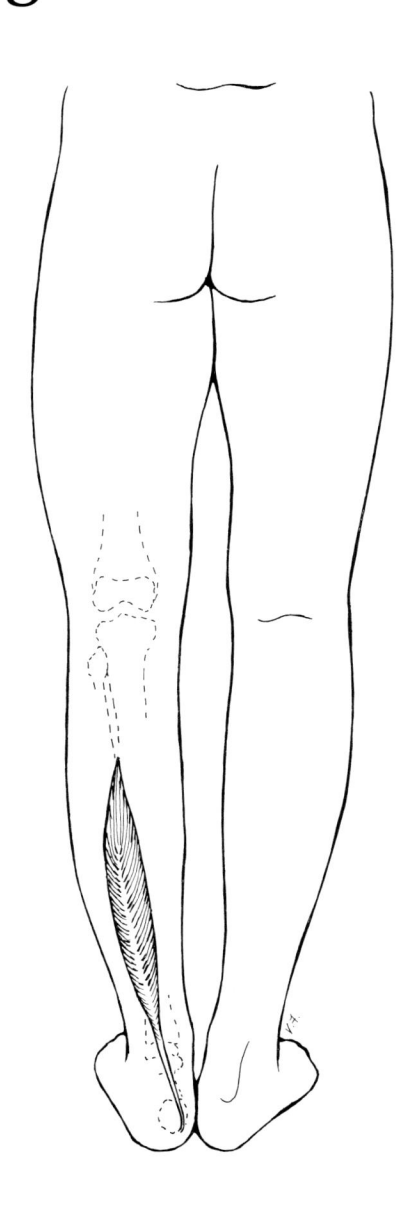

MEDIAL LEG

Flexor hallucis longus

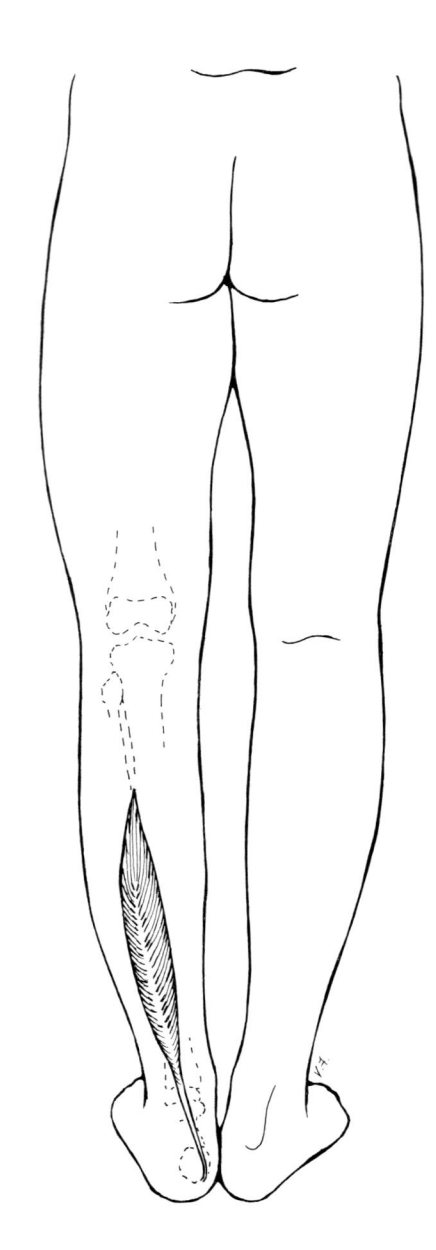

The flexor hallucis longus is deep to the soleus on the fibular side of the leg. It is shorter but thicker than the flexor digitorum longus.

Origin: Lower two-thirds of posterior fibula.
Insertion: Base of terminal phalanx of great toe.

MEDIAL LEG

Peroneal artery

Proximal pedicles

Distal pedicles

Blood supply

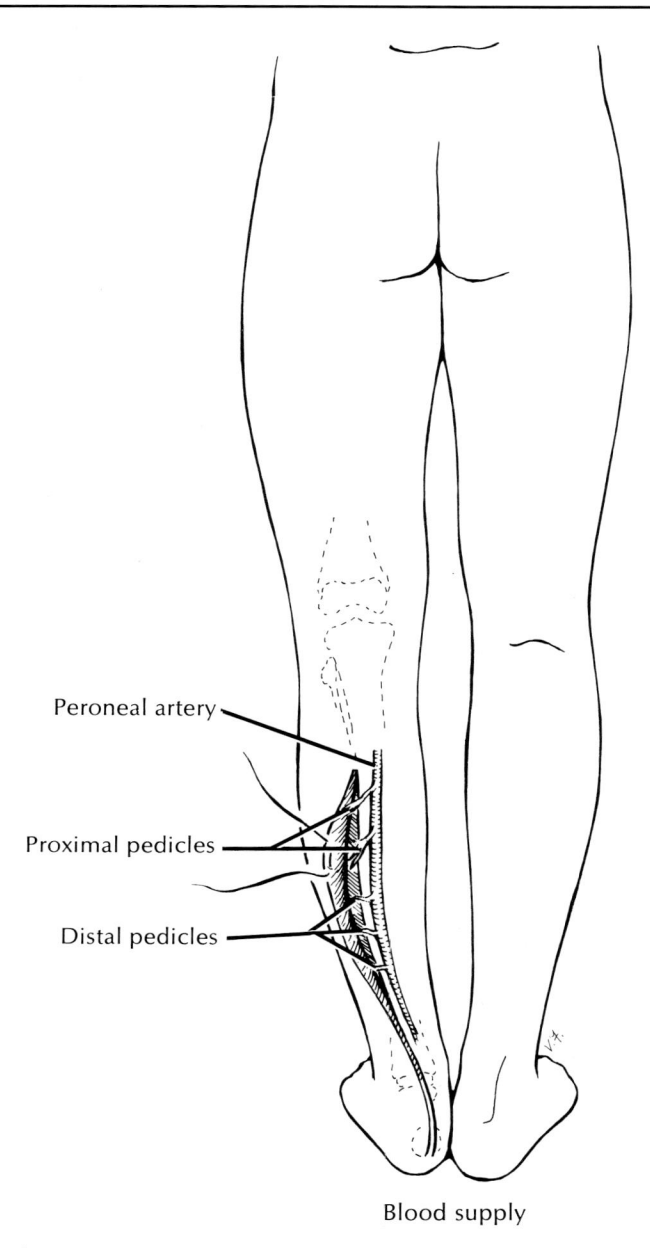

Nerve supply:

Motor—Branch of the posterior tibial nerve. The nerve supply enters deep on the medial surface and is a branch of the posterior tibial nerve. This nerve runs along the medial border of this muscle and enters the proximal third of the muscle.

Function: Flexion of the interphalangeal joint of the great toe. It is an expendable muscle, but the tendon may be spared by separating the muscle fibers from the tendon. (See flexor digitorum longus.) The distal muscle is then transposed, leaving the proximal muscle and tendon intact.

MEDIAL LEG

Flexor hallucis longus

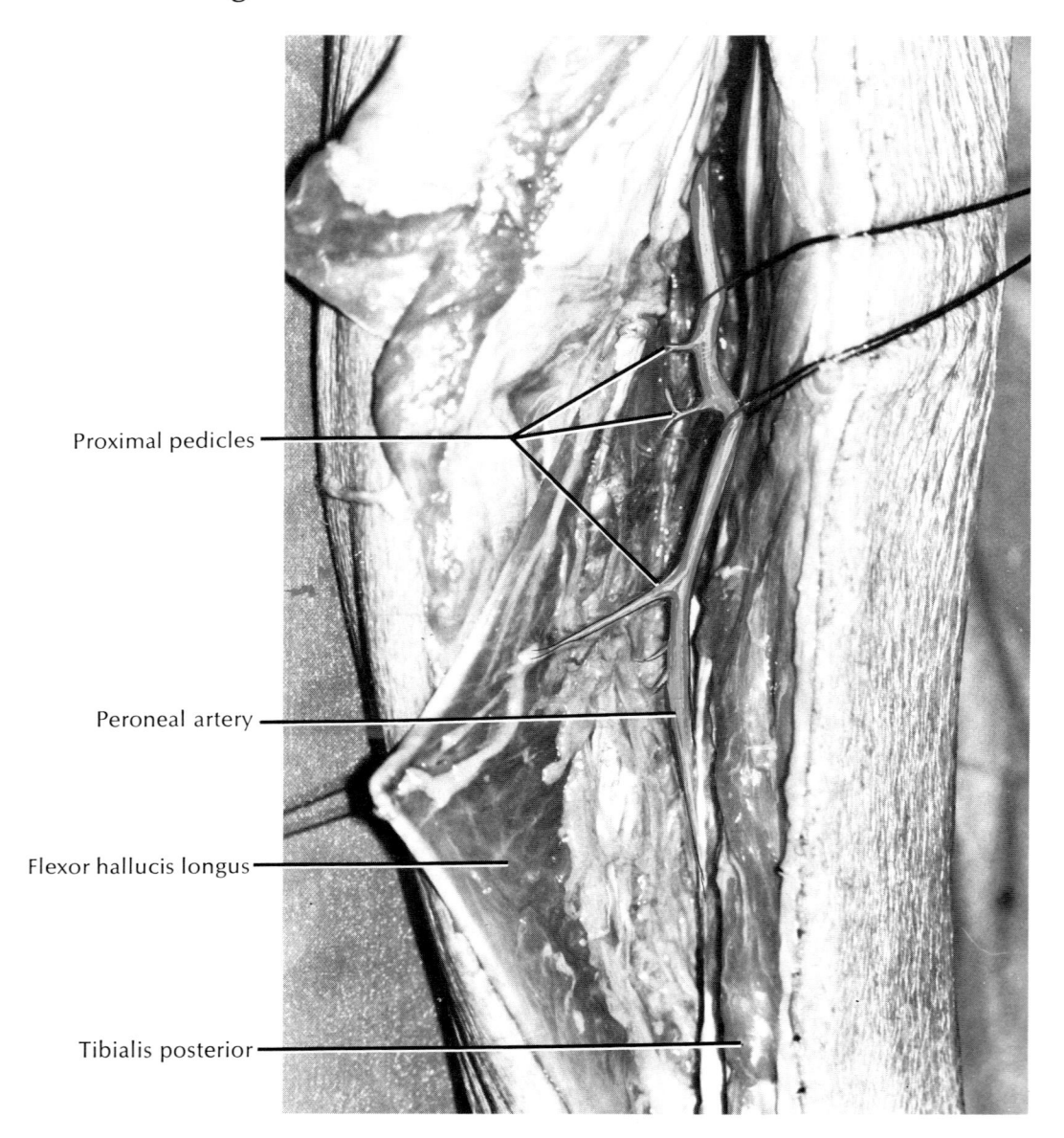

- Proximal pedicles
- Peroneal artery
- Flexor hallucis longus
- Tibialis posterior

Blood supply

This muscle has three or more segmental vascular pedicles that are based on the peroneal artery. The peroneal artery is a terminal branch of the popliteal artery running deep to the fibular head of the soleus, which it supplies through a direct branch. It then courses deep to or through the flexor hallucis longus toward the calcaneus.

MEDIAL LEG

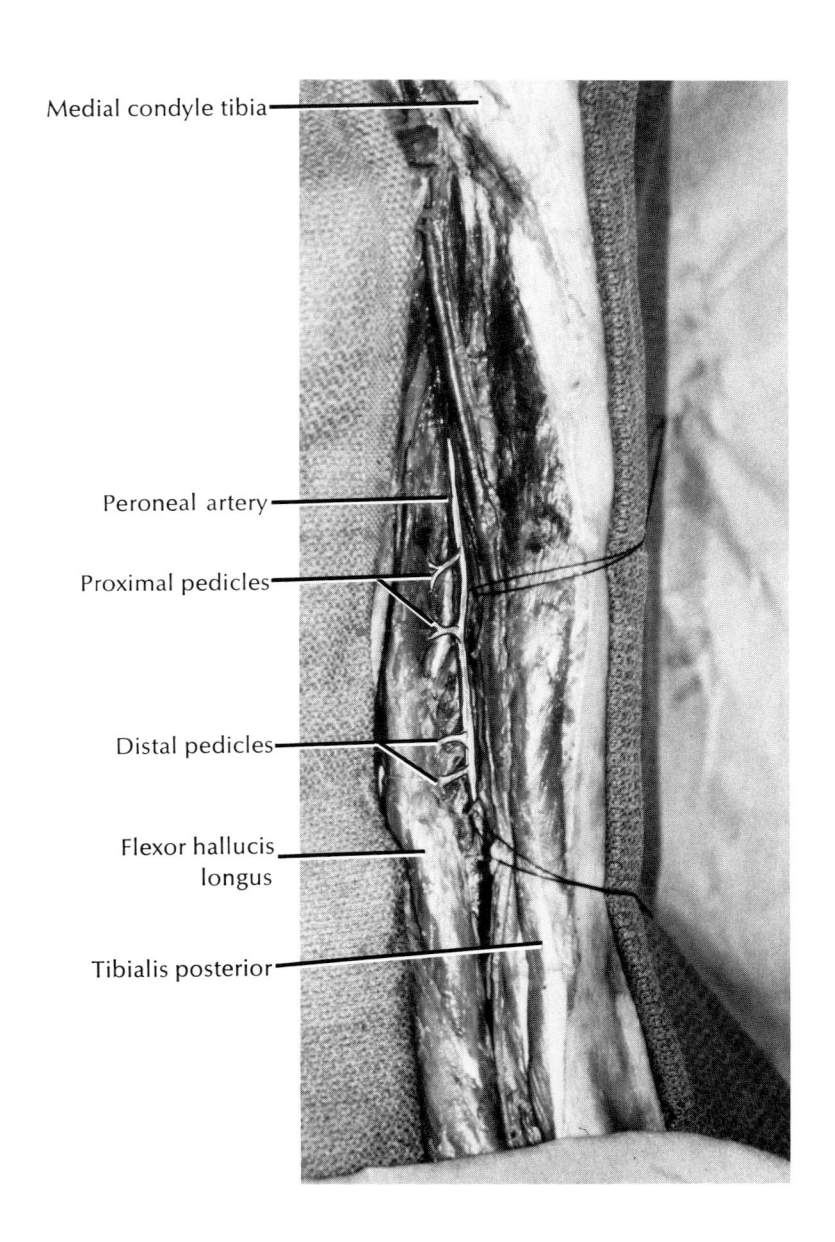

Medial condyle tibia

Peroneal artery

Proximal pedicles

Distal pedicles

Flexor hallucis longus

Tibialis posterior

MEDIAL LEG

Flexor hallucis longus

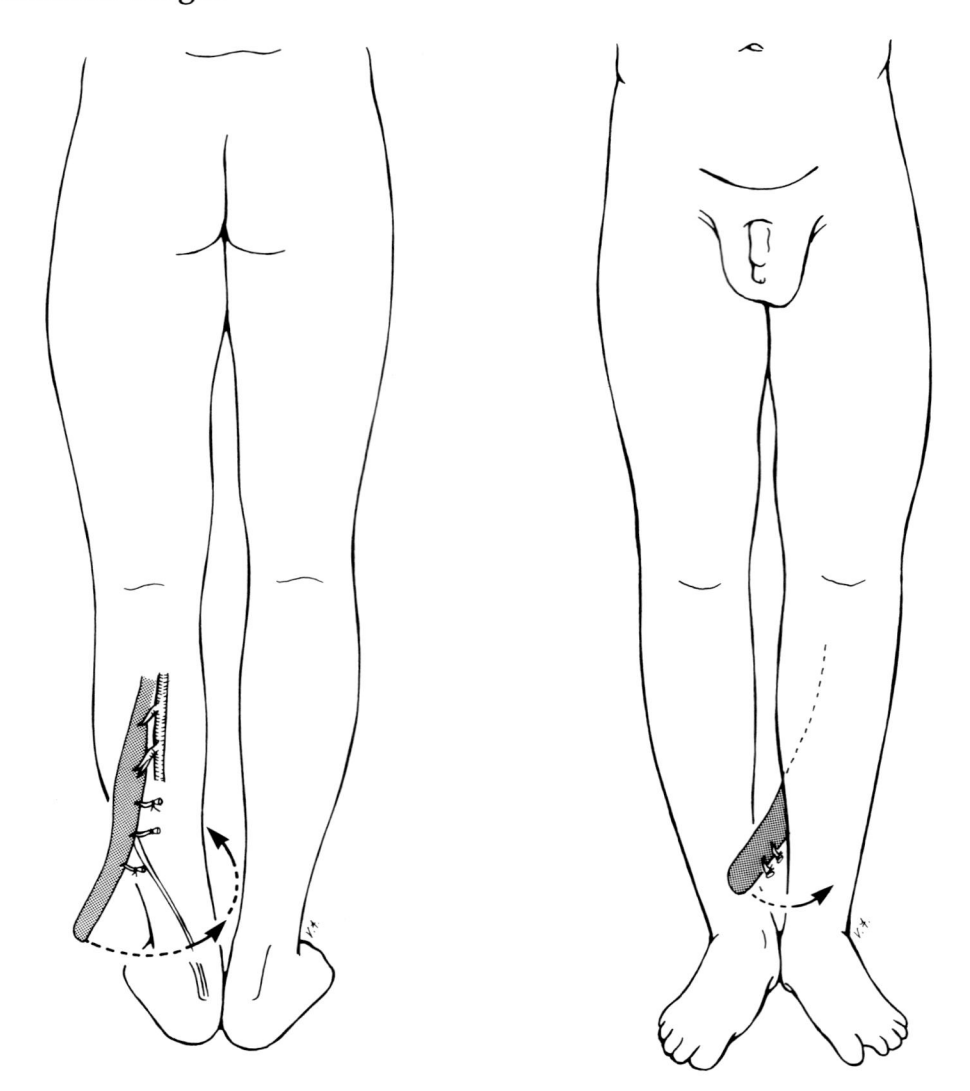

Arc of rotation

With the distal one or two pedicles divided, the distal portion of the muscle is separated from the tendon and is transposed laterally. The muscle will cover a small area below the arc of the soleus and flexor digitorum longus muscle. This muscle has limited usefulness when used alone, but is best used in combination with the other muscles in this group to cover a larger area of the distal tibia.

Based on the peroneal artery, the flexor hallucis longus has potential for free flap transfer as an innervated functional unit. The dissection would be difficult, but sacrifice of the peroneal artery would not significantly affect vascular supply to the foot.

MEDIAL LEG

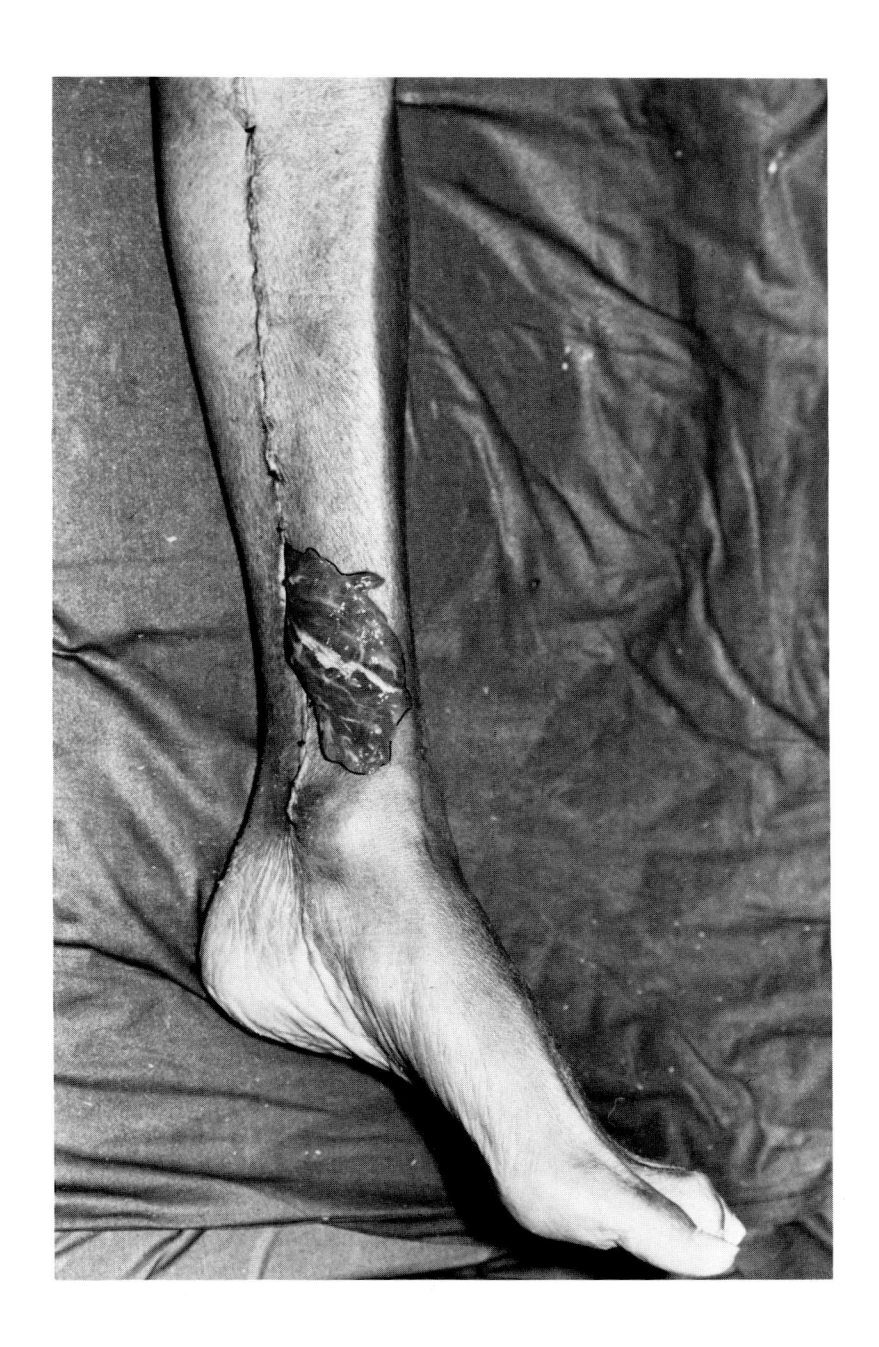

MEDIAL LEG

Flexor hallucis longus

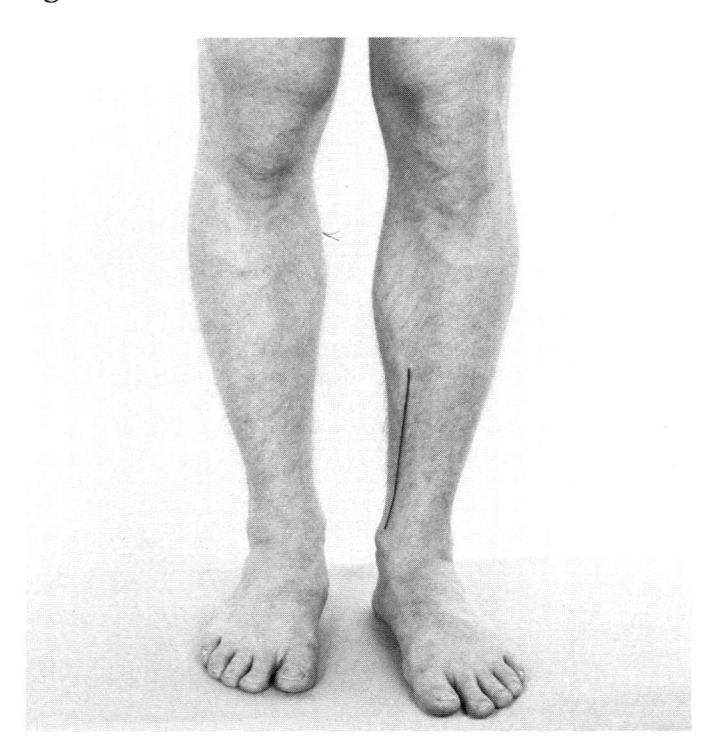

Elevation of flap

A medial incision similar to that for the soleus is made. The soleus is identified and retracted posteriorly. The flexor hallucis longus lies deep to the fibular side of the soleus and lateral to the posterior tibial artery and nerve. The peroneal artery is identified adjacent and lateral to the muscle or within the muscle. The muscle fibers are then dissected from the distal tendon for a distance of approximately 5 to 6 cm to gain enough muscle length to allow a lateral rotation to cover the distal tibia. It is necessary to leave the proximal one or two pedicles intact from the peroneal artery to ensure viability of this muscle following transposition. The donor defect may be closed primarily.

Precautions

- A small muscle, relatively difficult to dissect, with limited usefulness.
- The peroneal artery may run within this muscle and thus prevent its application as a muscle flap.
- The posterior tibial artery and nerve are lateral to the muscle in the distal leg.

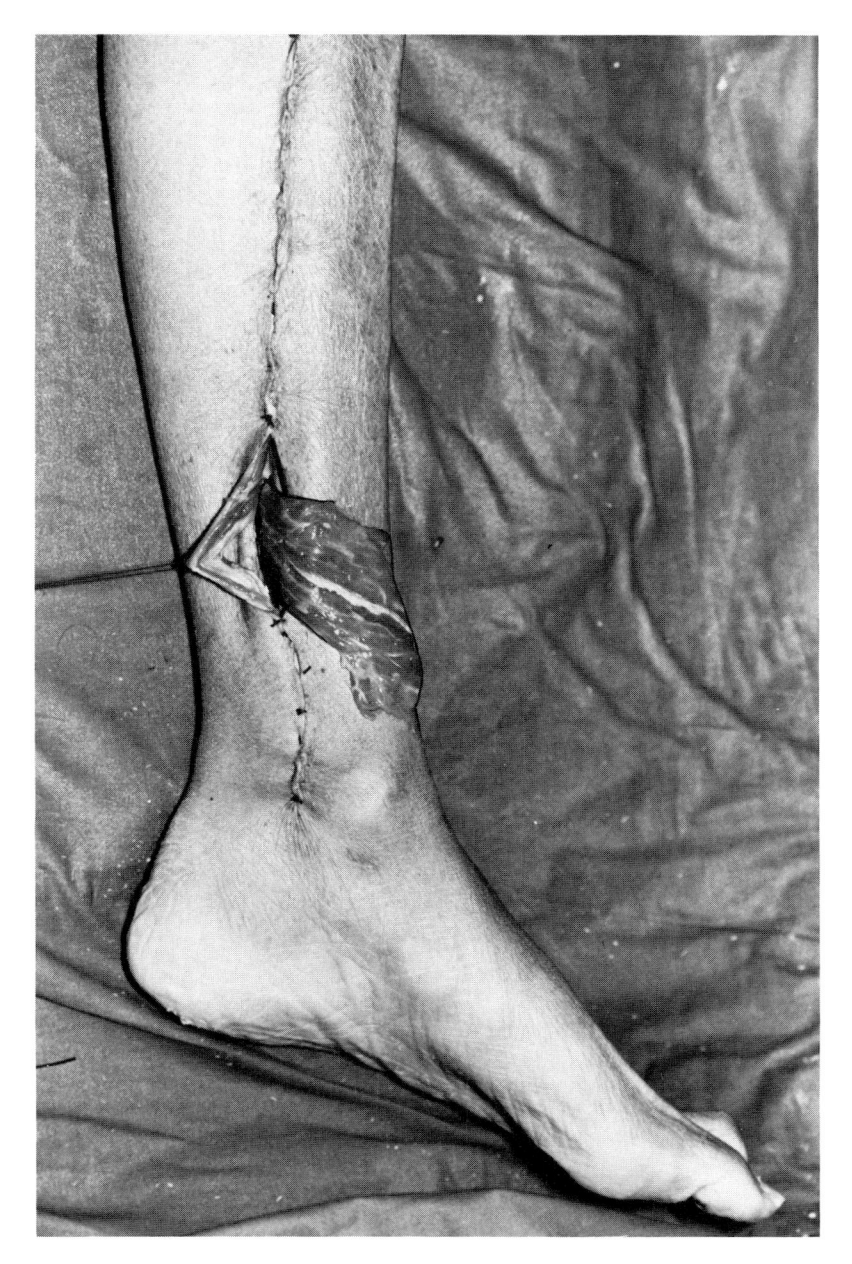

Function preserving elevation of muscle
(muscle belly separated from intact tendon)

SECTION FOUR
LATERAL LEG

Anterior group
 TIBIALIS ANTERIOR
 EXTENSOR DIGITORUM LONGUS
 EXTENSOR HALLUCIS LONGUS

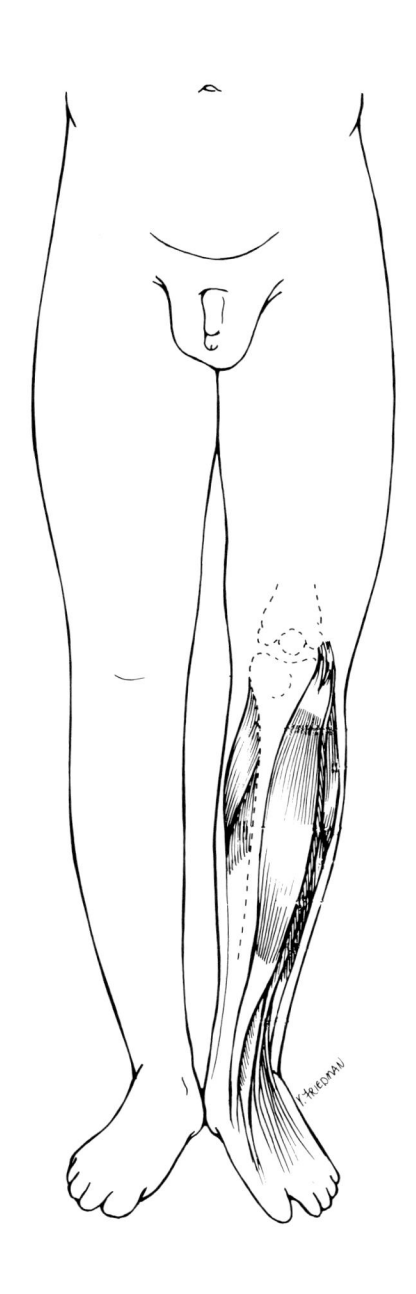

LATERAL LEG: ANTERIOR GROUP

- Gastrocnemius
- Peroneus longus
- Soleus
- Peroneus brevis
- Tibialis anterior
- Extensor digitorum longus
- Extensor hallucis longus

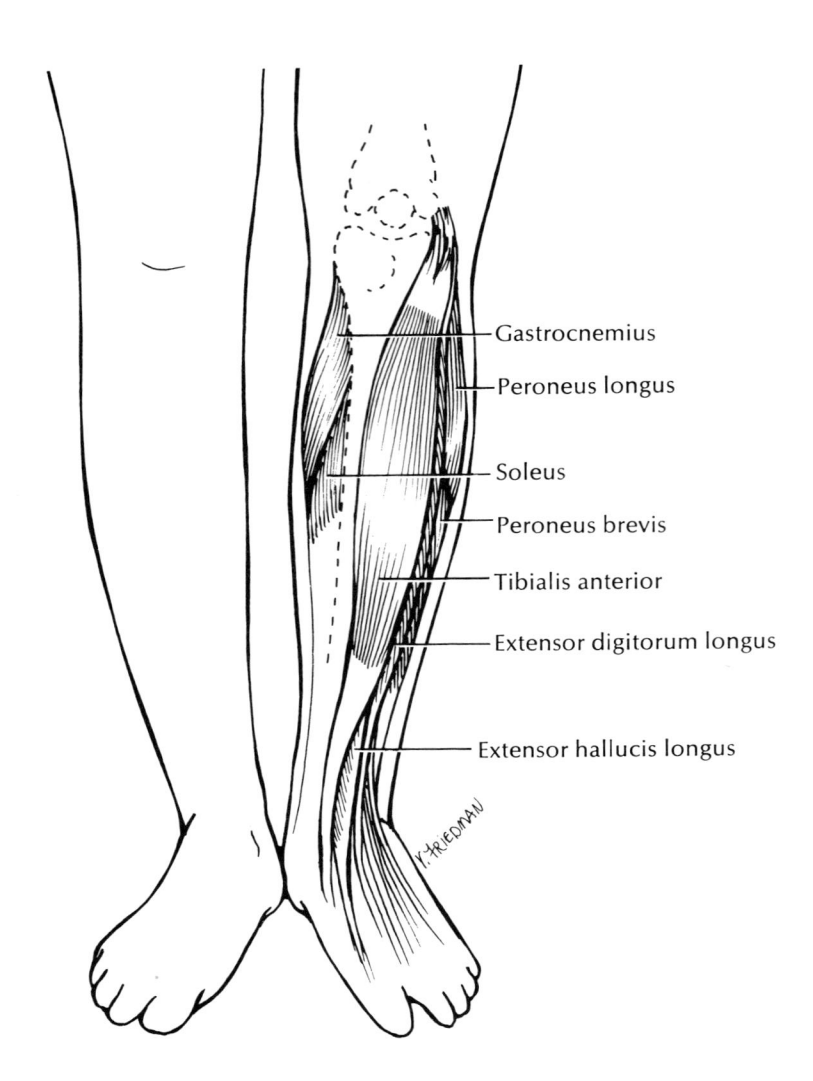

The anterior group of lateral leg muscles is approached through an anterolateral incision for use as transposition flaps. These muscles will cover the middle and lower thirds of the leg.

LATERAL LEG: ANTERIOR GROUP

- Tibialis anterior
- Peroneus longus
- Peroneus brevis
- Extensor hallucis longus
- Extensor digitorum longus

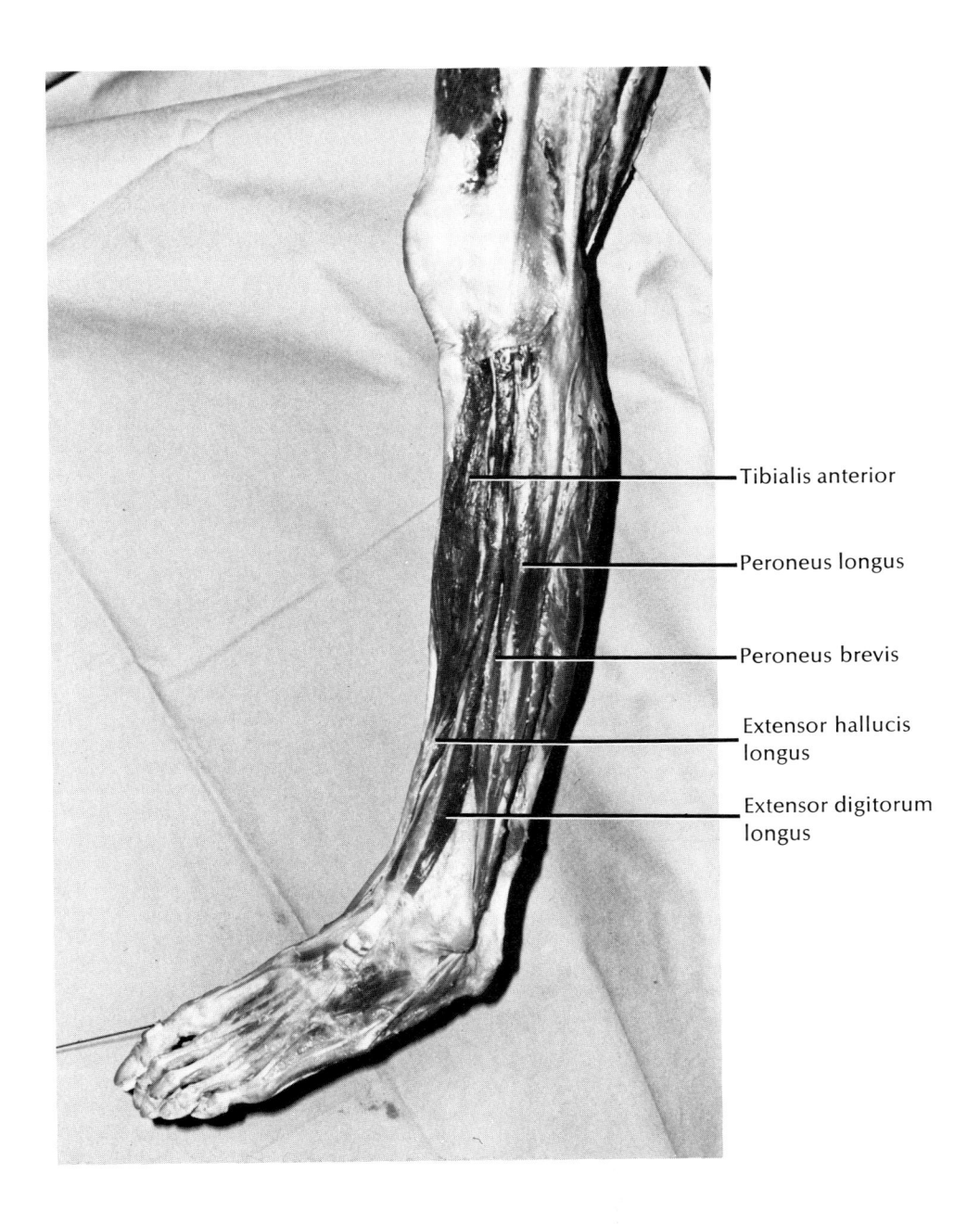

LATERAL LEG: ANTERIOR GROUP

- Tibialis anterior
- Extensor hallucis longus
- Extensor digitorum longus

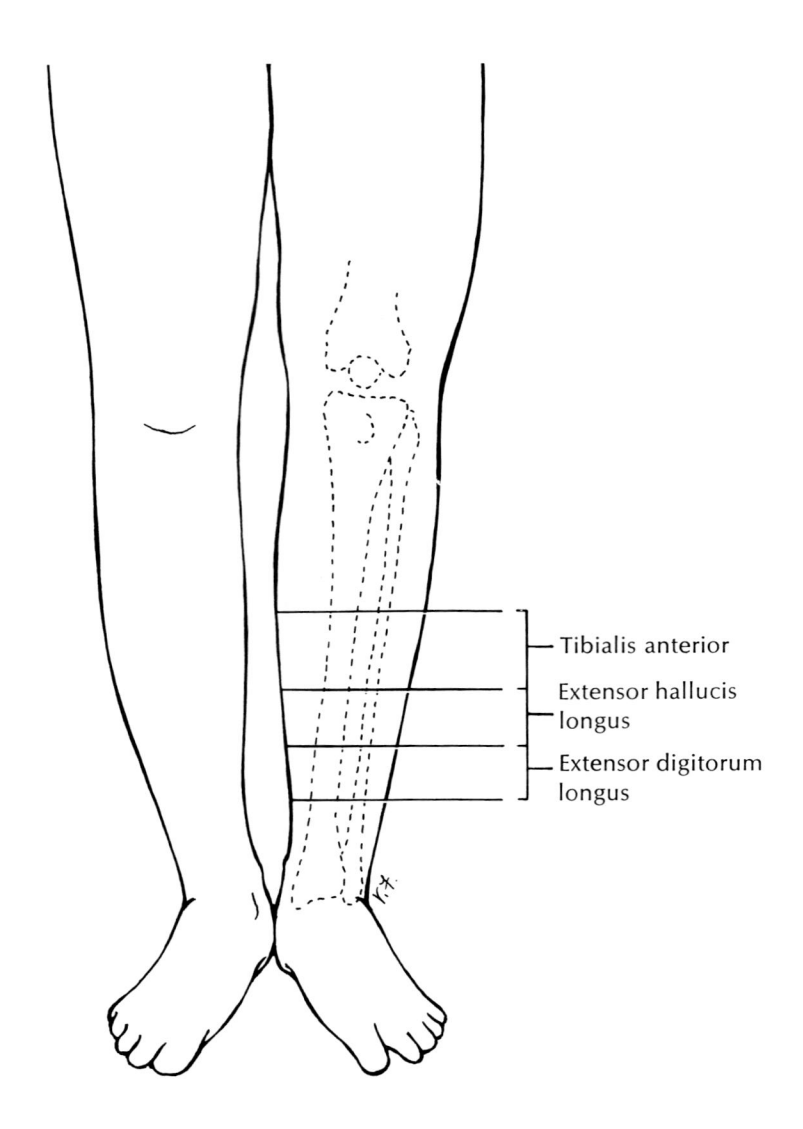

The tibialis anterior has the largest muscle belly, but has not been commonly used due to functional disability following use as a transposition muscle flap. New anatomic and clinical data confirm that the distal half of this group of muscles can be transposed with the tendon left intact with no noticeable loss of muscle function. The tibialis anterior muscle with tendon preservation is an excellent flap for tibial coverage in the lower leg. The extensor digitorum longus and extensor hallucis longus muscles are used primarily to extend inferior coverage of the transposed lateral leg muscles when the medial muscle groups are not available.

LATERAL LEG: ANTERIOR GROUP

Tibialis anterior

Extensor hallucis longus

Extensor digitorum longus

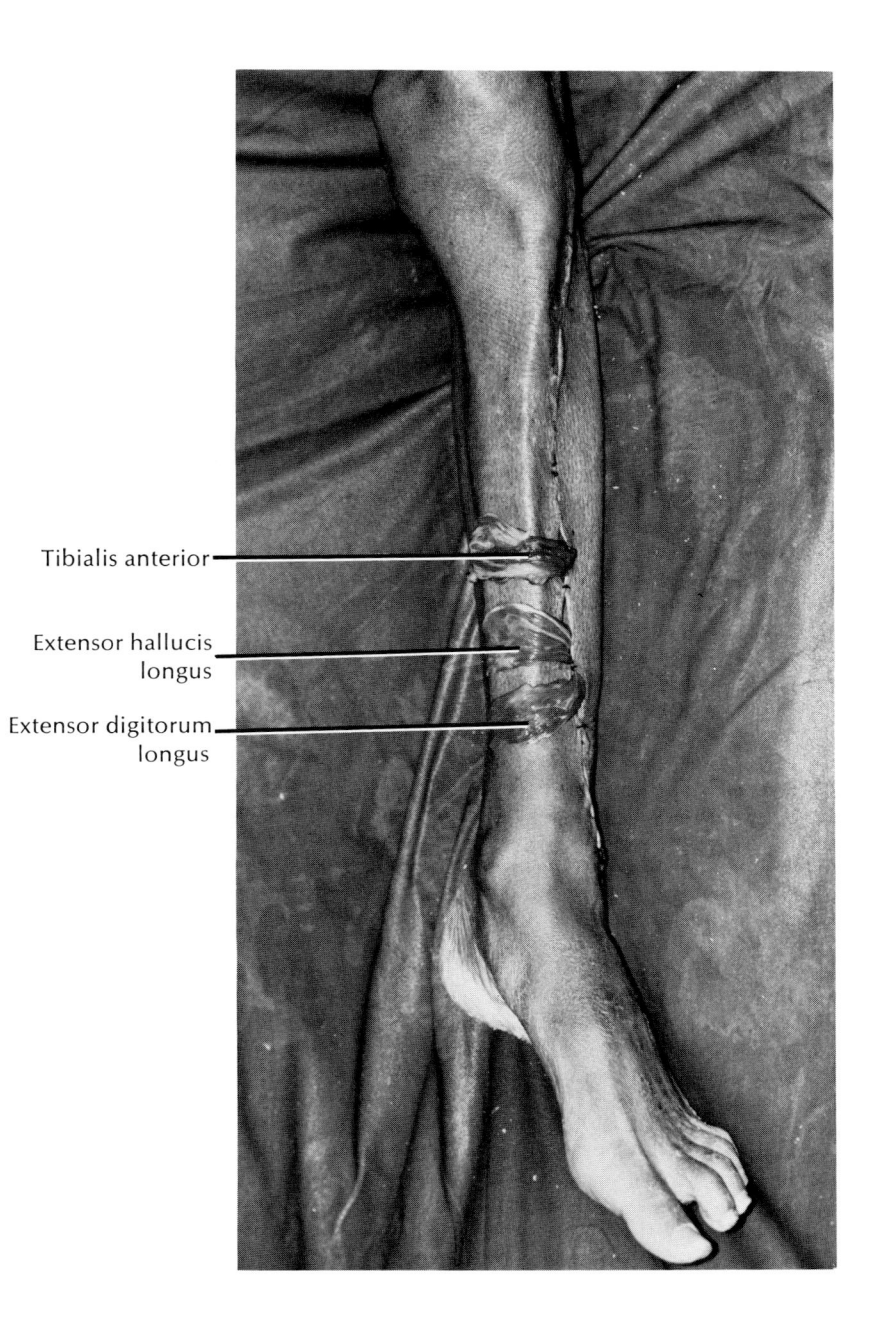

LATERAL LEG: ANTERIOR GROUP

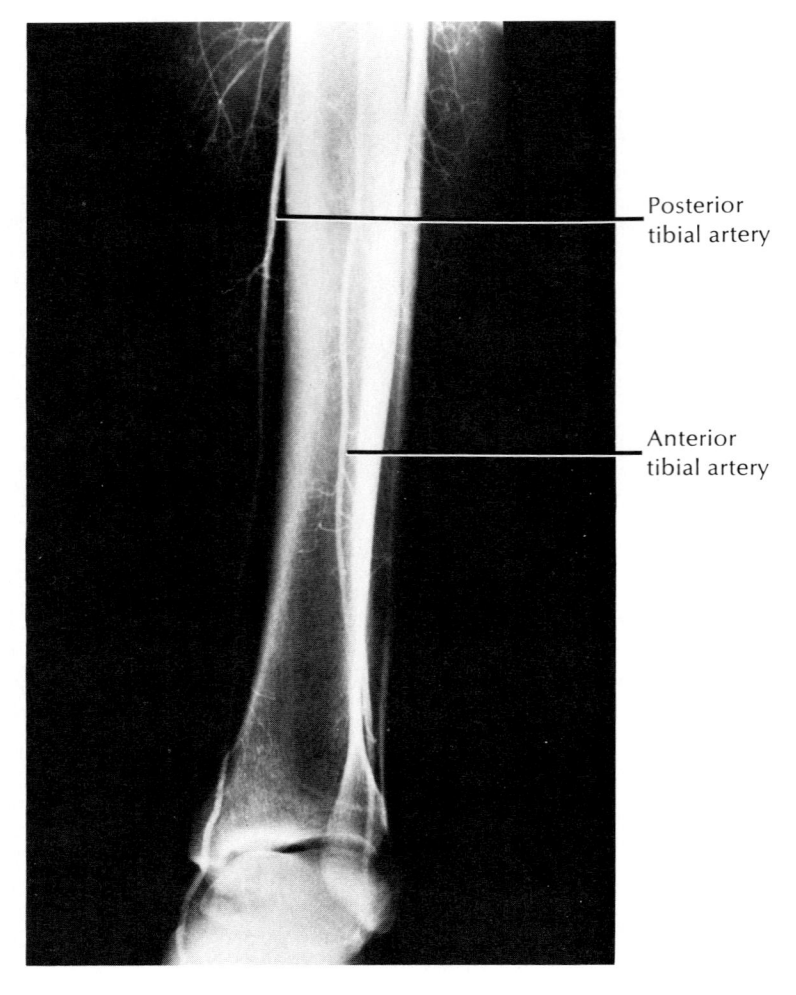

Angiogram of blood supply of lower leg

Blood supply

The blood supply to the lateral leg anterior muscle group is through the anterior tibial artery. This artery branches from the popliteal artery at the level of the popliteus muscle and courses between the two heads of the tibialis posterior and through the interosseous membrane. In the upper leg the anterior tibial artery courses between the tibialis anterior and extensor digitorum longus muscle. In the lower leg the anterior tibial artery courses between the tibialis anterior and extensor hallucis longus. The deep peroneal nerve courses with these vessels. During this course the anterior tibial artery supplies pedicles to the tibialis anterior, extensor digitorum longus, and extensor hallucis longus muscles.

LATERAL LEG: ANTERIOR GROUP

Tibialis anterior

Anterior tibial artery

Extensor digitorum longus
Extensor hallucis longus

Blood supply

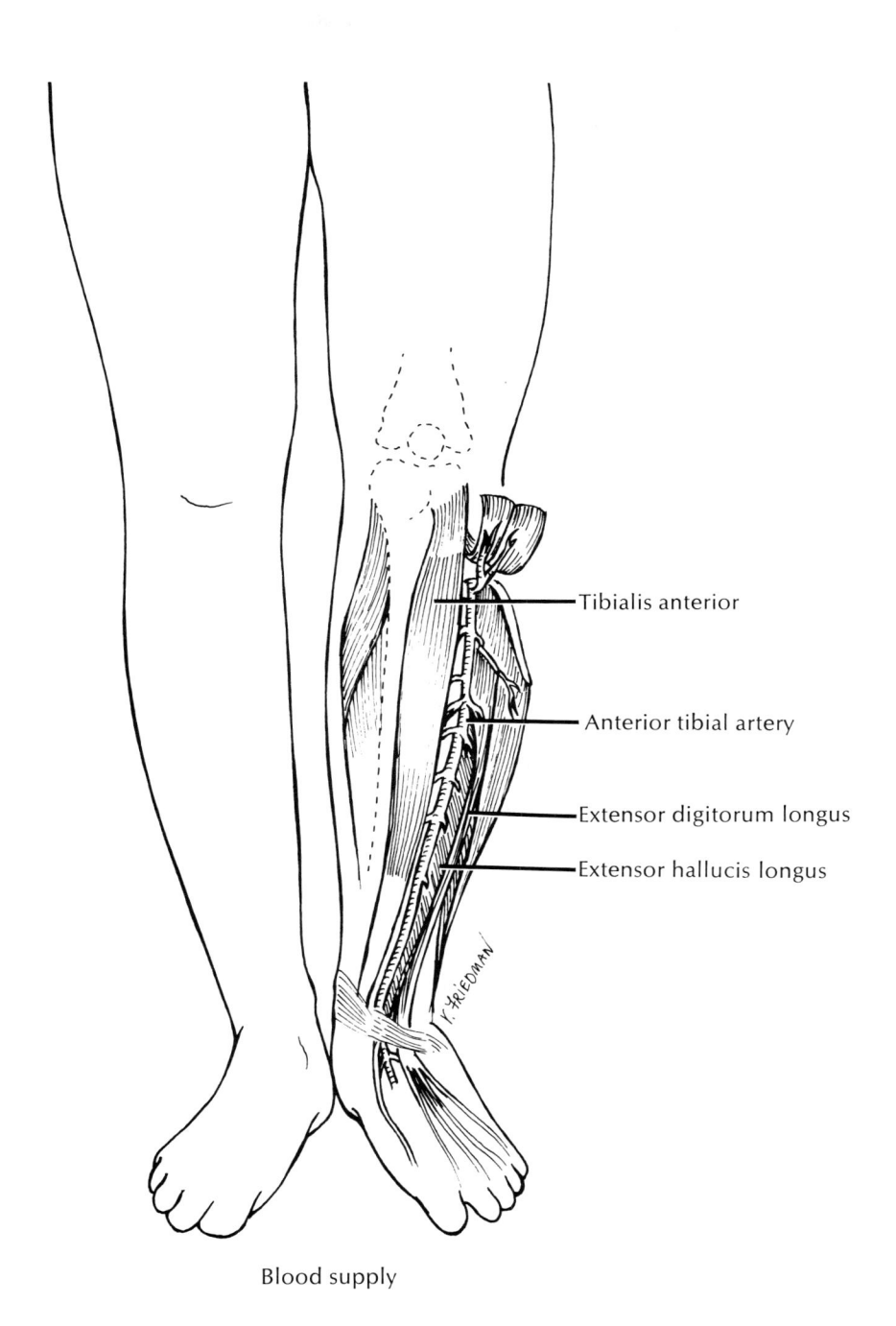

LATERAL LEG: ANTERIOR GROUP

Tibialis anterior

MUSCLE FLAP
MUSCULOCUTANEOUS FLAP

Application

Coverage of:

MIDDLE THIRD OF LEG

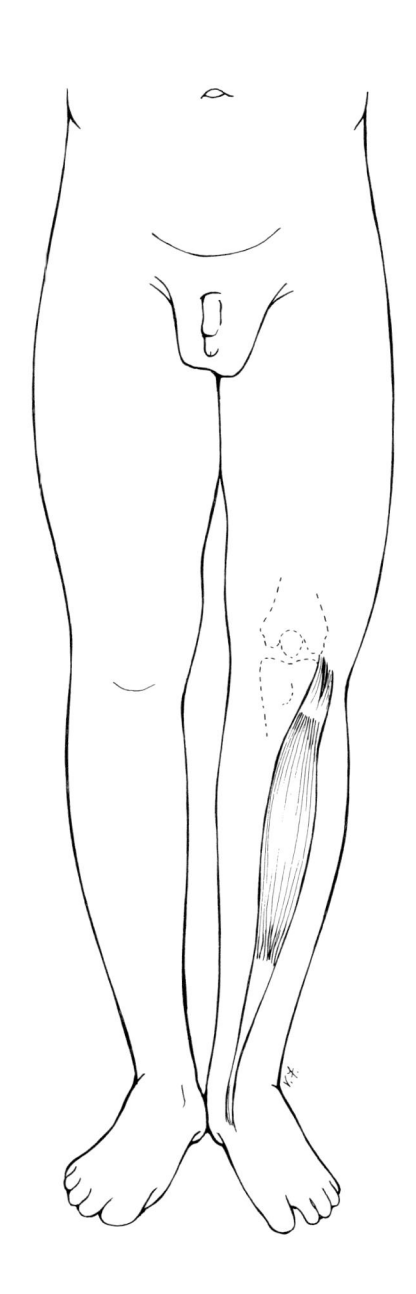

LATERAL LEG: ANTERIOR GROUP

Tibialis anterior

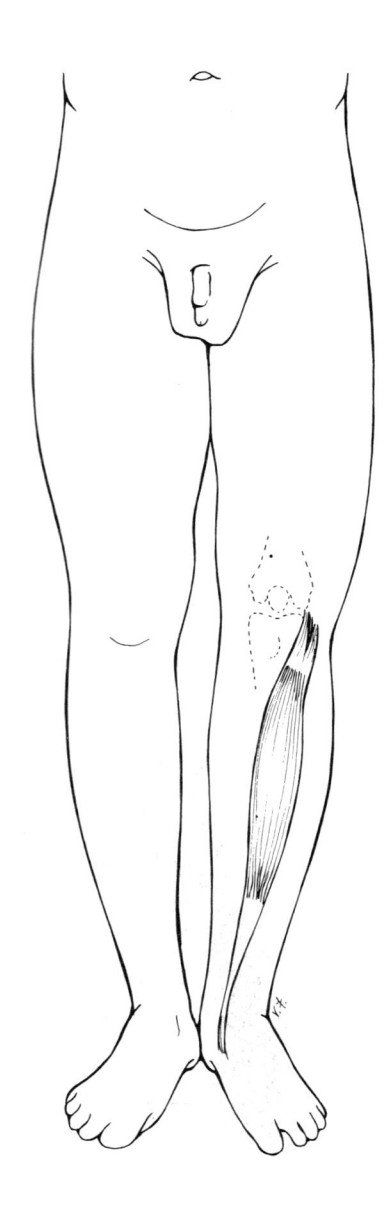

The large tibialis anterior leg muscle is located adjacent and lateral to the tibia. The exterior digitorum longus and extensor hallucis longus form its lateral border. The posterior muscle belly has a close relationship to the anterior tibial artery and the deep peroneal nerve.

Origin: Lateral condyle of tibia, upper lateral surface of tibia, interosseus membrane.
Insertion: Medial cuneiform bone, base of first metatarsal bone.

LATERAL LEG: ANTERIOR GROUP

Blood supply

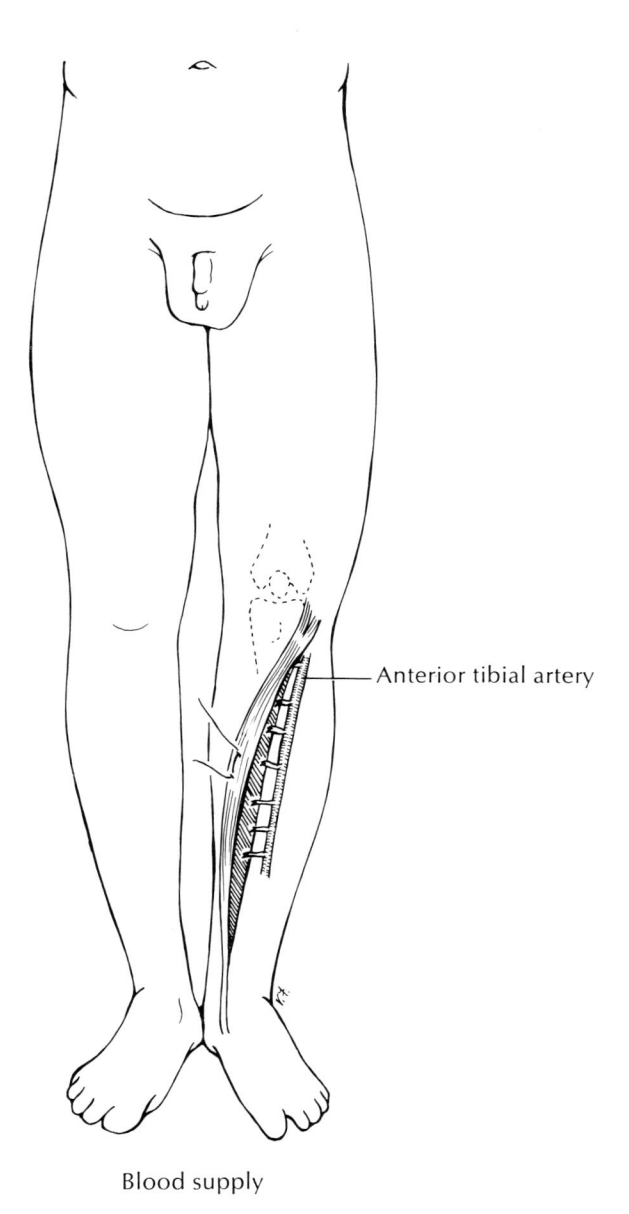

Nerve supply:

    Motor — Branches of deep peroneal nerve (anterior tibial). This motor nerve branch enters the proximal posterior aspect of the muscle belly.

Function: This is not an expendable muscle in an ambulatory patient. Foot dorsiflexion and inversion could be impaired despite presence of the extensor hallucis longus and extensor digitorum longus muscles. When the tendon is preserved, it appears that tibialis anterior function has been maintained despite muscle transposition.

LATERAL LEG: ANTERIOR GROUP

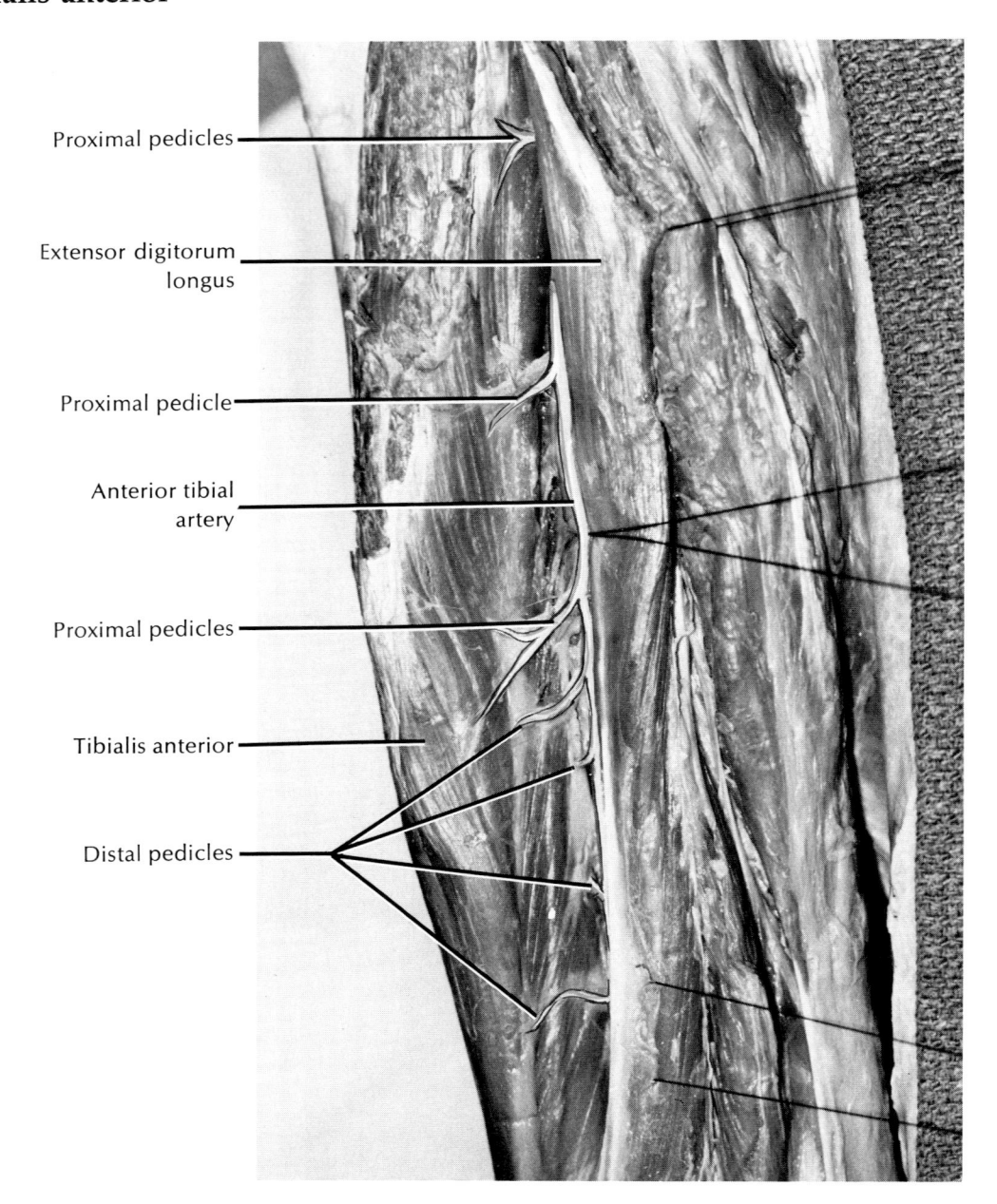

Tibialis anterior

- Proximal pedicles
- Extensor digitorum longus
- Proximal pedicle
- Anterior tibial artery
- Proximal pedicles
- Tibialis anterior
- Distal pedicles

Blood supply

There are six to eight muscular arterial branches from the anterior tibial artery to the tibialis anterior muscle belly. The muscle will survive division of the inferior pedicles. The proximal third of the muscle belly receives three to four pedicles, which will sustain this muscle following transposition.

LATERAL LEG: ANTERIOR GROUP

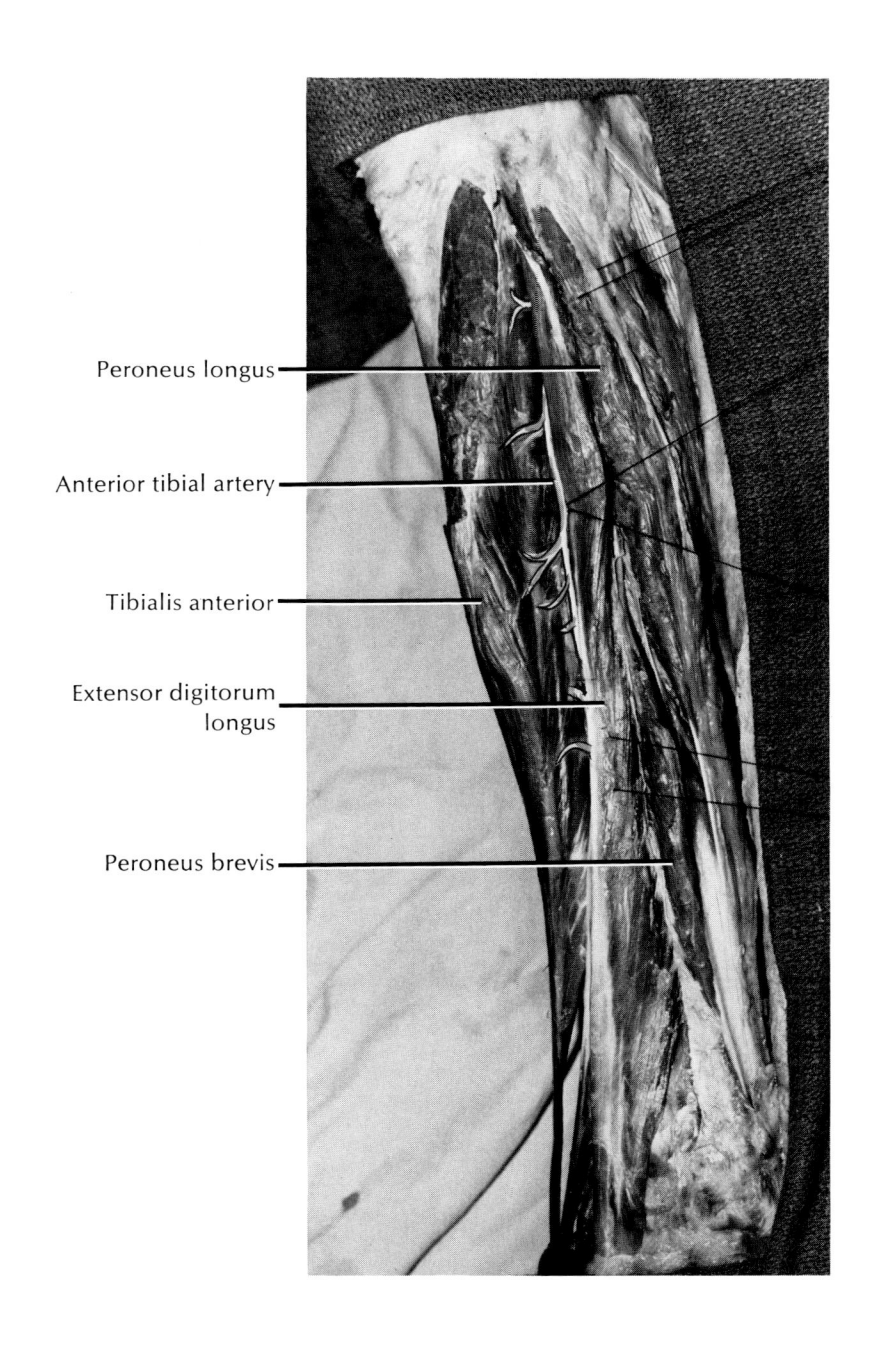

- Peroneus longus
- Anterior tibial artery
- Tibialis anterior
- Extensor digitorum longus
- Peroneus brevis

LATERAL LEG: ANTERIOR GROUP

Tibialis anterior

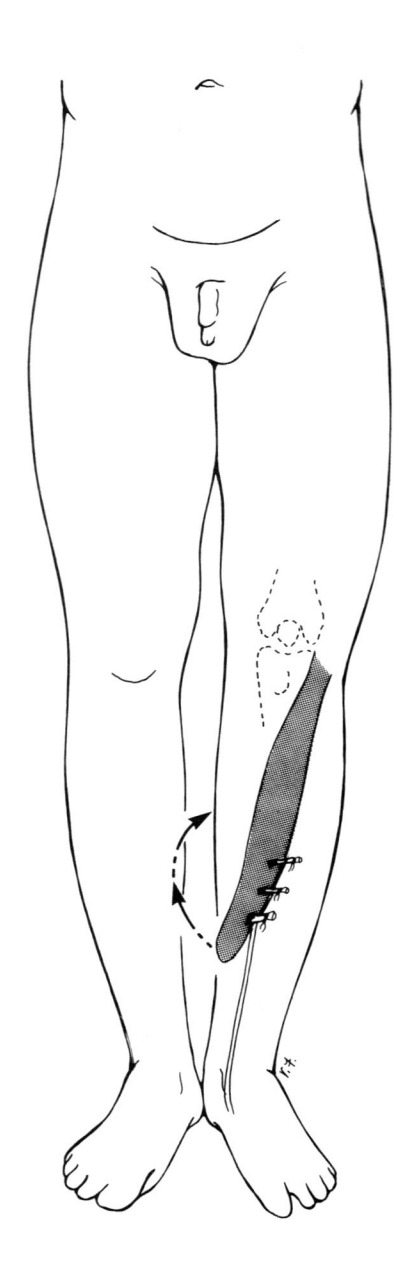

Arc of rotation

The tibialis anterior muscle has a point of rotation 5 to 10 cm inferior to the level of the tibial tubercle. This muscle may be transposed in a medial direction to cover the middle third of exposed tibia.

LATERAL LEG: ANTERIOR GROUP

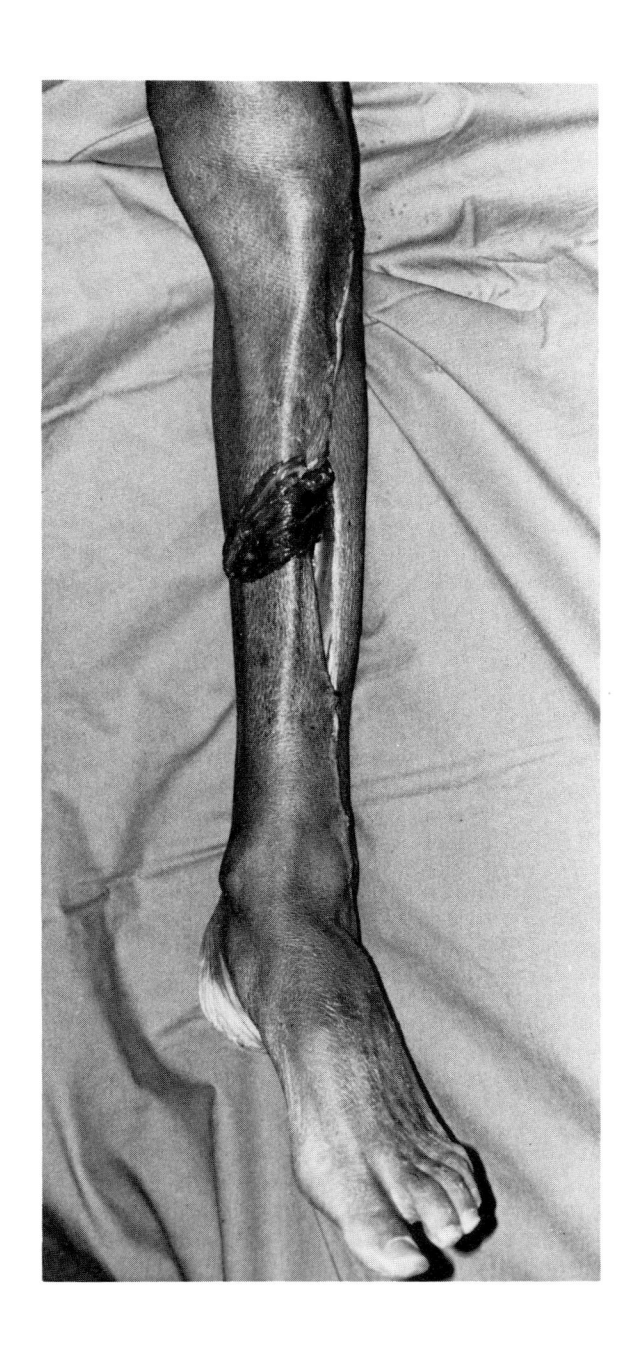

LATERAL LEG: ANTERIOR GROUP

Tibialis anterior

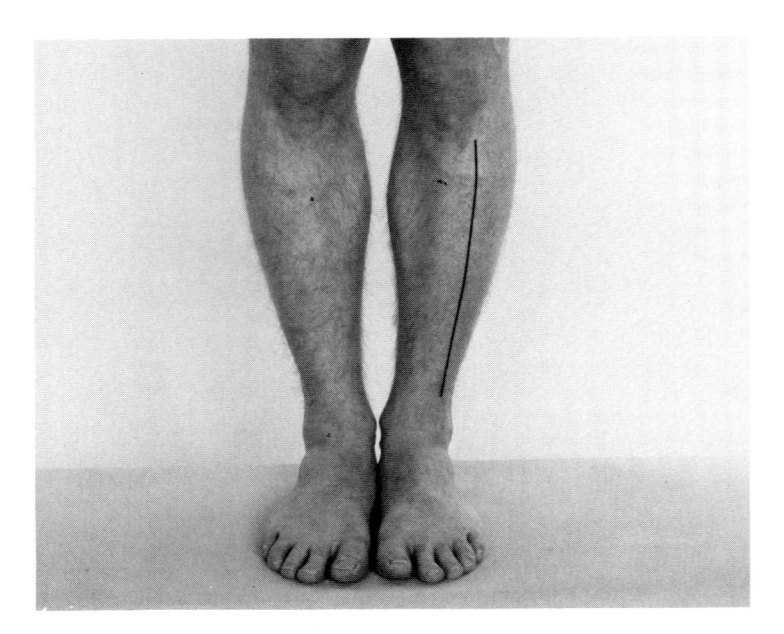

Skin incision

Elevation of flap

The tibialis anterior is directly exposed by a vertical incision lateral to the tibia. The tibialis anterior tendon is divided inferiorly, and the muscle is elevated off the anterior tibial artery, deep peroneal nerve, and extensor hallucis longus muscle. Vascular pedicles are divided to a level 5 to 10 cm inferior to the tibial tubercle, depending on the length of desired muscle transposition. The donor area may be closed primarily.

The muscle flap may be elevated with the tendon intact. This will preserve tibialis anterior function and allow use of this muscle with less functional disability. The arc of rotation is slightly decreased, but the close proximity of the muscle to the tibia eliminates the need for a long arc of rotation.

The cutaneous territory of this muscle is generally not useful for transposition, since the donor defect may leave exposure of the anterior tibial artery and deep peroneal nerve.

Precautions

- Functional disability following transposition of the entire unit may occur.
- The posterior muscle belly is closely related to the anterior tibial artery and deep peroneal nerve.

LATERAL LEG: ANTERIOR GROUP

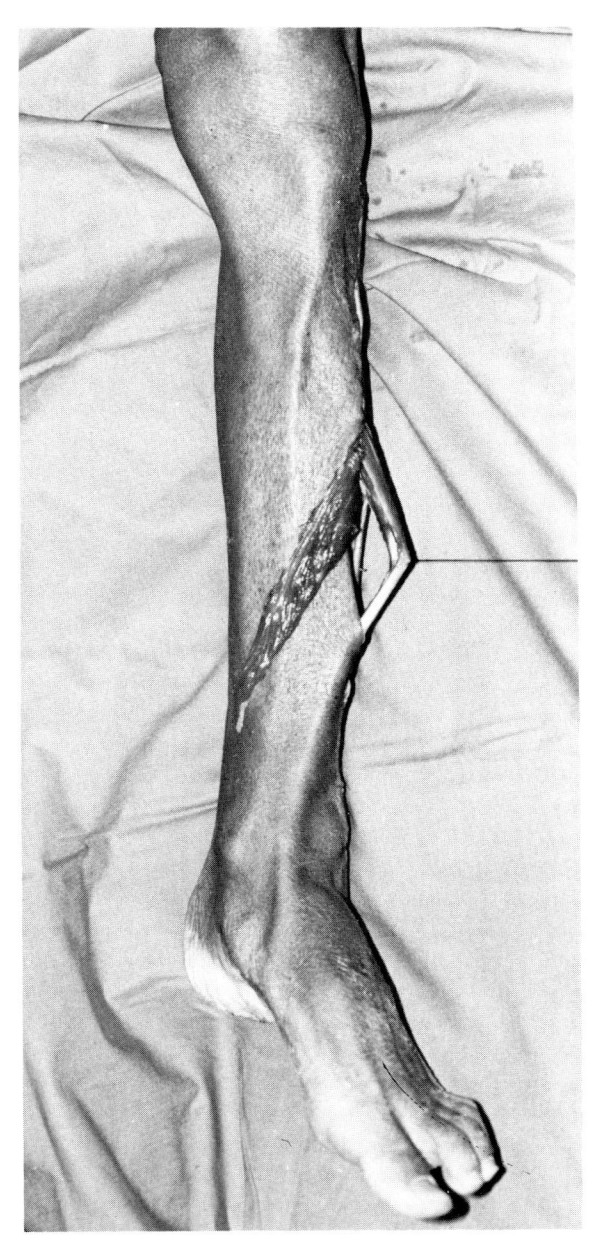

Function preserving elevation of muscle
(muscle belly separated from intact tendon)

Extensor digitorum longus

MUSCLE FLAP

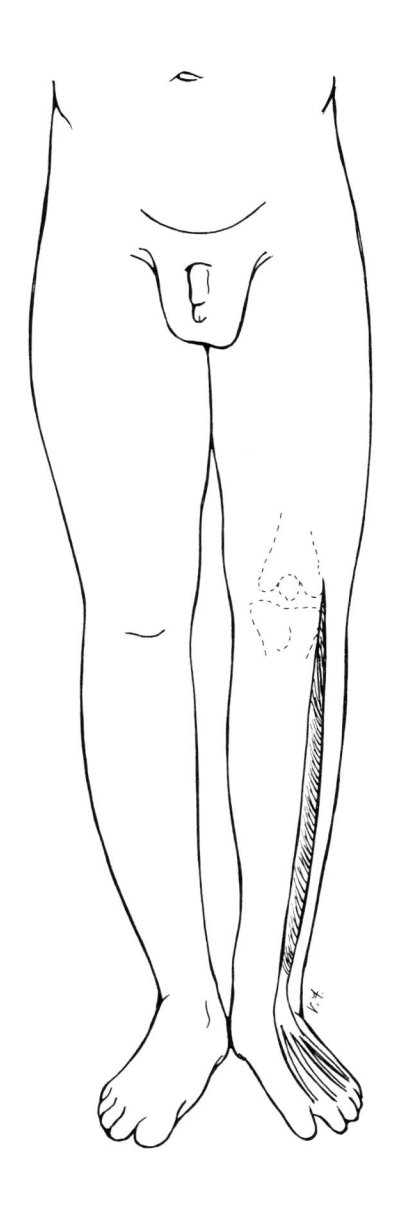

Applications

Coverage of:
LOWER THIRD OF LEG

Free flap:
FUNCTIONAL MUSCLE

LATERAL LEG: ANTERIOR GROUP

Extensor digitorum longus

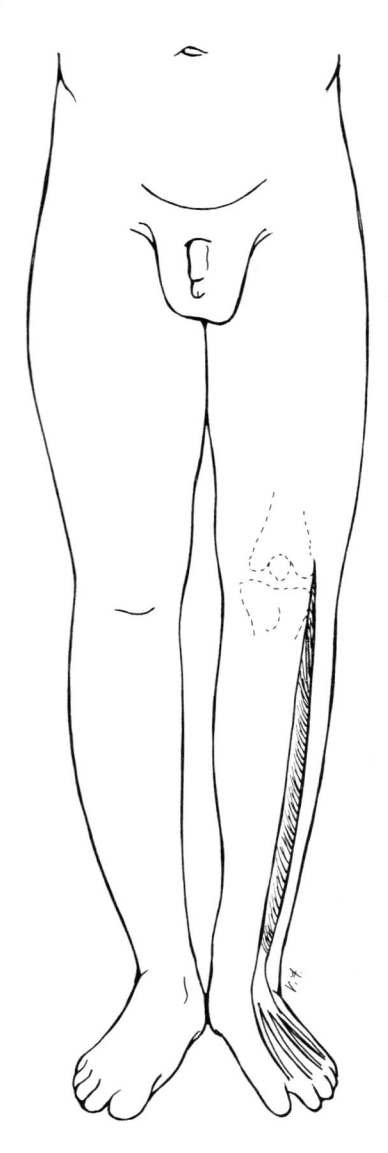

The long, thin extensor digitorum longus muscle is bordered medially and anteriorly by the tibialis anterior and extensor hallucis longus muscles. It has a posterolateral relationship with the peroneal muscles.

> **Origin:** Lateral condyle of tibia, anterior surface of fibula interosseous membrane.
> **Insertion:** Middle and distal phalanges of four lateral toes.

LATERAL LEG: ANTERIOR GROUP

Anterior tibial artery

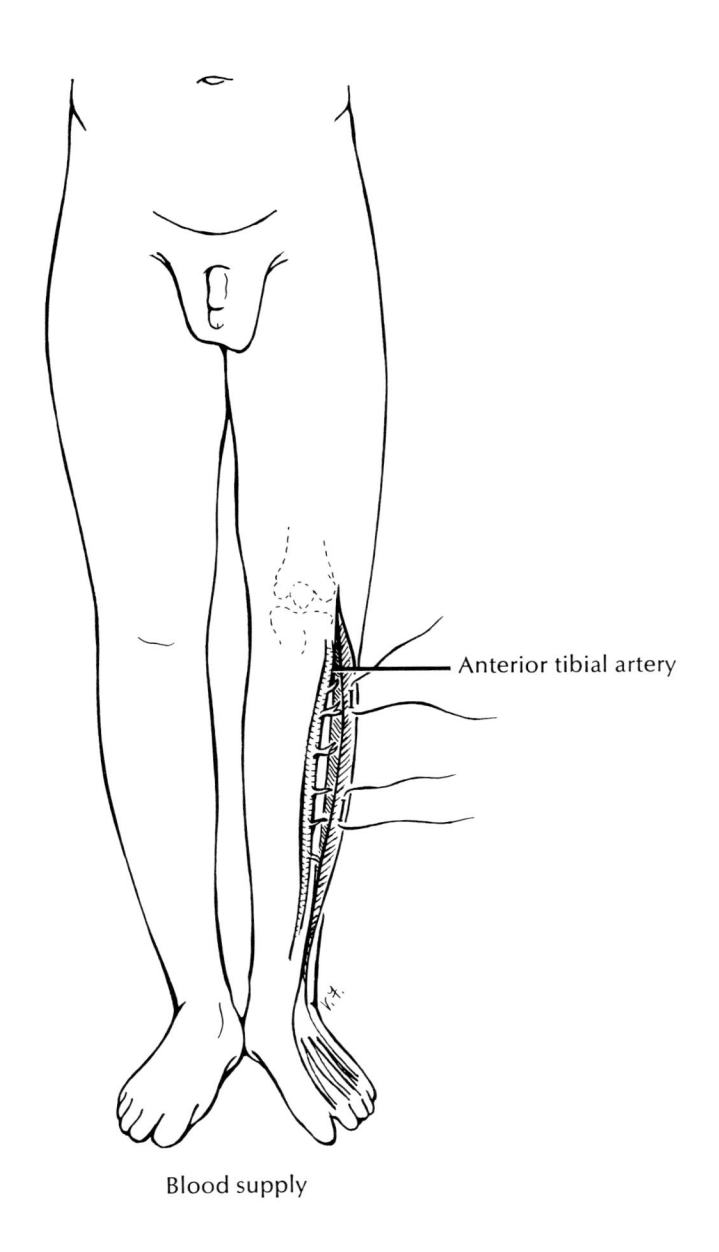

Blood supply

Nerve supply:
  Motor—Deep peroneal. The motor nerve enters the posterior proximal portion of the muscle.
Function: This is an expendable muscle. The function of phalangeal extension would be preserved by the extensor digitorum brevis. (There is loss of extension of the distal phalanx.) Dorsiflexion would be preserved unless the remaining anterior muscle groups are transposed or injured.

LATERAL LEG: ANTERIOR GROUP

Extensor digitorum longus

Tibialis anterior

Anterior tibial artery

Proximal pedicles

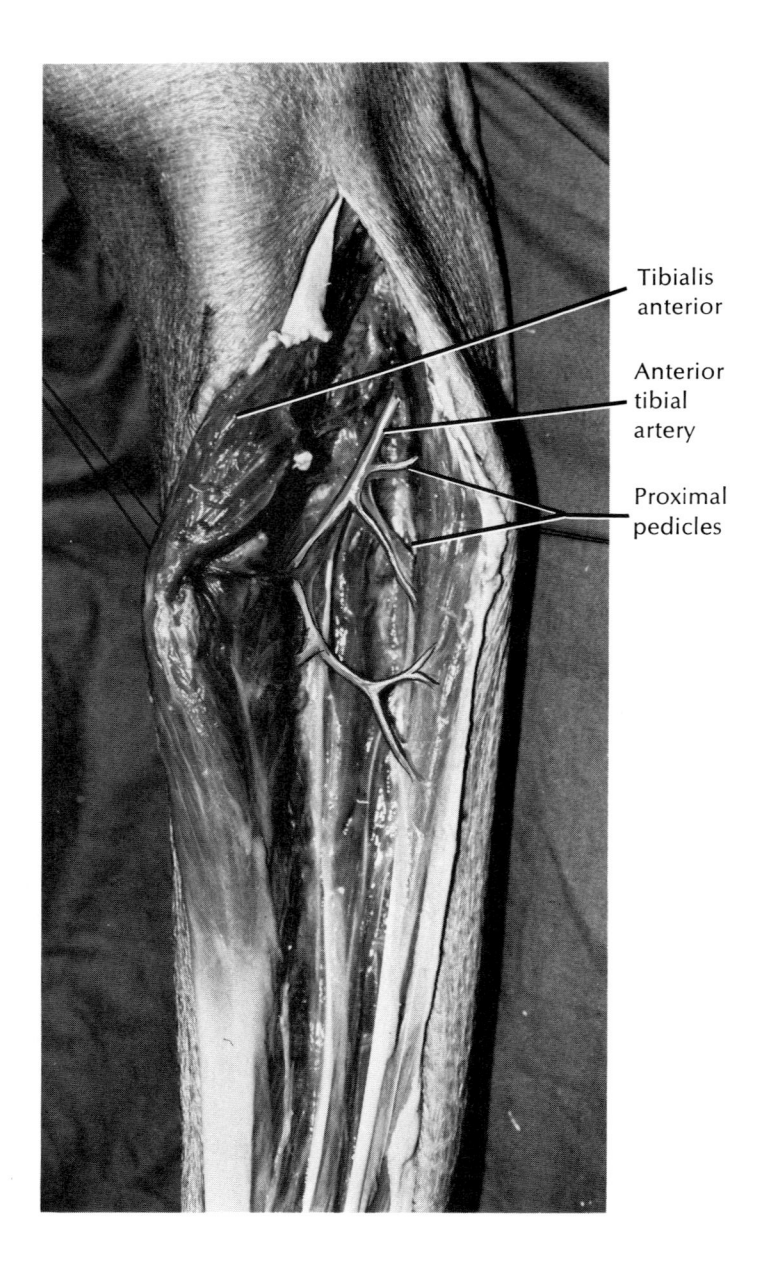

Blood supply

There are six to eight vascular pedicles to the extensor digitorum longus muscle from the anterior tibial artery. These pedicles enter the posterior muscle belly. The proximal pedicles above the mid tibia must be preserved during muscle transposition.

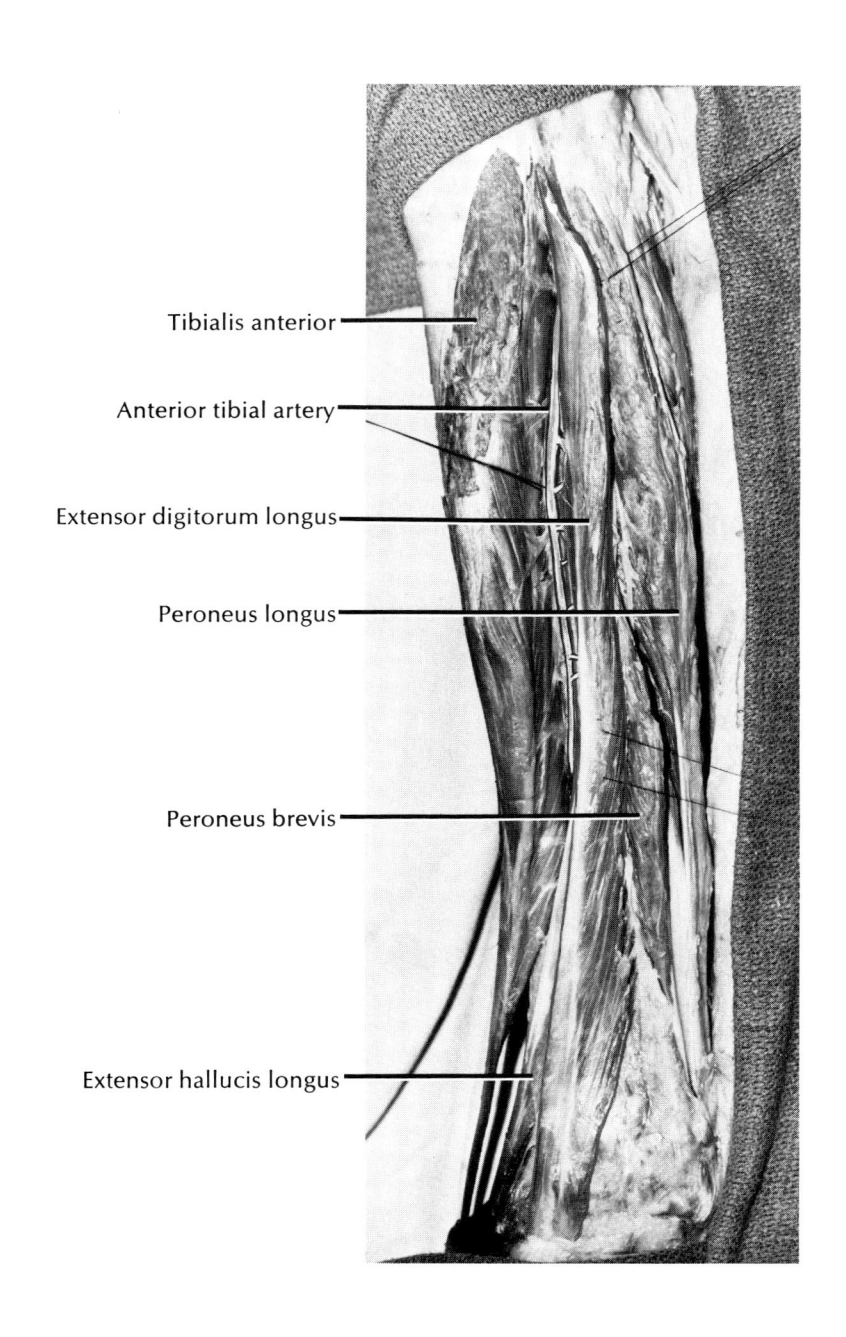

LATERAL LEG: ANTERIOR GROUP

Tibialis anterior

Anterior tibial artery

Extensor digitorum longus

Peroneus longus

Peroneus brevis

Extensor hallucis longus

LATERAL LEG: ANTERIOR GROUP

Extensor digitorum longus

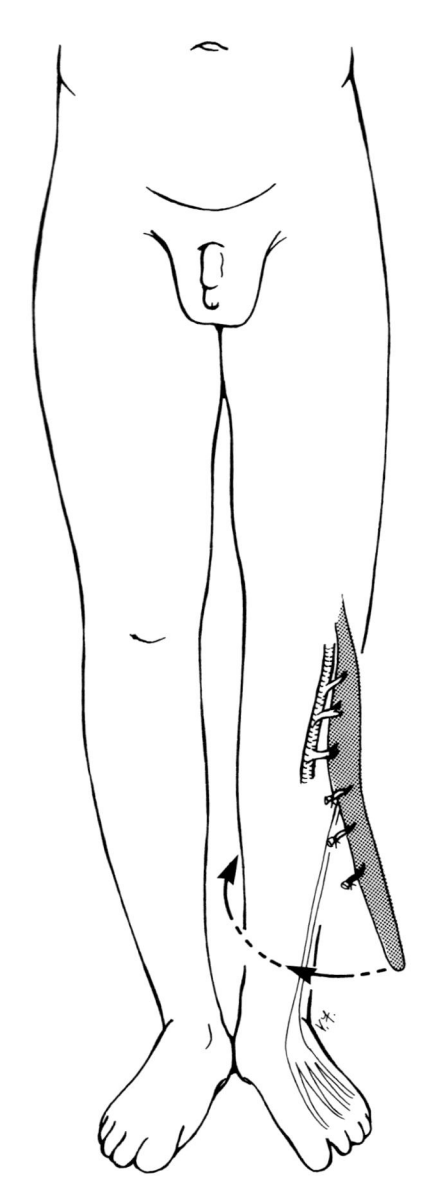

Arc of rotation

The extensor digitorum longus muscle has a point of rotation in the mid portion of the lower leg. This muscle has a small arc of rotation, which will cover a portion of the lower anterior tibia following medial transposition.

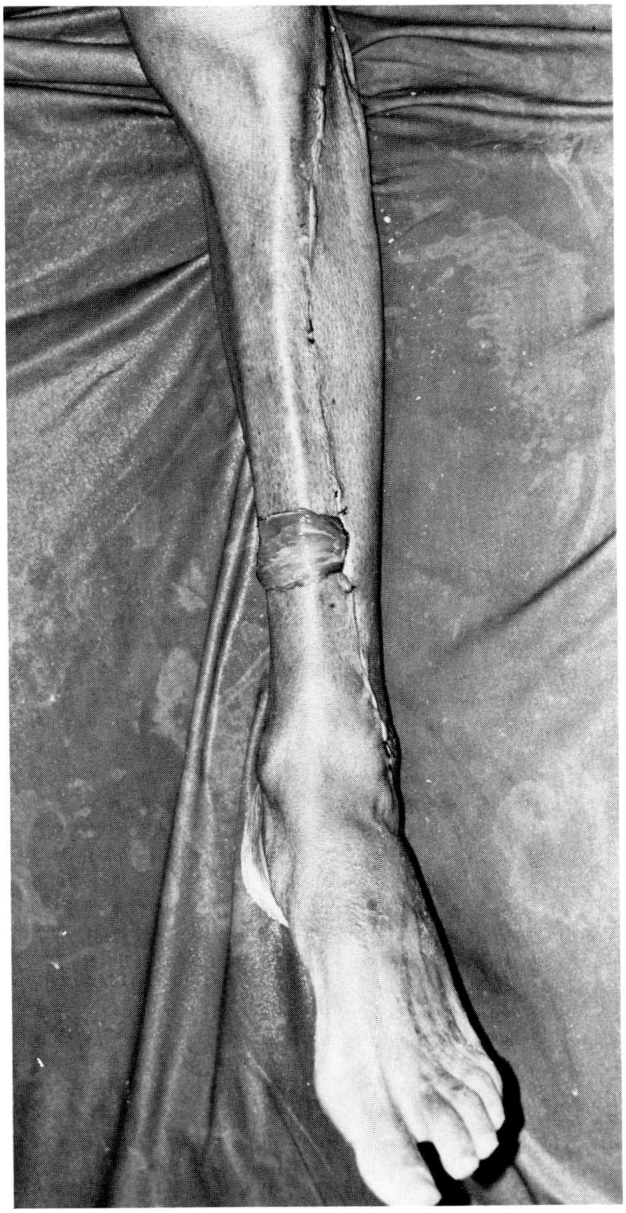

LATERAL LEG: ANTERIOR GROUP

Extensor digitorum longus

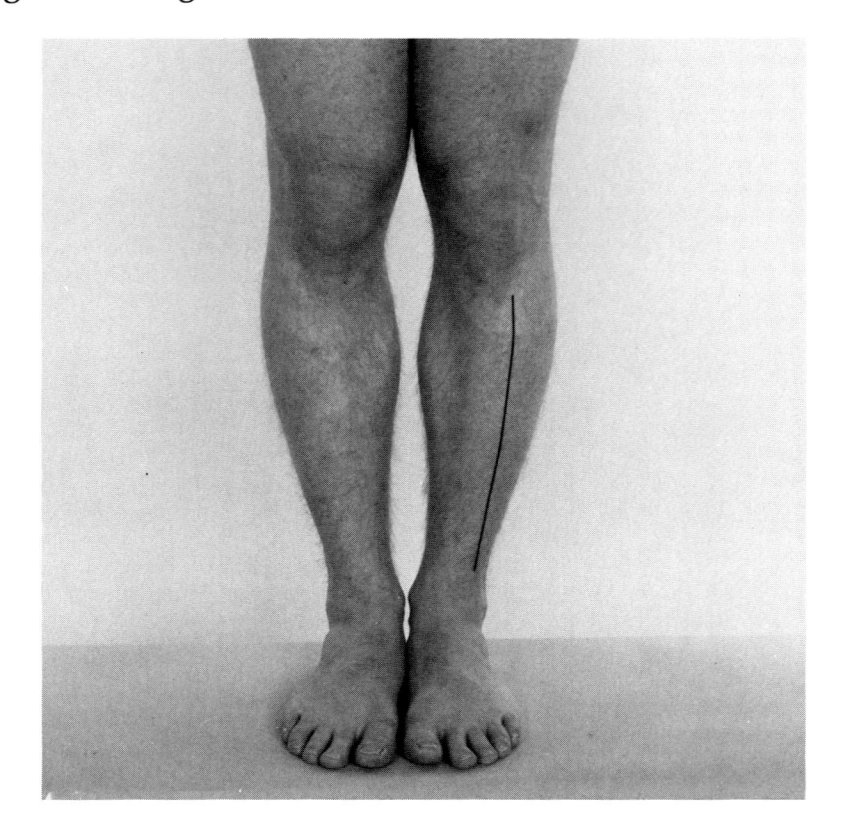

Elevation of flap

The extensor digitorum longus is approached through an anterolateral vertical incision. The tendon is located at the level of the inferior extensor retinaculum between the extensor hallucis longus tendon and the peroneal tendon. The tendon is divided and the flap elevated to the mid tibial level. Two or three vascular pedicles must be divided to reach this level. Based on anatomic studies, the distal muscle belly may be dissected from the tendon to this same level, allowing muscle transposition and preserving function via the intact tendon. The donor region may be closed primarily.

The extensor digitorum longus can be used in free transfer as a functional motor unit. The dissection may necessitate sacrifice of the proximal anterior tibial artery. The required extensive dissection would not be warranted unless conventional muscle transfers are unavailable.

Precautions

- The muscle belly is small and useful only in conjunction with transposition of other lateral muscles when medial muscles are unavailable for tibial coverage.
- To avoid potential weakness in toe extension, the tendon may be preserved.
- The superficial peroneal nerve has a close relationship to the proximal lateral muscle belly.

Extensor hallucis longus

MUSCLE FLAP

Applications

Coverage of:
LOWER THIRD OF LEG

Free flap:
FUNCTIONAL MUSCLE

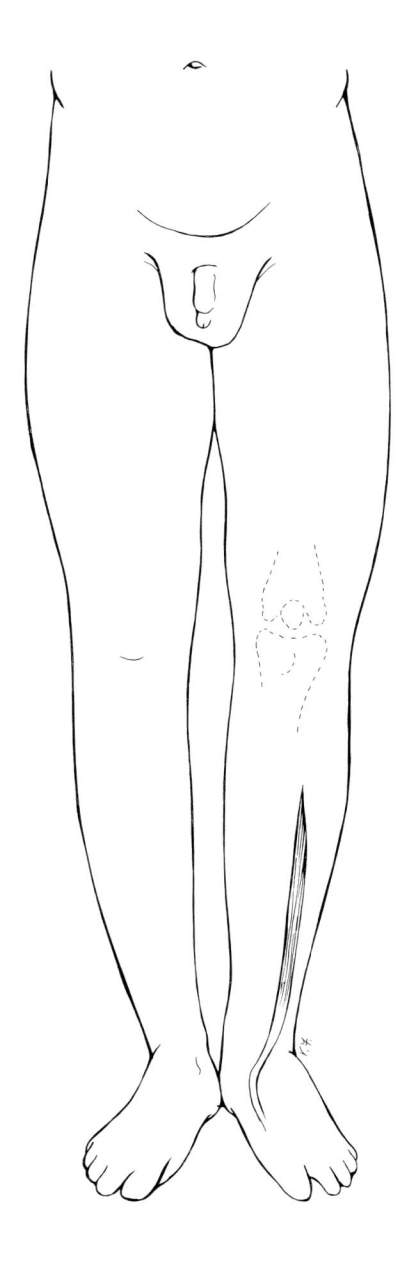

LATERAL LEG: ANTERIOR GROUP

Extensor hallucis longus

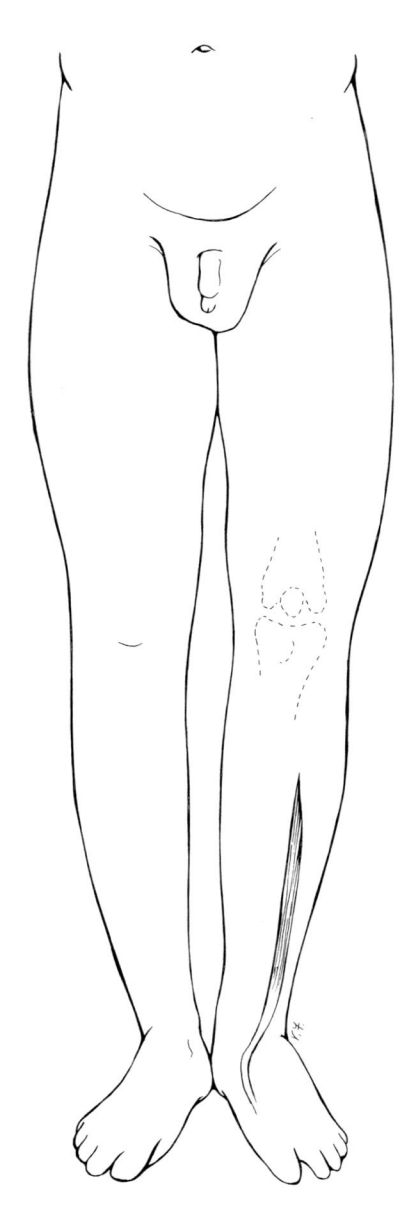

The extensor hallucis longus is a short, narrow muscle located in the anterior leg. The muscle is adjacent to the tibialis anterior muscle anteromedially and the anterior tibial artery and deep peroneal nerve posteromedially. The muscle borders the extensor digitorum longus muscle laterally. At the level of the inferior extensor retinaculum, the tendon crosses anterior to the dorsalis pedis artery and deep peroneal nerve.

Origin: Mid anterior fibula and interosseous membrane.
Insertion: Base of distal phalanx of great toe.

LATERAL LEG: ANTERIOR GROUP

Anterior tibial artery

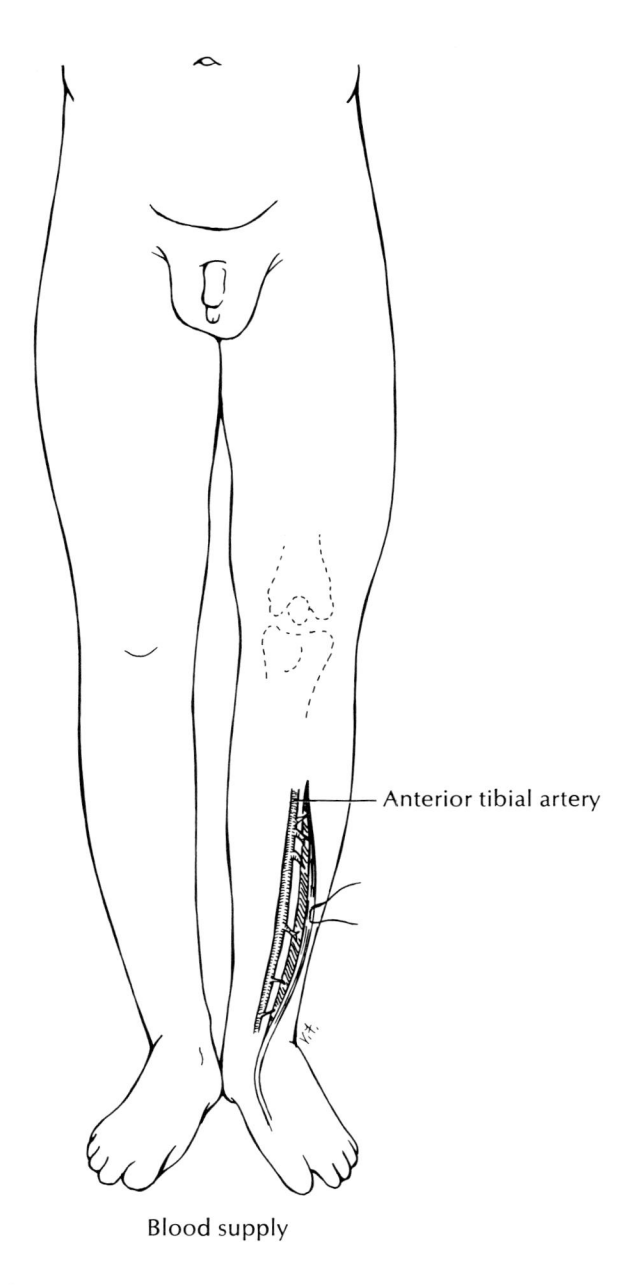

Blood supply

Nerve supply:
   Motor—Muscular branch of deep peroneal nerve. The motor branches enter the proximal and middle thirds of the posterior muscle belly.

Function: This is an expendable muscle. The function of extension of the great toe will be maintained at the proximal phalanx by the extensor digitorum brevis. Foot dorsiflexion will be preserved by the remaining anterior leg musculature.

LATERAL LEG: ANTERIOR GROUP

Extensor hallucis longus

Blood supply

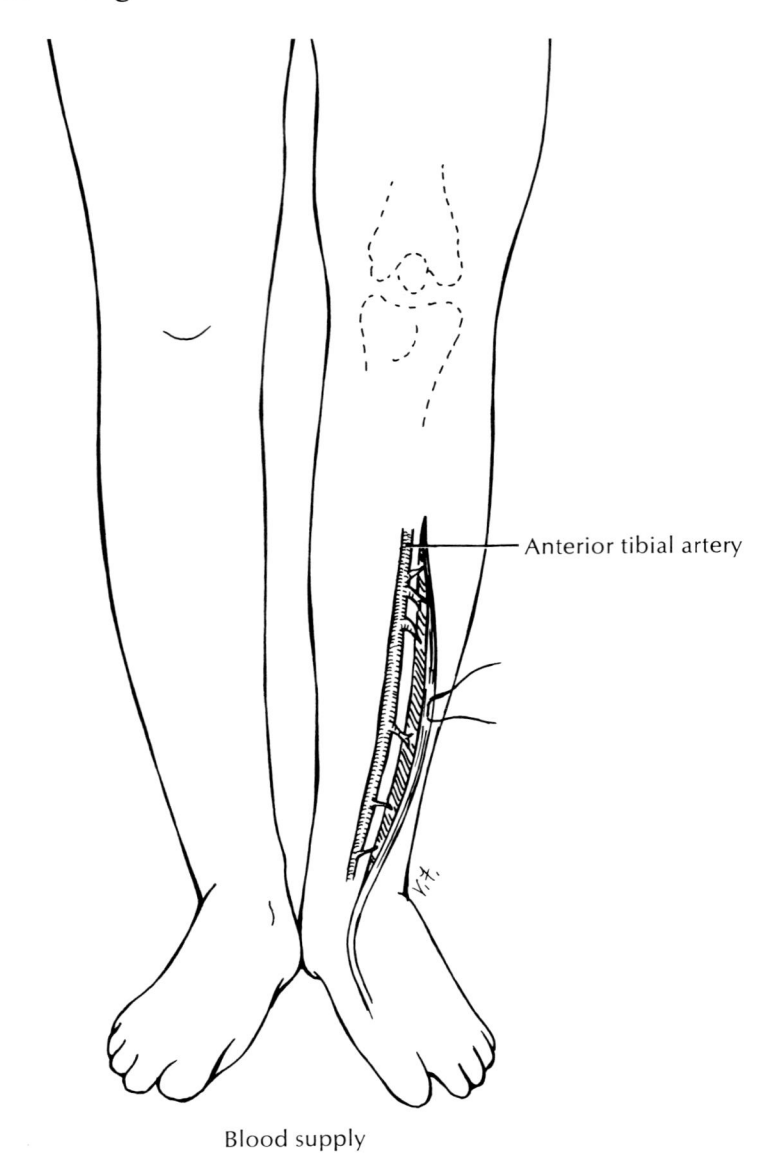

Blood supply

This muscle receives five to six arterial pedicles from the anterior tibial artery. These pedicles enter the medial border of the muscle. The proximal two or three pedicles must be preserved for muscle transposition.

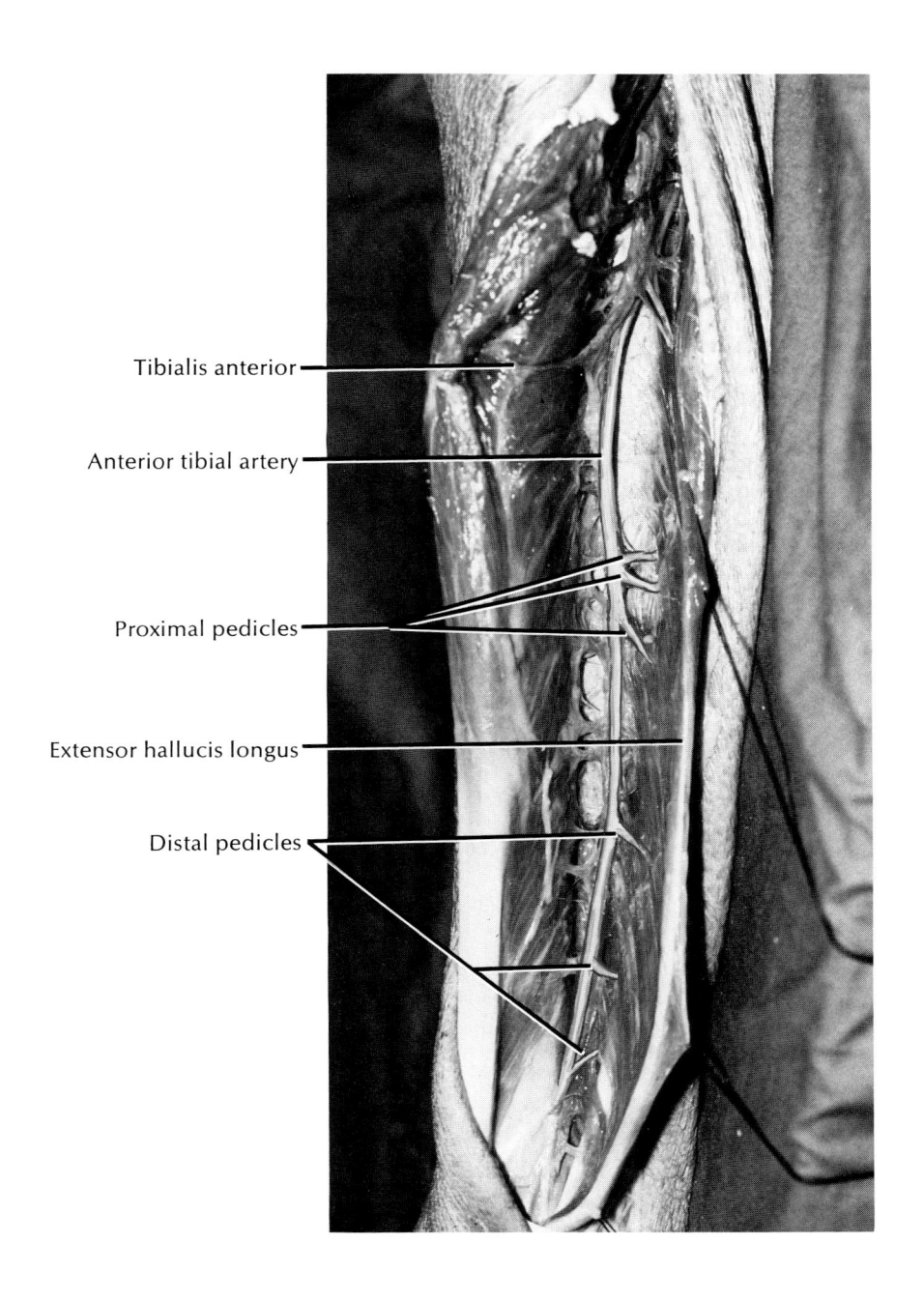

LATERAL LEG: ANTERIOR GROUP

- Tibialis anterior
- Anterior tibial artery
- Proximal pedicles
- Extensor hallucis longus
- Distal pedicles

LATERAL LEG: ANTERIOR GROUP

Extensor hallucis longus

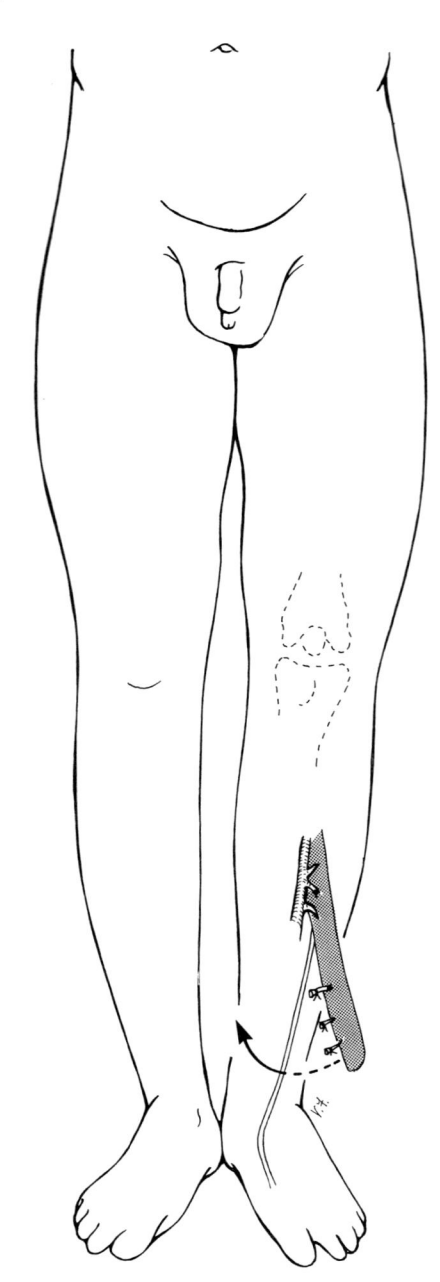

Arc of rotation

The extensor hallucis longus has a point of rotation in the mid lower leg. This small muscle will cover a proximal portion of the distal third of the tibia following medial transposition.

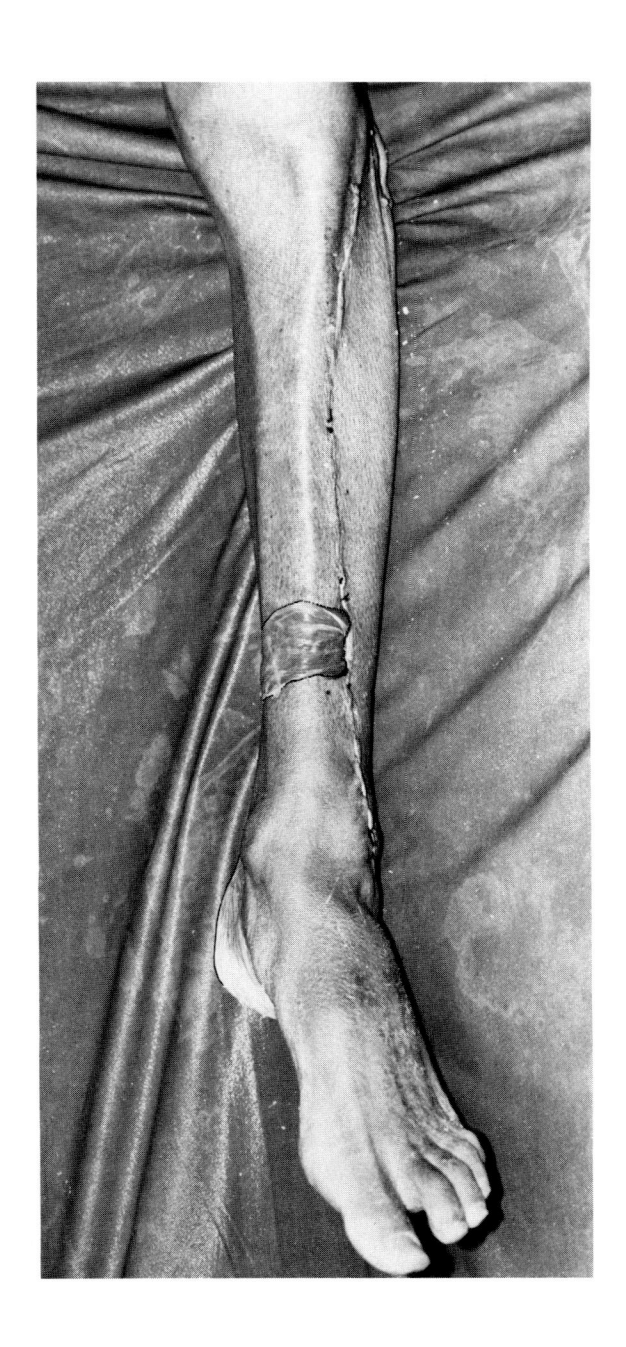

LATERAL LEG: ANTERIOR GROUP

LATERAL LEG: ANTERIOR GROUP

Extensor hallucis longus

Elevation of flap

The extensor hallucis longus muscle is approached through an anterolateral vertical incision. The tendon is located at the level of the inferior extensor retinaculum between the tibialis anterior and extensor digitorum longus tendons. The distal portion of the anterior tibial artery is immediately posterior to the extensor hallucis tendon. Following tendon division, the muscle is then elevated proximally from its tibial origin with ligation of two or three inferior pedicles. The flap is elevated to the mid lower leg, leaving the proximal two or three pedicles intact. Based on anatomic studies, the distal muscle belly may be dissected from the tendon to this level, allowing muscle transposition and preserving muscle blood flow and function via the intact tendon. The donor area may be closed primarily following muscle transposition.

The extensor hallucis longus muscle can be used in free transfer as a functional motor unit. The dissection may necessitate sacrifice of the mid anterior tibial artery. The required extensive dissection would not be warranted unless conventional muscle transfers were not available.

Precaution

- The medial posterior belly has a close relationship to the anterior tibial artery and the deep peroneal nerve.

LATERAL LEG: ANTERIOR GROUP

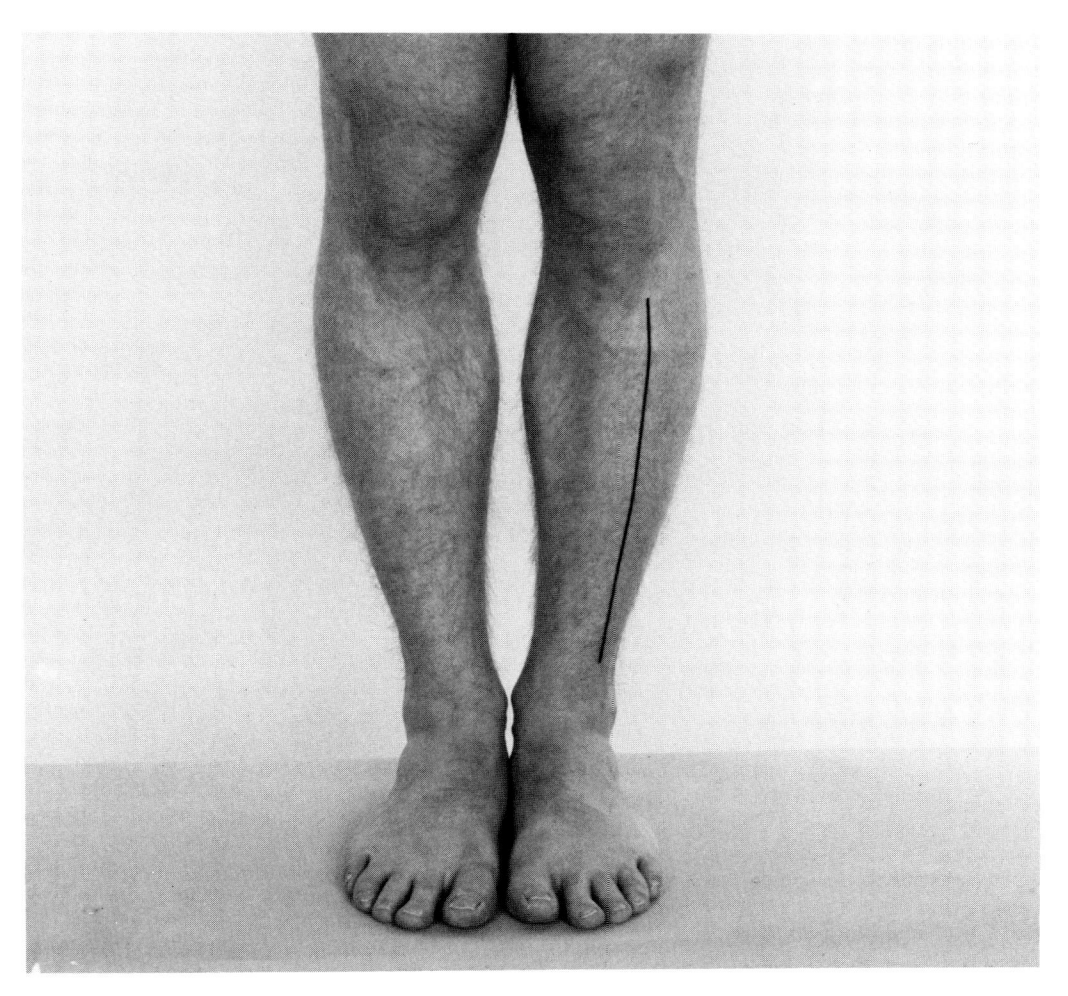

Skin incision

LATERAL LEG

Posterior group
PERONEUS LONGUS
PERONEUS BREVIS

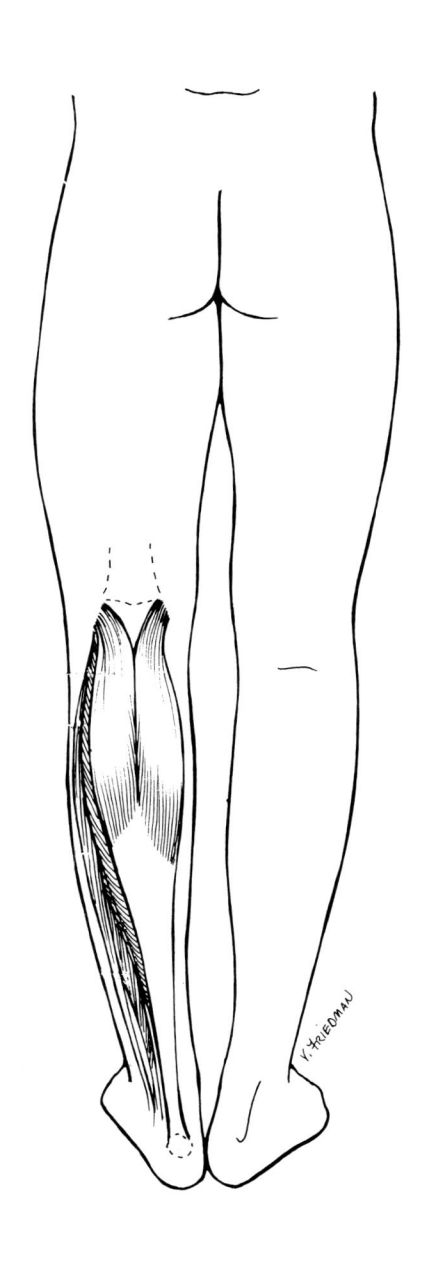

LATERAL LEG: POSTERIOR GROUP

- Gastrocnemius
- Soleus
- Peroneus longus
- Peroneus brevis

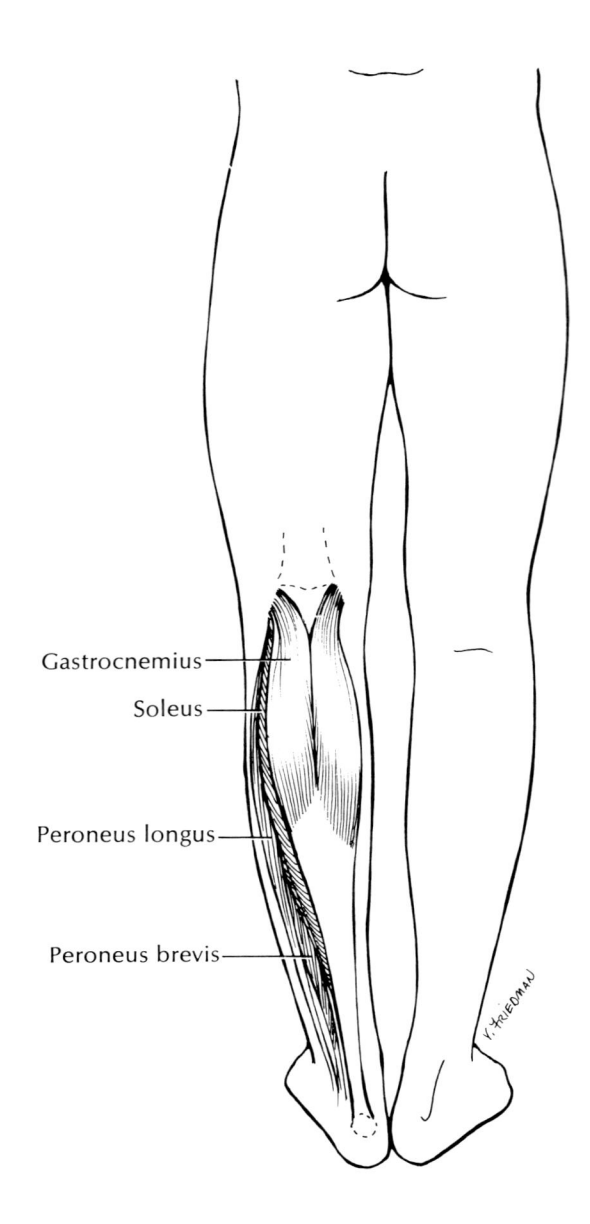

The posterior group of lateral leg muscles is located behind the extensor digitorum longus. These are small muscles that are seldom used in reconstructive surgery. However, for small defects or if the soleus is not available, this group of muscles is an alternative for coverage of the middle third of the leg. These muscles, however, must reach across the extensor digitorum longus and tibialis anterior to cover the tibia.

The blood supply to these muscles is based on perforating branches of the peroneal artery.

LATERAL LEG: POSTERIOR GROUP

Peroneus longus
Peroneus brevis

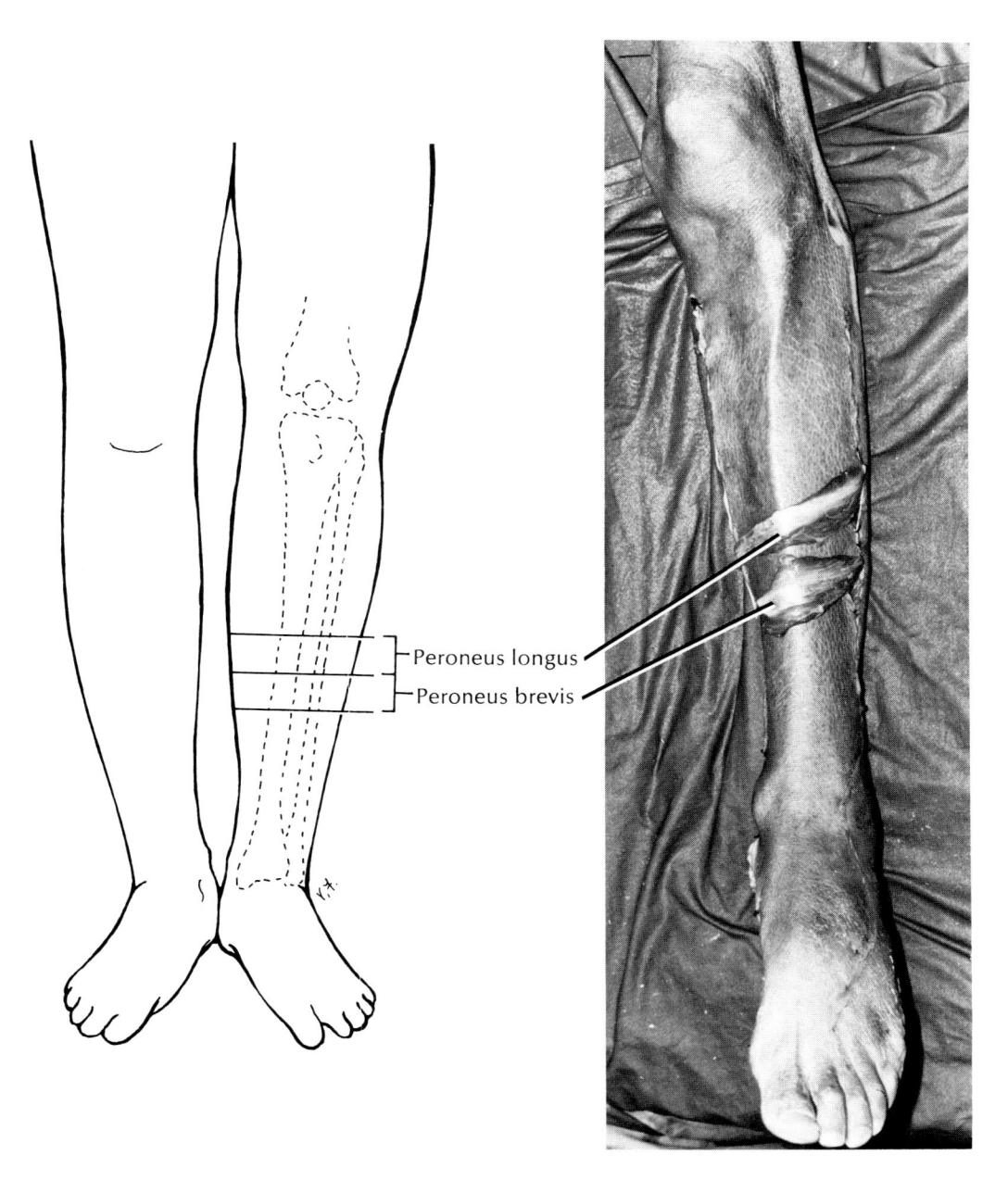

Areas of coverage

239

Peroneus longus

MUSCLE FLAP

Application

Coverage of:
SMALL DEFECTS OF MIDDLE THIRD OF LEG

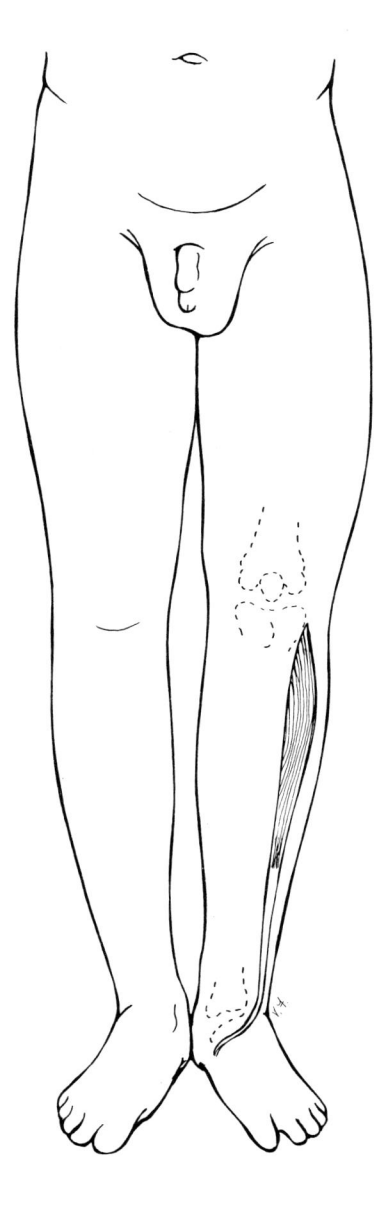

LATERAL LEG: POSTERIOR GROUP

Peroneus longus

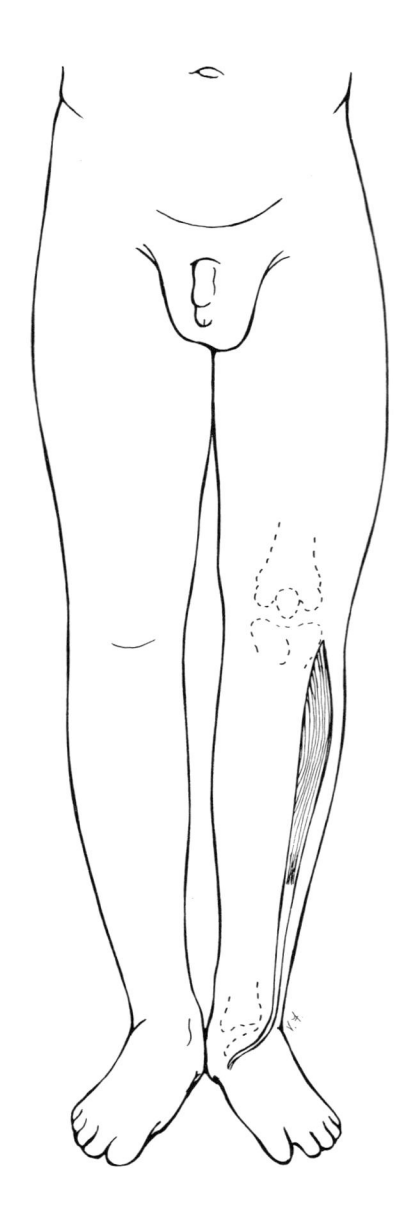

The peroneus longus is the larger of the two muscles in the posterior group of lateral leg muscles and lies superficial to the peroneus brevis behind the extensor digitorum longus.

Origin: Head and upper two-thirds of the lateral fibula.
Insertion: Lateral side of the base of the first metatarsal and the medial cuneiform.

LATERAL LEG: POSTERIOR GROUP

- Superficial femoral artery
- Peroneal artery

Blood supply

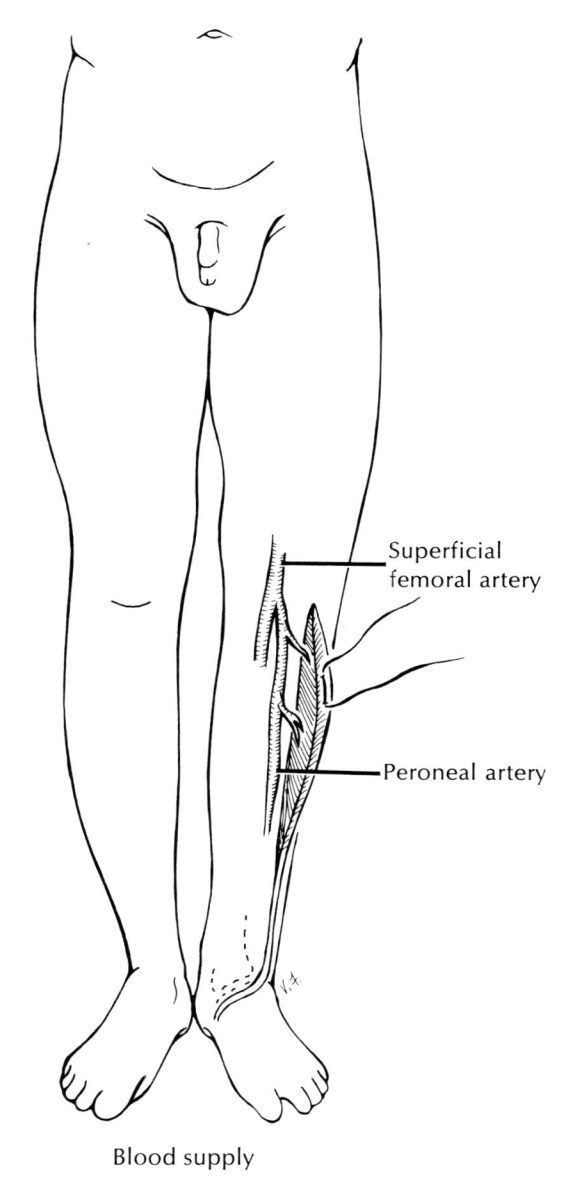

Nerve supply: The common peroneal nerve winds around the neck of the fibula to lie deep to the peroneus longus. It then divides into the superficial peroneal and deep peroneal nerves (anterior tibial nerve). The superficial peroneal nerve innervates the peroneus longus near its origin on the deep surface.

Function: This muscle everts and plantar flexes the foot. The peroneus longus, peroneus brevis, and peroneus tertius (if present) are the evertors of the foot. Foot eversion is preserved if only one muscle is used. The disability from loss of foot eversion is minimal.

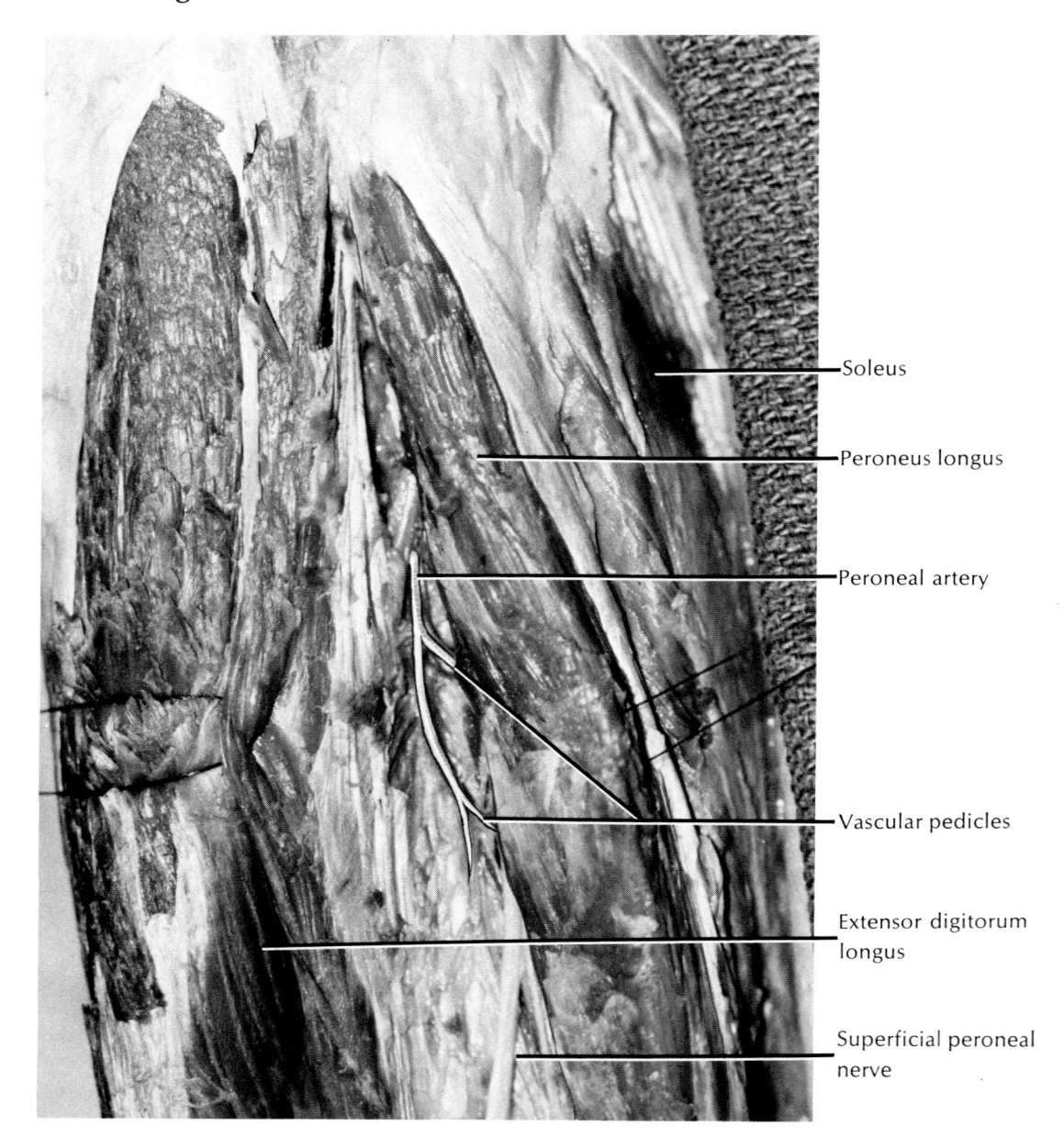

LATERAL LEG: POSTERIOR GROUP

Peroneus longus

- Soleus
- Peroneus longus
- Peroneal artery
- Vascular pedicles
- Extensor digitorum longus
- Superficial peroneal nerve

Blood supply

The peroneal artery, a branch of the popliteal artery, runs anterior to or within the flexor hallucis longus muscle. The peronei lie anterior to the flexor hallucis longus, and muscular branches from the peroneal artery course anteriorly to supply the peroneus longus. These are three or more vessels that enter the muscle proximally.

LATERAL LEG: POSTERIOR GROUP

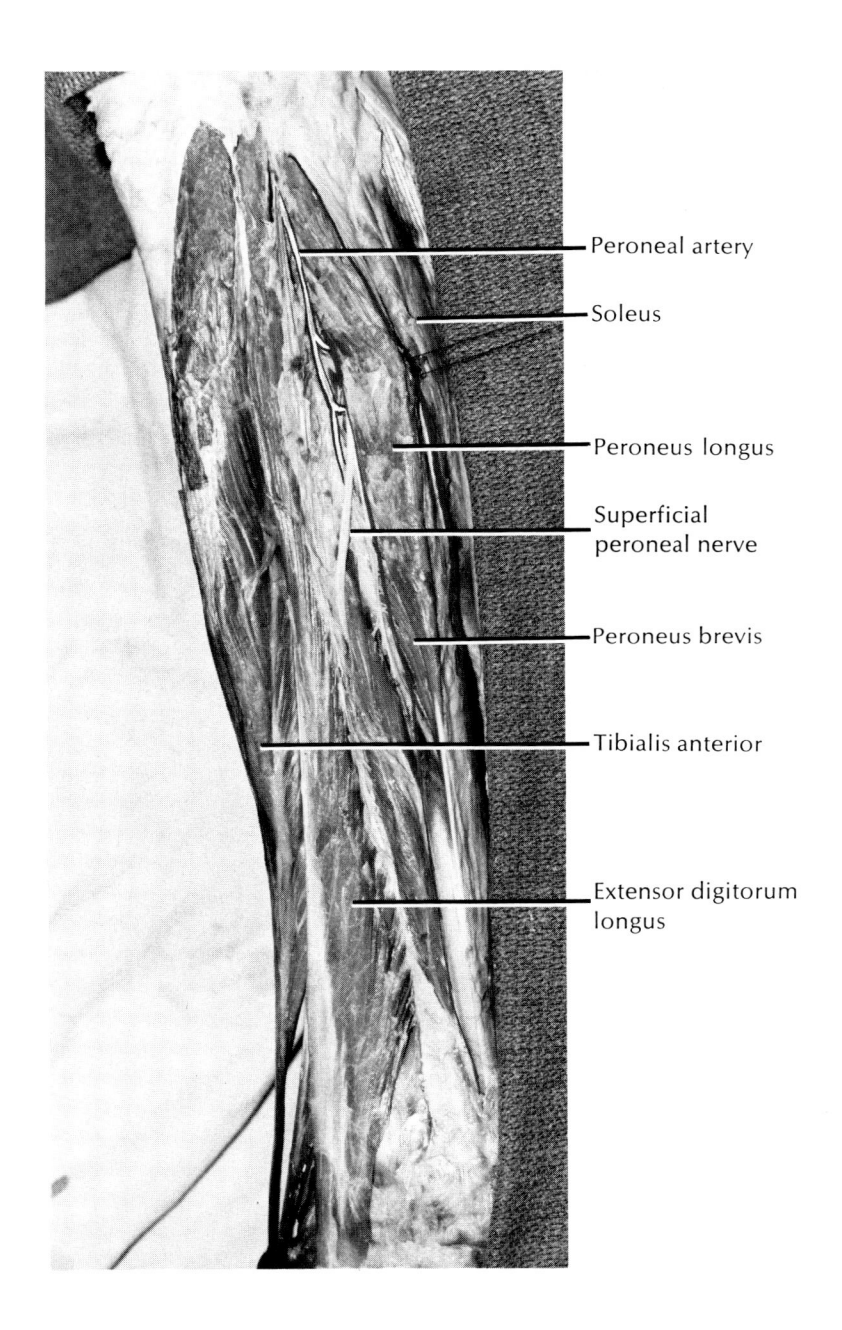

- Peroneal artery
- Soleus
- Peroneus longus
- Superficial peroneal nerve
- Peroneus brevis
- Tibialis anterior
- Extensor digitorum longus

LATERAL LEG: POSTERIOR GROUP

Peroneus longus

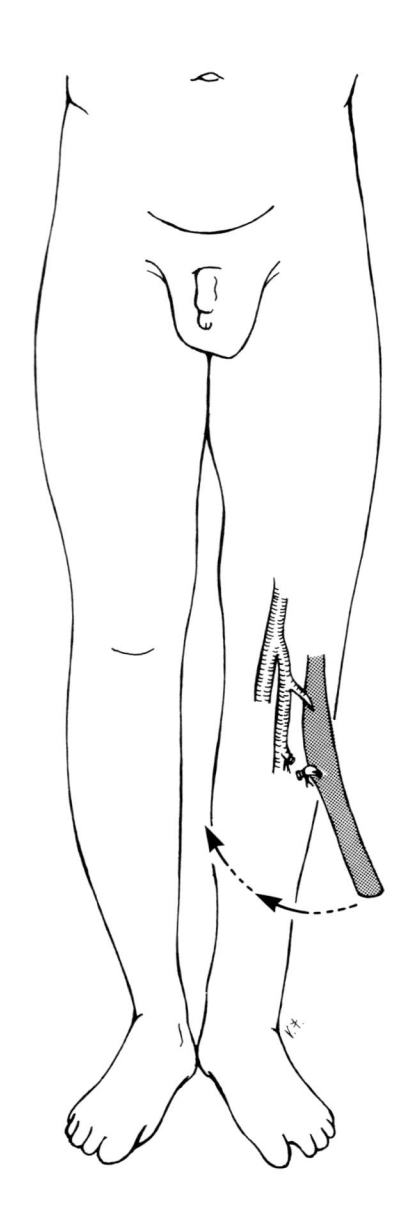

Arc of rotation

The point of rotation of this muscle is proximal. The distal vascular pedicle may safely be divided, then the muscle can be transposed from lateral to medial to cover a small area in the upper part of the middle third of the leg.

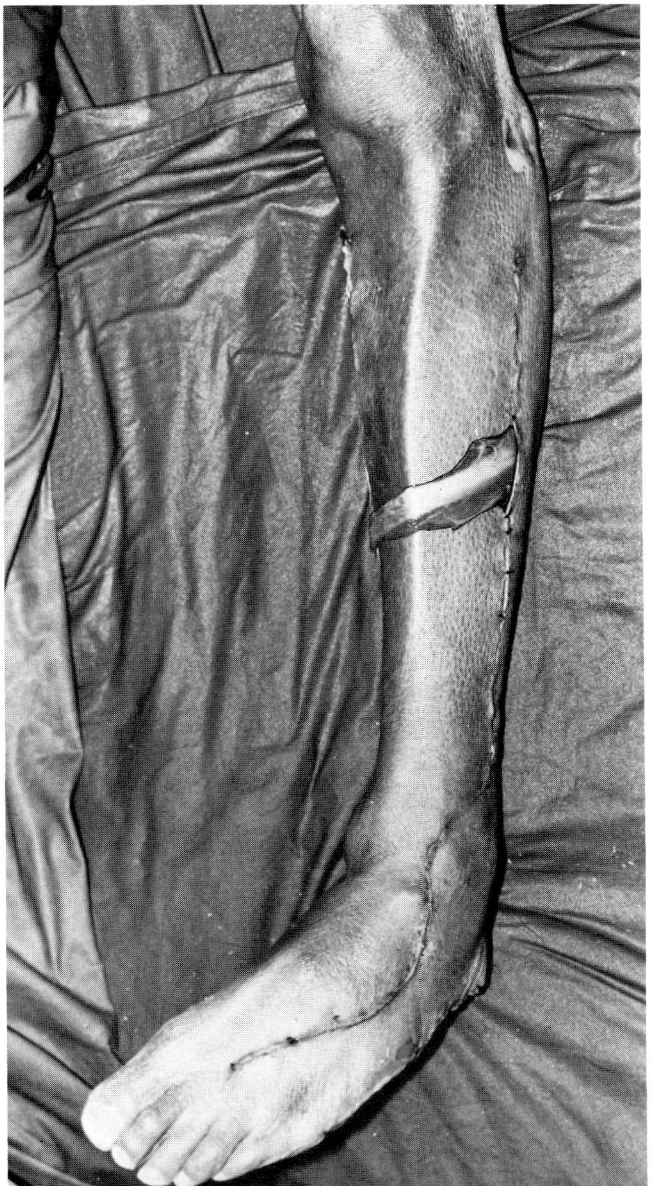

LATERAL LEG: POSTERIOR GROUP

Peroneus longus

Elevation of flap

An incision is made on the anterolateral aspect of the leg (the peroneus longus is the longer and more superficial of the two peroneal muscles), covering almost all but the tendon of the peroneus brevis. The muscle lies between the extensor digitorum longus anteriorly and the flexor hallucis longus and soleus posteriorly. The muscle is identified, and the tendon is divided just above the lateral malleolus. The flap is then elevated upward, exposing the peroneus brevis muscle. If necessary, the most distal vascular pedicle is divided and the muscle transposed from lateral to medial.

Precautions

- This is a small muscle that will only cover a small part of the middle third of the leg, and should only be considered if the soleus is not available.
- If both peroneal muscles are used, foot eversion is lost.

LATERAL LEG: POSTERIOR GROUP

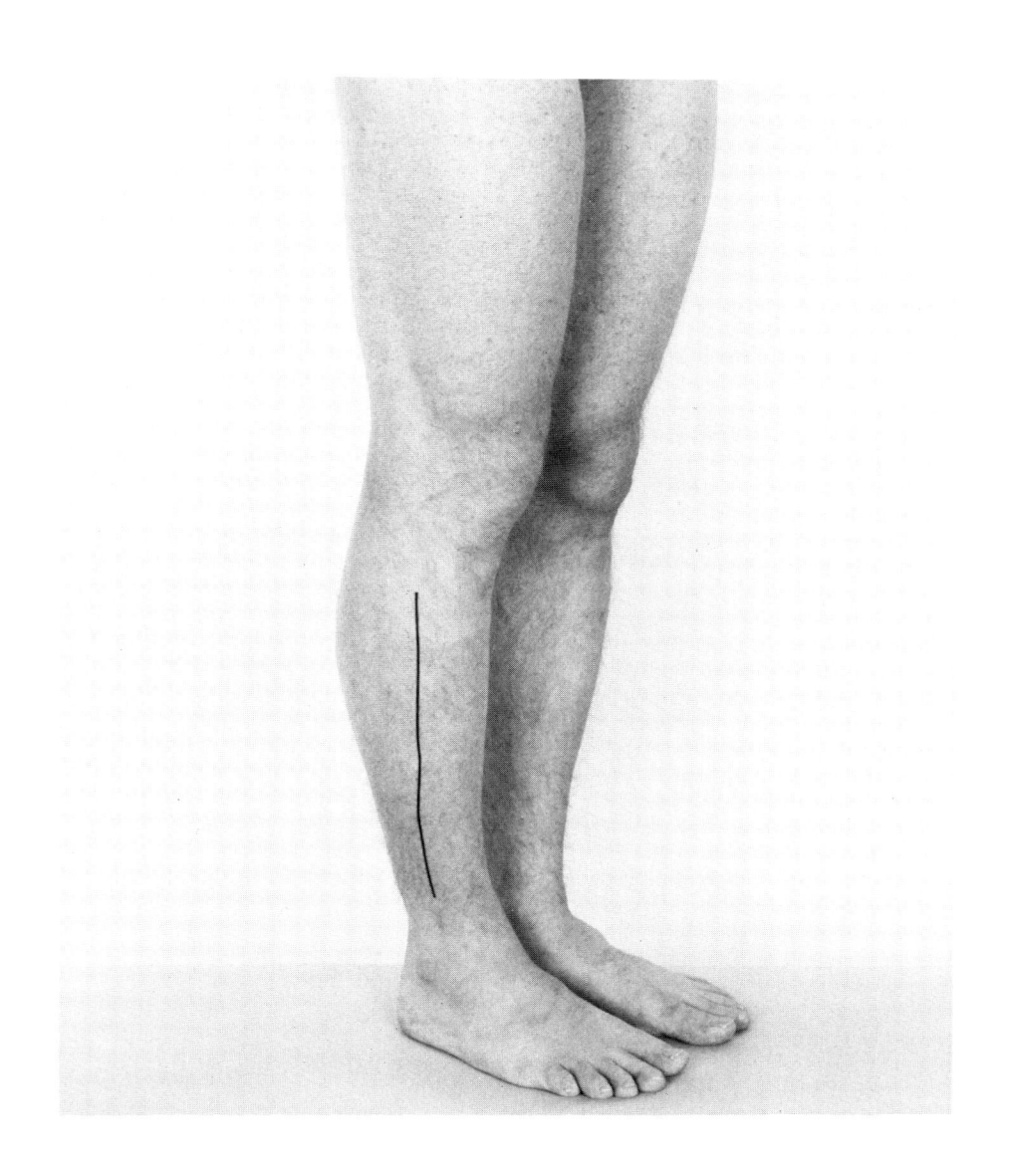

Peroneus brevis

MUSCLE FLAP

Application

Coverage of:

LOWER PORTION OF MIDDLE THIRD OF LEG
LOWER THIRD OF LEG

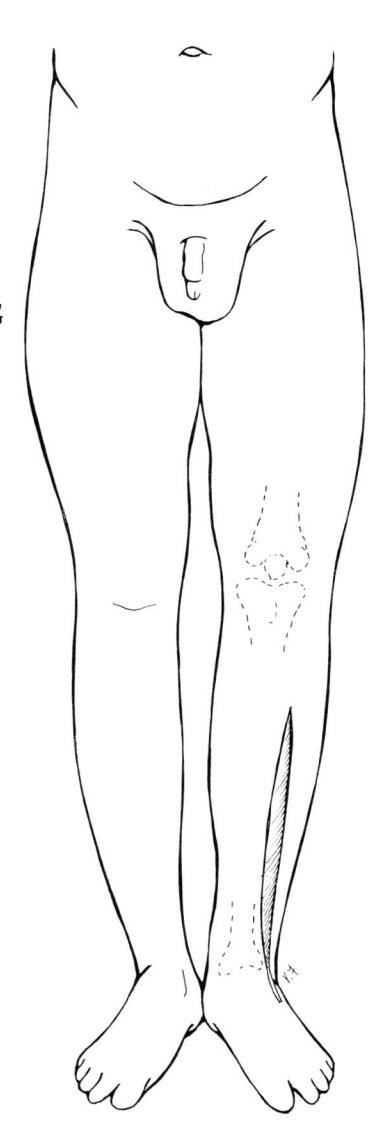

LATERAL LEG: POSTERIOR GROUP

Peroneus brevis

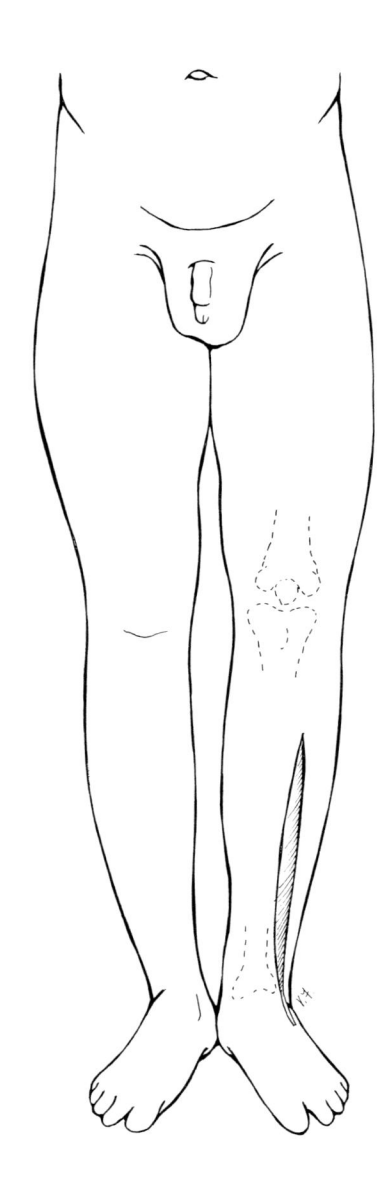

The peroneus brevis is the smaller of the two peroneal muscles and lies below and deep to the peroneus longus. It lies between the extensor digitorum longus anteriorly and the flexor hallucis longus posteriorly.

Origin: Lower third of the lateral fibula below the origin of the peroneus longus.
Insertion: Fifth metatarsal.

LATERAL LEG: POSTERIOR GROUP

Blood supply

— Peroneal artery

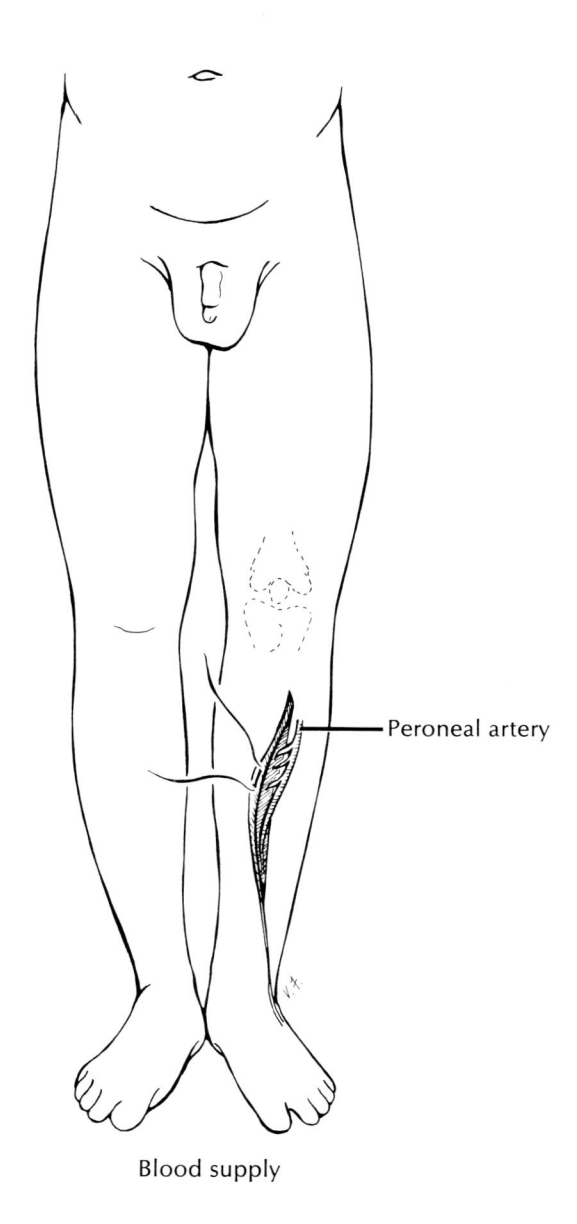

Nerve supply: The muscle is innervated by the superficial peroneal nerve, the motor nerve entering the proximal part of the muscle.

Function: This muscle plantar flexes and everts the foot. If both peroneal muscles are used, foot eversion is lost. Preservation of one muscle will not result in any appreciable loss of foot eversion.

LATERAL LEG: POSTERIOR GROUP

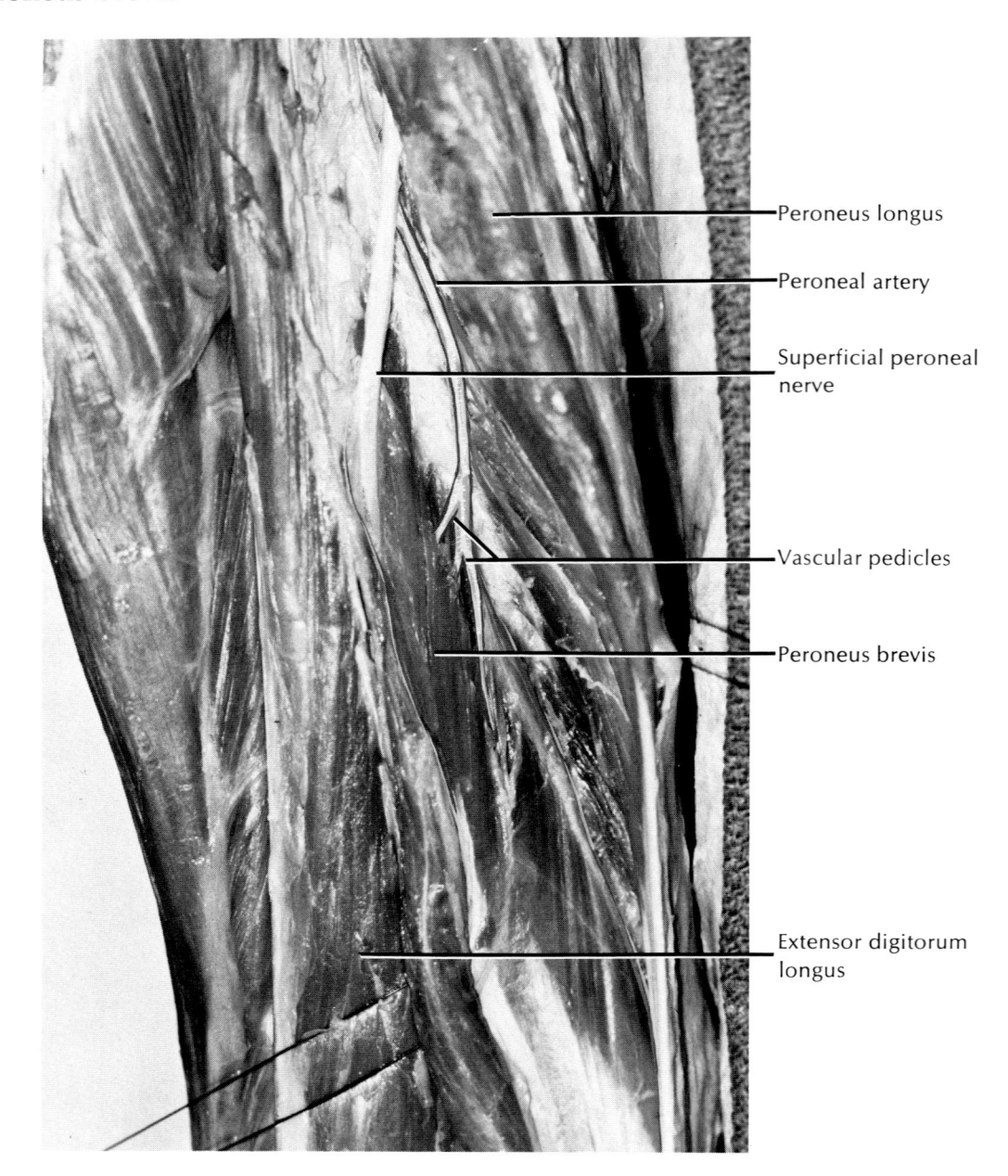

Peroneus brevis

- Peroneus longus
- Peroneal artery
- Superficial peroneal nerve
- Vascular pedicles
- Peroneus brevis
- Extensor digitorum longus

Blood supply

The peroneal artery runs anterior to or within the flexor hallucis longus muscle and sends two or more branches anteriorly to the peroneus brevis. These vascular pedicles enter the proximal part of the muscle.

LATERAL LEG: POSTERIOR GROUP

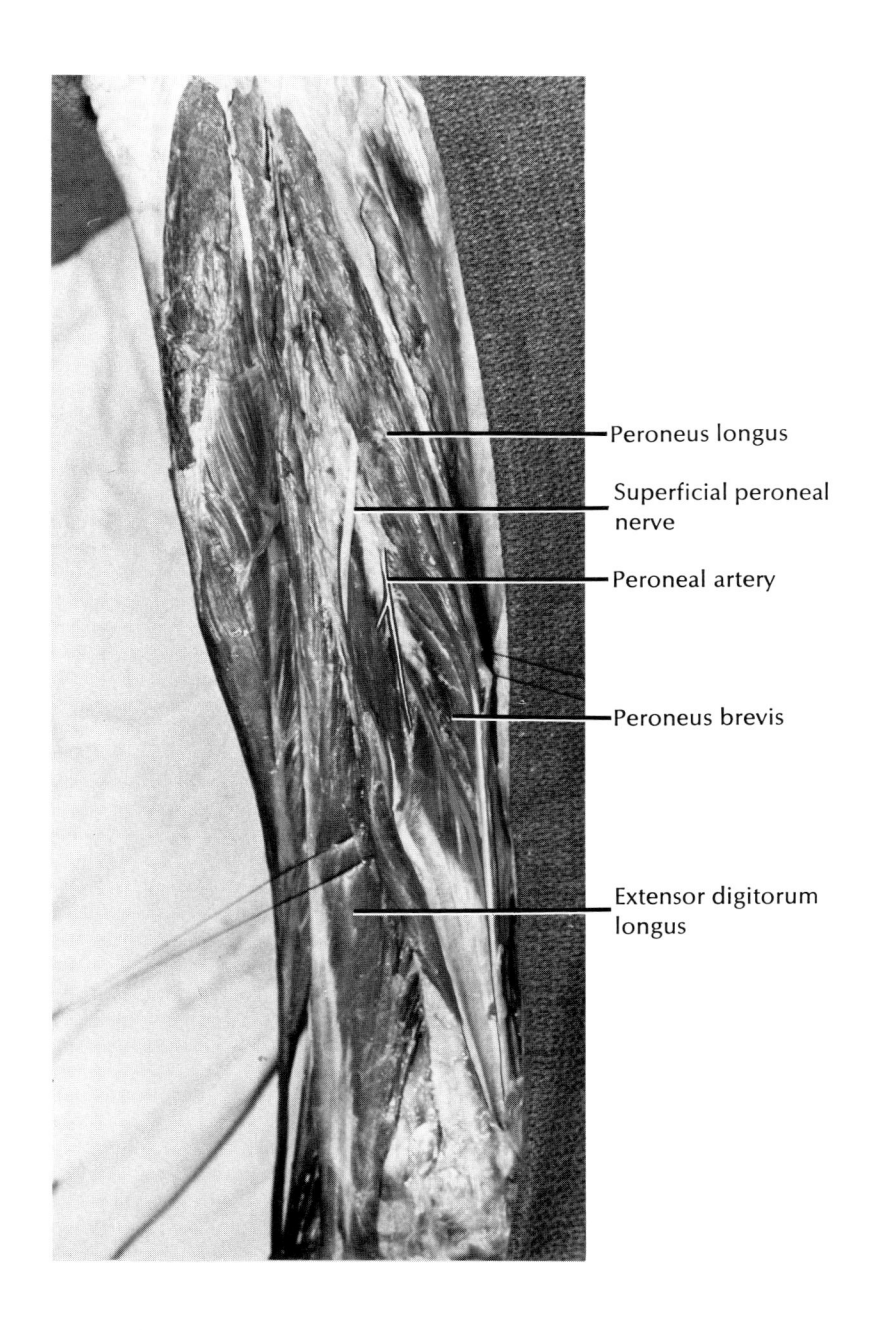

LATERAL LEG: POSTERIOR GROUP

Peroneus brevis

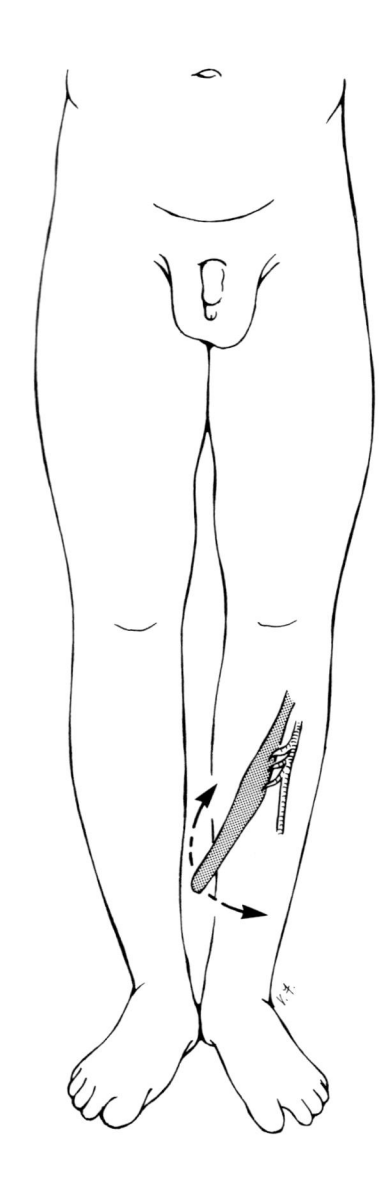

Arc of rotation

The point of rotation of the muscle is the proximal vascular pedicles. The most distal pedicle can be safely divided and the muscle elevated and transposed to cover a small area in the lower part of the middle third of the leg.

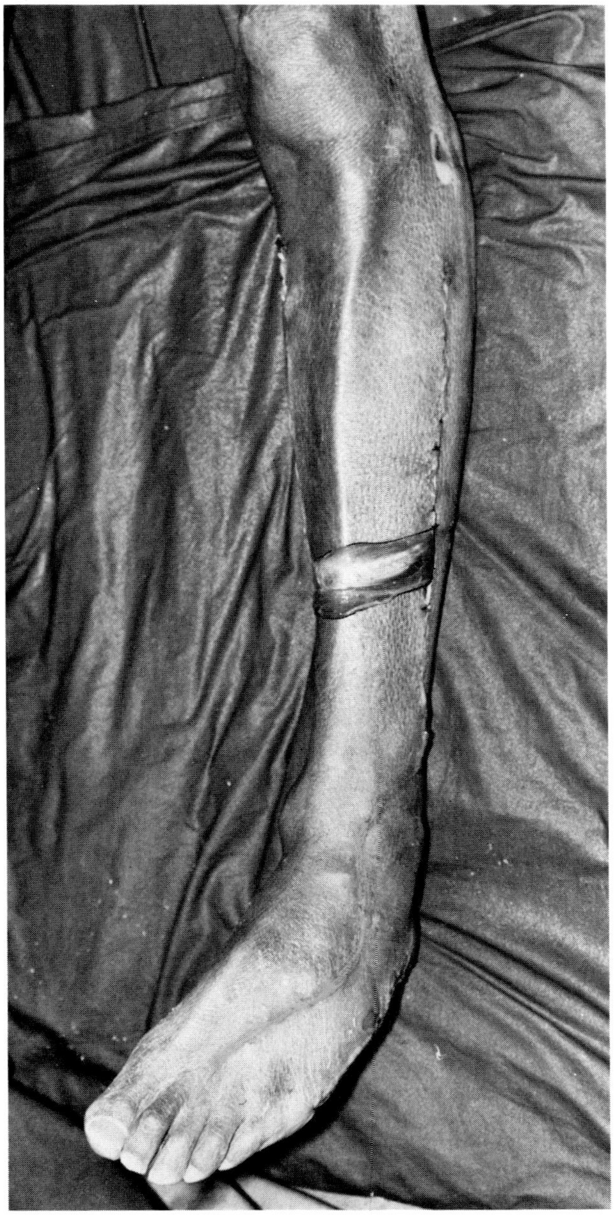

LATERAL LEG: POSTERIOR GROUP

Peroneus brevis

Elevation of flap

An incision on the anterolateral aspect of the leg is made. The peroneus longus is the most superficial muscle in this area and covers all but the tendon of the peroneus brevis. The peroneus brevis is deep to the peroneus longus, with the extensor digitorum longus lying anterior and the flexor hallucis longus posterior to it. The muscle is identified, the tendon is divided, and the muscle is dissected upward. The most distal vascular pedicle may be divided and the muscle transposed from lateral to medial over the extensor digitorum longus and tibialis anterior to cover the tibia.

Precautions

- A small muscle that is not the flap of choice for coverage of the middle third of the leg. Rather, it is an alternative if the soleus is not available.
- Foot eversion is lost if both peroneal muscles are used.

LATERAL LEG: POSTERIOR GROUP

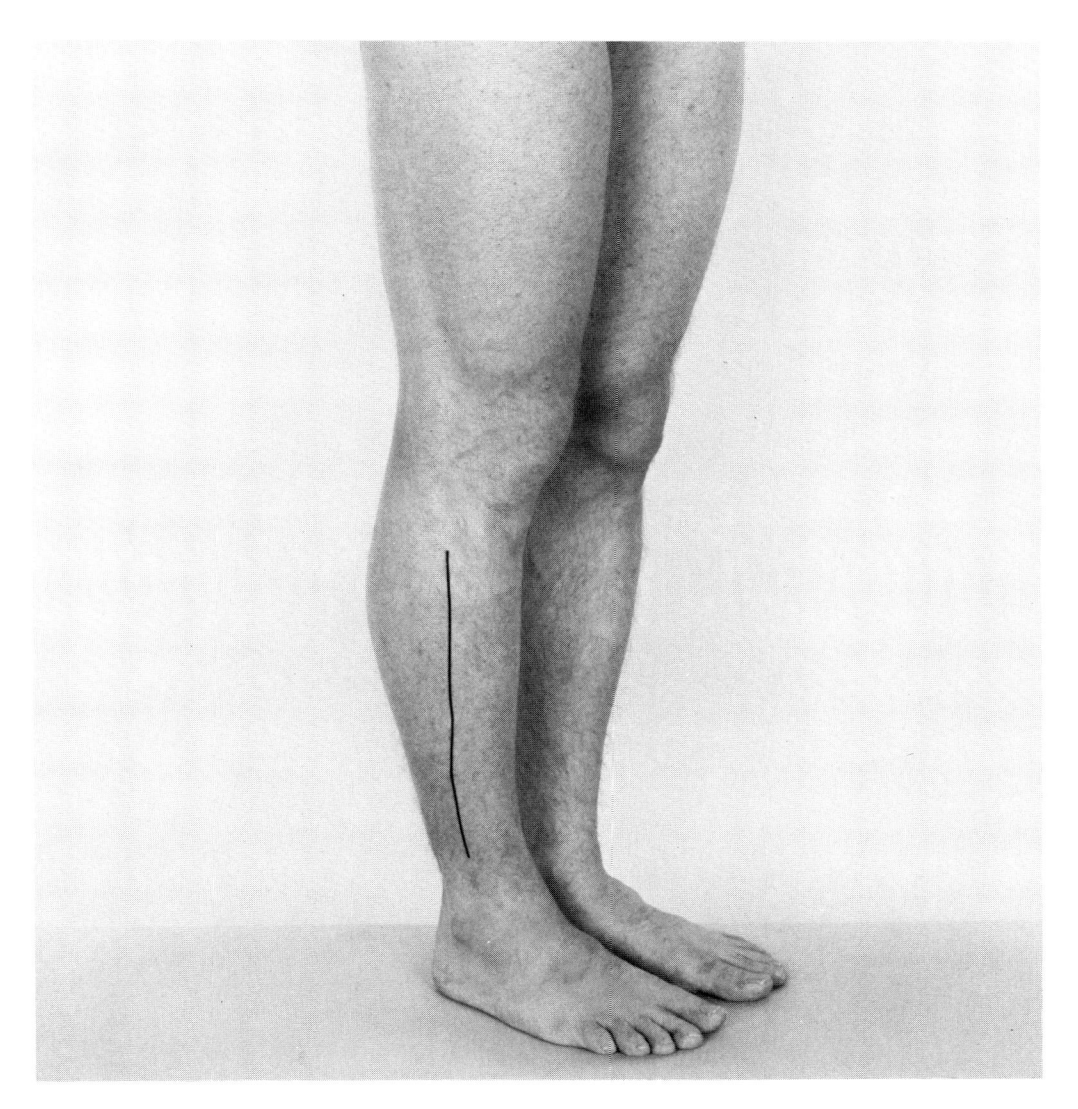

Skin incision

LATERAL LEG: POSTERIOR GROUP

Peroneus brevis

Coverage of exposed fibula and plate

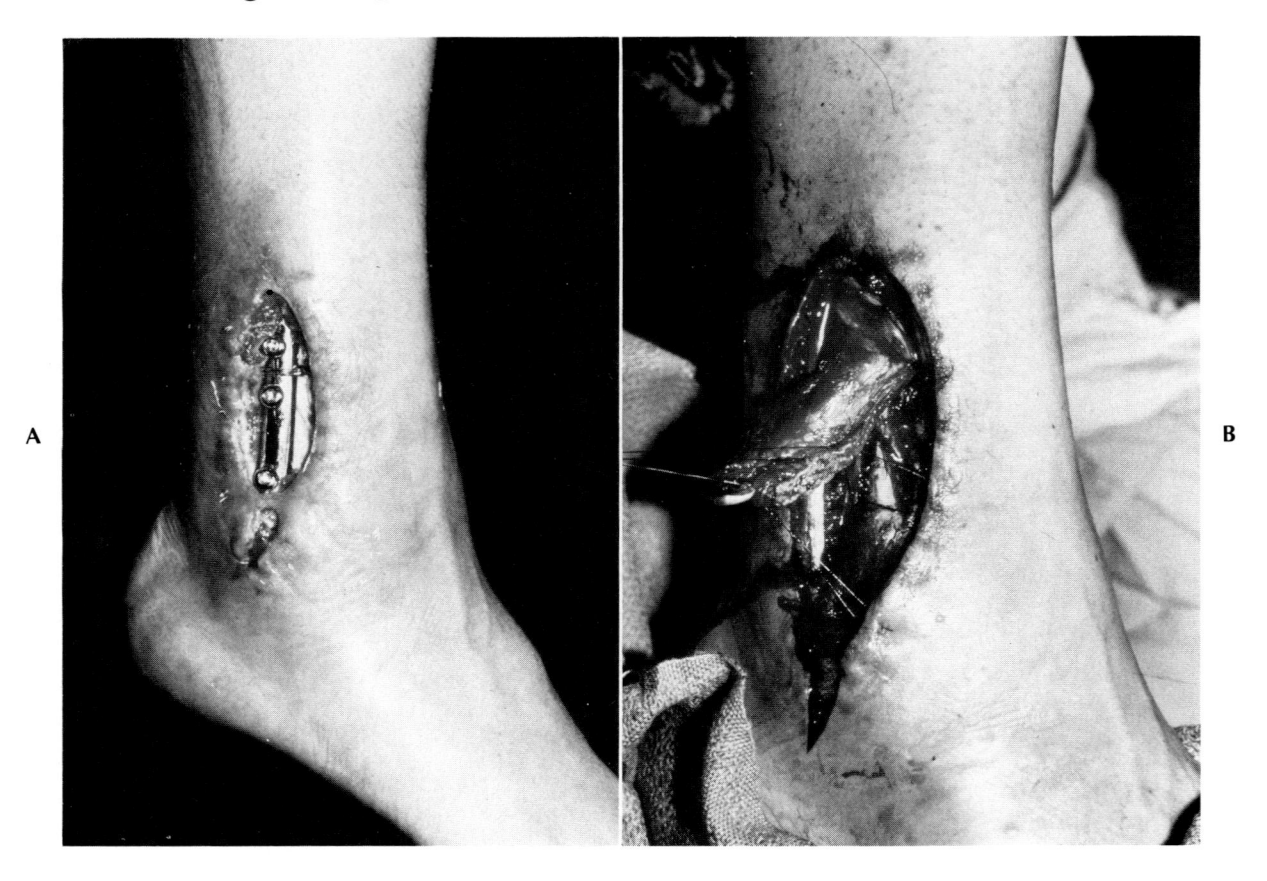

A Exposed fibula and plate.
B Peroneus brevis muscle elevated.

LATERAL LEG: POSTERIOR GROUP

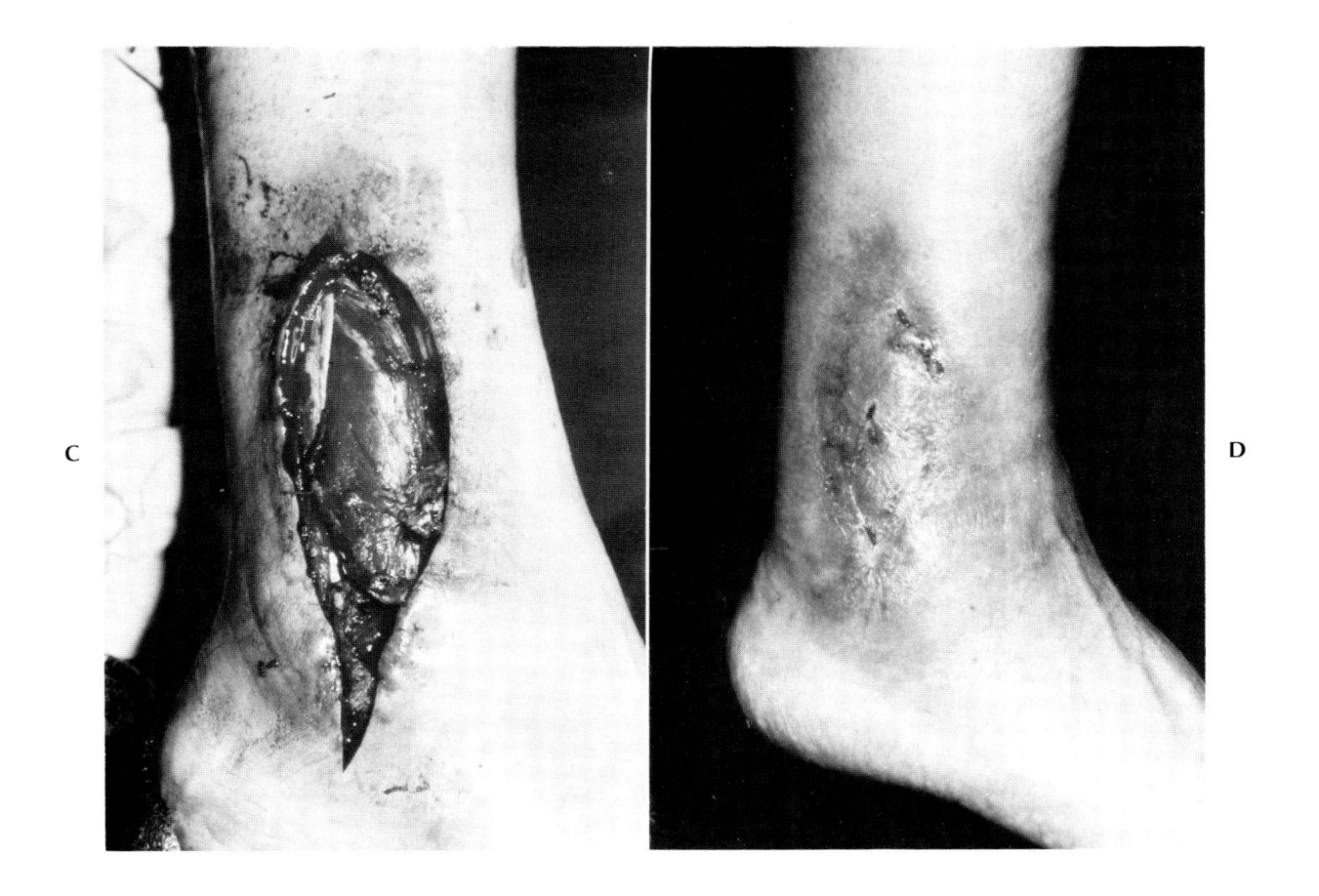

C Muscle transposed over exposed plate.
D Final result. Muscle was covered with split skin. Single-stage coverage of exposed bone and plate.

SECTION FIVE
FOOT

Abductor hallucis

Flexor digitorum brevis

Abductor digiti minimi

Extensor digitorum brevis

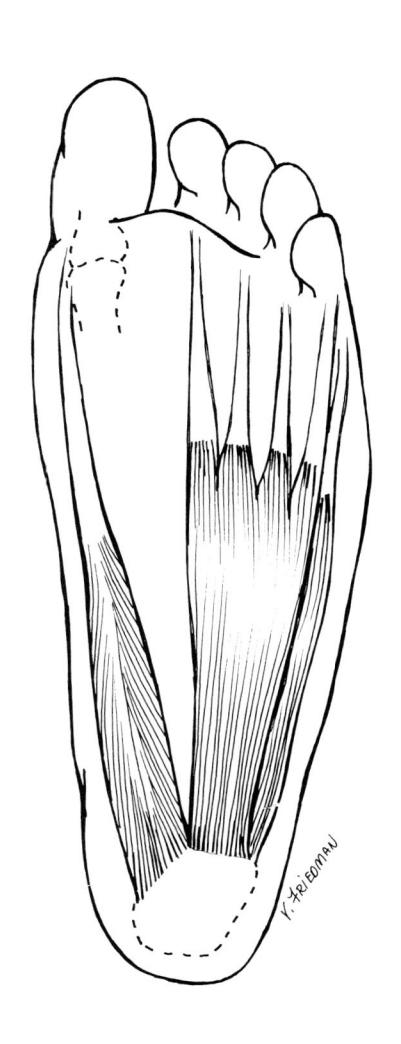

263

FOOT

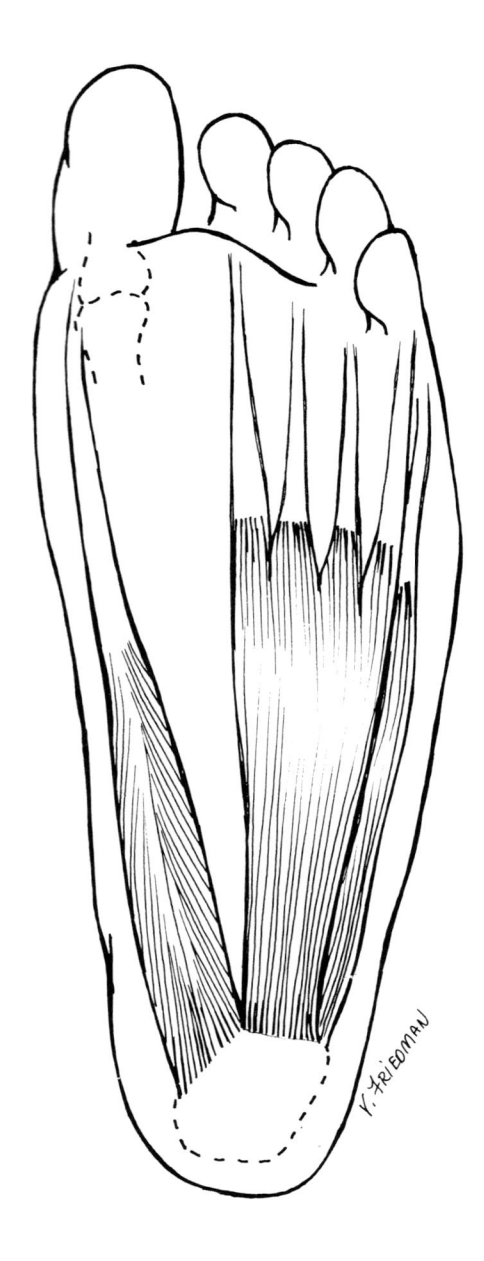

This group of small foot muscles has proximal dominant vascular pedicles, allowing an arc of rotation to the ankle. Thus these muscles can be used to cover defects on the heel and medial and lateral ankle. The extensor digitorum brevis is included in this section because it is a useful muscle for distant free functional muscle transfer.

FOOT

Abductor hallucis

Flexor digitorum brevis

Abductor digiti minimi

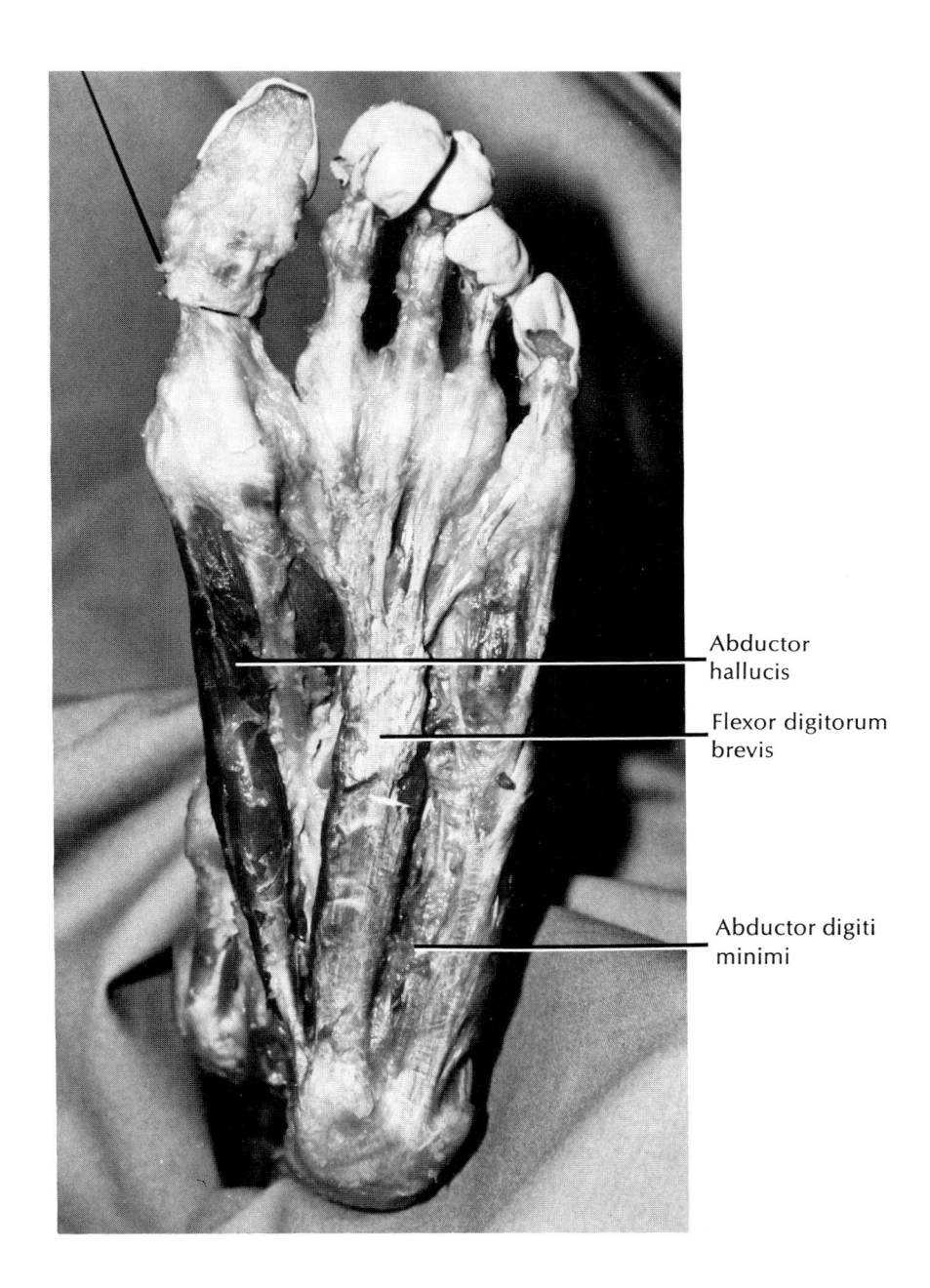

FOOT

Labels on angiogram:
- Dorsalis pedis artery
- Plantar arch
- Posterior tibial artery
- Medial plantar artery
- Lateral plantar artery

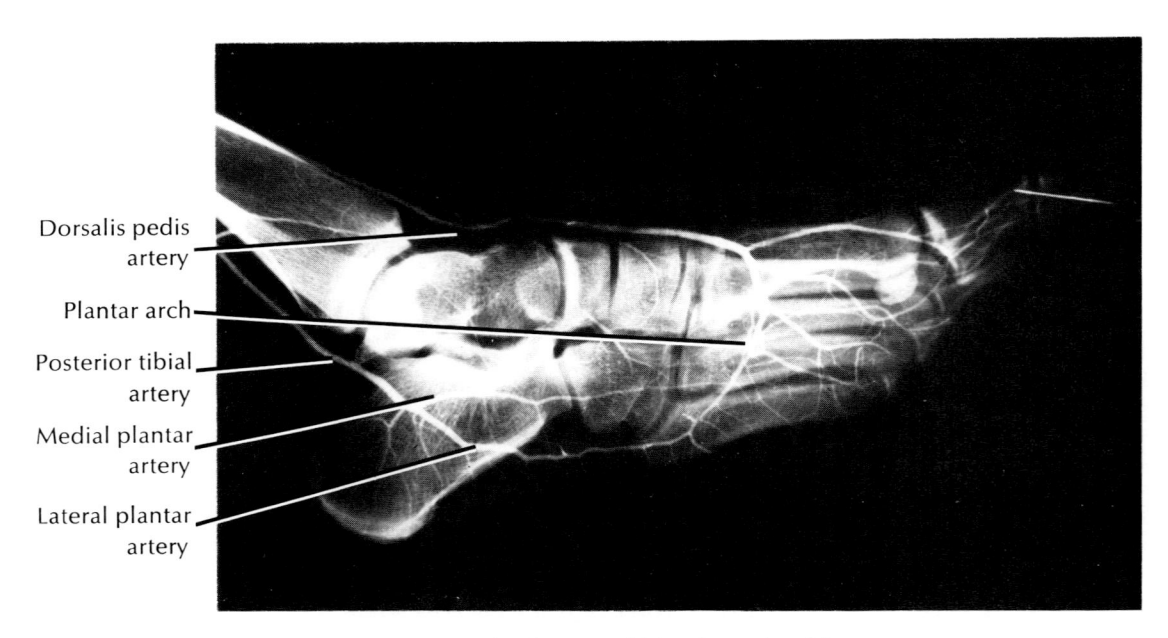

Angiogram of blood supply of foot

Blood supply

The plantar muscles of the foot receive their blood supply from the posterior tibial artery. At the ankle the posterior tibial artery courses between the medial malleolus and the calcaneal tuberosity. Posterior to the origin of the abductor hallucis muscle, the posterior tibial artery divides into the medial and lateral plantar arteries. The medial plantar artery courses between the abductor hallucis muscle and the flexor digitorum brevis. This artery contributes branches into the proximal muscle bellies of the abductor hallucis and flexor digitorum brevis muscles. The artery terminates in the great toe. The lateral plantar artery courses beneath the flexor digitorum brevis muscle and superficial to the quadratus plantae muscle. It courses anteriorly along the abductor digiti minimi muscle. This artery contributes branches to the proximal muscle bellies of the flexor digitorum brevis and abductor digiti minimi muscles. The artery then courses deep to the plantar fascia and superficial to the abductor digiti minimi in the groove between the abductor digiti minimi and the flexor digitorum brevis. The artery then terminates in an oblique course medial to the first interosseous space to unite with the deep plantar branch of the dorsalis pedis artery to form the plantar arch. The plantar arch has perforating and plantar metatarsal branches.

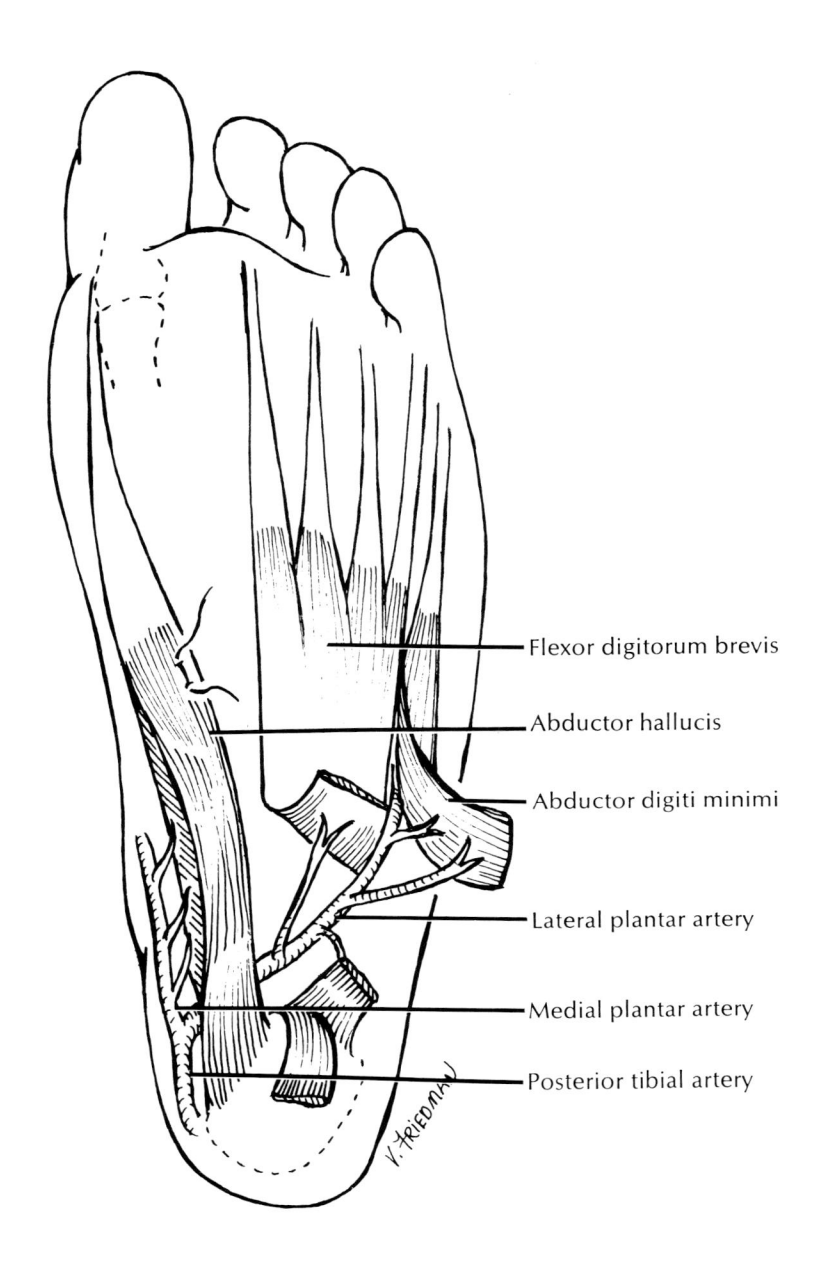

- Flexor digitorum brevis
- Abductor hallucis
- Abductor digiti minimi
- Lateral plantar artery
- Medial plantar artery
- Posterior tibial artery

The dorsalis pedis artery passes beneath the first tendon of the extensor digitorum brevis. This artery has an anteromedial relationship with the tendon of the extensor hallucis longus and an anterolateral relationship with the remaining extensor digitorum brevis muscle bellies. At the level of the navicular bone, the lateral tarsal artery, a branch of the dorsalis pedis artery, supplies the extensor digitorum brevis muscle.

Abductor hallucis

MUSCLE FLAP

Application

   Coverage of:
     MEDIAL ANKLE

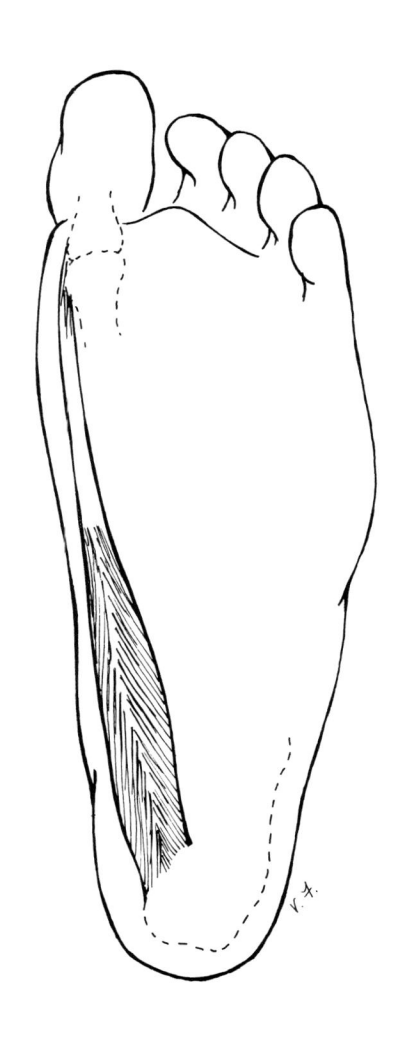

FOOT

Abductor hallucis

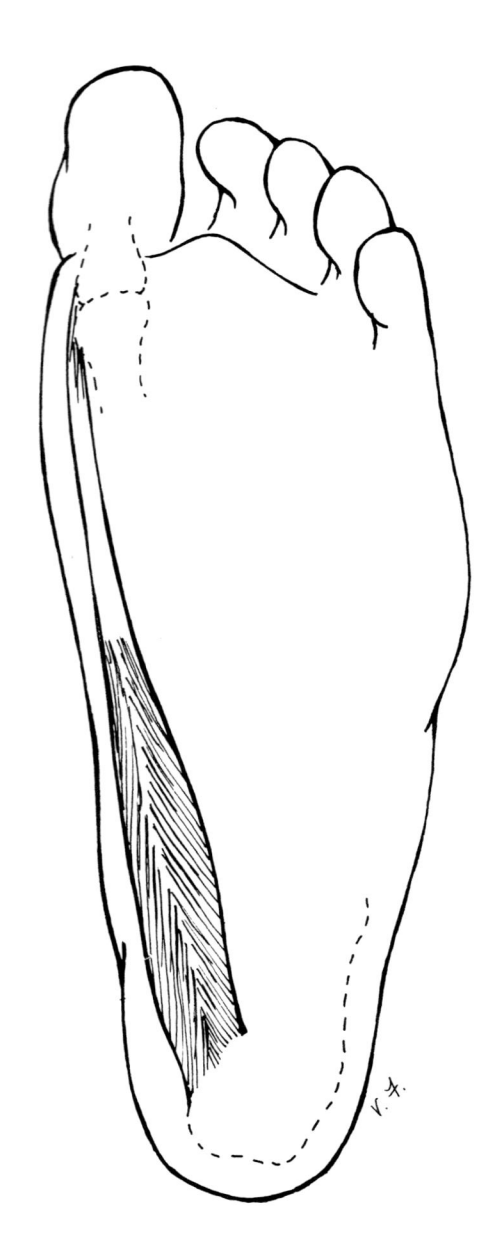

The abductor hallucis is a long, thin muscle located in the medial foot. It lies between the flexor digitorum brevis muscle and flexor hallucis longus tendon laterally and the bones forming the medial border of the foot.

Origin: Calcaneus, plantar aponeurosis, adjacent intermuscular septum.
Insertion: Medial base of proximal phalanx of the great toe.

FOOT

Medial plantar artery

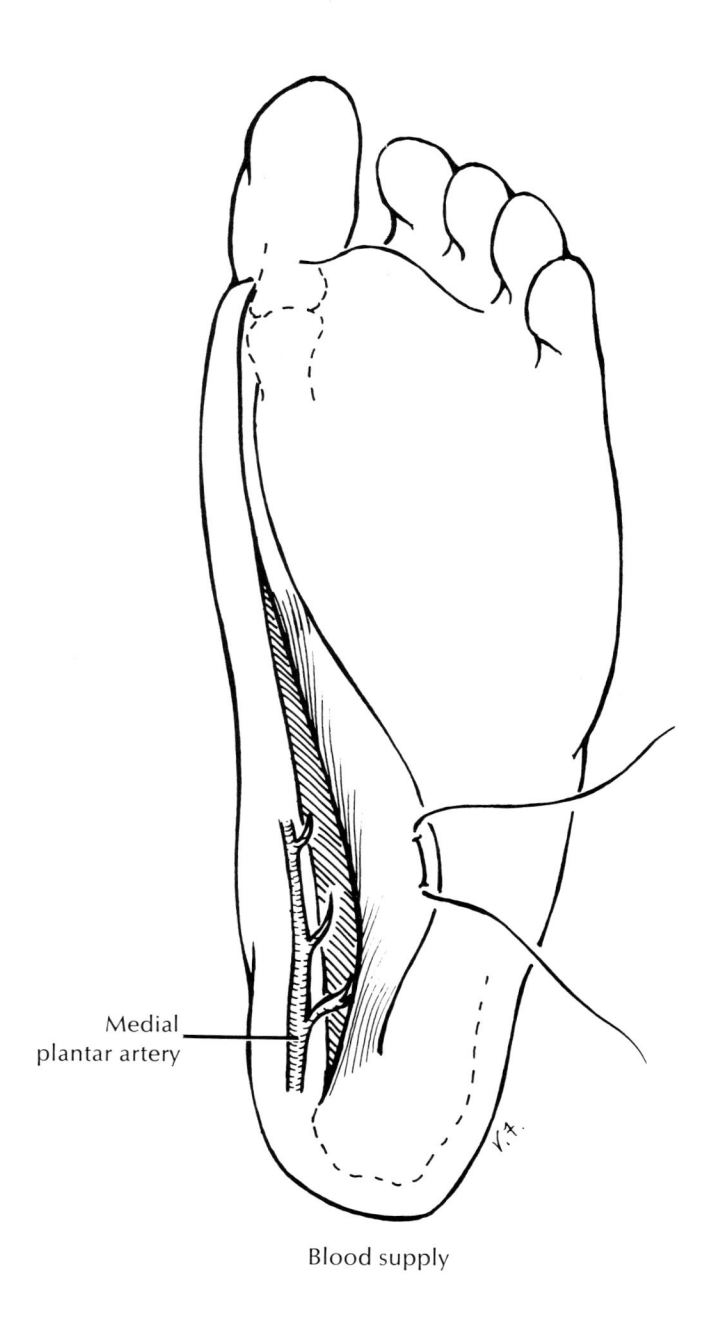

Blood supply

Nerve supply:
  Motor — Branch of medial plantar nerve. The motor branch enters the muscle proximally with the dominant vascular pedicles.
Function: This is an expendable muscle. The function of great toe abduction is lost following transposition of this muscle.

FOOT

Abductor hallucis

Medial plantar artery

Abductor hallucis

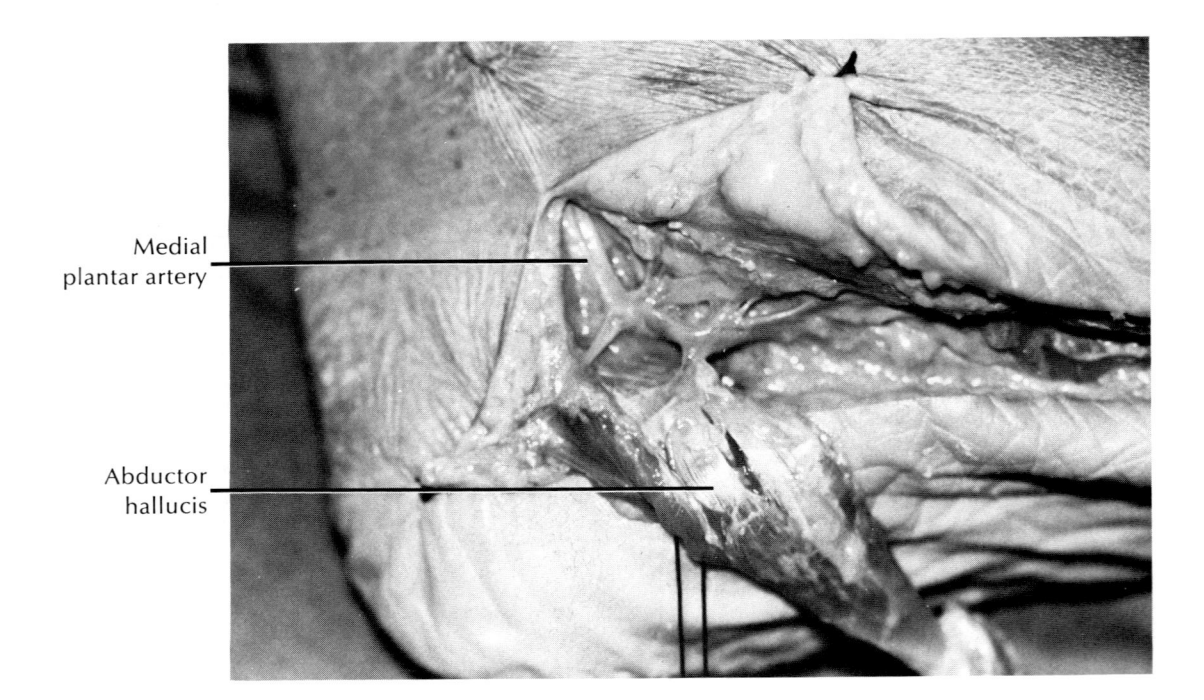

Blood supply

The medial plantar artery supplies three or four pedicles to the posterior muscle belly as it courses between the abductor hallucis muscle and the flexor digitorum brevis and flexor hallucis brevis muscles.

Medial plantar artery

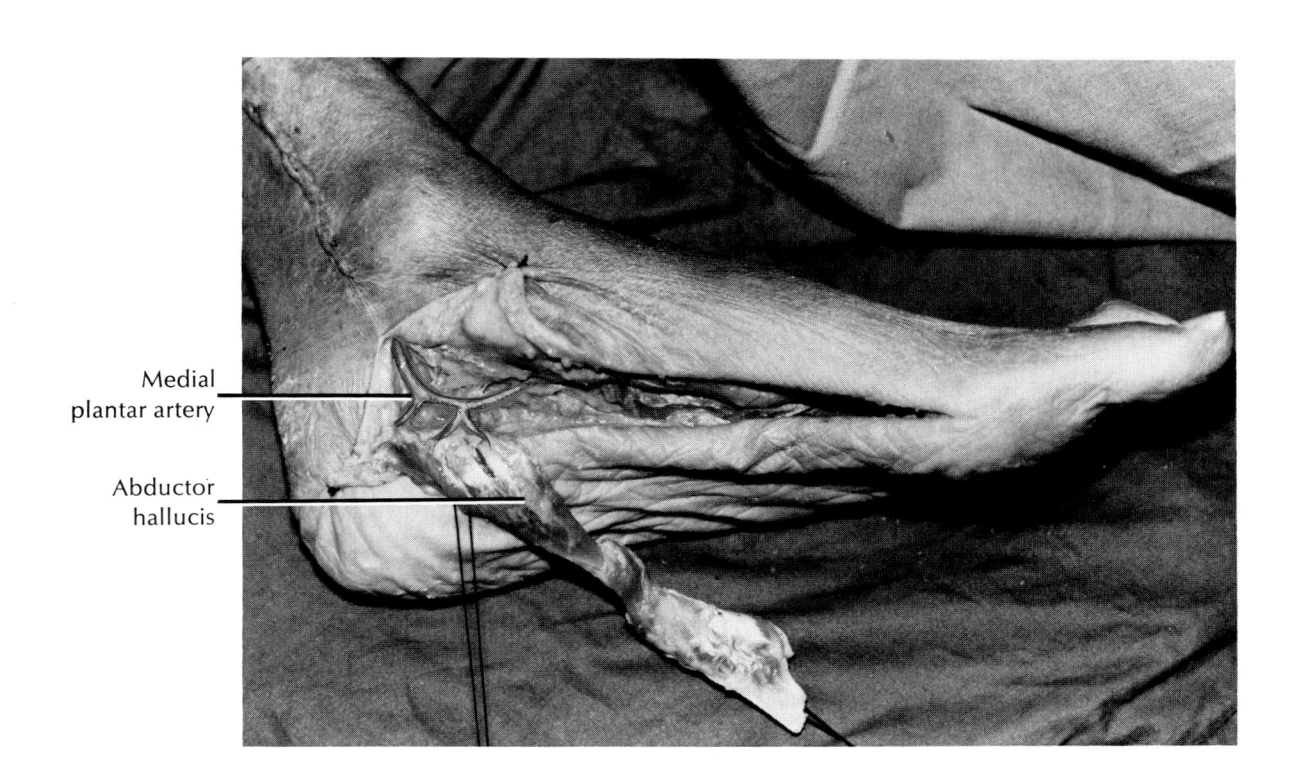

Abductor hallucis

FOOT

Abductor hallucis

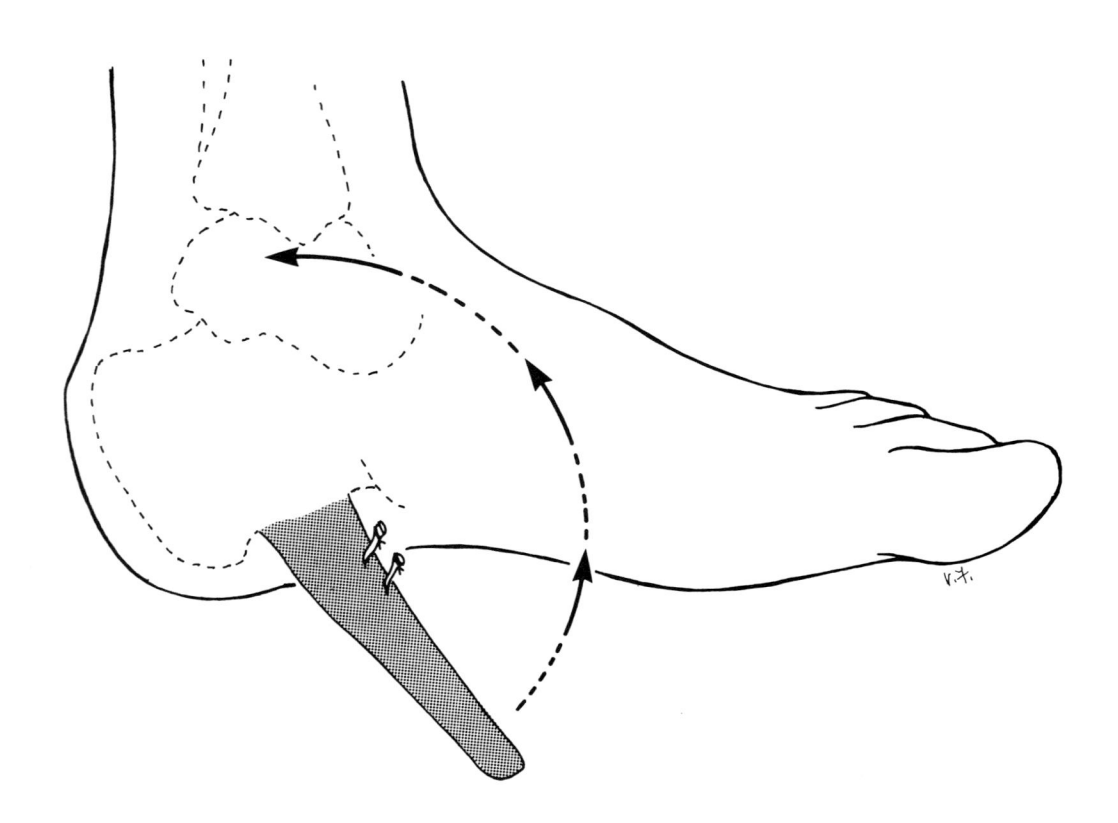

Arc of rotation

This muscle has a superior arc of rotation that will reach the medial malleolus. The muscle will cover defects in the medial ankle and medial foot.

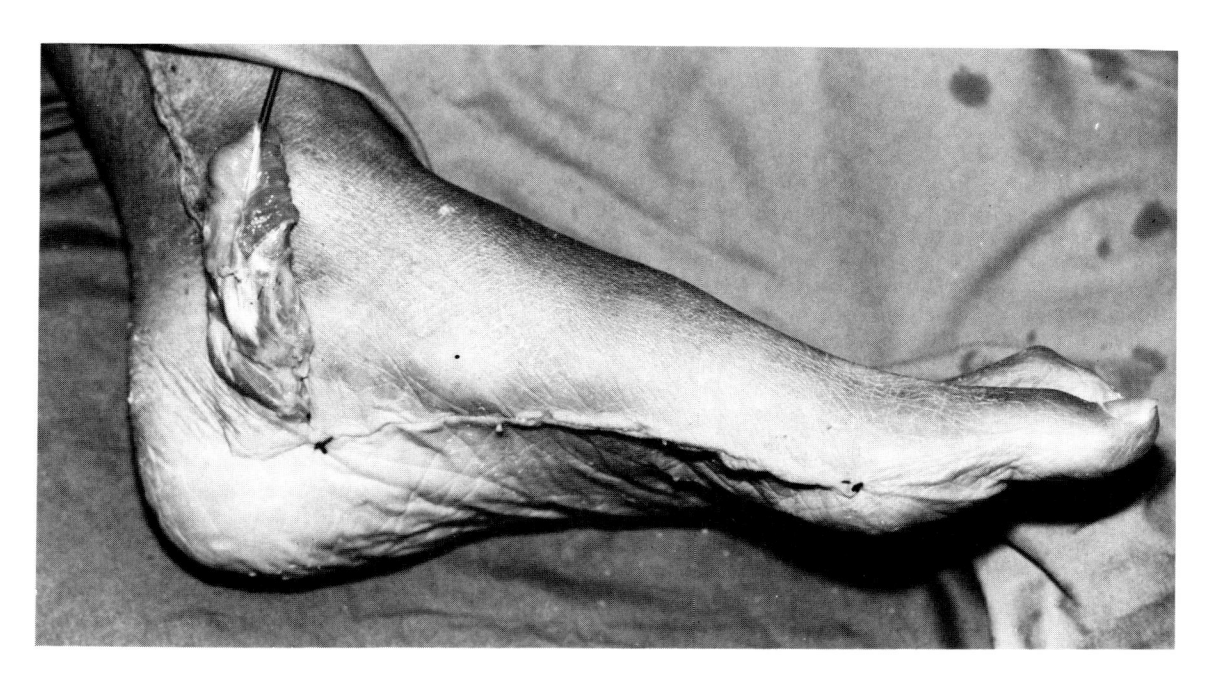

Coverage of medial malleolus

FOOT

Abductor hallucis

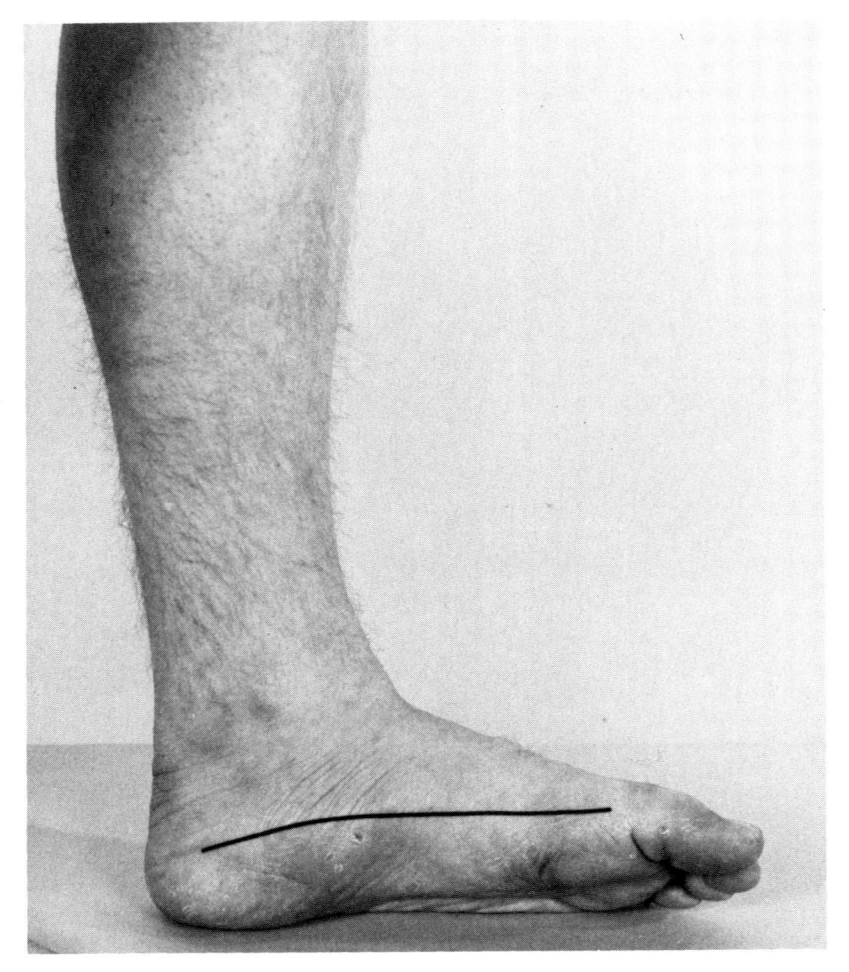

Skin incision

Elevation of flap

The muscle is approached through a medial foot incision. The muscle is identified distally at the level of the medial first metatarsal bone. The tendon is divided and the muscle retracted medially. The distal pedicles to the muscle from the adjacent medial plantar artery are divided to the level of the distal medial malleolus approximately 4 cm from the posterior aspect of the calcaneus. At this level the proximal two or three pedicles to the muscle are mobilized and preserved to allow an adequate safe arc of rotation. The donor defect can be closed directly.

The muscle is generally not used as a musculocutaneous unit, since the donor defect would require skin grafts, leaving an unacceptable defect.

Precautions

- The tendon of the flexor hallucis longus muscle is anterolateral to the muscle and should be preserved.
- The sensory branch of the medial plantar nerve to the great toe has a close relationship to the tendon of the abductor hallucis muscle.

Flexor digitorum brevis

MUSCLE FLAP
FREE FLAP

Applications

Coverage of:

HEEL

Free functional muscle transfer

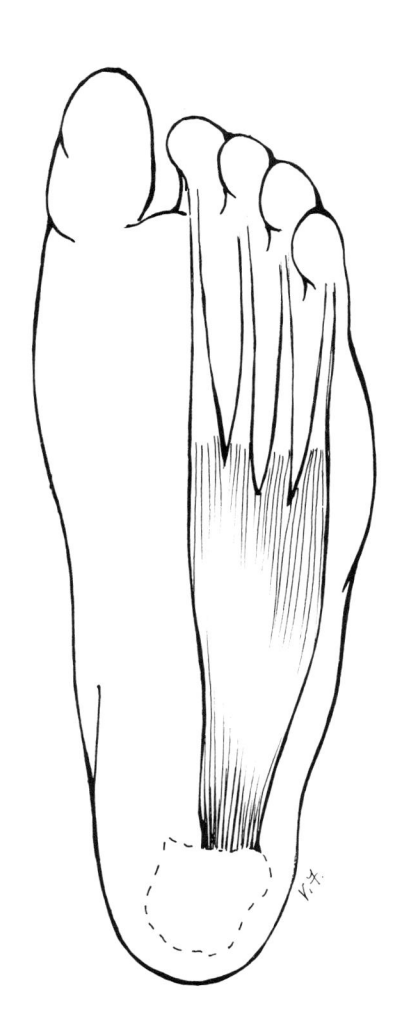

FOOT

Flexor digitorum brevis

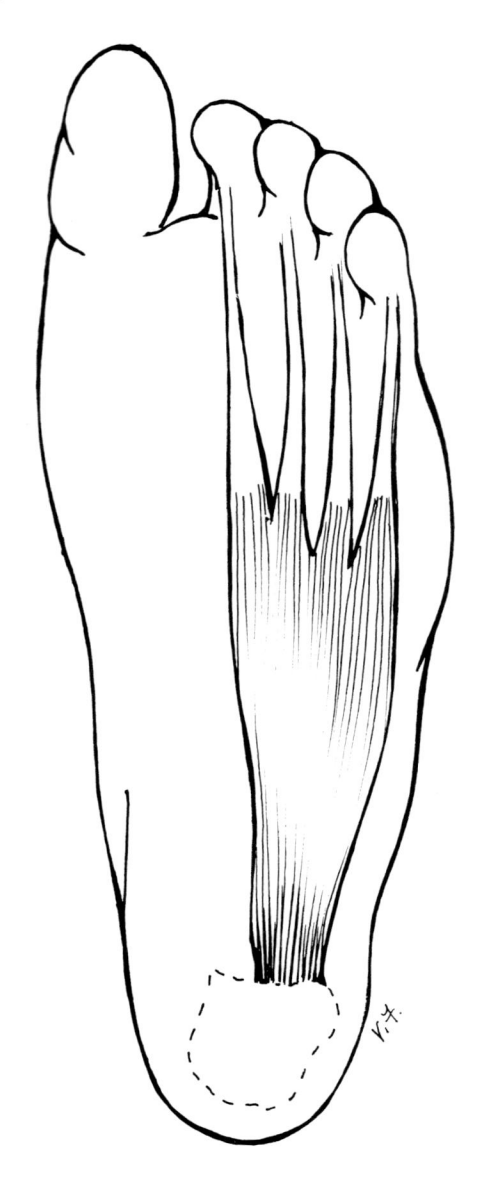

The flexor digitorum brevis is located in the central plantar aspect of the foot. The muscle is deep to the plantar aponeurosis and superficial to the quadratus plantar muscle. It is bordered by the abductor hallucis medially and the abductor digiti minimi laterally.

Origin: Medial process of calcaneus, plantar aponeurosis, intermuscular septa.
Insertion: Middle phalanx of the second, third, fourth, and fifth toes.

FOOT

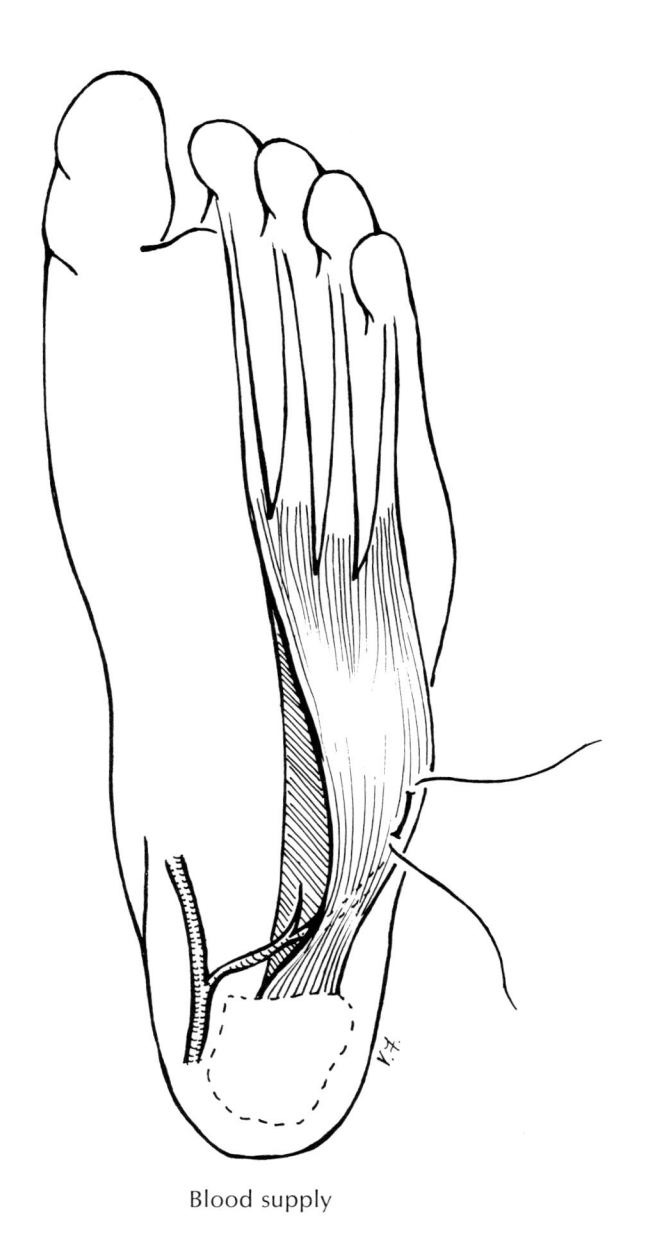

Blood supply

Nerve supply:
    Motor—Branch of medial plantar nerve. This motor nerve branch enters the proximal posteromedial muscle belly adjacent to the vascular pedicles.

Function: This is an expendable muscle. The function of toe flexion is maintained by the flexor digitorum longus muscle.

FOOT

Flexor digitorum brevis

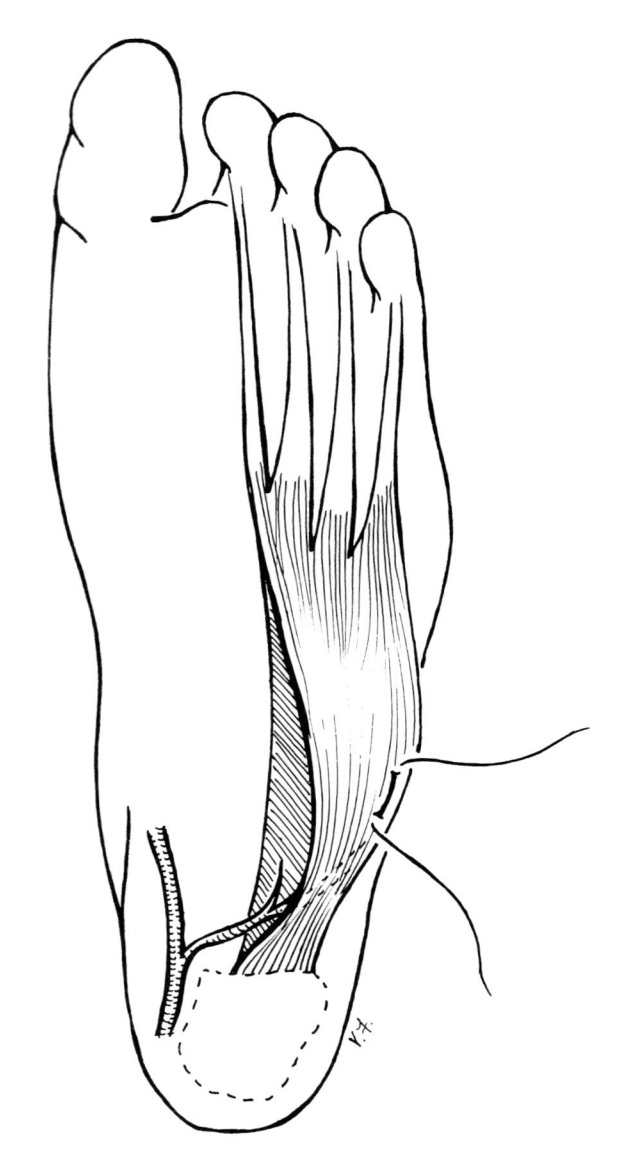

Blood supply

The muscle receives its blood supply from both the medial and lateral plantar arteries. The medial plantar artery gives pedicles to the medial mid muscle belly. The lateral plantar artery courses between the flexor digitorum brevis and the quadratus plantae muscles. In the region of the proximal muscle belly, the lateral plantar artery has pedicles both to the flexor digitorum brevis and to the anterior quadratus plantae.

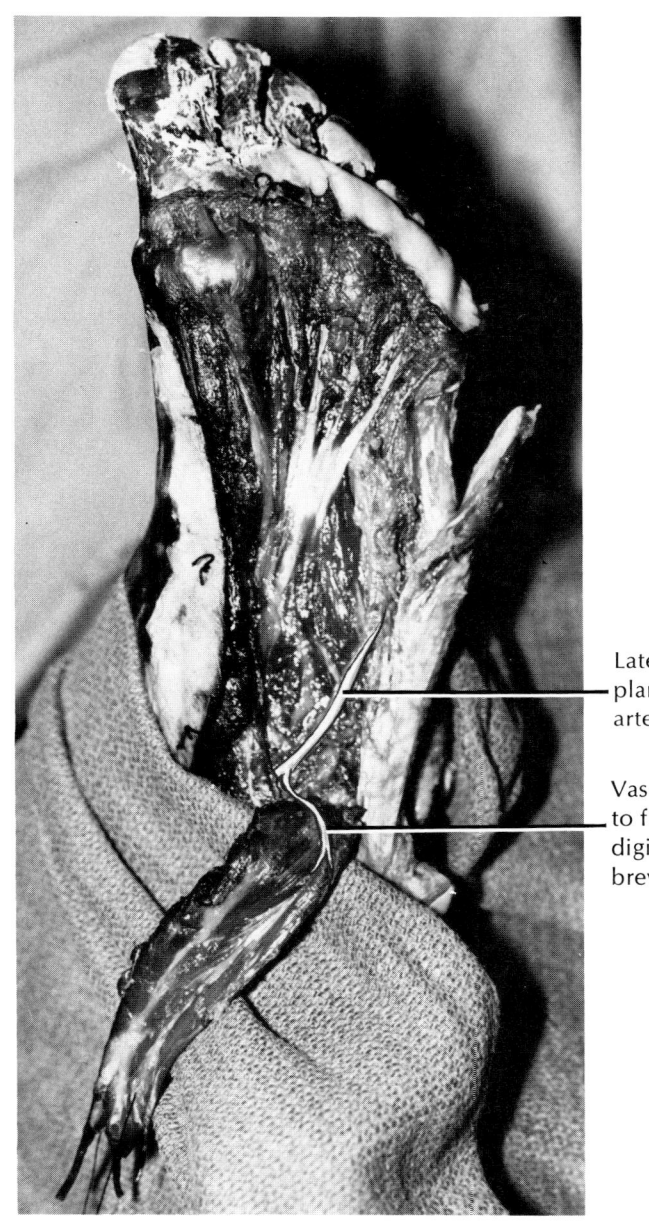

FOOT

Lateral plantar artery

Vascular pedicle to flexor digitorum brevis

FOOT

Flexor digitorum brevis

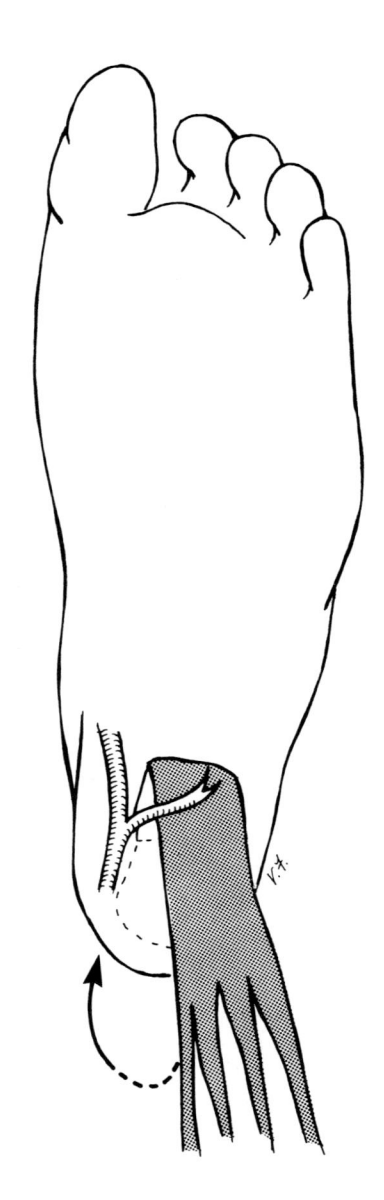

Arc of rotation

After division of the tendons of the flexor digitorum brevis and the distal pedicles to the muscle from the medial plantar artery, this muscle has a posterior arc of rotation. The muscle transposition arc will allow coverage of heel defects with exposed calcaneus.

FOOT

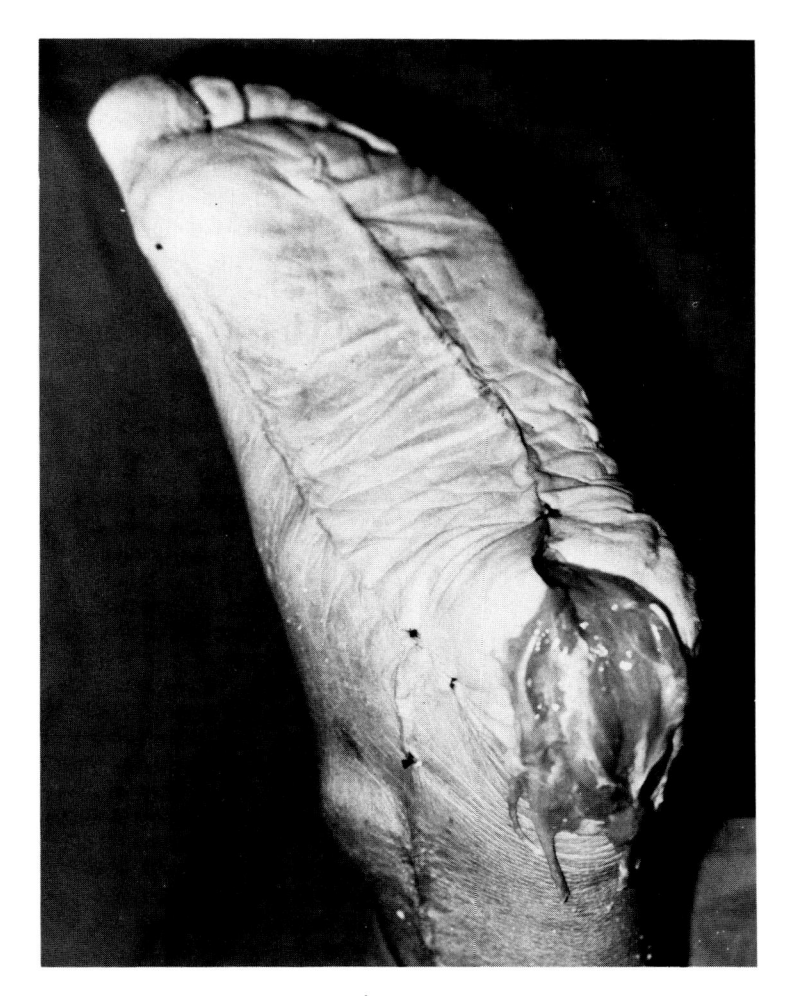

Heel coverage

FOOT

Flexor digitorum brevis

Elevation of flap

This muscle is located by a midline incision on the plantar aspect of the foot. The plantar aponeurosis is also divided in the midline of the foot and reflected medially and laterally. The tendons of the flexor digitorum brevis are located and divided. The muscle is then elevated toward its origin. During this blunt dissection, medial pedicles from the medial plantar artery must be divided. The lateral plantar artery is visualized and reflected posteriorly with the muscle flap at the border of the calcaneus. The lateral plantar artery does not need to be divided, as it extends laterally to the abductor digiti minimi muscle. Since the muscle is generally used to cover exposed calcaneus, it is helpful to lower the posterior projection of the calcaneus with a bony excision using an osteotome. This increases the posterior reach of the flexor digitorum brevis muscle flap and allows the thicker portion of the muscle to be located in the weight-bearing area. The tendons are sutured to the Achilles tendon or adjacent soft tissue to maintain flap transposition. A suction drainage catheter is left in the donor area, which is closed primarily.

Although a musculocutaneous unit is feasible, the donor defect would not close primarily and may not be acceptable for weight bearing.

This muscle unit may be transferrred to a distant site as a functional motor unit. The lateral plantar artery can be taken with the unit without vascular compromise to the foot.

Precautions

- The tendons of the flexor digitorum brevis must be distinguished from the tendons of the flexor digitorum longus during flap elevation.
- The medial plantar nerve is located deep to the muscle belly and should be preserved.
- If the calcaneus bone is not lowered by partial excision, the muscle will not reach the posterior aspect of the heel.

FOOT

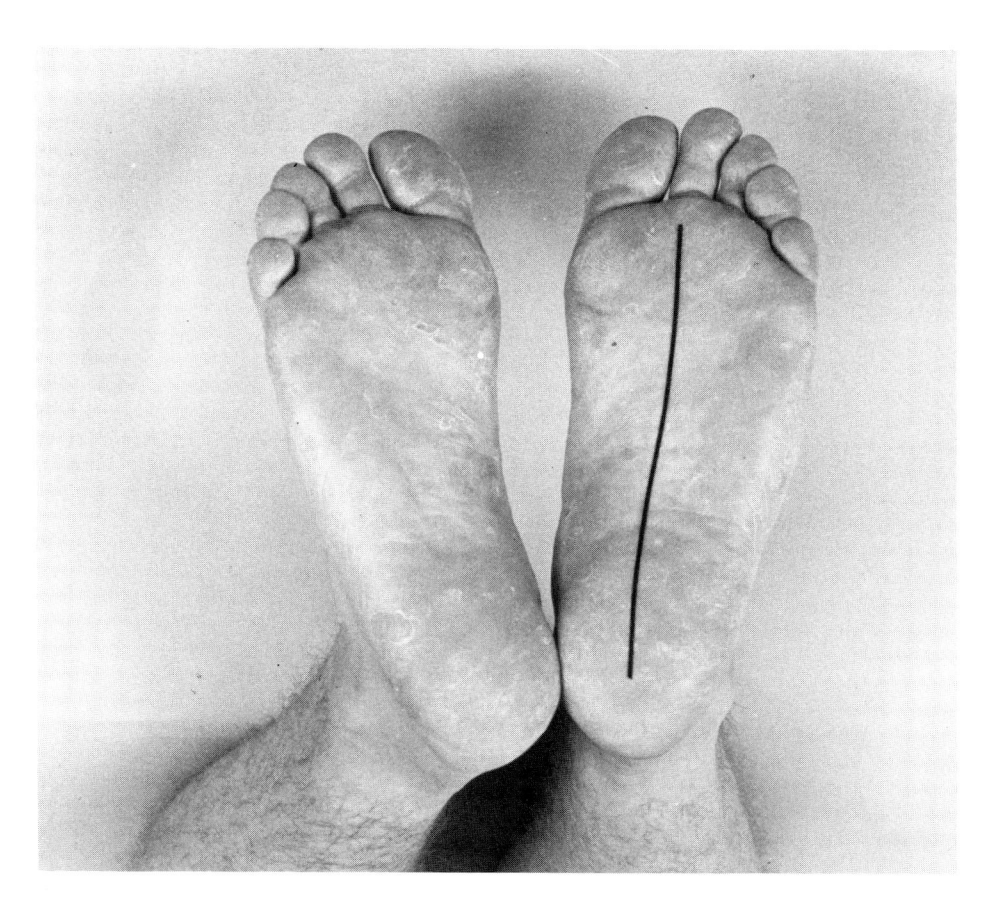

Skin incision

FOOT

Flexor digitorum brevis

Coverage of heel

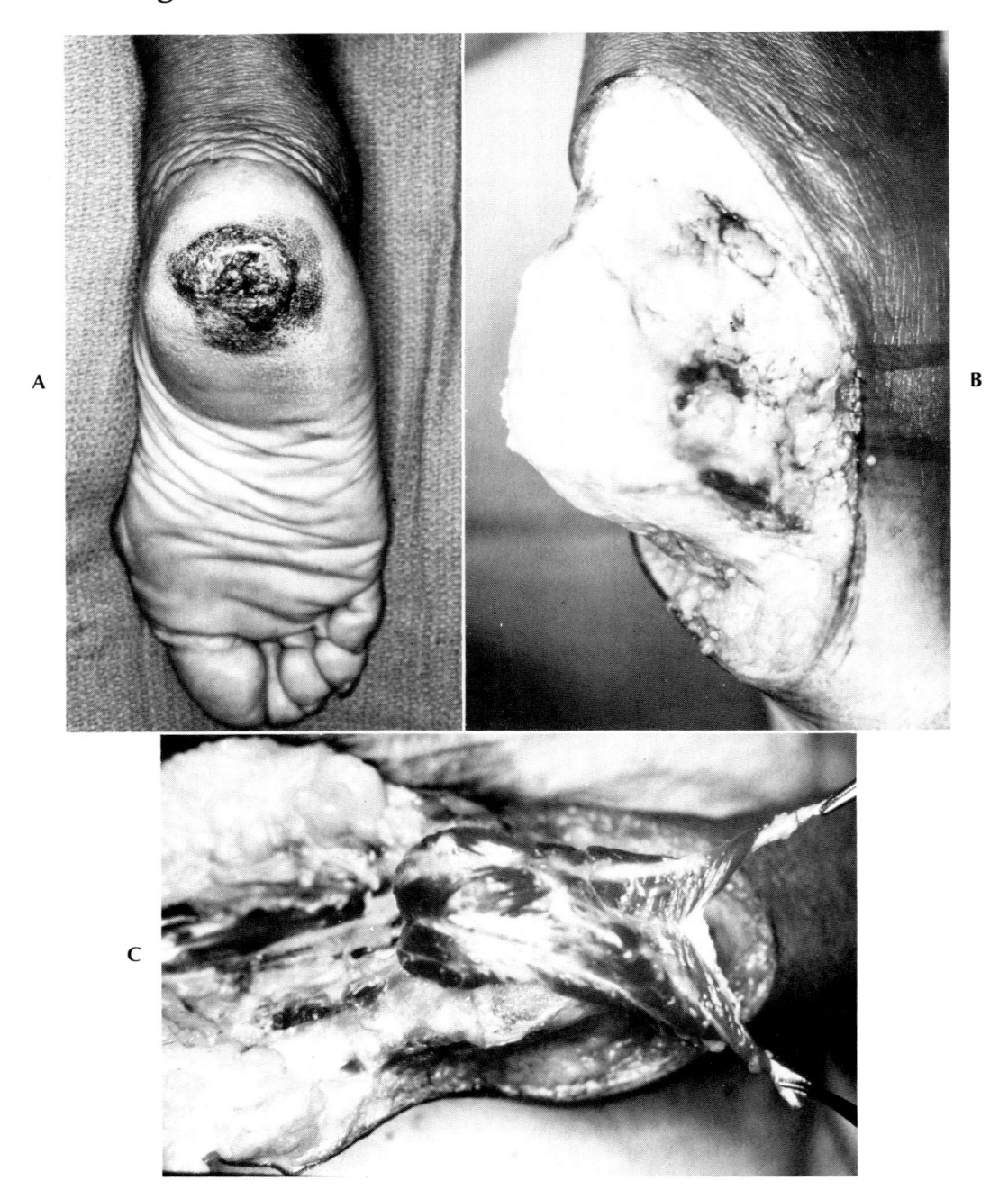

A   Level V melanoma of heel.

B   Wide local excision with exposure of calcaneus in ambulatory patient.

C   Flap elevated; calcaneus partially excised.

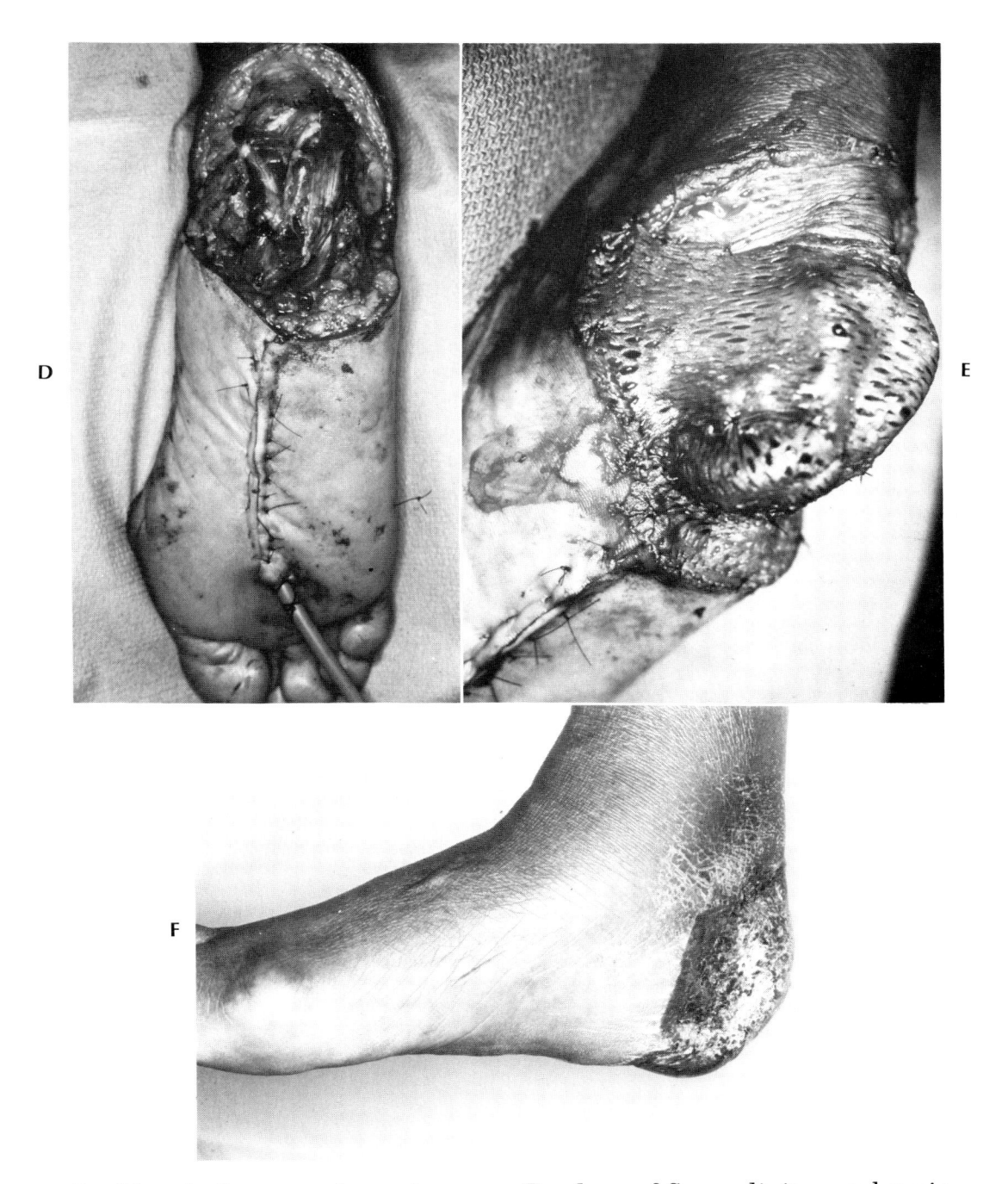

D   Muscle flap covering calcaneus. Tendons of flexor digitorum brevis sutured into Achilles tendon.

E   Split-skin graft over muscle.

F   Long-term result in ambulatory patient, following single-stage reconstruction of heel.

Abductor digiti minimi

MUSCLE FLAP
MUSCULOCUTANEOUS FLAP

Application

   Coverage of:

     LATERAL ANKLE

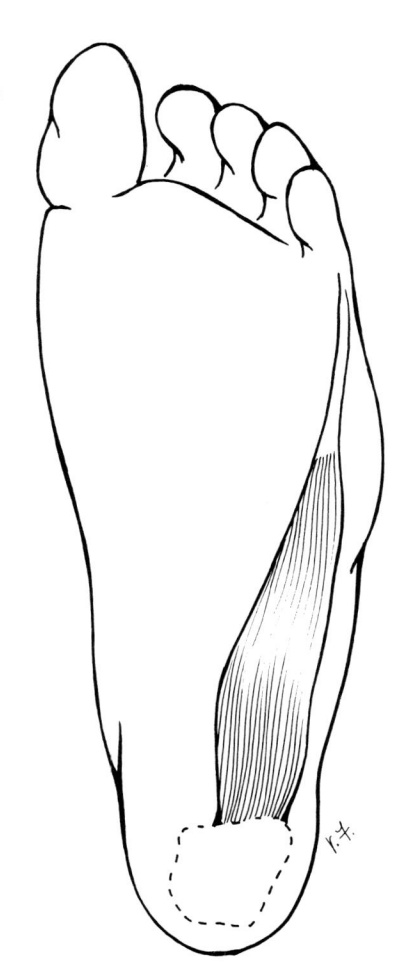

FOOT

Abductor digiti minimi

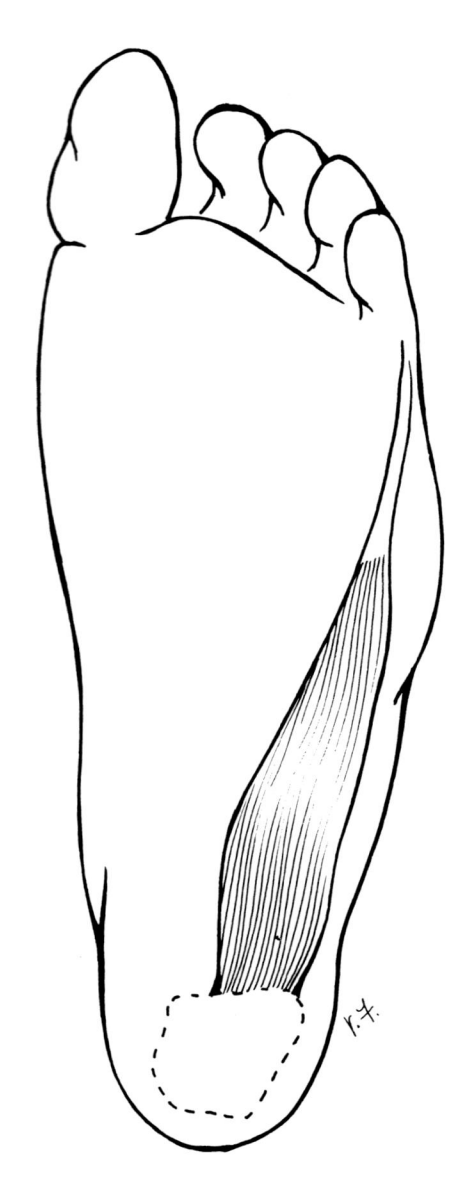

The abductor digiti minimi muscle is located on the lateral border of the foot. The muscle lies between the flexor digitorum brevis muscle medially and the bones of the lateral border of the foot. The muscle is very thin and stretches over the plantar surface of the fifth metatarsal bone.

Origin: Lateral and medial processes of the tuberosity of the calcaneus, adjacent intermuscular septum.

FOOT

Blood supply

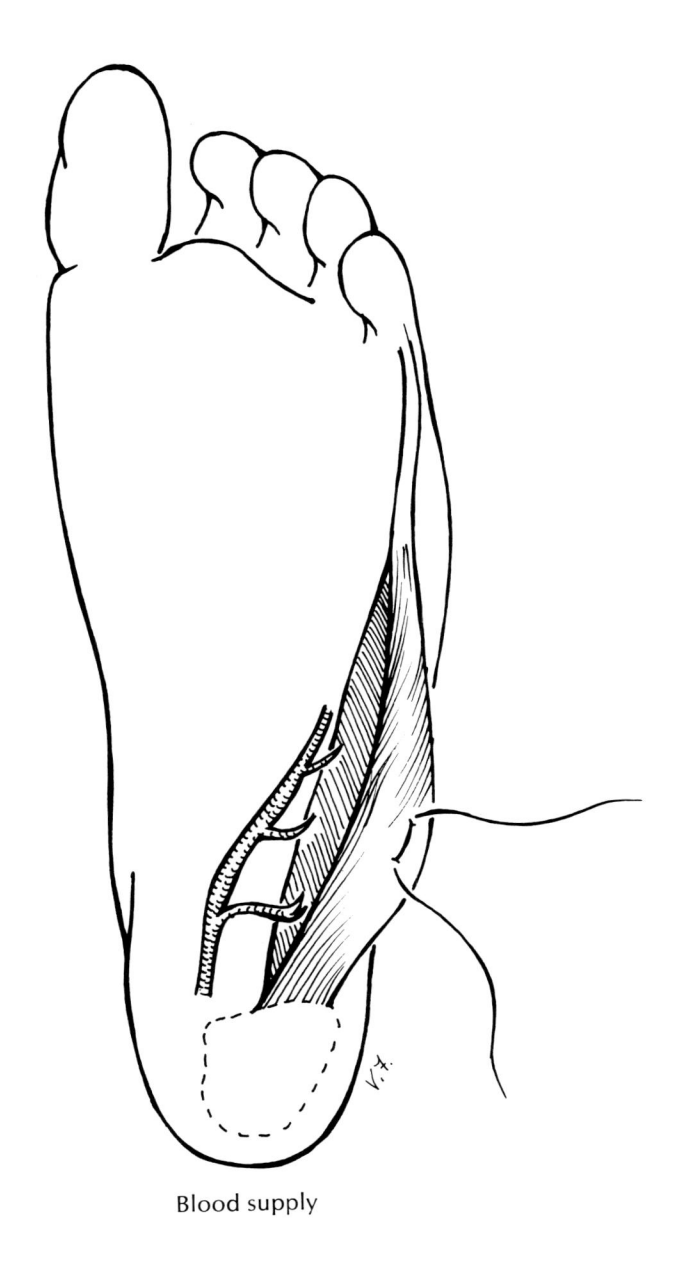

Insertion: Lateral aspect of the base of the proximal phalanx of the little toe.
Nerve supply:
 Motor — Branches of lateral plantar nerve.
 Sensory — Distal branches of sural nerve.
Function: This is an expendable muscle. The function of little toe abduction is lost following flap transposition.

FOOT

Abductor digiti minimi

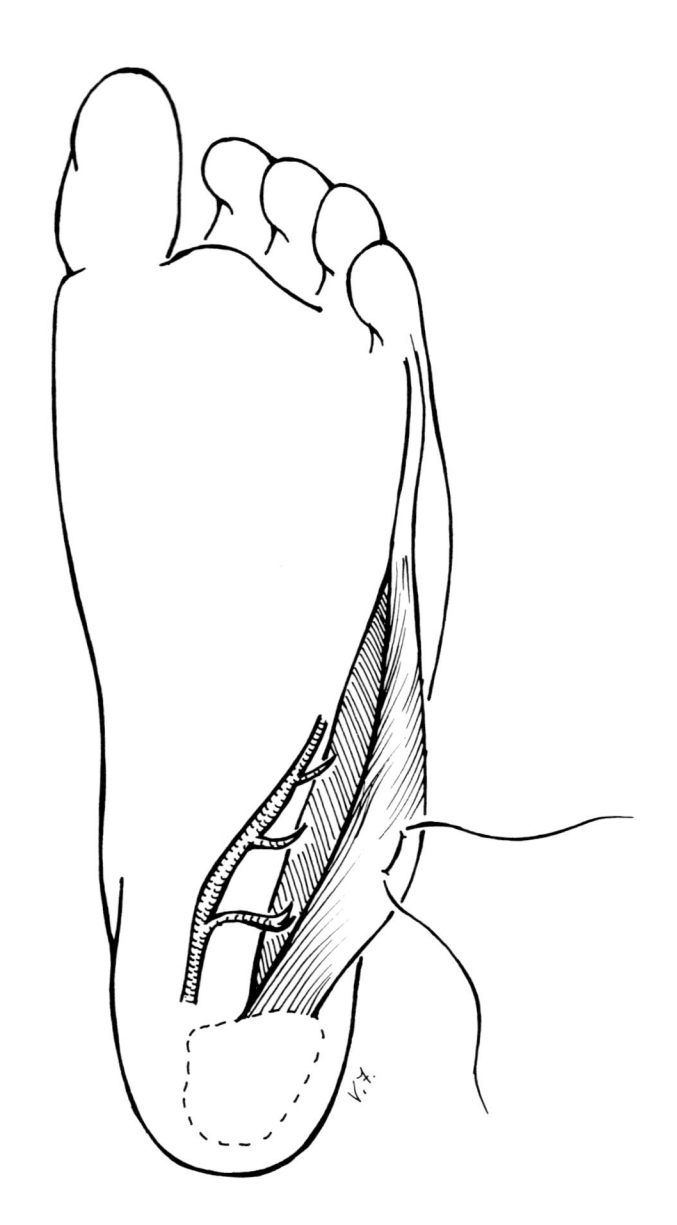

Blood supply

The muscle receives two or three branches from the lateral plantar artery into the posteromedial proximal muscle belly.

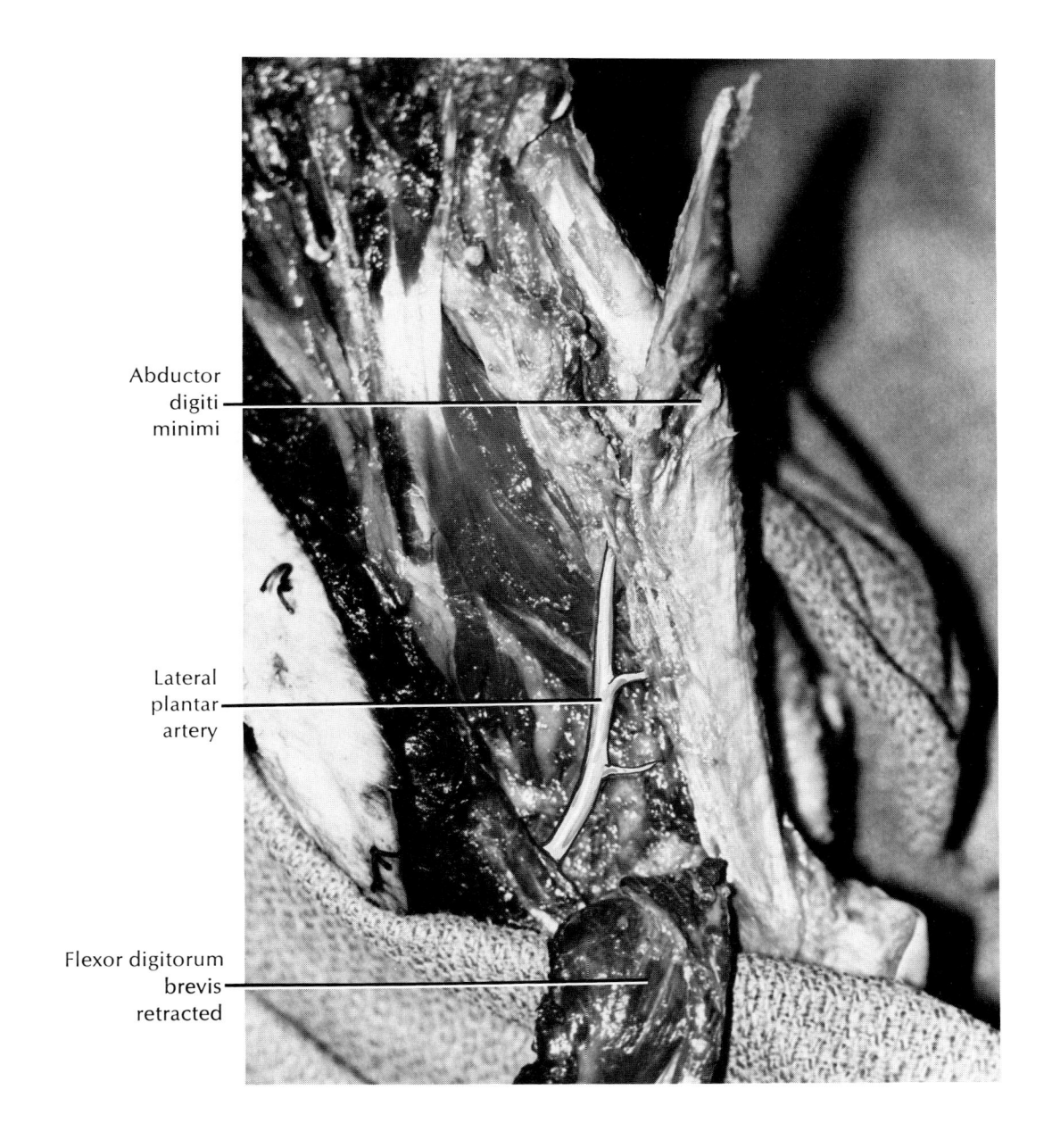

FOOT

- Abductor digiti minimi
- Lateral plantar artery
- Flexor digitorum brevis retracted

FOOT

Abductor digiti minimi

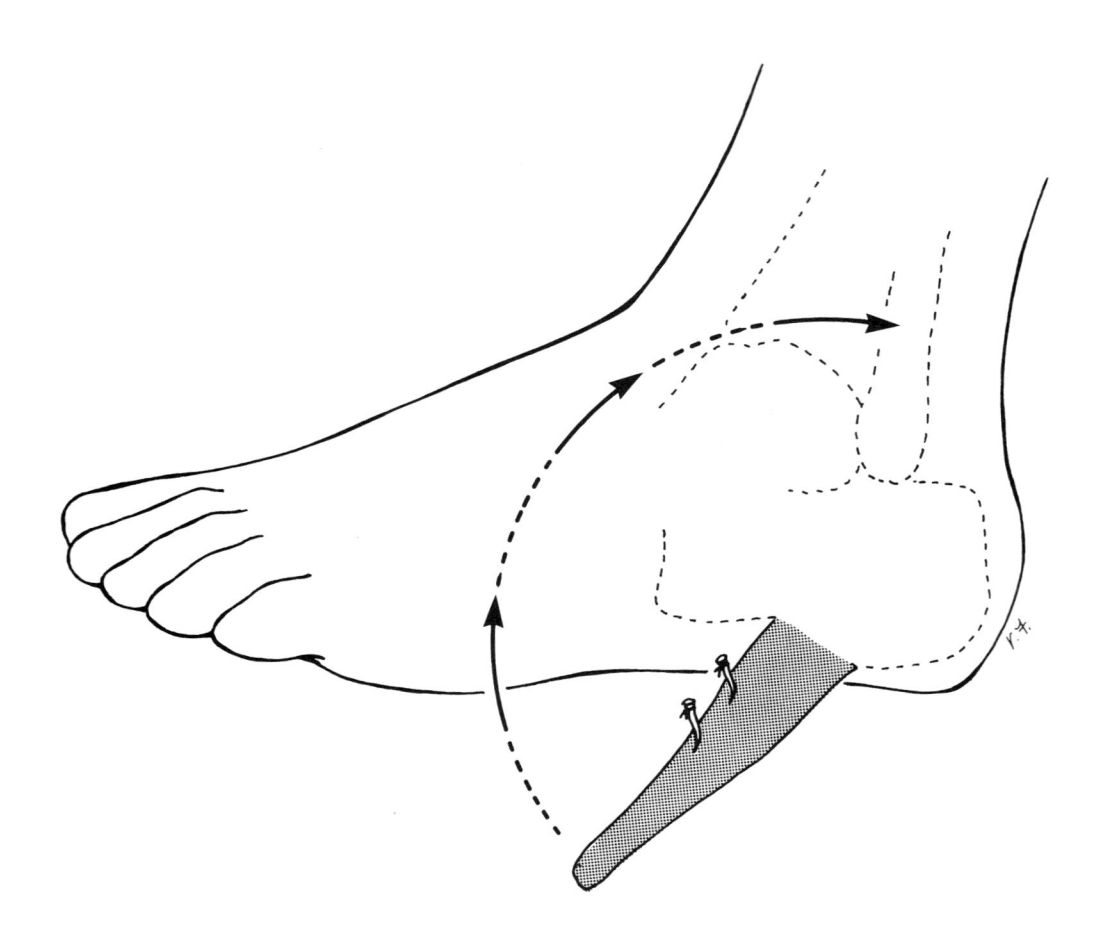

Arc of rotation

The muscle has a short arc of rotation from the lateral foot. The muscle, following posterior or superior rotation, will cover defects on the lateral ankle, including exposed distal Achilles tendon and fibula.

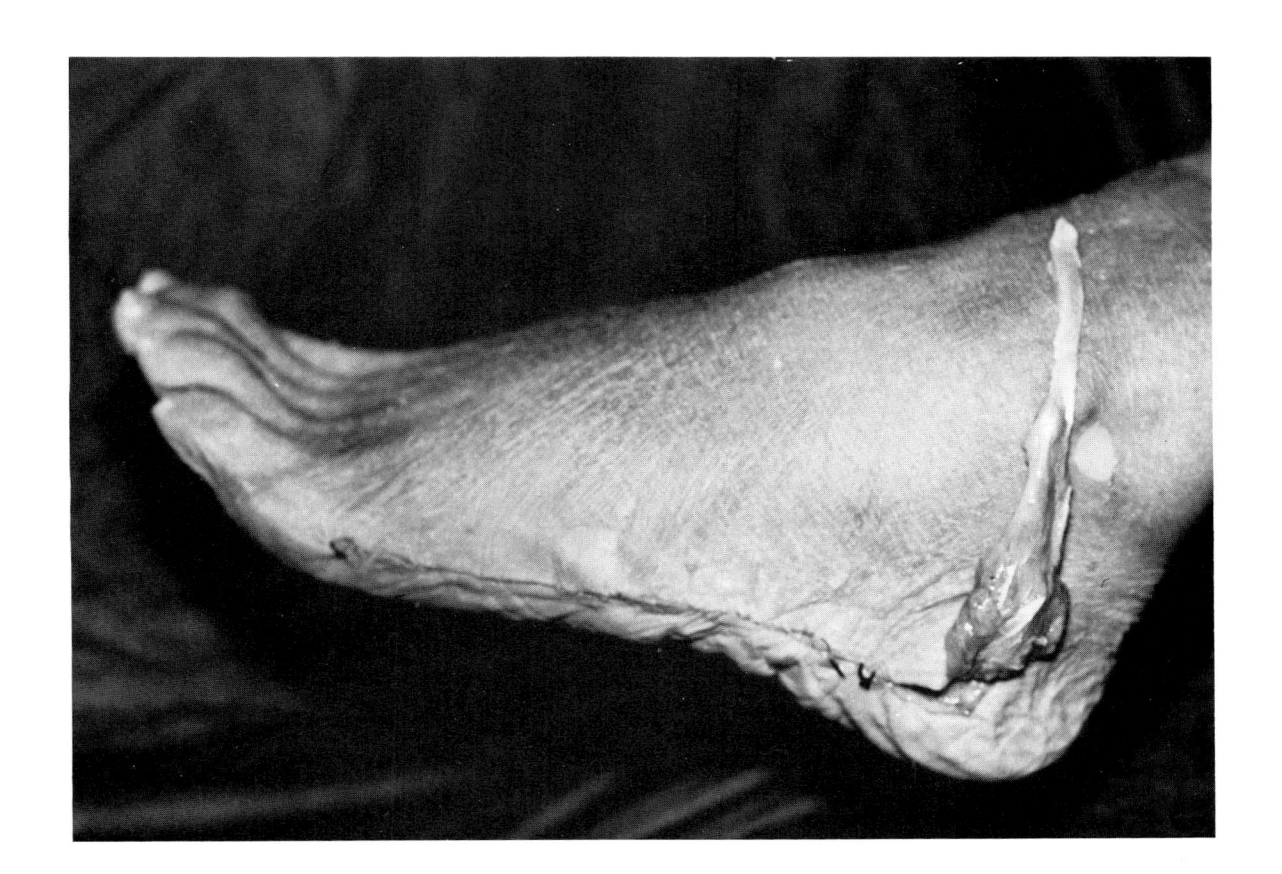

FOOT

FOOT

Abductor digiti minimi

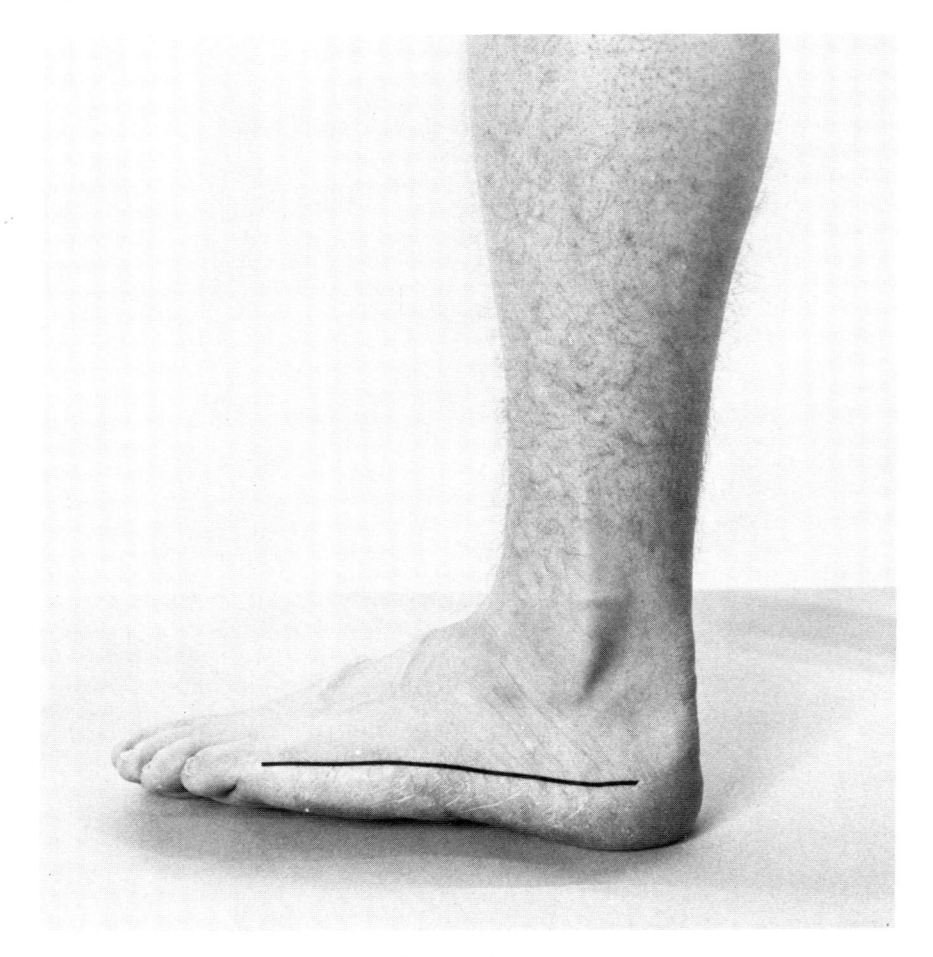

Skin incision

Elevation of flap

The muscle is exposed through an incision on the lateral foot. The muscle is located adjacent to its insertion and elevated proximally. The muscle must be elevated from its attachments to the base of the fifth metatarsal bone. The muscle is very thin at this level. One or two distal pedicles from the lateral plantar artery are divided in the region of the cuboid bone 4 to 5 cm from the lateral malleolus. Care must be taken to preserve the lateral tarsal artery and its proximal pedicles to the muscle. The muscle is transposed posteriorly to cover defects in proximity to the lateral malleolus and Achilles tendon. The donor defect can be closed primarily.

This unit can be transferred as a musculocutaneous flap incorporating the cutaneous territory of the sural nerve. In this case the donor defect would require skin grafting.

Precautions

- The muscle is very thin in the region of the head of the metatarsal, and the distal muscle circulation can be damaged during the dissection in this region.
- The lateral plantar artery as it crosses the posterior aspect of the flexor digitorum brevis is the point of rotation of this flap. Tension must be avoided, especially since the flap is small and generally inadequate for large defects of the ankle.
- The skin-grafted donor defect following musculocutaneous flap transposition may be unstable in the region of the projecting fifth metatarsal head.

Extensor digitorum brevis

FREE FLAP

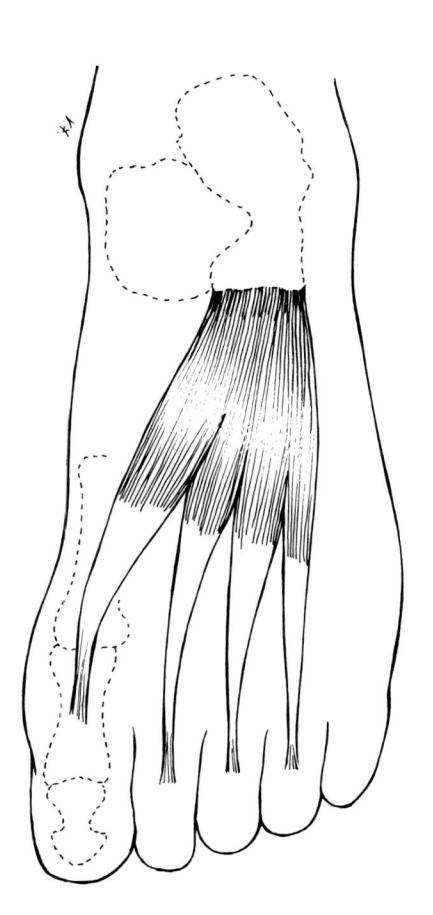

Application

Free functional muscle transfer

FOOT

Extensor digitorum brevis

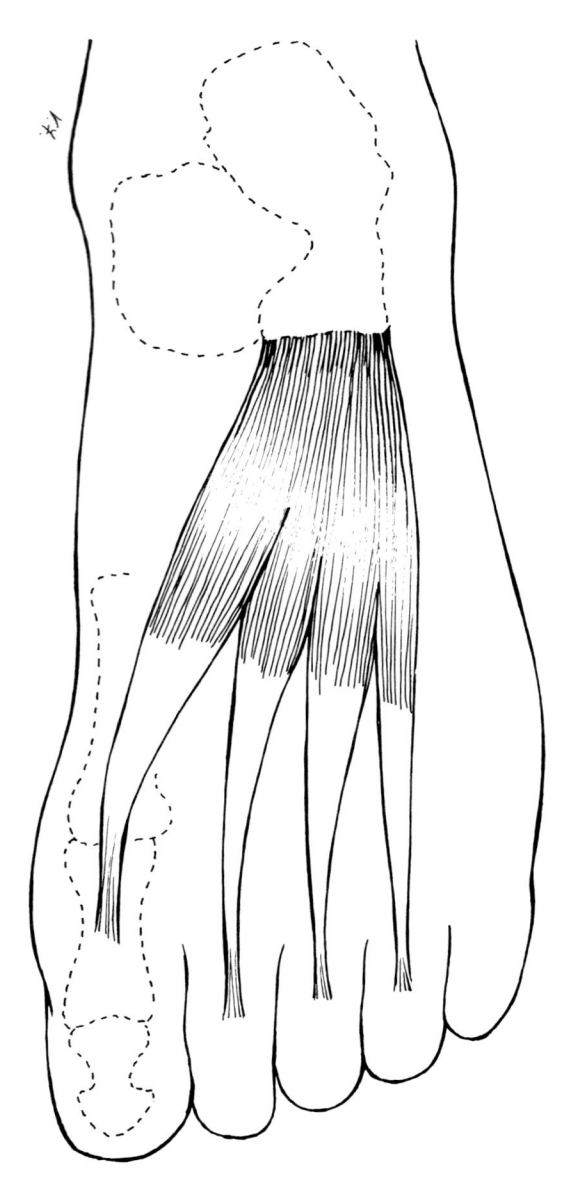

The extensor digitorum brevis, a broad, thin muscle, is located on the dorsum of the foot. The muscle is deep to the tendons of the extensor digitorum longus muscle and superficial to the tarsal bones.

Origin: Distal and lateral portions of the calcaneus, lateral talocalcaneal ligament.
Insertion: First muscle slip (extensor hallucis brevis): base of proximal phalanx of great toe; second through fourth muscle slips: lateral sides of extensor digitorum longus tendons.

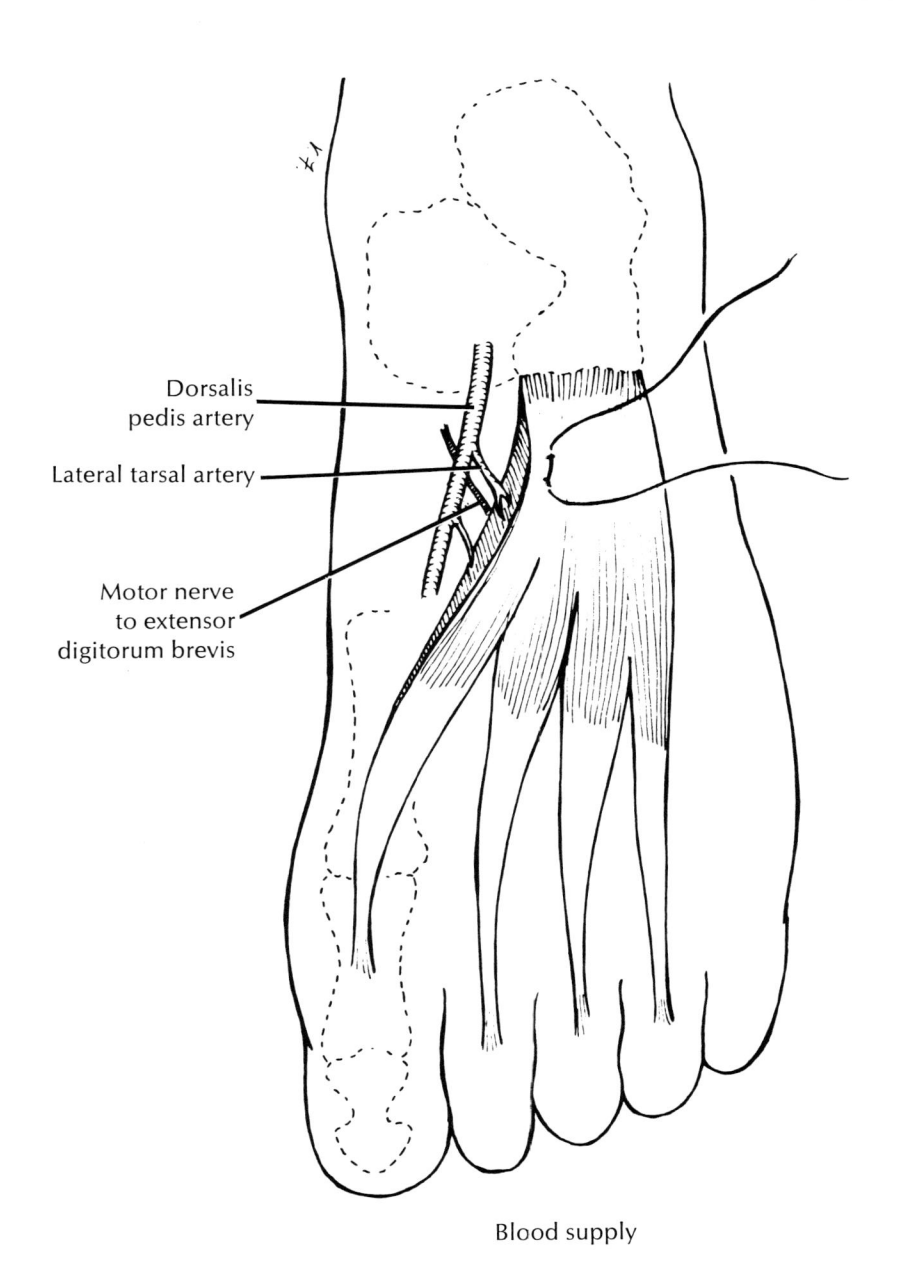

FOOT

- Dorsalis pedis artery
- Lateral tarsal artery
- Motor nerve to extensor digitorum brevis

Blood supply

Nerve supply: Branch of deep peroneal nerve. The nerve enters the proximal posterior muscle belly in close relationship to the vascular pedicle.

Function: This is an expendable muscle. The function of extension of the four medial toes will be maintained by the extensor digitorum longus and extensor hallucis longus muscles.

FOOT

Extensor digitorum brevis

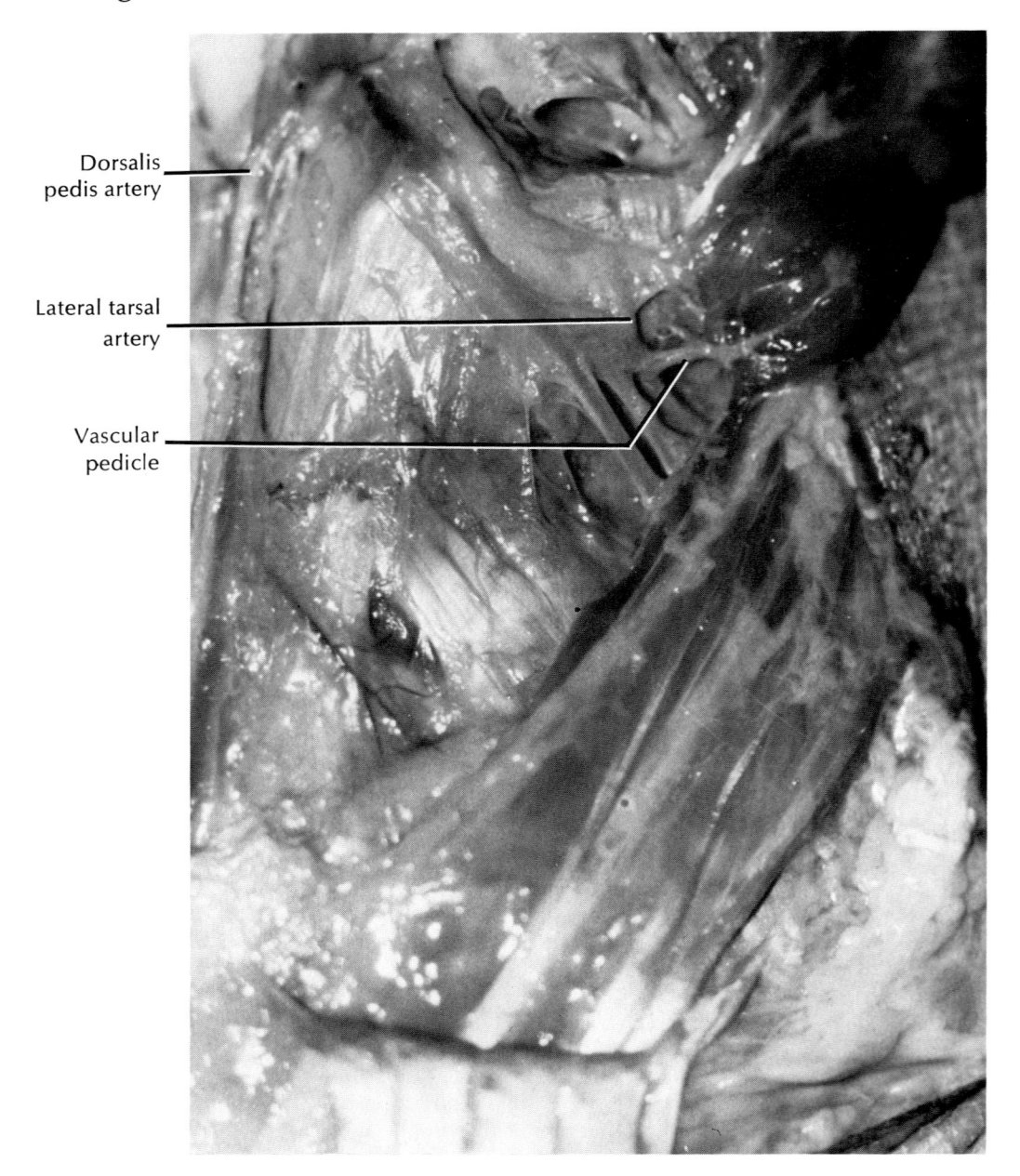

- Dorsalis pedis artery
- Lateral tarsal artery
- Vascular pedicle

Blood supply

The lateral tarsal artery, a branch of the dorsalis pedis artery, gives two branches to the posteromedial muscle belly. This vessel enters the proximal aspect of the muscle.

FOOT

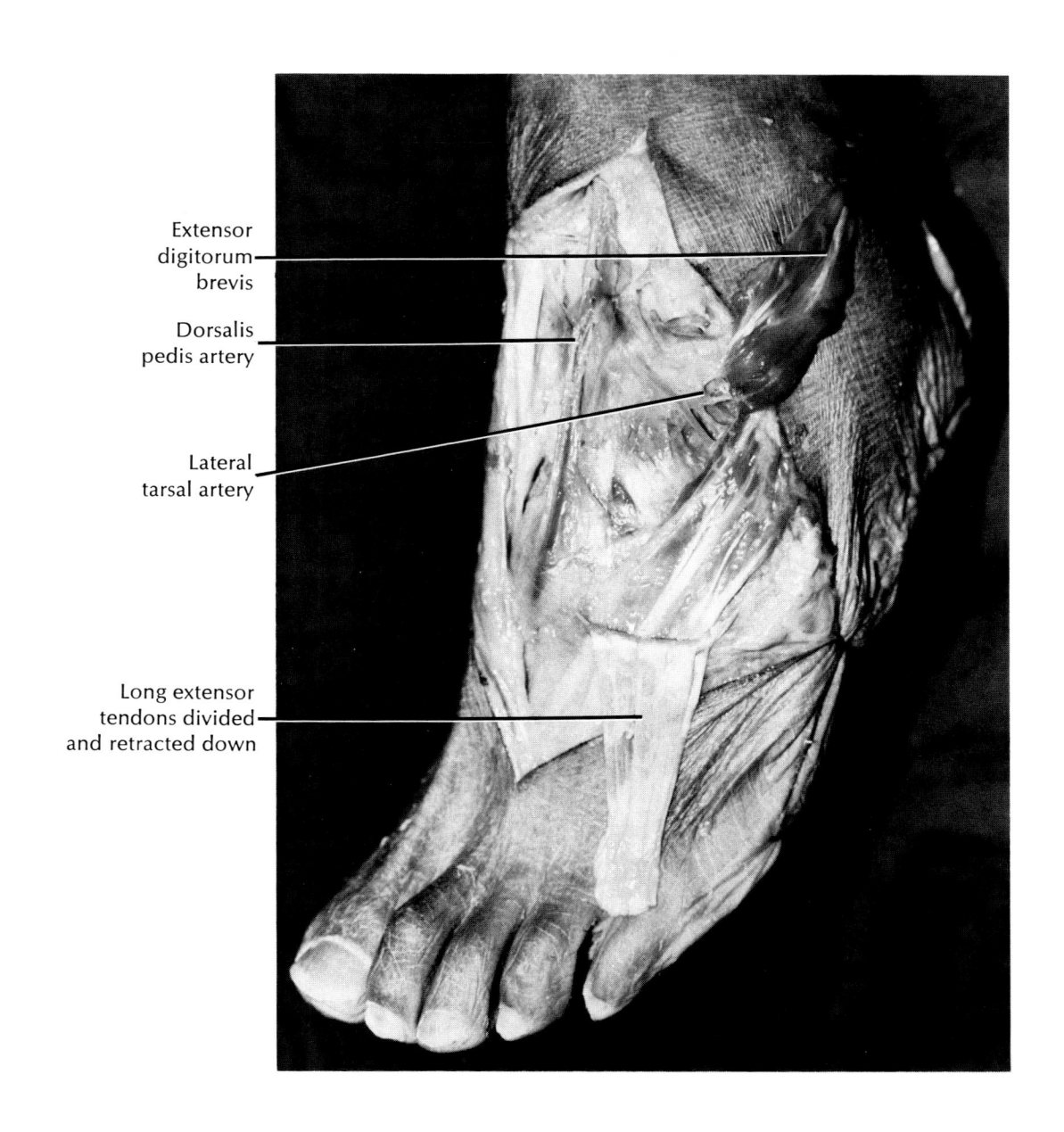

Extensor digitorum brevis

Dorsalis pedis artery

Lateral tarsal artery

Long extensor tendons divided and retracted down

FOOT

Extensor digitorum brevis

Arc of rotation

This muscle has a small arc to the medial and lateral foot. The muscle is not adequate in size to warrant use in muscle transposition. However, this unit is included to present its suitability for transfer as a free functional muscle flap.

Elevation of flap

This muscle can be located by a dorsal medial incision on the foot. The long extensor tendons are retracted laterally, and the distal tendons of the extensor digitorum brevis are divided close to their insertion to maintain length. The muscle is elevated proximally. Adjacent to the insertion of the first slip to the great toe (occasionally known as the extensor hallucis brevis), the lateral tarsal artery is identified. With preservation of this vessel, the origin of the muscle is detached. The dorsalis pedis artery and associated venae comitantes are similarly elevated. The proximal dorsalis pedis artery and veins are dissected to predetermined required length for free flap transfer. Care must be taken during this flap elevation to include a vein that drains the extensor digitorum brevis muscle in conjunction with the lateral plantar artery. The motor nerve to the muscle is isolated adjacent to the vascular pedicle and dissected to its junction with the deep peroneal nerve. If further nerve length is required, interfascicular separation can be accomplished under microscopic control. This functional muscle unit is then ready for transfer by microsurgical techniques.

FOOT

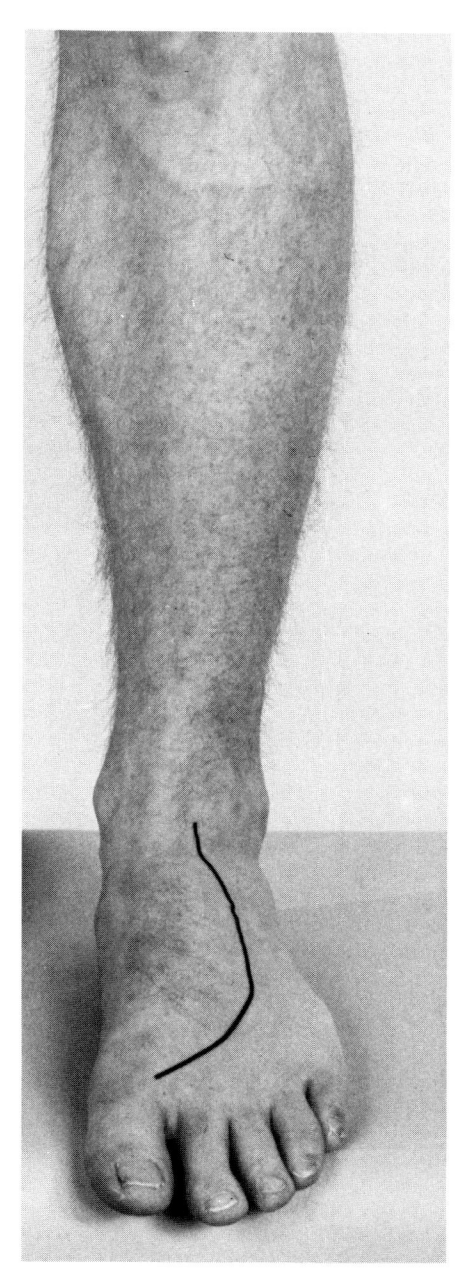

Skin incision

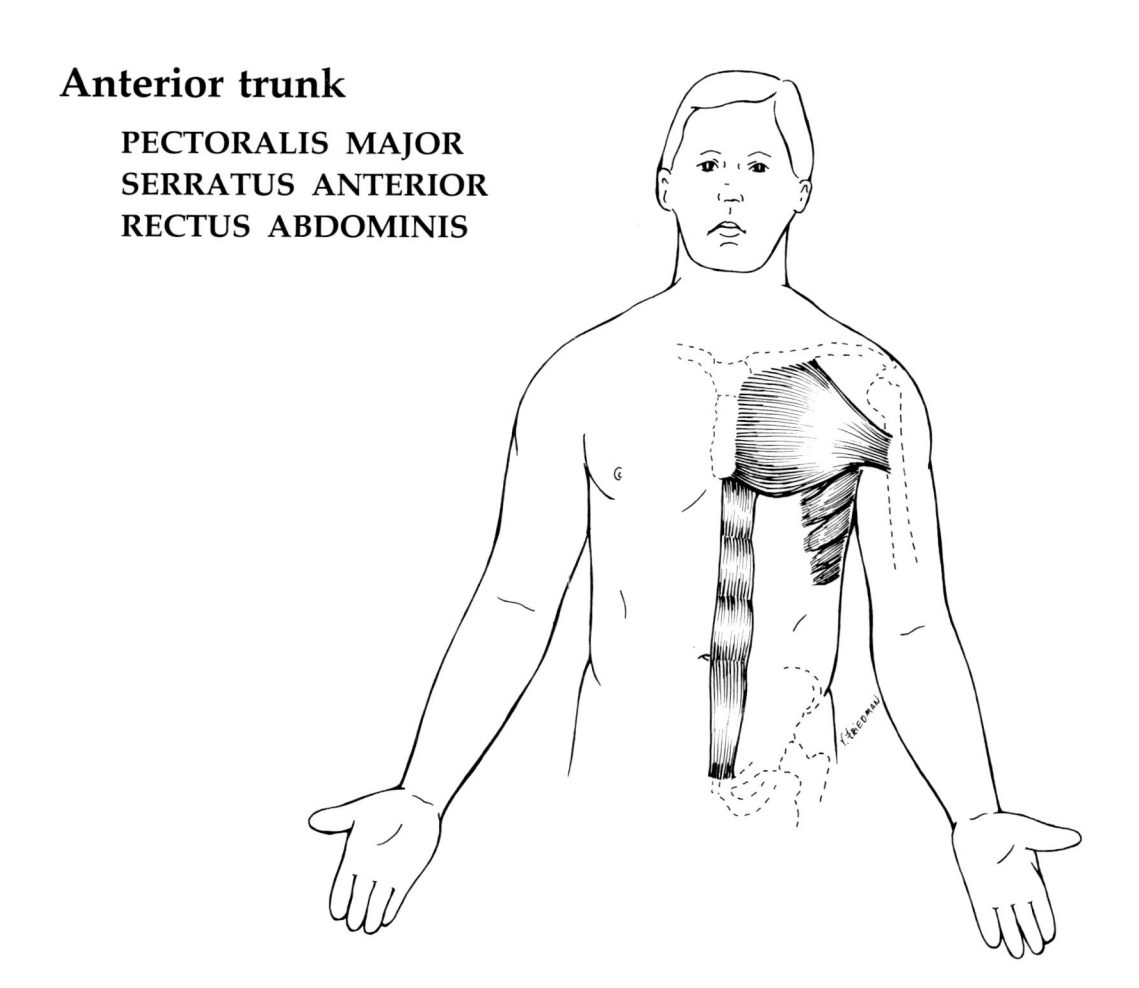

SECTION SIX
TRUNK

Anterior trunk
PECTORALIS MAJOR
SERRATUS ANTERIOR
RECTUS ABDOMINIS

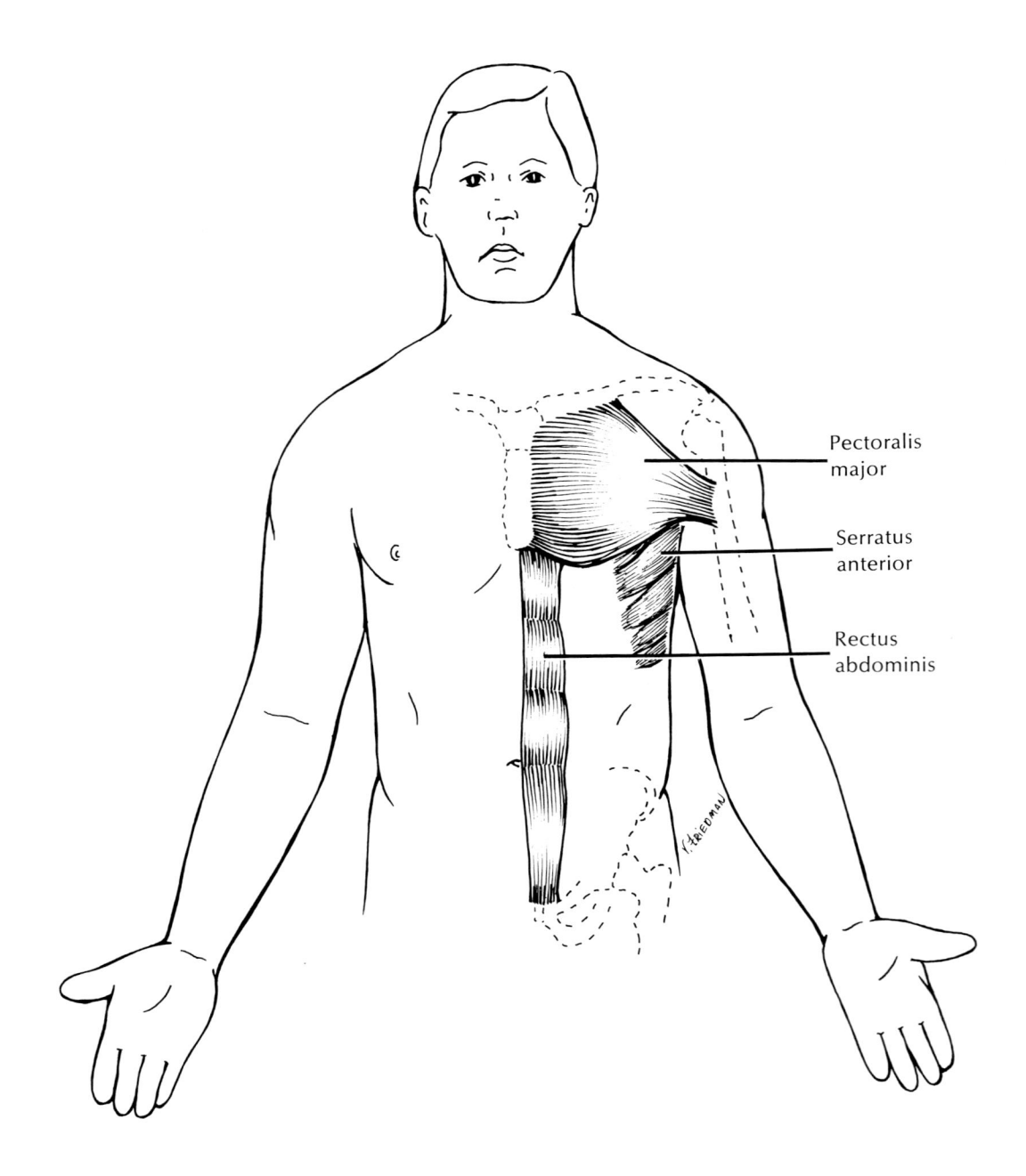

TRUNK: ANTERIOR TRUNK

- Pectoralis major
- Serratus anterior
- Rectus abdominis

The anterior trunk muscles are an extremely useful group of muscles and musculocutaneous flaps, with great potential for reconstruction of the trunk, head, and neck.

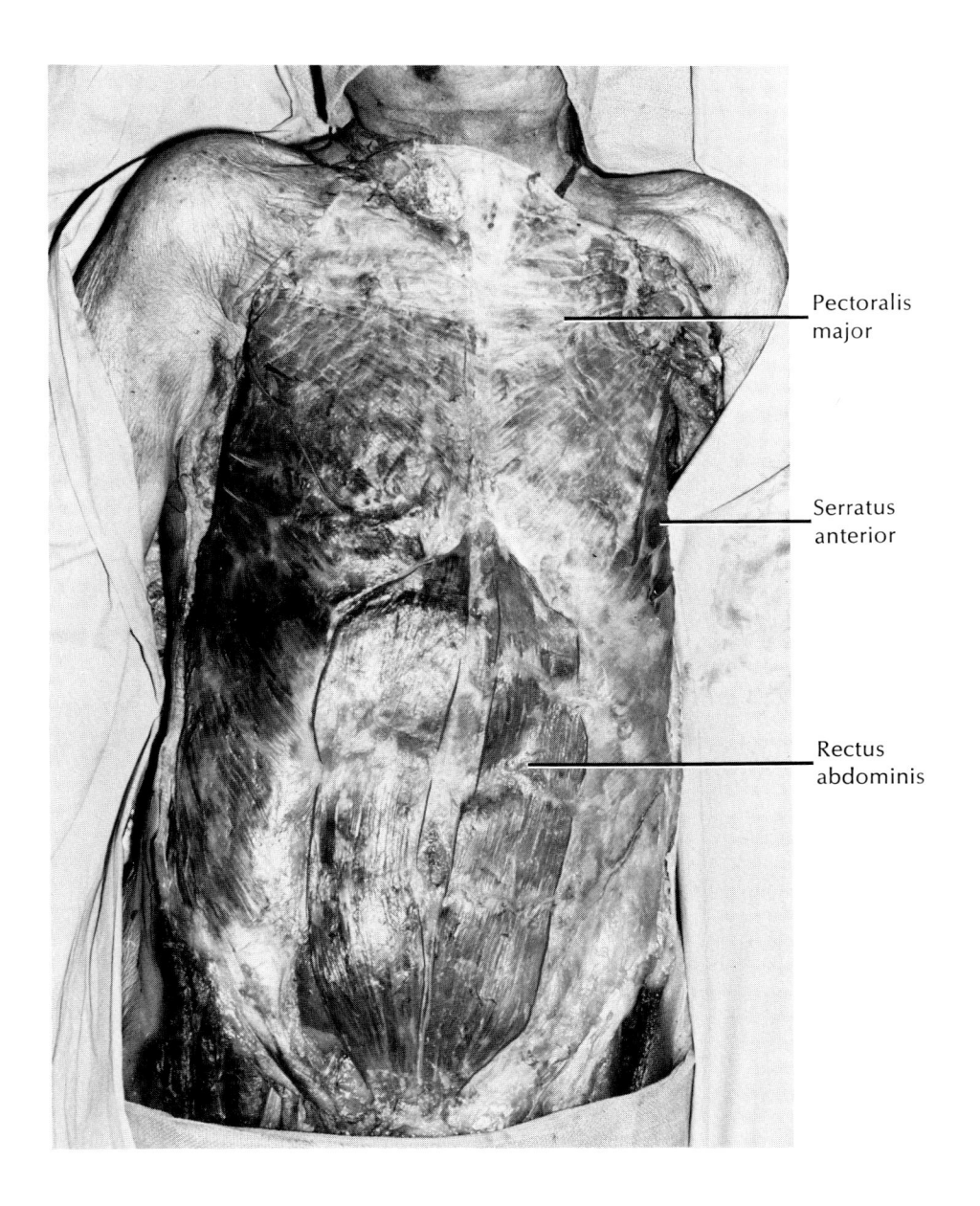

TRUNK: ANTERIOR TRUNK

Pectoralis major

Serratus anterior

Rectus abdominis

TRUNK: ANTERIOR TRUNK

Axillary artery

Subscapular artery

Pectoral artery

Thoracodorsal artery

Latissimus dorsi muscle

Lateral thoracic artery

Serratus anterior

Pectoralis major

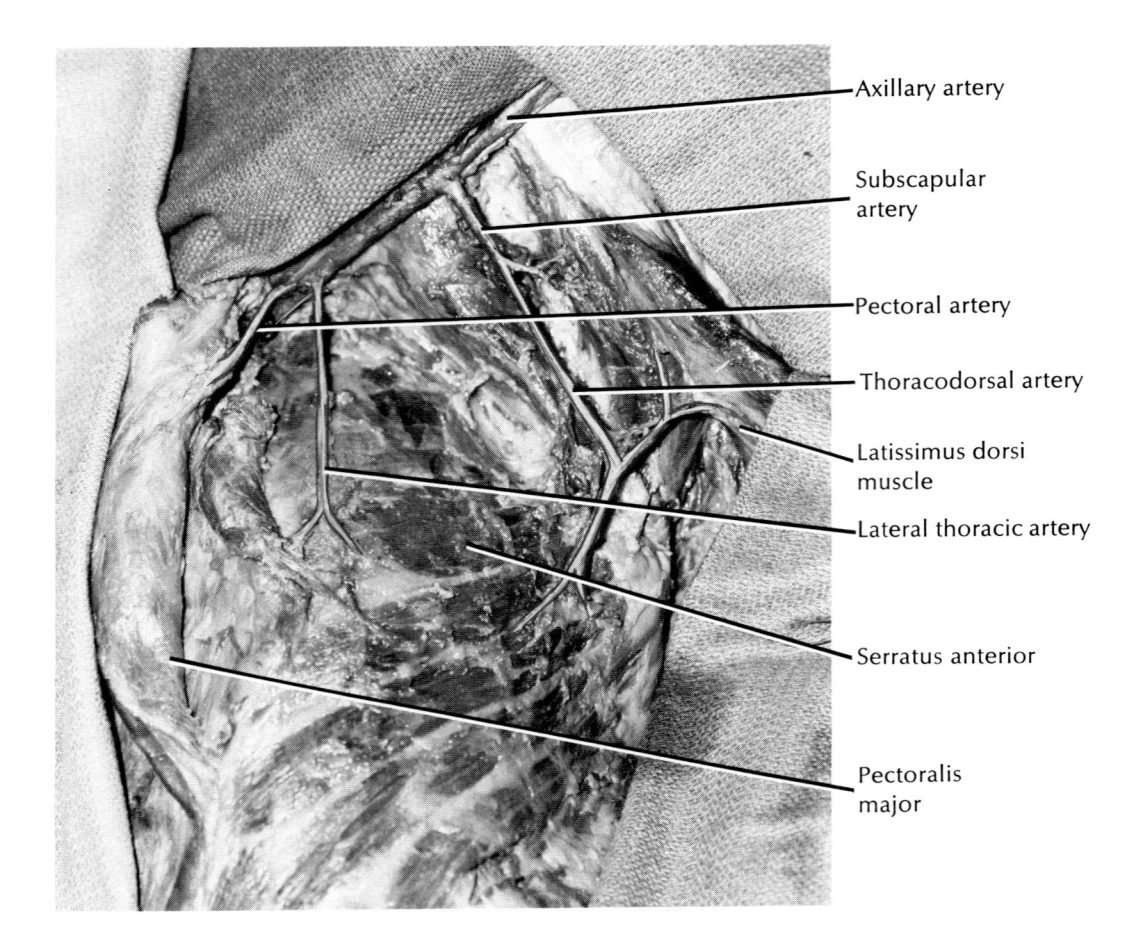

Blood supply

The blood supply to the pectoralis major and serratus anterior is based on the axillary artery. The rectus abdominis is based on the internal mammary and inferior epigastric arteries. The internal mammary artery also supplies minor segmental pedicles to the pectoralis major. As the blood supply of the latissimus dorsi is based on the axillary artery, it will be discussed in this section too.

The three branches of the axillary artery, which are significant in terms of these muscles, are the thoracoacromial (pectoralis major), lateral thoracic (serratus anterior), and the subscapular (latissimus

TRUNK: ANTERIOR TRUNK

Blood supply

- Thoracoacromial artery
- Axillary artery
- Subscapular artery
- Circumflex scapular artery
- Thoracodorsal artery
- Lateral thoracic artery
- Superior epigastric artery
- Inferior epigastric artery

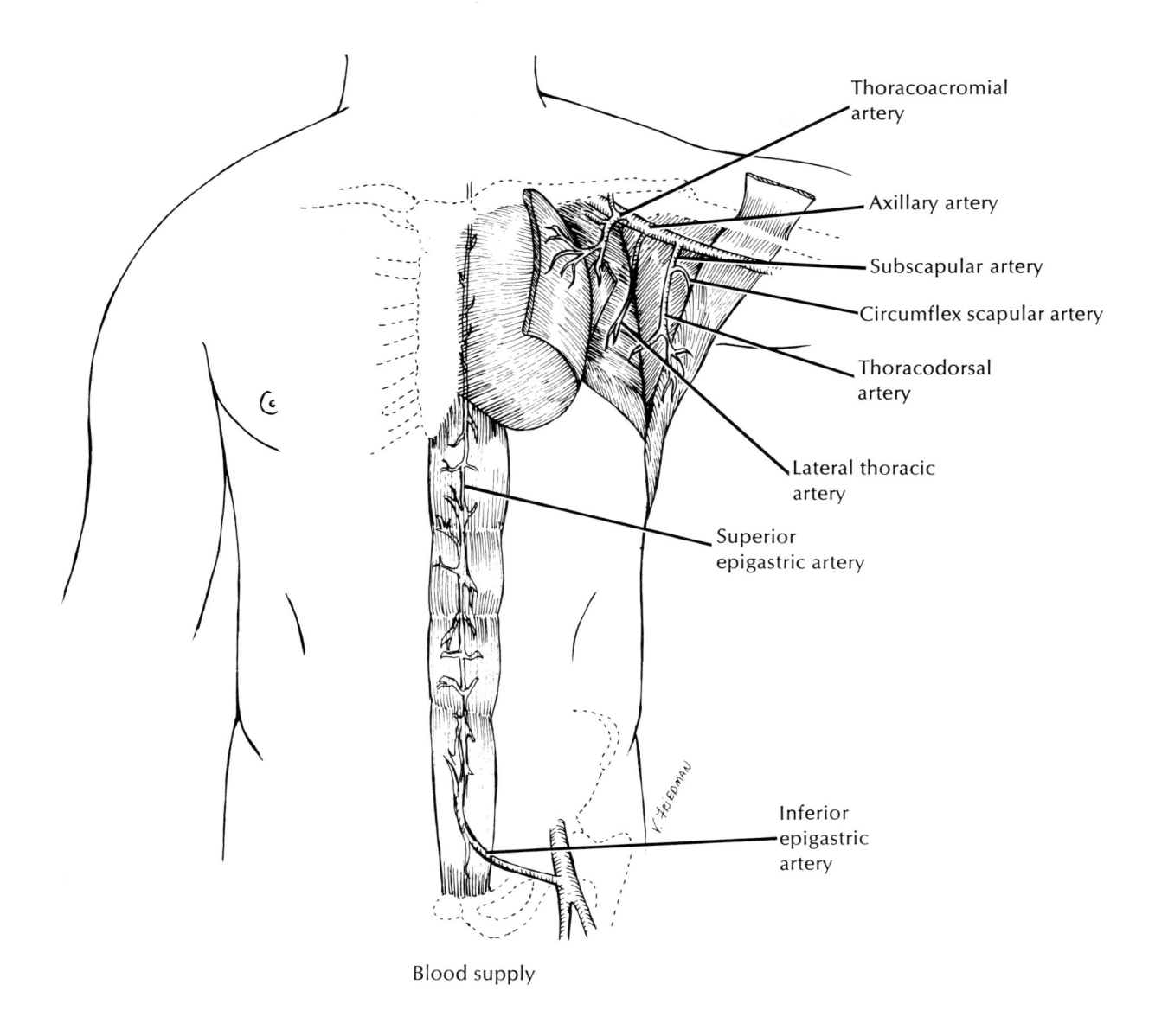

dorsi) arteries. The first two are branches of the second part of the axillary artery, and the subscapular is a branch of the third part of the artery.

The thoracoacromial artery is a short vessel that branches anteriorly from the axillary artery deep and medial to the pectoralis minor. It immediately divides, giving a large pectoral branch that is the major vascular pedicle to the pectoral muscles.

The lateral thoracic artery branches from the lower aspect of the axillary artery, behind the pectoralis major, and along the lower border of the pectoralis minor. The vessel follows the lateral border of the pec-

TRUNK: ANTERIOR TRUNK

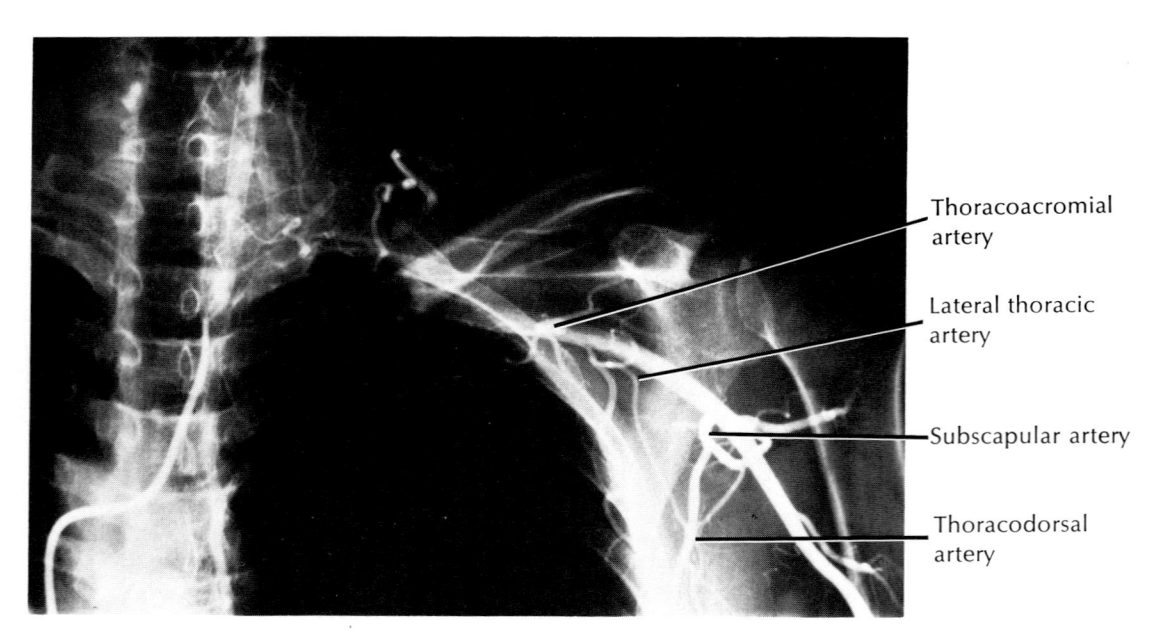

Angiogram of blood supply of anterior trunk

toralis minor muscle toward the chest wall, where it becomes the major pedicle to the serratus anterior. It has minor branches to the pectoral muscles.

The subscapular artery branches from the lower aspect of the axillary artery and is the largest branch of the artery. It arises at the distal border of the subscapularis muscle. This short (5 cm) vessel divides into two terminal branches, the circumflex scapular artery and the thoracodorsal artery. The circumflex scapular artery is larger and courses posteriorly. The thoracodorsal artery is the continuation of the subscapular artery and courses distally along the anterior border of the latissimus dorsi muscle, accompanied by the thoracodorsal nerve, to enter the latissimus dorsi muscle as its major vascular pedicle. The thoracodorsal artery has one or two branches that cross the axilla to supply the serratus anterior.

The internal mammary artery is a branch of the first part of the subclavian artery. It runs distal behind the costal cartilages close to the lateral margin of the sternum. In each interspace it has a perforating branch that passes through the intercostal muscles to supply the pectoralis major and the overlying skin. The internal mammary artery divides into the musculophrenic and superior epigastric arteries in the sixth intercostal space. The superior epigastric artery continues between the sternal and costal leaves of the diaphragm to enter the sheath of the rectus abdominis. The vessel first lies between the muscle and the posterior sheath, and then enters the muscle to supply it and the overlying skin.

The inferior epigastric artery branches from the external iliac artery just above the inguinal ligament; it curves along the medial margin of the internal inguinal ring and up through the transversalis fascia, through the posterior rectus sheath, and finally into the rectus abdominis muscle to supply the muscle and overlying skin. Terminal branches of this vessel have anastomotic connections with the terminal branches of the internal mammary artery.

Pectoralis major

MUSCLE FLAP
MUSCULOCUTANEOUS FLAP
FREE FLAP

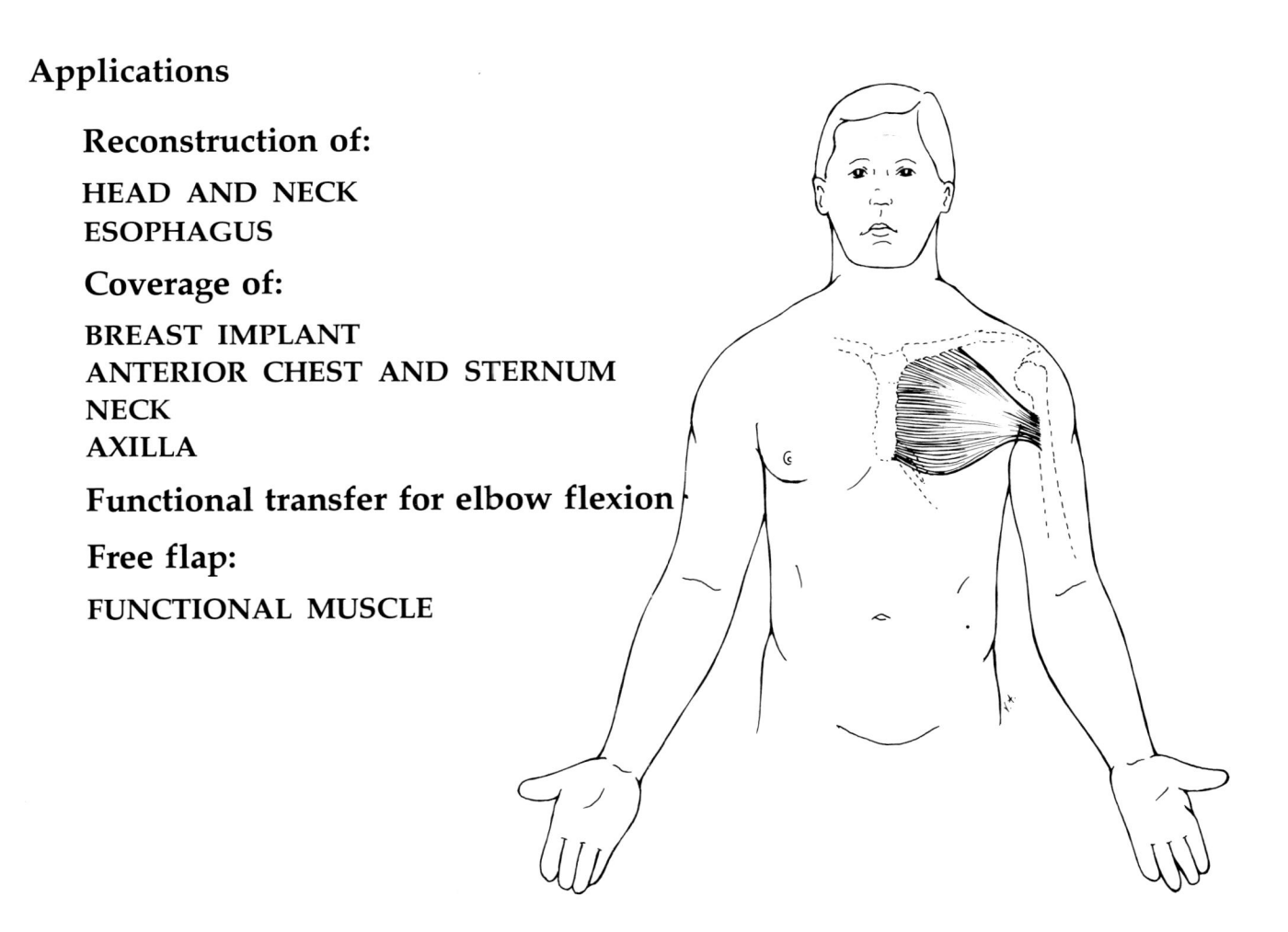

Applications

Reconstruction of:

HEAD AND NECK
ESOPHAGUS

Coverage of:

BREAST IMPLANT
ANTERIOR CHEST AND STERNUM
NECK
AXILLA

Functional transfer for elbow flexion

Free flap:

FUNCTIONAL MUSCLE

TRUNK: ANTERIOR TRUNK

Pectoralis major

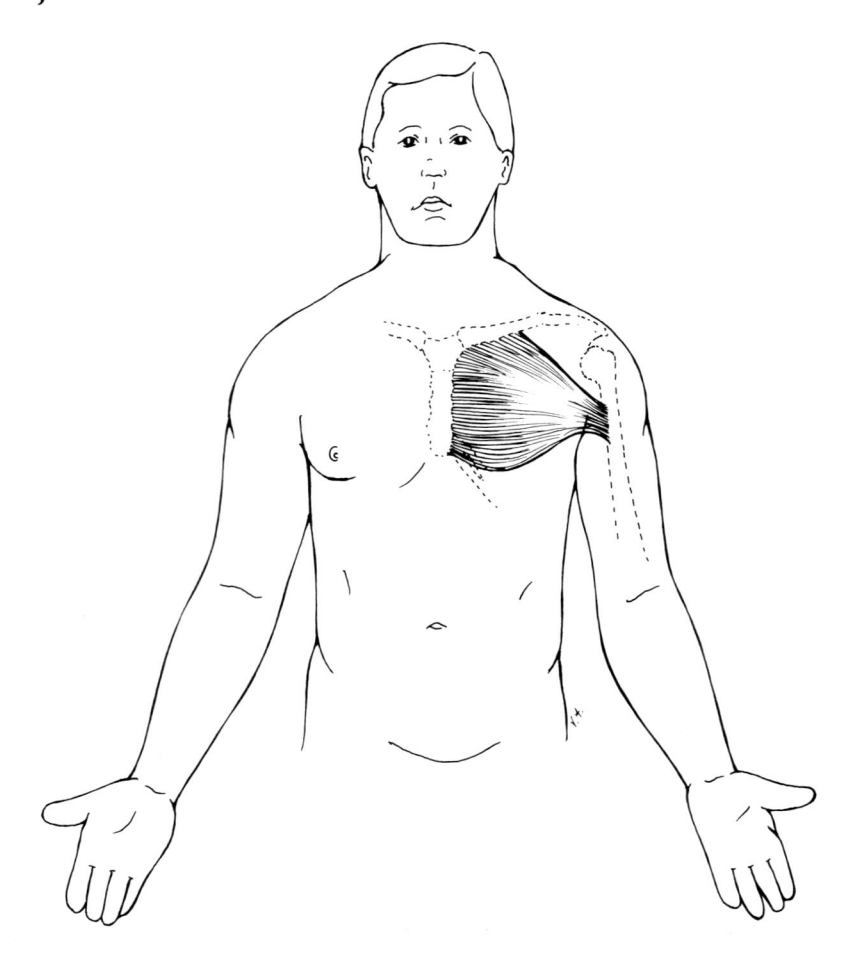

The pectoralis major is a broad, flat, fan-shaped muscle lying superficially on the anterior chest wall. Deep to the pectoralis major are the pectoralis minor, serratus anterior, and the intercostal muscles. The upper border of the muscle is related to the clavicle and deltoid, and the lower border is related to the rectus abdominis and the external oblique.

- **Origin:** The pectoralis major has two origins, the clavicular and sternal. The clavicular origin is from the medial half of the clavicle, and the sternal head is from the anterior half of the sternum down to the seventh rib, with slips from the first seven ribs.
- **Insertion:** The fibers converge into a flat tendon to insert into the lateral lip of the bicipital groove of the humerus.

TRUNK: ANTERIOR TRUNK

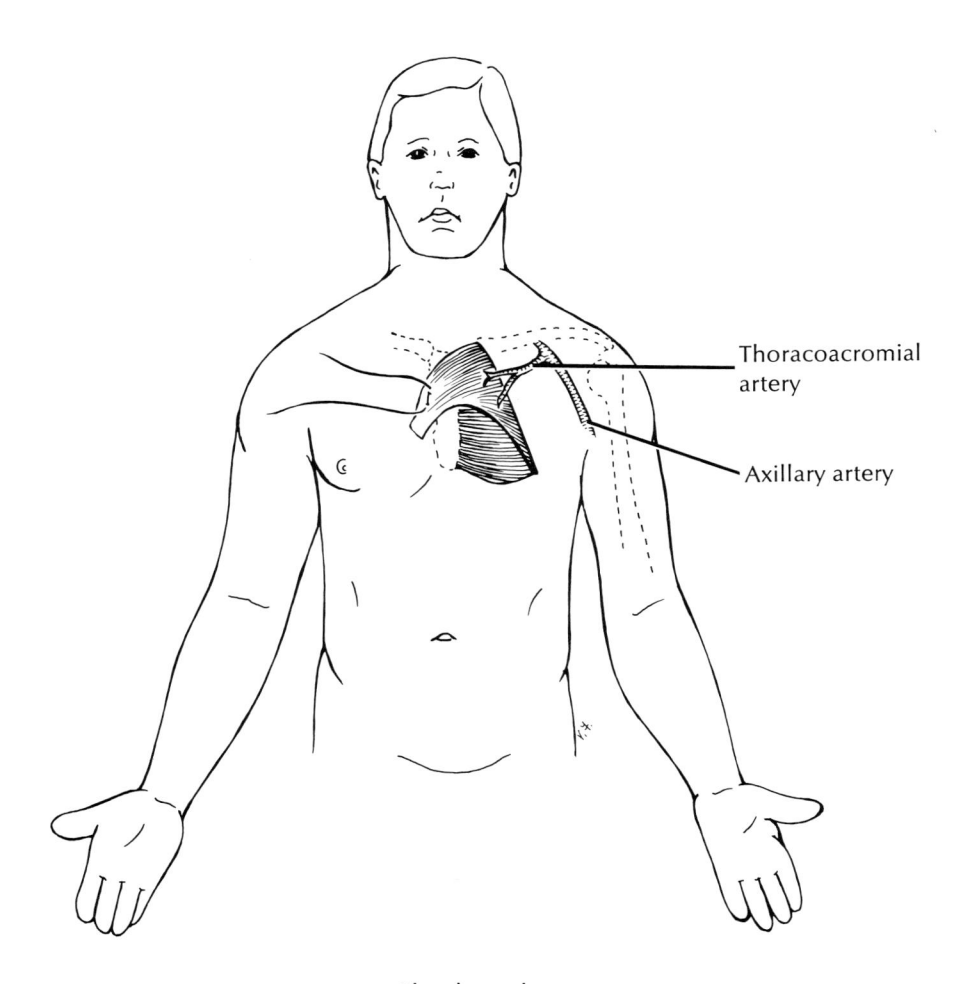

Blood supply

Nerve supply: The medial and lateral pectoral nerves innervate this muscle. These are direct branches from the lateral and medial cords of the brachial plexus in the axilla. The nerves enter the muscle with the major vascular pedicle.

Function: The pectoralis major is an adductor and medial rotator of the arm. Some disability results from loss of pectoralis muscle function. However, as this is such a versatile unit for reconstruction, its use is justified.

TRUNK: ANTERIOR TRUNK

Pectoralis major

Thoracoacromial artery

Axillary artery

Pectoral branches of thoracoacromial artery

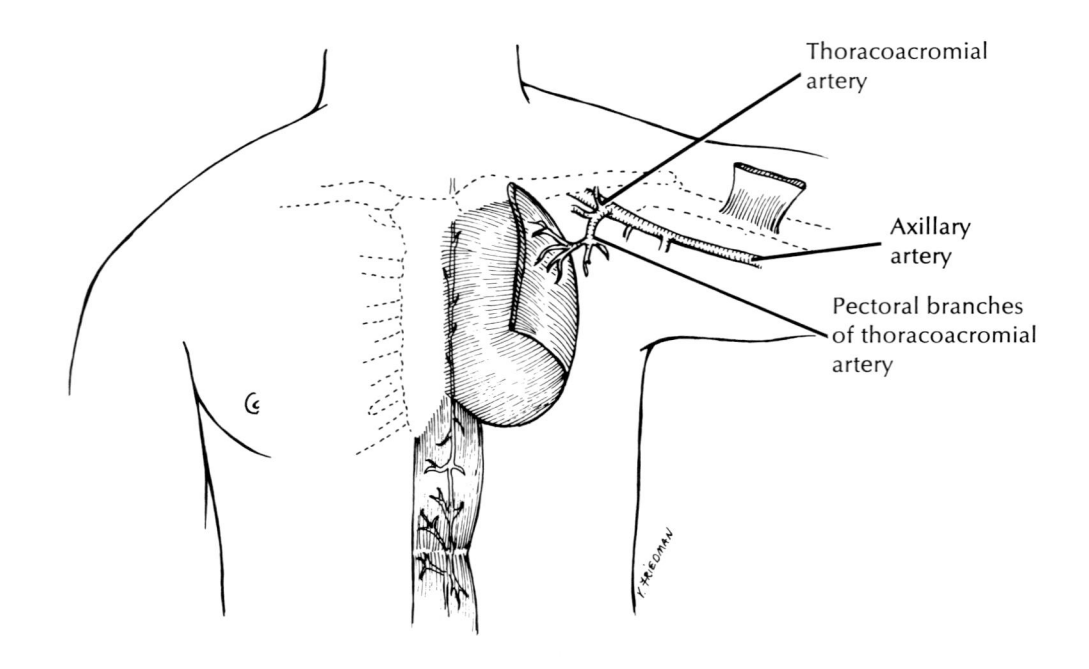

Blood supply

The pectoralis major muscle is supplied by the dominant major pedicle from the thoracoacromial artery and by several segmental minor pedicles, the perforating branches of the internal mammary artery.

The thoracoacromial artery is a short anterior branch of the second part of the axillary artery. It branches off deep to the pectoralis minor muscle and almost immediately divides at the upper edge of the pectoralis minor to give a large pectoral branch. The thoracoacromial vessel then continues laterally and divides into deltoid and acromial branches. The pectoral branch then divides into a small branch to supply the pectoralis minor, a small clavicular branch, and a large branch that becomes the major pedicle of the overlying pectoralis major muscle. The vessel enters the deep surface of the muscle at the level of the upper border of the pectoralis minor.

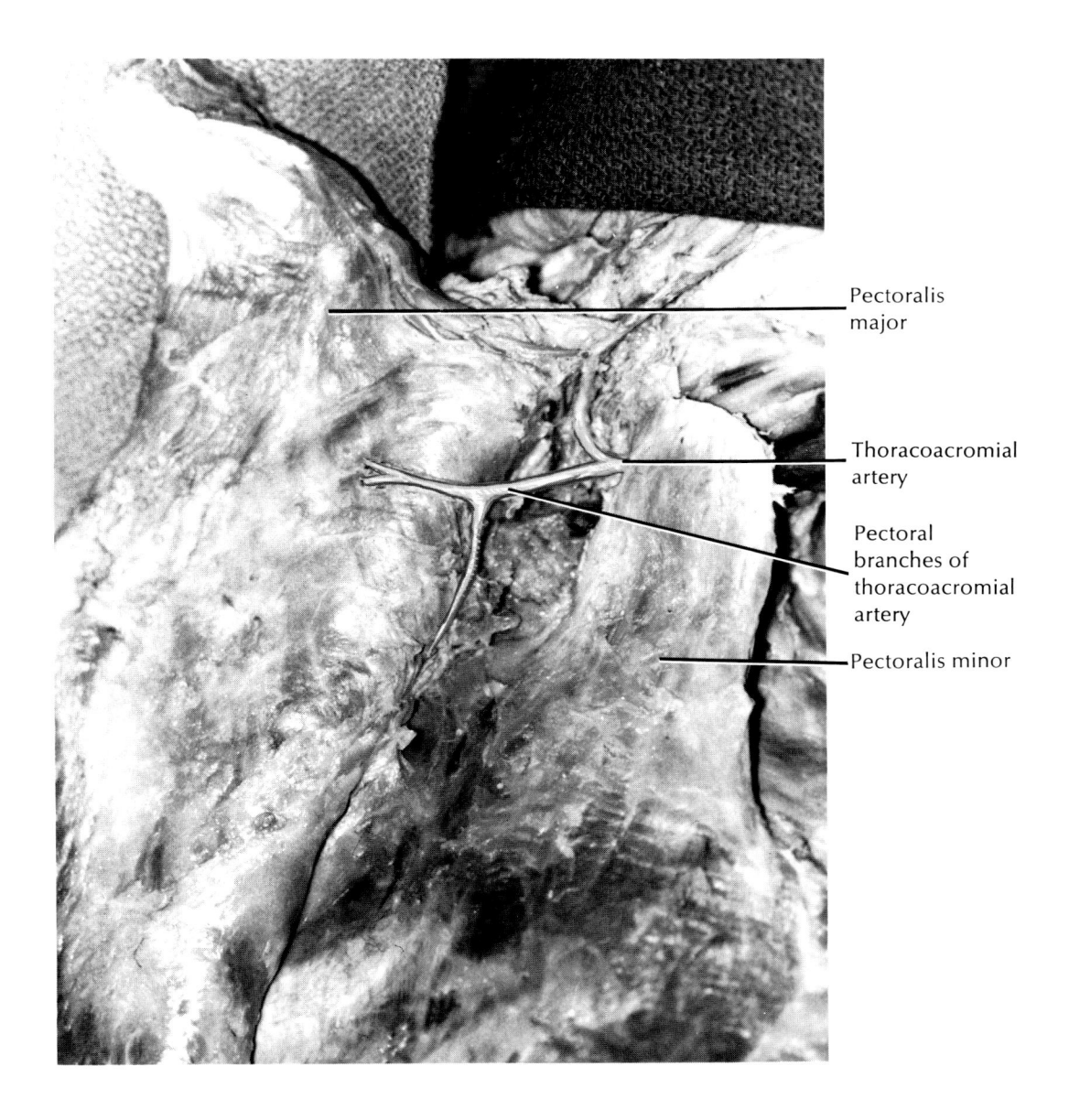

TRUNK: ANTERIOR TRUNK

Pectoralis major

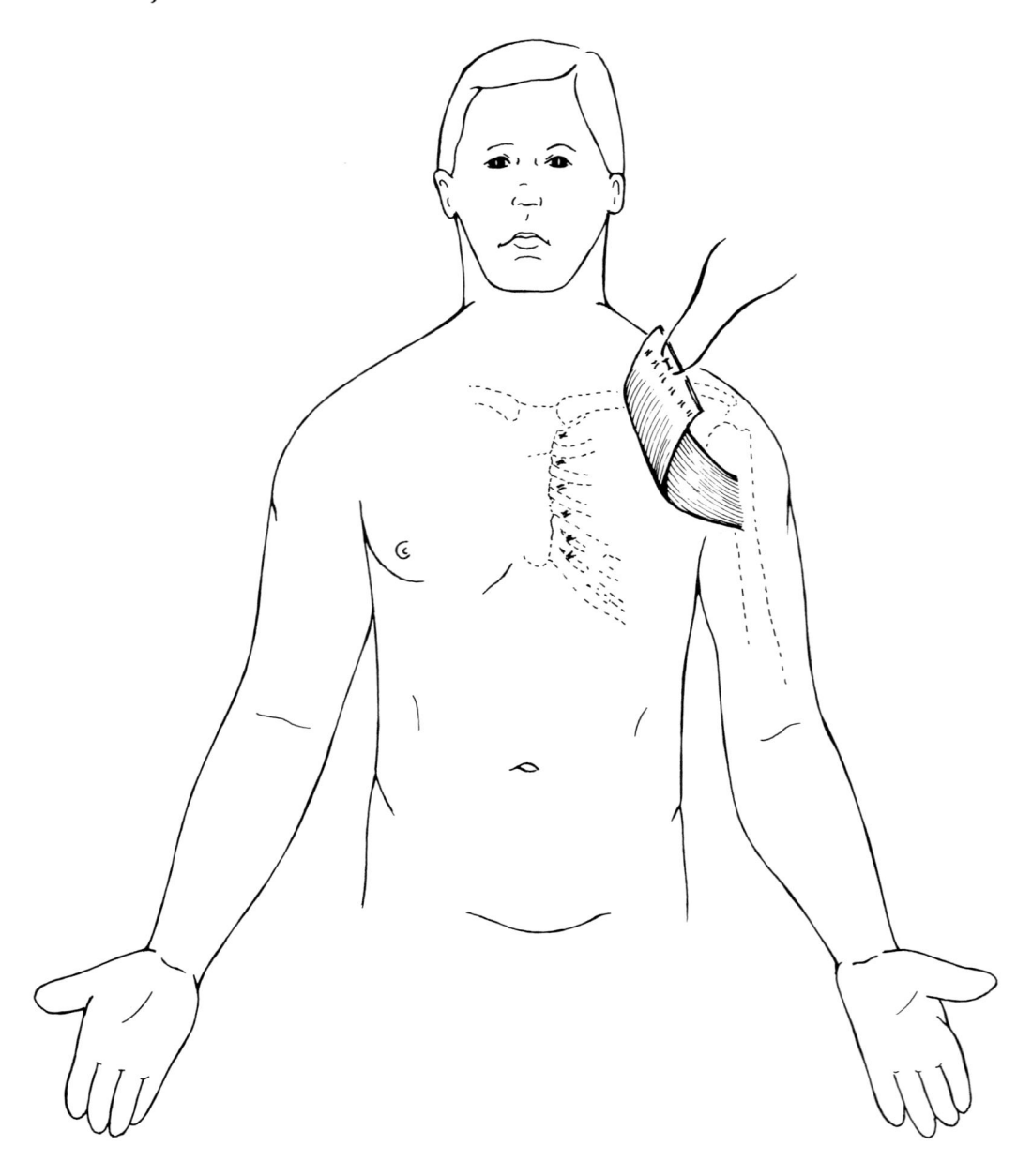

   The internal mammary artery, a branch of the first part of the subclavian artery, runs behind the cartilage of the upper six ribs close to the lateral border of the skin, and has perforating branches through the intercostal spaces to supply the pectoralis and the overlying skin. These perforating branches, which are variable in size, are the basis of the deltopectoral flap and contribute in part to the pectoralis major musculocutaneous flap. These may support a medially based pectoralis flap if the major pedicle is divided.

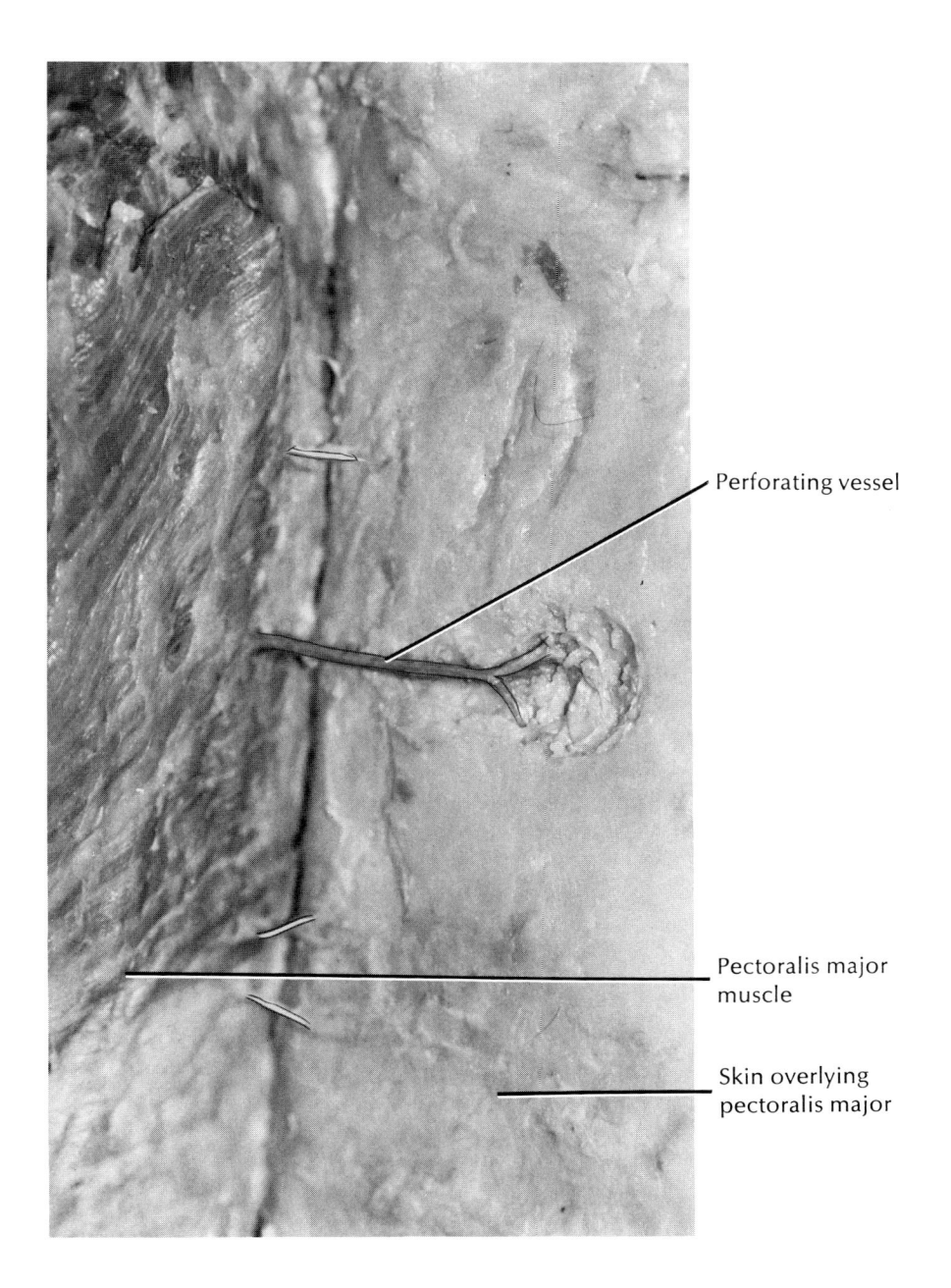

[Image: Dissection photograph with labels "Perforating vessel", "Pectoralis major muscle", and "Skin overlying pectoralis major"]

Through the intercostal perforators and another group of perforating musculocutaneous vessels through the clavicular head of the pectoralis major muscle, the entire area of overlying skin or parts of it can be elevated as a musculocutaneous island based on the major pedicle of the muscle.

TRUNK: ANTERIOR TRUNK

Pectoralis major

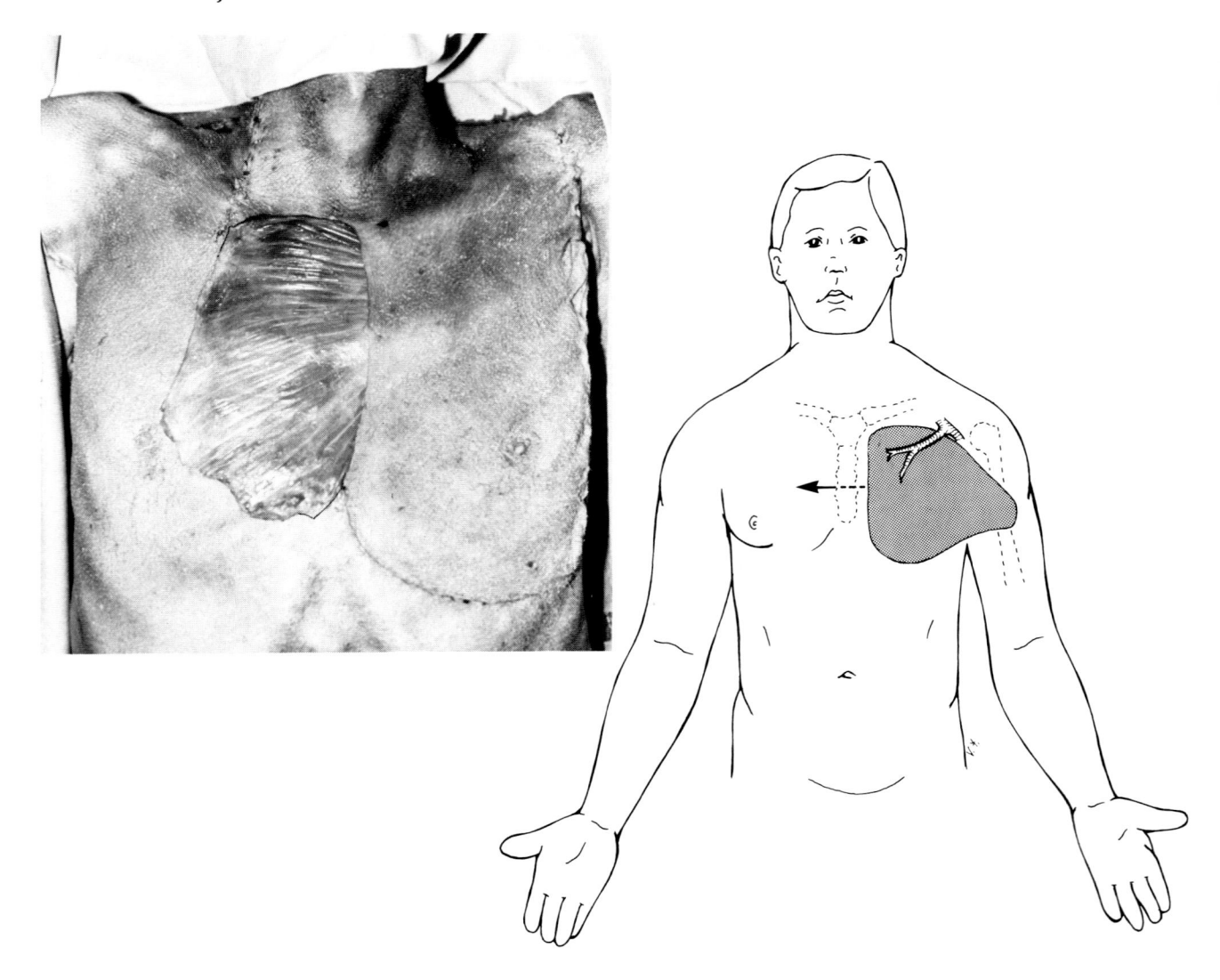

Arc of rotation

With the origin and insertion divided, and based on the major vascular pedicle in the apex of the axilla, this unit has a useful arc of rotation as a muscle or musculocutaneous flap.

It will easily advance medially as a muscle or musculocutaneous flap to cover the entire sternum and extend to the costochondral cartilage on the opposite side. If only the sternum is to be covered, division of the origin and elevation of the muscle off the chest wall may be all that is required, but for further advancement it is necessary to divide the insertion through an axillary incision, and for a musculocutaneous flap the inferior border of the skin territory along the sixth intercostal space will have to be incised.

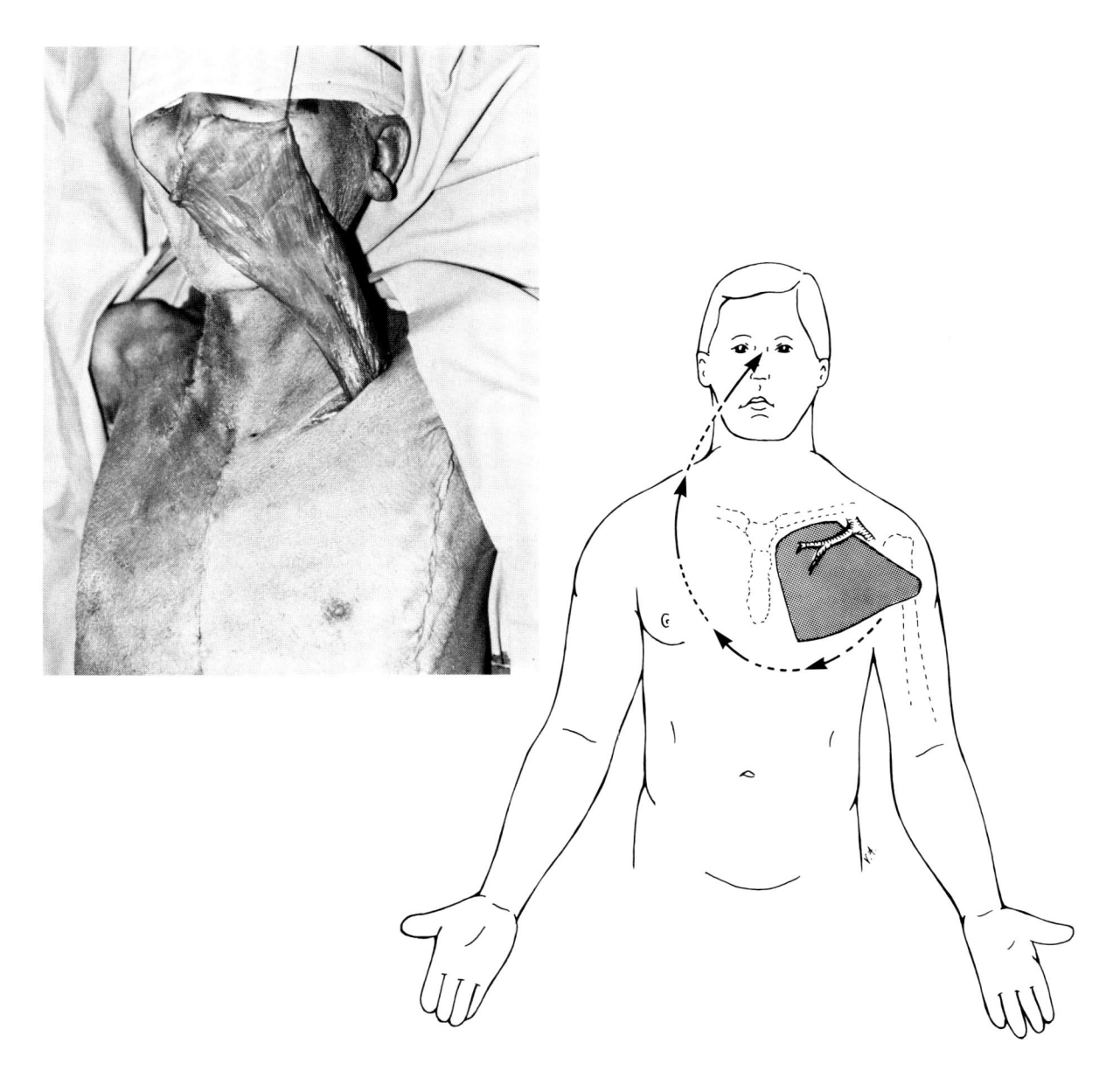

Either as a muscle flap or as an island musculocutaneous flap, the arc of rotation will easily reach the midface, where it is an extremely useful unit for reconstruction of the head and neck.

The muscle can be easily transposed into the axilla.

The donor defects for the muscle flap and for small musculocutaneous flaps can be closed directly, but larger islands will require skin grafting.

TRUNK: ANTERIOR TRUNK

Pectoralis major

Coverage of sternum

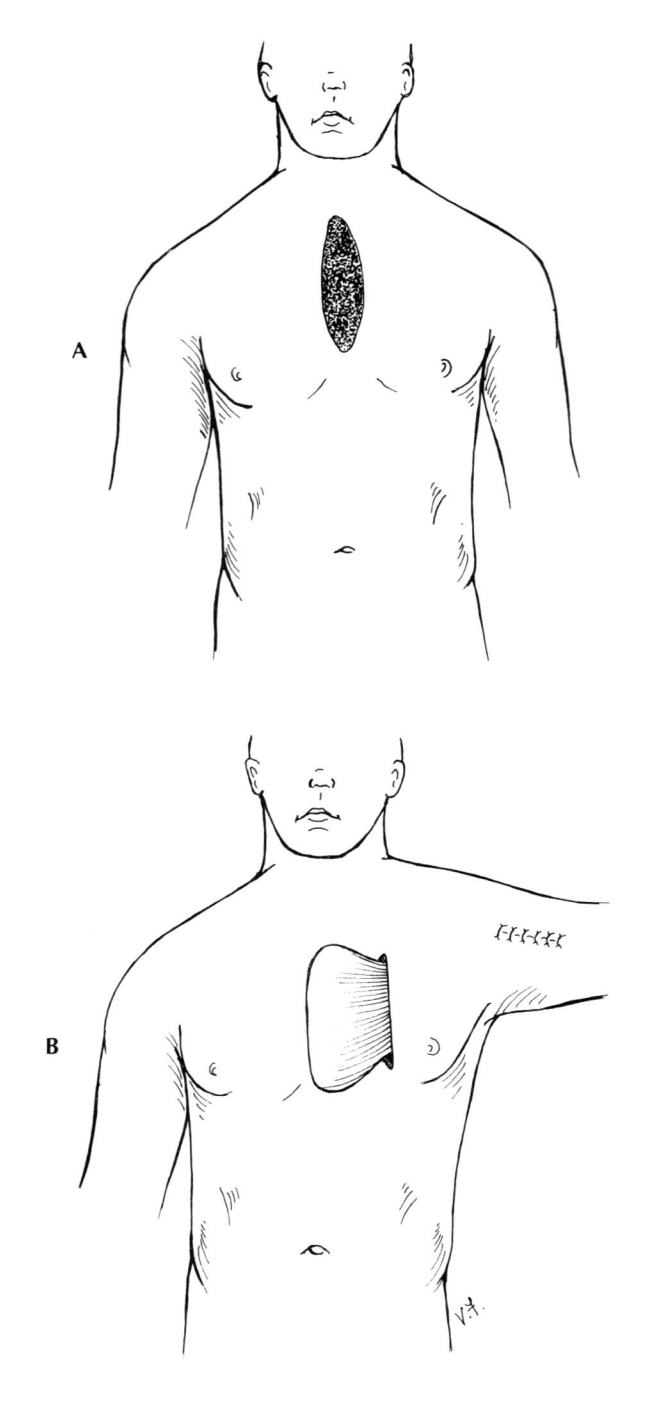

A and **B** Medial transposition for sternal cover.

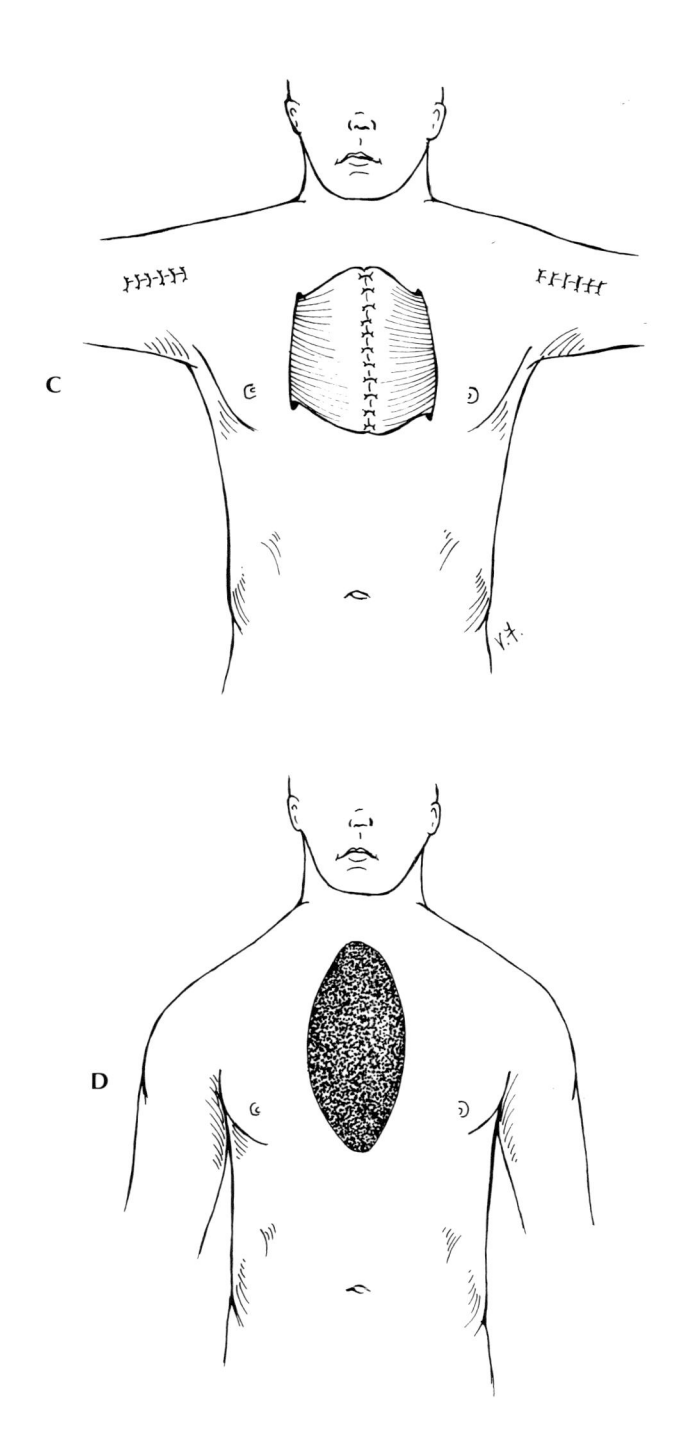

C and D Bilateral medial advancement for closure of larger defects.

TRUNK: ANTERIOR TRUNK

Pectoralis major

Elevation of flap

As an island musculocutaneous flap, the skin island is first outlined. With the arm abducted to 90°, the incision is made in the anterior axillary fold over the pectoralis major. The muscle is identified and dissected to its insertion into the lateral lip of the bicipital groove. It is divided close to the insertion, and the tendon is retracted medially to expose the pectoralis minor muscle. The major vascular pedicle enters the muscle at the level of the upper medial border of the pectoralis minor. The pedicle is identified and preserved. If necessary, the pedicle can be lengthened by identifying and ligating the branch to the pectoralis minor and the acromial and deltoid branches. The skin island is then incised and the margin of the skin island is temporarily sutured to the underlying muscle to protect the musculocutaneous perforators. Skin flaps are then raised above and below the island to expose the origin of the muscle. The origin is divided and the muscle is elevated off the chest wall. The perforating vessels from the internal mammary artery are identified and ligated during this dissection.

The muscle or musculocutaneous unit may be advanced medially to provide sternal coverage. In this case it is not necessary to identify the vascular pedicle. The muscle with or without the overlying skin is freed from the chest wall through a sternal incision; the clavicular origin is divided through the same incision. By division of the muscle origin, advancement of this flap is facilitated and the reach of the muscle increased.

Anatomic dissection demonstrates vascular connections between the muscle and its origin from the ribs. Each rib may be included with the muscle for both soft tissue and mandible reconstruction.

Precautions

- This is a reliable and useful flap, but its usefulness as a musculocutaneous flap in the female may be limited by the breast.
- Use of this flap as a musculocutaneous flap will alter nipple position.
- During the dissection to identify the vascular pedicle, especially with division of branches to lengthen the pedicle, care should be taken to avoid injury to the axillary vein.

TRUNK: ANTERIOR TRUNK

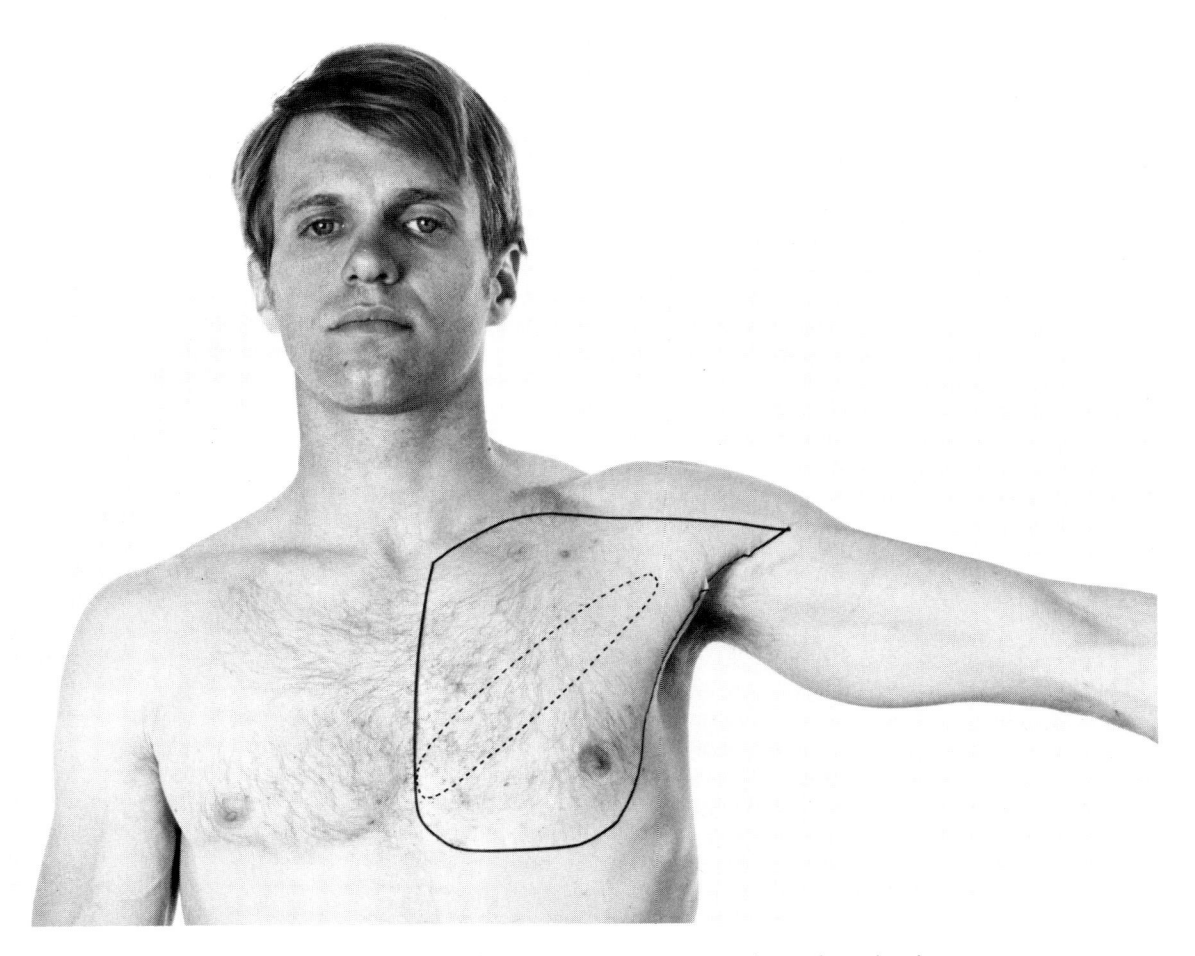

Solid line outlines muscle margin Dotted line outlines skin island

TRUNK: ANTERIOR TRUNK

Pectoralis major

Musculocutaneous flap for closure of sternal defect

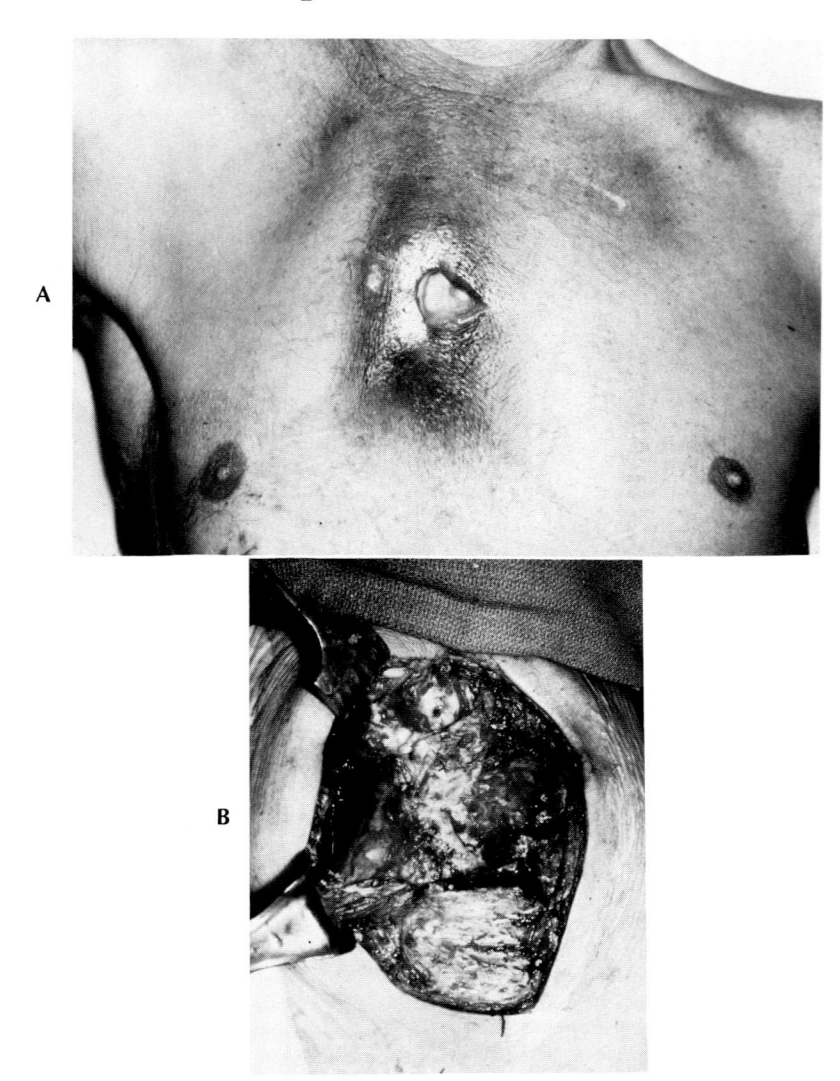

A   Radiation necrosis ulcer over sternum.
B   Resultant defect after wide excision of ulcer and sternum.

TRUNK: ANTERIOR TRUNK

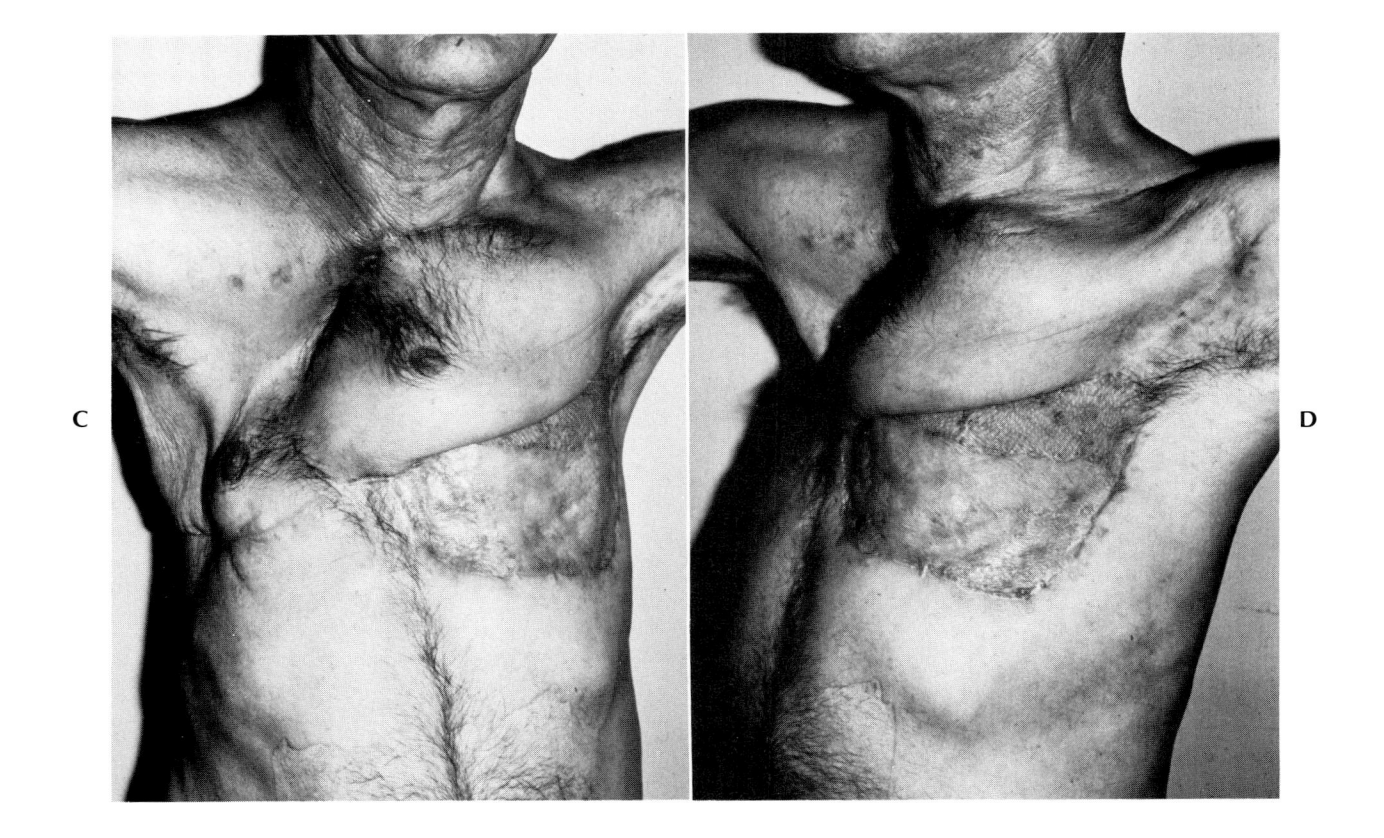

C Pectoralis major musculocutaneous flap incorporating entire overlying cutaneous territory. Note abnormal nipple position.

D Secondary defect was grafted. Single-stage reconstruction of anterior chest wall.

TRUNK: ANTERIOR TRUNK

Pectoralis major

Musculocutaneous island flap for head and neck reconstruction

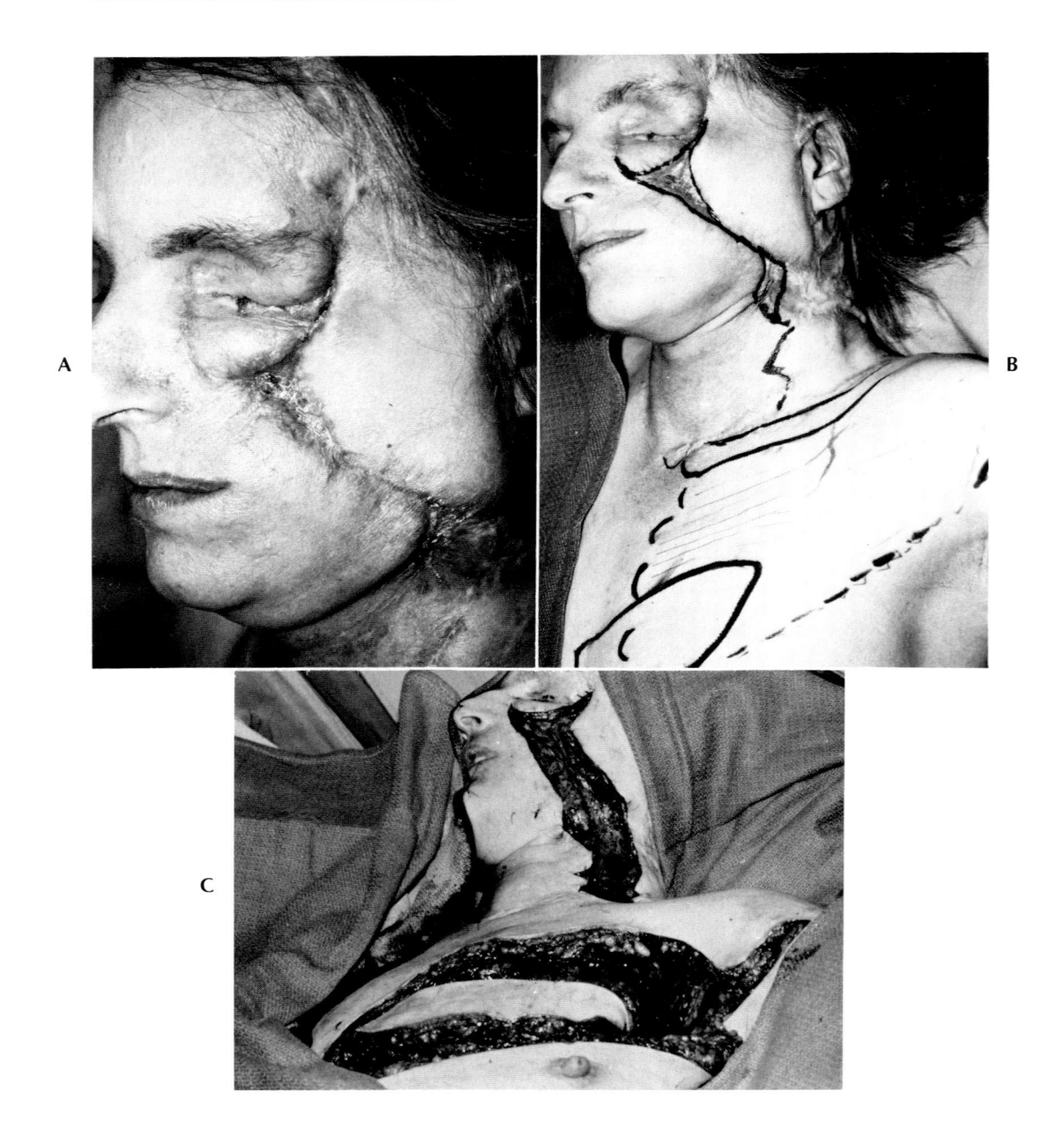

A Soft tissue deformity of cheek secondary to shotgun wound.

B Flap and island of skin outlined.

C Skin island and muscle elevated. Facial scars excised and released.

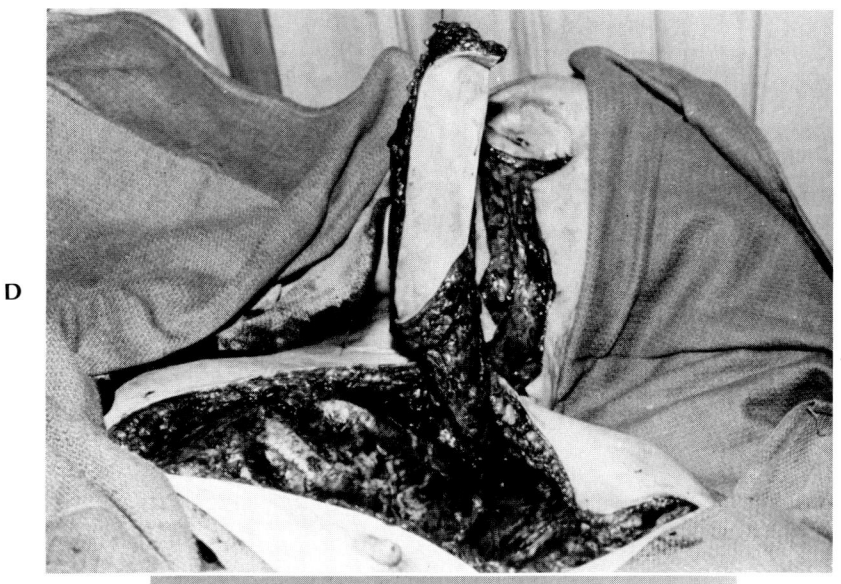

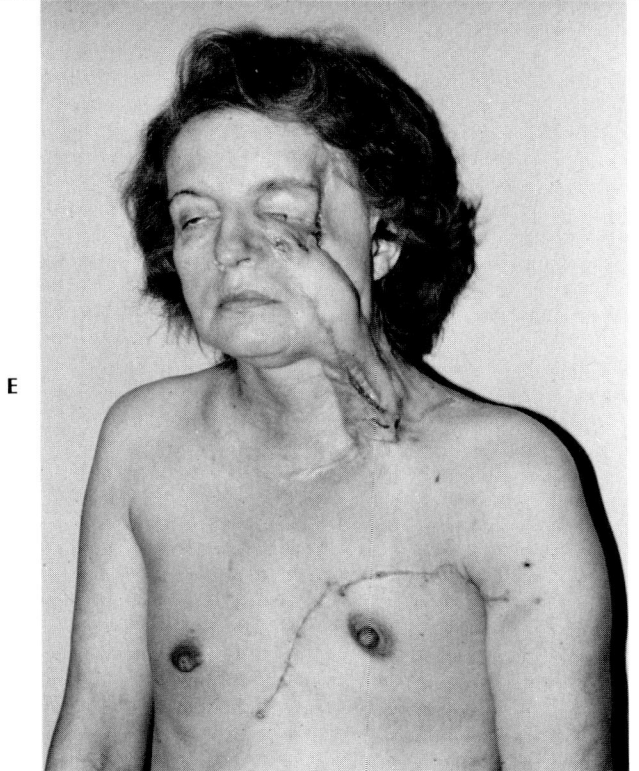

D   Muscle and skin island rotated superiorly.

E   Healed flap. Donor defect closed primarily with minimal elevation of nipple position.

TRUNK: ANTERIOR TRUNK

Pectoralis major

Head and neck reconstruction

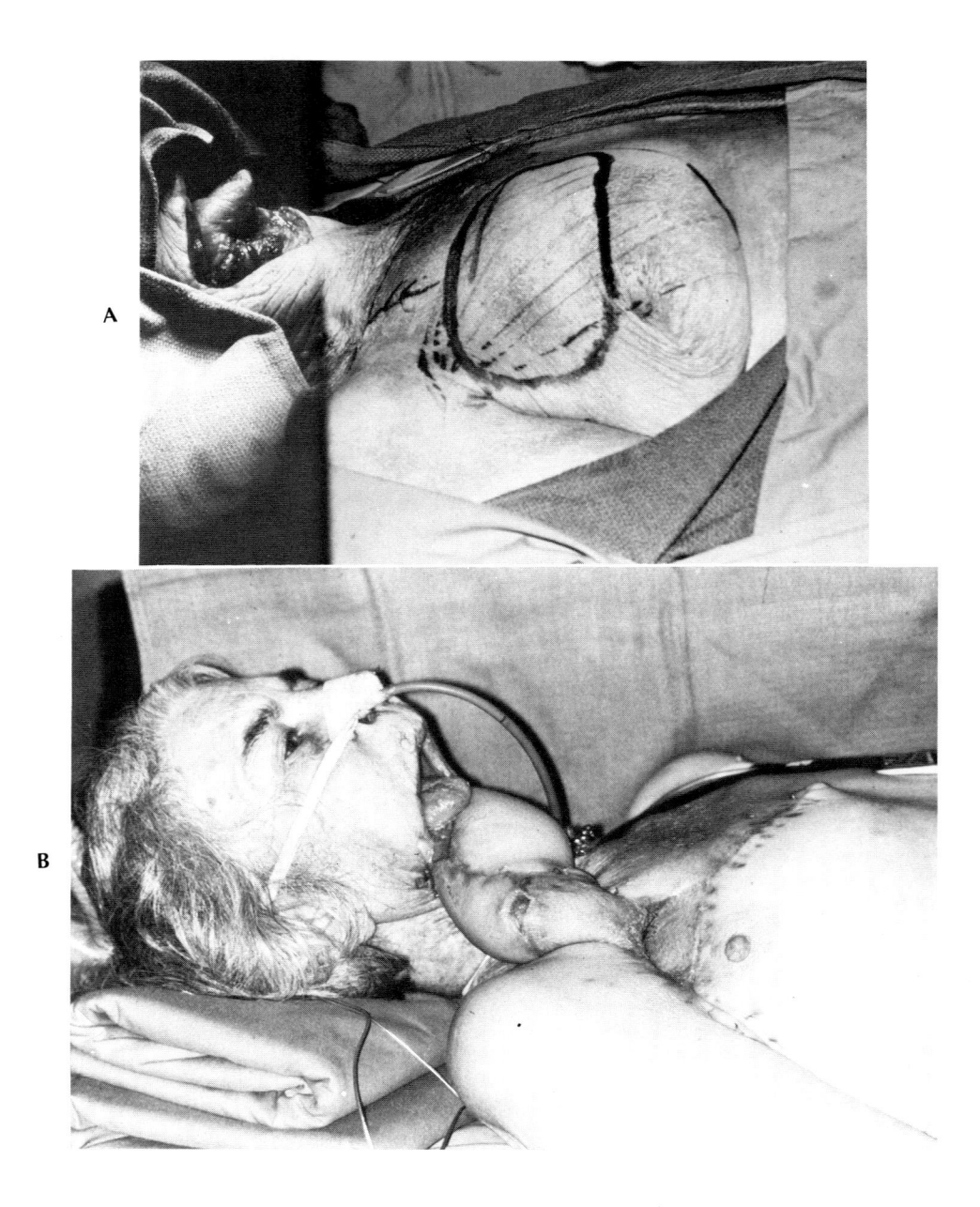

A Composite resection for carcinoma of floor of mouth. Pectoralis major musculocutaneous flap designed.

B Pectoralis major musculocutaneous flap. Donor defect closed primarily.

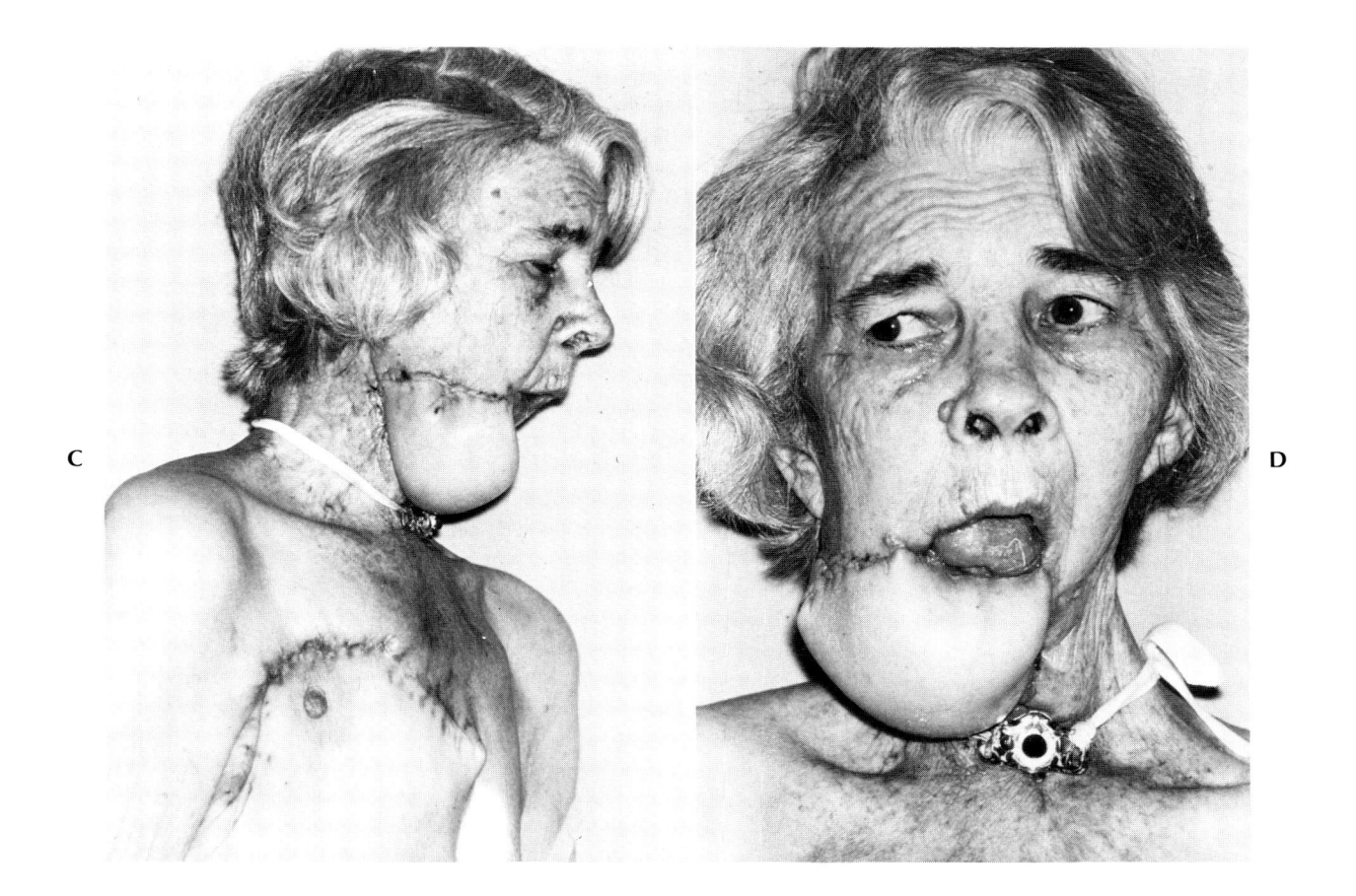

C Pectoralis major musculocutaneous flap inset.
D Reconstruction of floor of mouth and chin with pectoralis major musculocutaneous island flap.

Serratus anterior

MUSCLE FLAP
MUSCULOCUTANEOUS FLAP
FREE FLAP

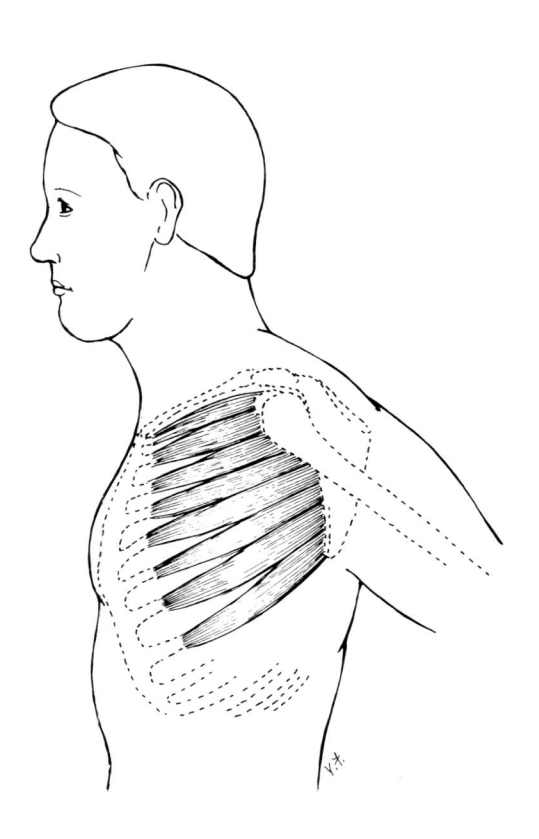

Applications

Reconstruction of:
BREAST

Coverage of:
CHEST WALL
NECK
AXILLA
BREAST IMPLANT

Free flap:
DISTANT COVERAGE

TRUNK: ANTERIOR TRUNK

Serratus anterior

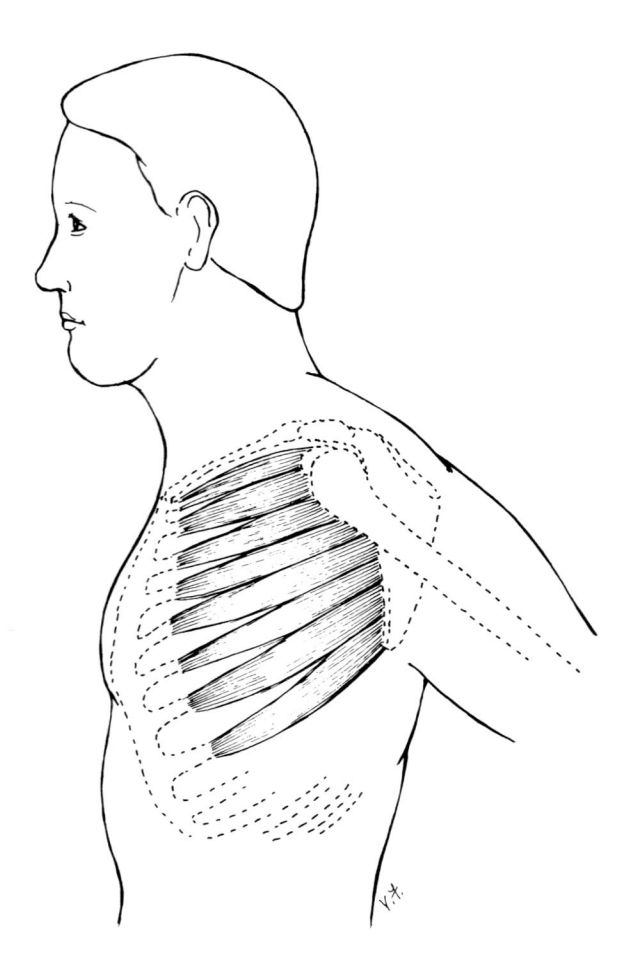

The serratus anterior, a thin muscle, is located on the anterolateral chest between the anterior axillary line and the border of the scapula. The muscle has a close relationship on its outer surface with the pectoralis minor muscle anteriorly and the latissimus dorsi muscle and scapula posteriorly. The costal surface has contact with the ribs and intercostal muscles.

Origin: Outer surface of the upper eight or nine ribs.
Insertion: Costal surface of the vertebral border of scapula.

TRUNK: ANTERIOR TRUNK

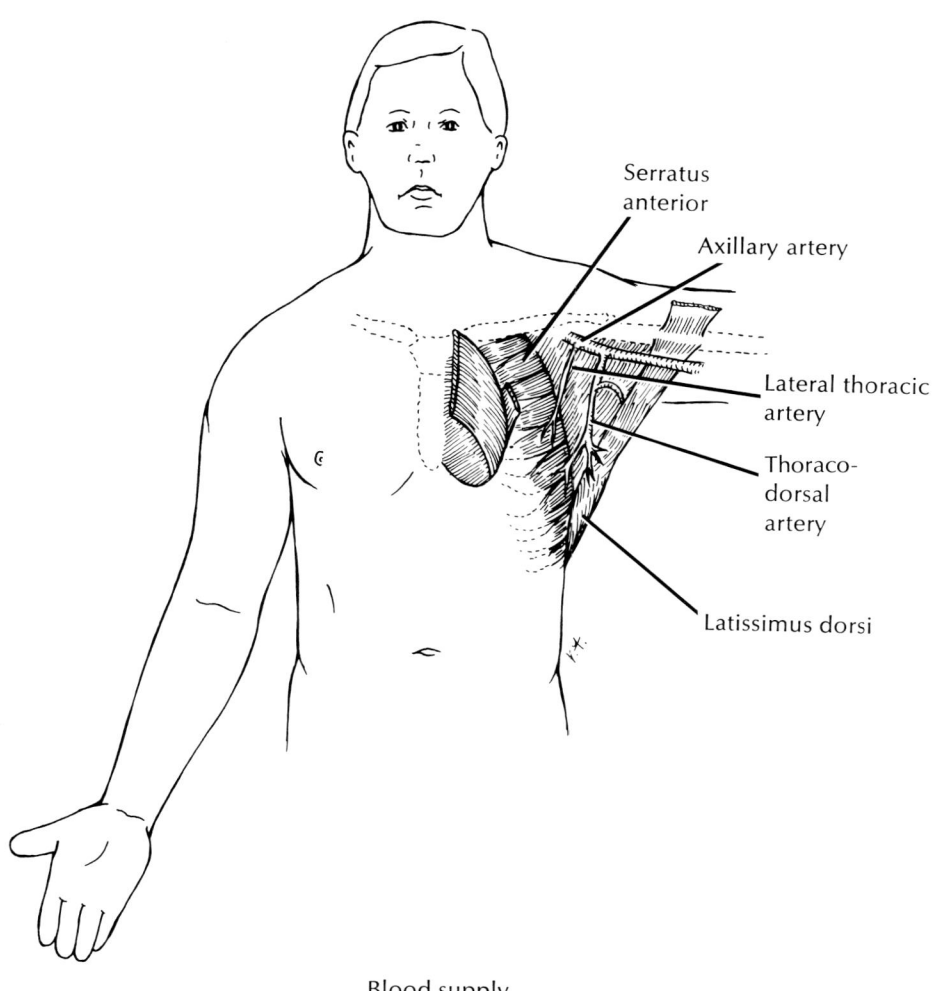

Blood supply

Nerve supply:
    Motor — Long thoracic nerve.
    Sensory — Intercostobrachial nerve.
Function: This is an expendable muscle. However, with the loss of scapular adduction, there may be objectionable winging of the scapula.

TRUNK: ANTERIOR TRUNK

Serratus anterior

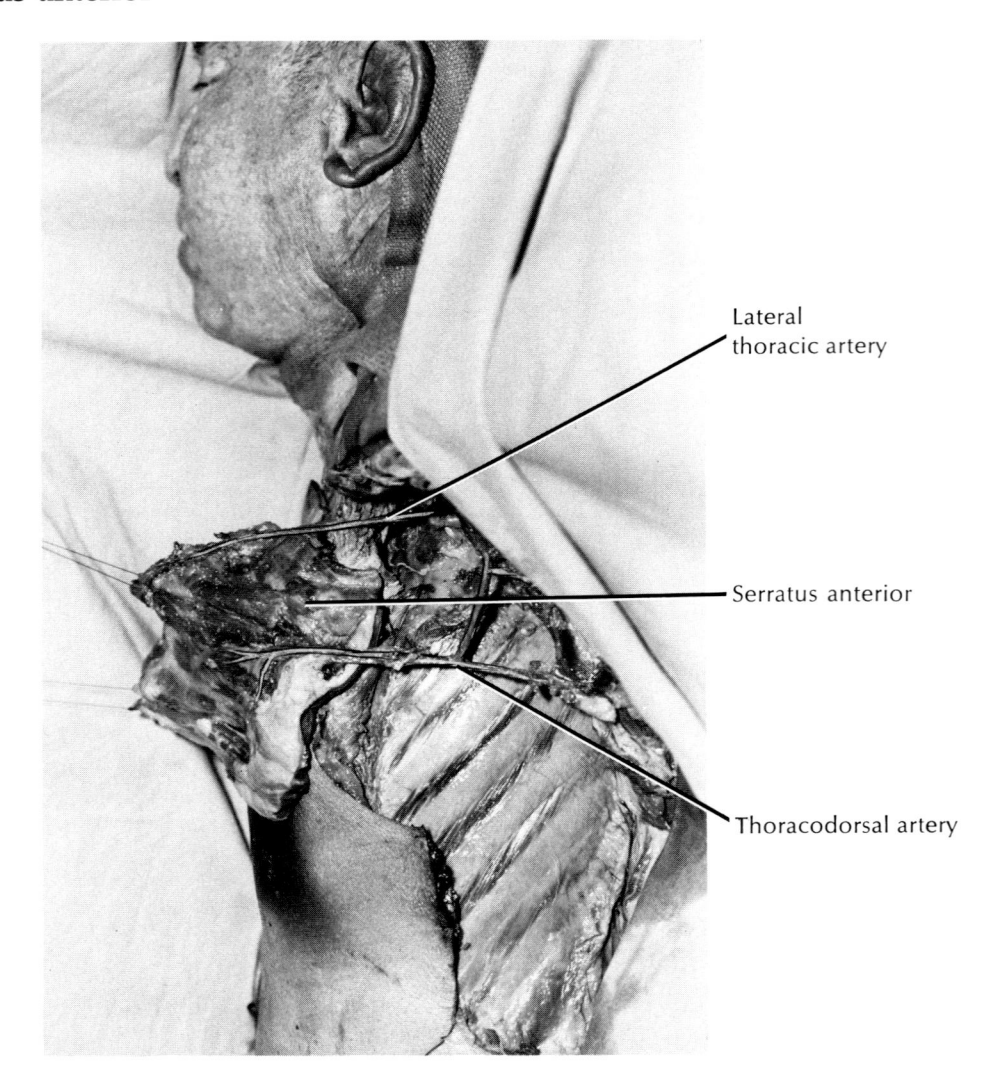

Blood supply

This muscle receives a dual blood supply that enters the proximal superior muscle within the axilla. The long thoracic artery, the second major branch of the axillary artery, enters directly onto the lateral surface of the muscle. The artery then courses inferiorly with multiple branches into the muscle.

The second major blood supply to this muscle is via the thoracodorsal artery. Within the axilla, this artery has a branch entering the posterolateral aspect of the muscle before the artery terminates in the latissimus dorsi muscle.

TRUNK: ANTERIOR TRUNK

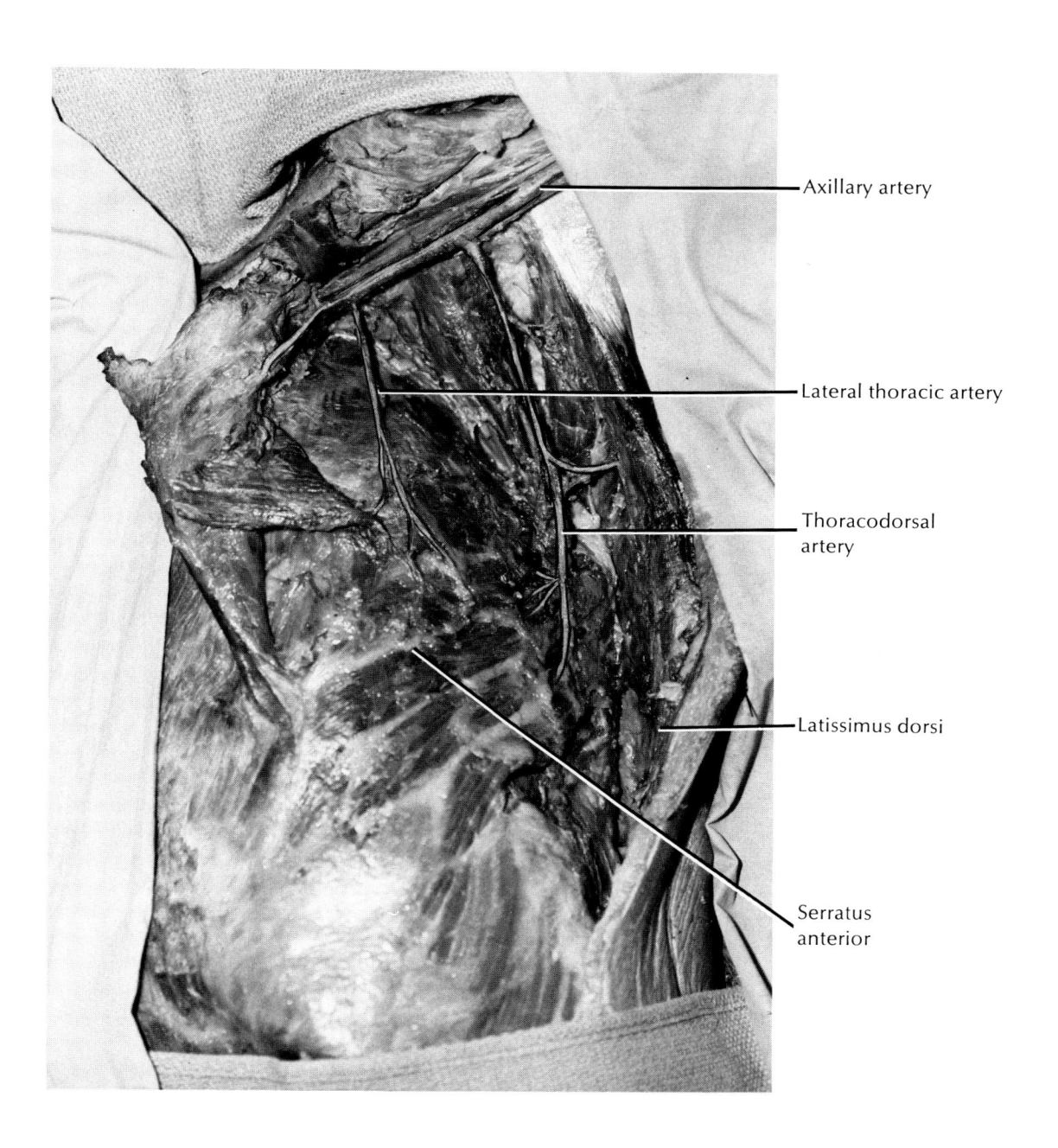

- Axillary artery
- Lateral thoracic artery
- Thoracodorsal artery
- Latissimus dorsi
- Serratus anterior

TRUNK: ANTERIOR TRUNK

Serratus anterior

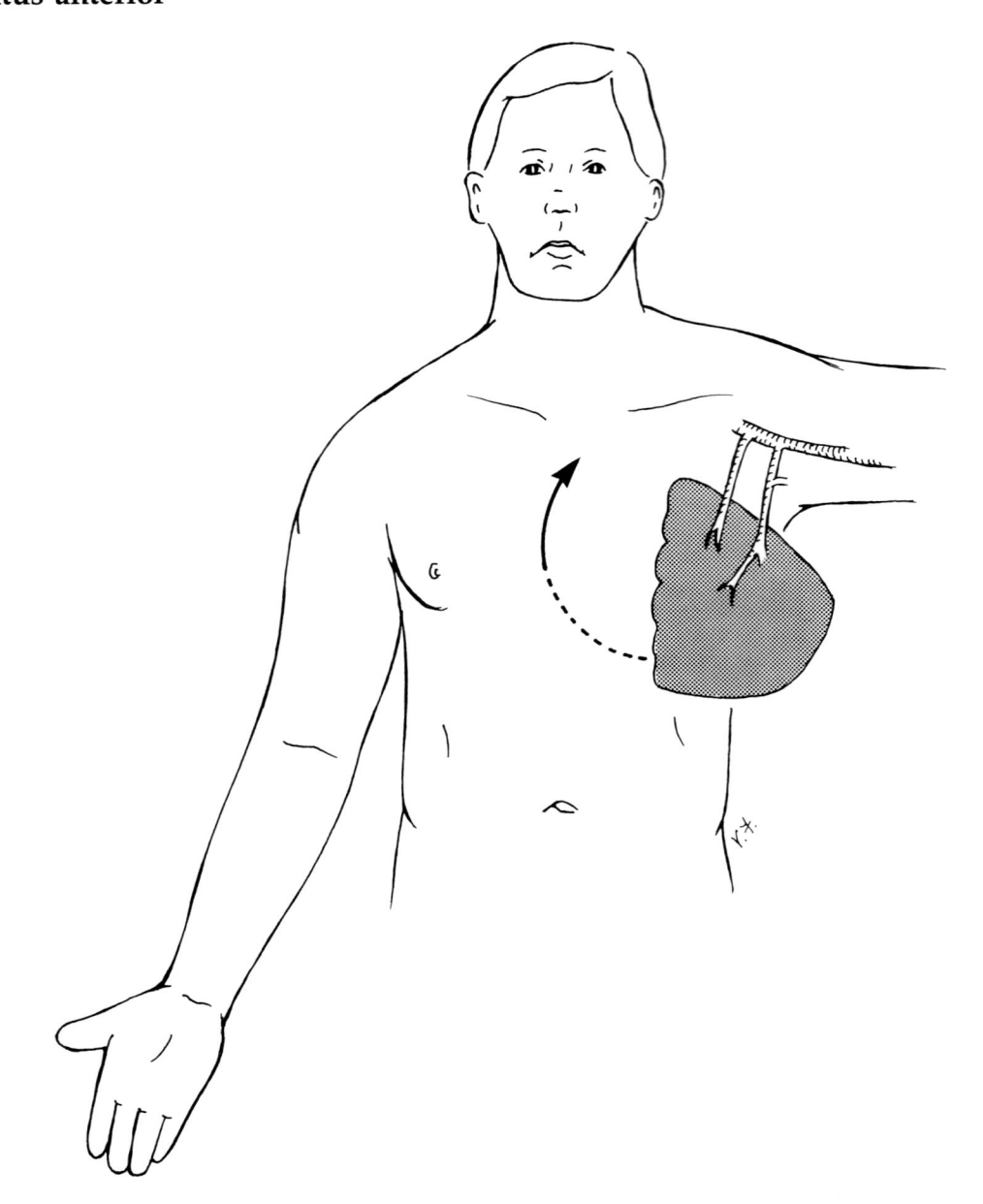

Arc of rotation

Following release of both the origin and insertion, this muscle has a wide arc of rotation based in the axilla on the long thoracic artery and the posterosuperior arterial branch from the thoracodorsal artery. This arc is primarily directed anteriorly over the chest and neck. This arc is increased by division of the posterosuperior arterial branch from the thoracodorsal artery.

TRUNK: ANTERIOR TRUNK

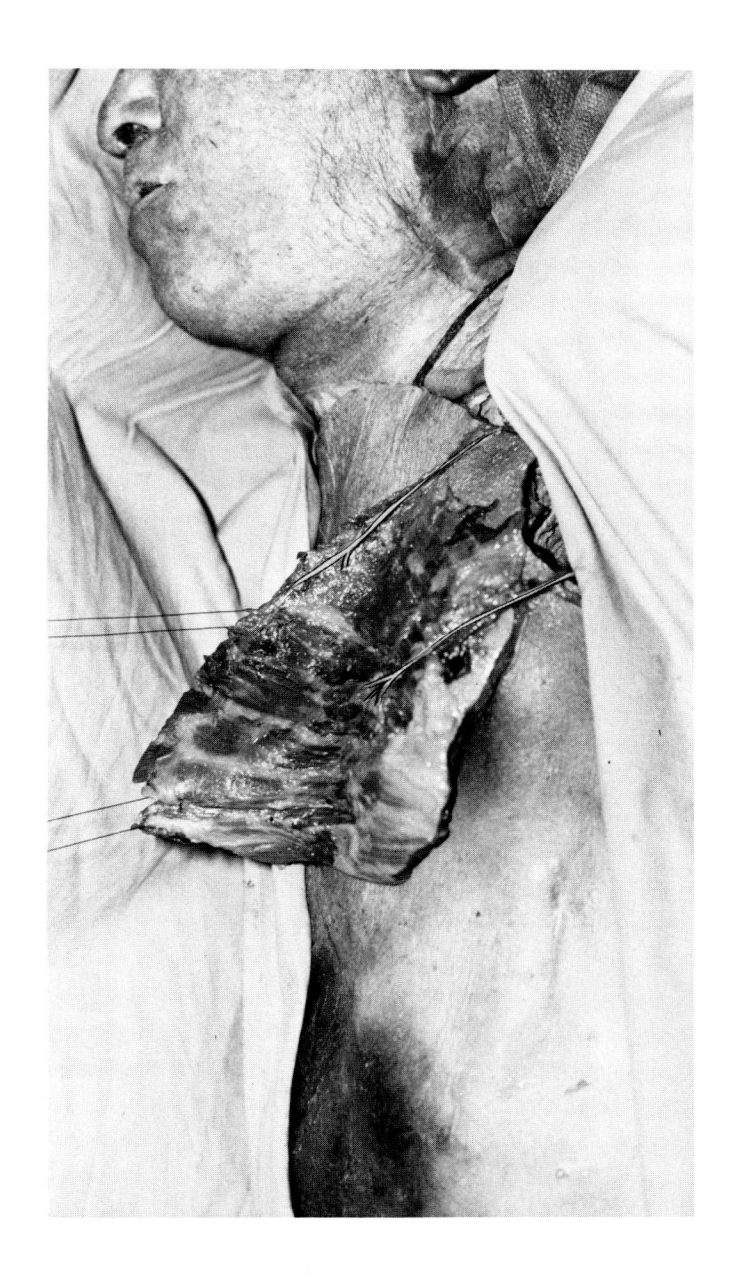

TRUNK: ANTERIOR TRUNK

Serratus anterior

Elevation of flap

This muscle is best approached through a vertical incision in the mid axillary line. However, the muscle may be adequately visualized through any anterolateral chest incision. The muscle is elevated from its costal relationship inferiorly by blunt dissection. The origin is detached from the ribs along the anterior axillary line. In the region of the pectoralis minor muscle as it crosses the serratus muscle in the axilla, the long thoracic artery is located close to the muscle origin. If a large muscle flap is needed, this proximal medial dissection must be performed with the long thoracic artery under visualization. The muscle may be advanced anteriorly with the insertion intact to provide anterior chest coverage. This maneuver may be useful in providing coverage for an exposed breast implant or for lateral implant coverage in breast reconstruction. However, only after division of the insertion of the muscle along the vertebral border of the scapula does the muscle have a wide anterior arc of rotation. For a greater length in this arc, the muscle can be rotated as an island flap on the long thoracic artery. The donor area can be closed primarily.

The muscle has a cutaneous territory between the pectoralis major and the latissimus dorsi that can be carried with the muscle in an anterior transposition flap. With this limited cutaneous territory the donor area can be closed primarily.

This unit has the potential for free transfer either as a muscle or musculocutaneous flap. The long thoracic artery makes a long pedicle suitable for microsurgical free transfer. A rib at the muscle origin could be transferred with this unit when required for reconstructive problems.

Precautions

- Use of this muscle in transfer transposition or free flaps will result in winged scapula deformity.
- Until more clinical experience with this muscle unit is gained, the use of this muscle in anterior chest reconstruction is not indicated when the latissimus dorsi muscle is available.

TRUNK: ANTERIOR TRUNK

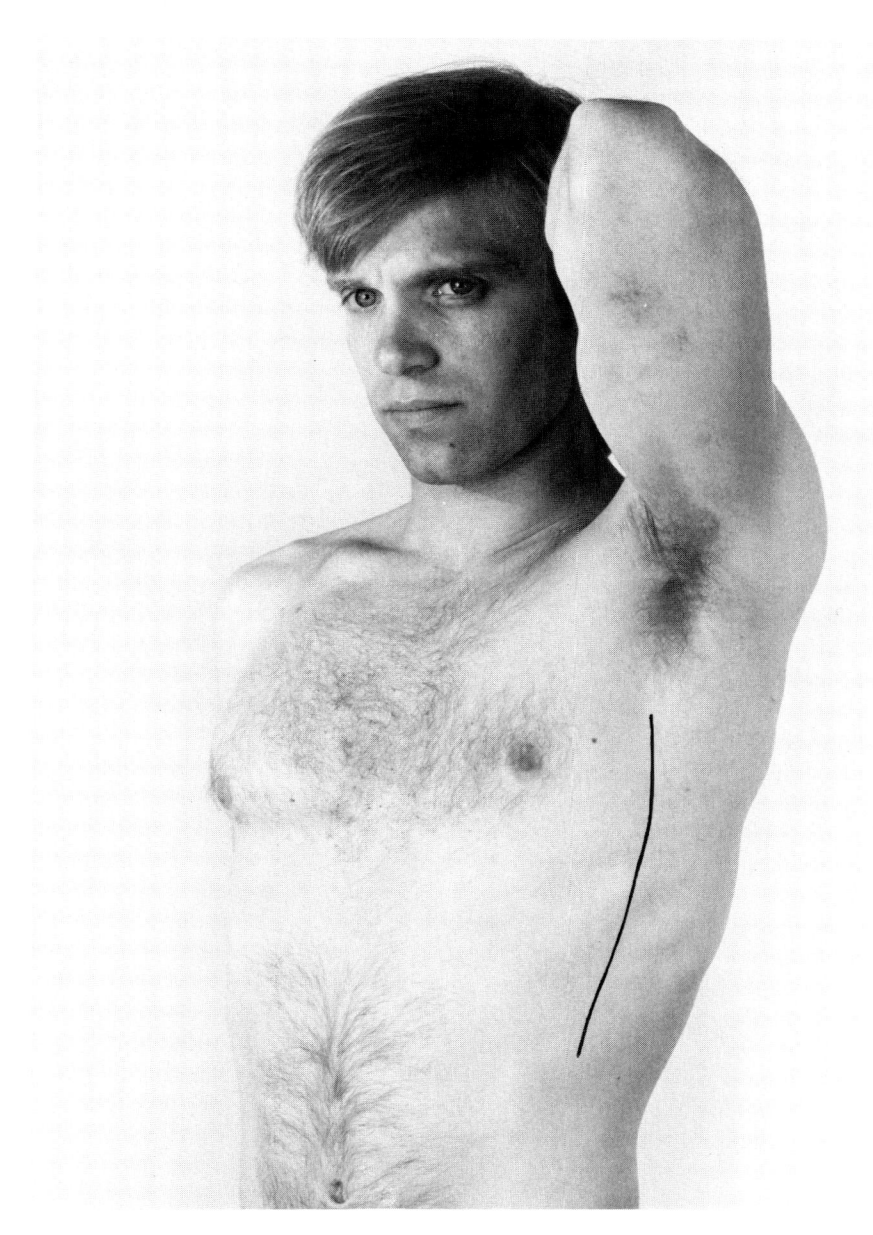

Skin incision

Rectus abdominis

MUSCLE FLAP
MUSCULOCUTANEOUS FLAP
FREE FLAP

Applications

Reconstruction of:

ABDOMINAL WALL
CHEST WALL

Coverage of:

THORAX
ABDOMINAL WALL
GROIN

Free flap:

DISTANT COVERAGE

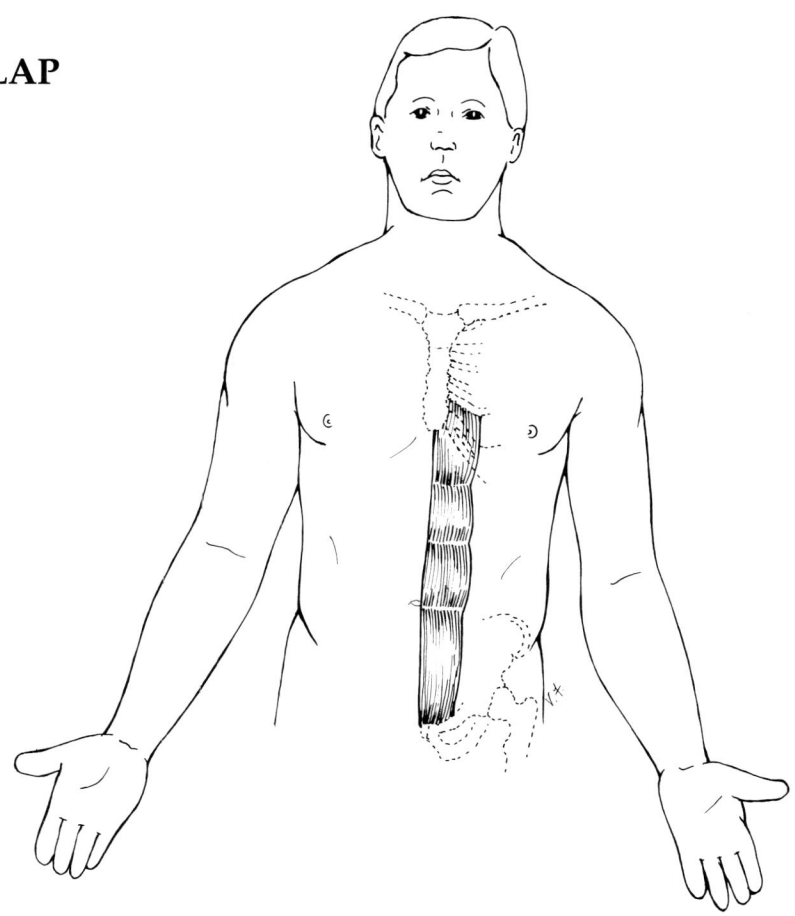

TRUNK: ANTERIOR TRUNK

Rectus abdominis

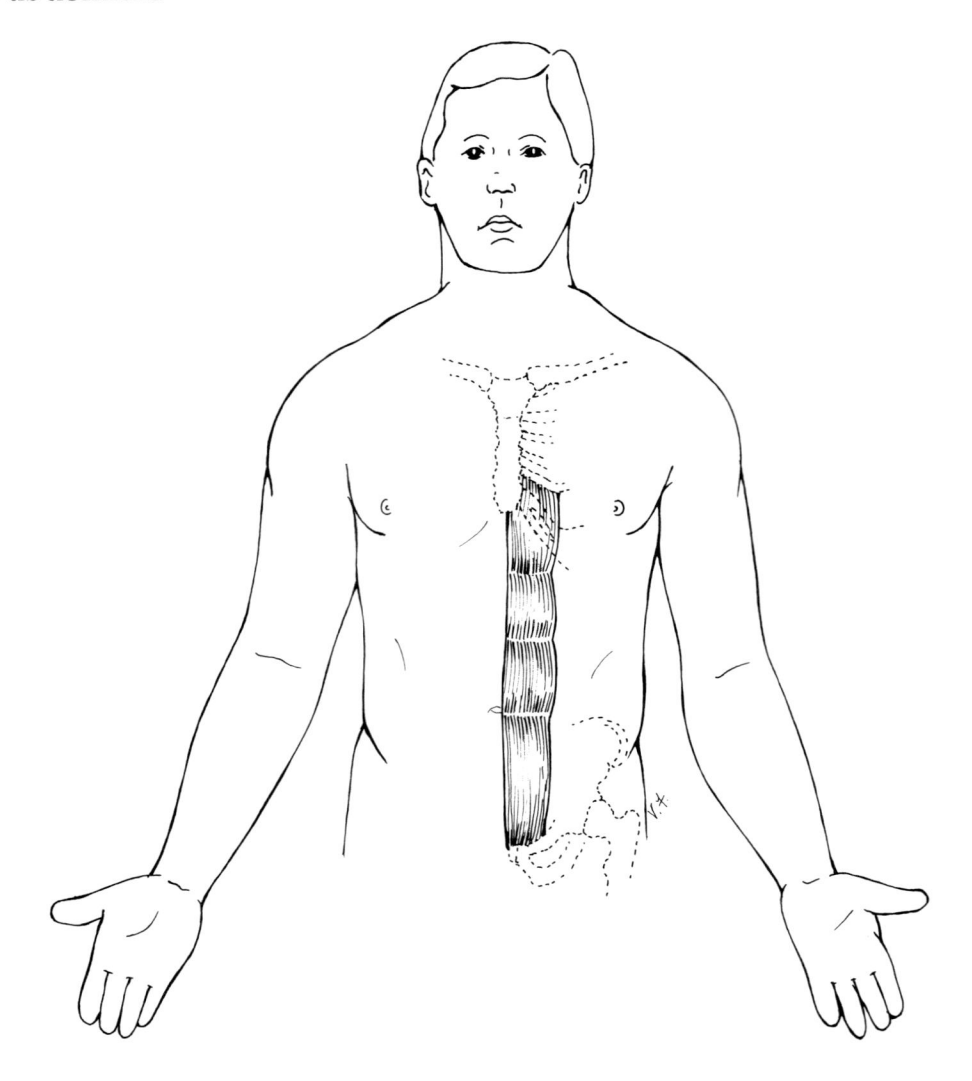

The rectus abdominis is a long, flat muscle that is traversed by three tendinous intersections. It is ensheathed by the anterior and posterior rectus sheaths, and extends the length of the anterior abdominal wall.

- **Origin:** The muscle has two heads of origin: a lateral head from the pubic crest and a medial head from the front of the symphysis pubis. The muscle is broader at its insertion.
- **Insertion:** The muscle is inserted by three slips into the costal cartilage of the fifth, sixth, and seventh ribs. These slips lie deep to the pectoralis major muscle.

TRUNK: ANTERIOR TRUNK

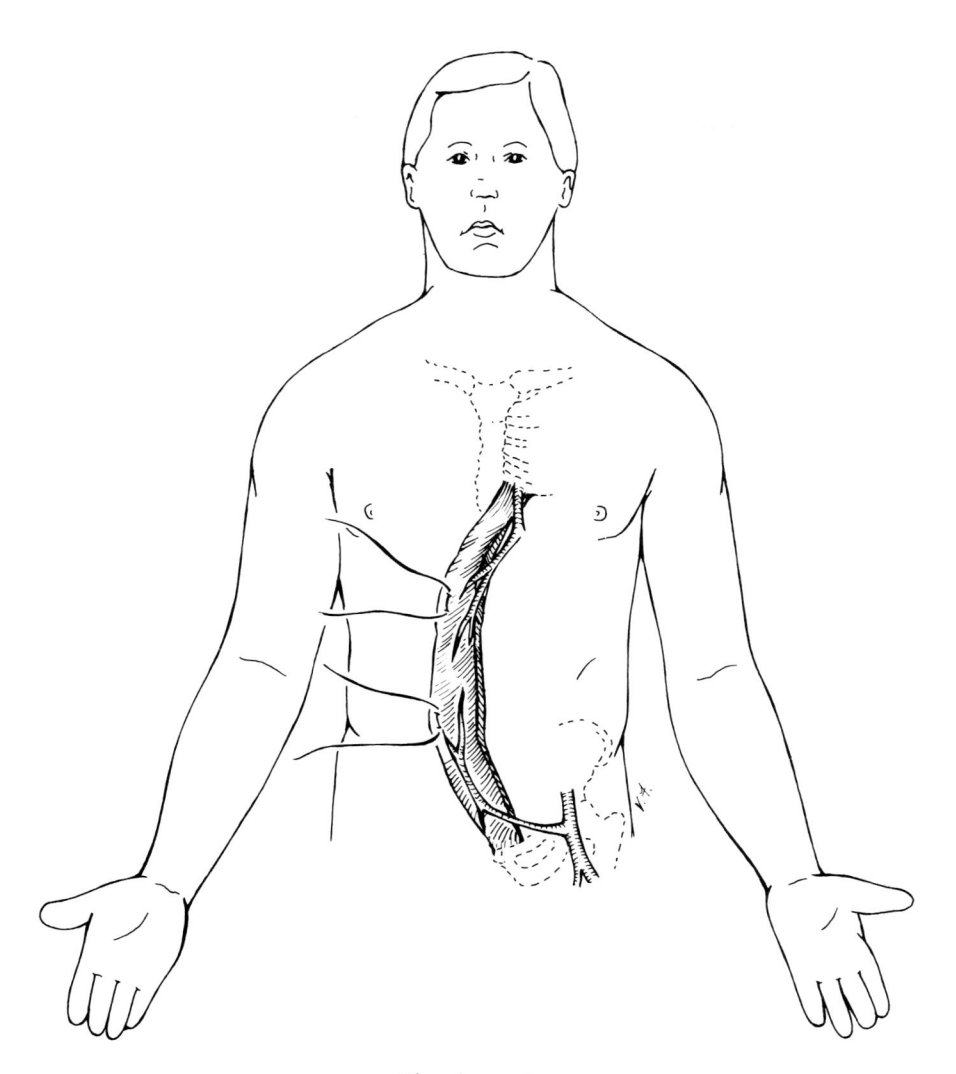

Blood supply

Nerve supply: The nerve supply to this muscle is segmental. Motor branches from the seventh through twelfth intercostal nerves innervate the muscle on its deep surface.

Function: The rectus abdominis muscle flexes the vertebral column and tightens the abdominal wall. It is a relatively expendable muscle.

TRUNK: ANTERIOR TRUNK

Rectus abdominis

- Superior epigastric artery
- Rectus abdominis
- Inferior epigastric artery

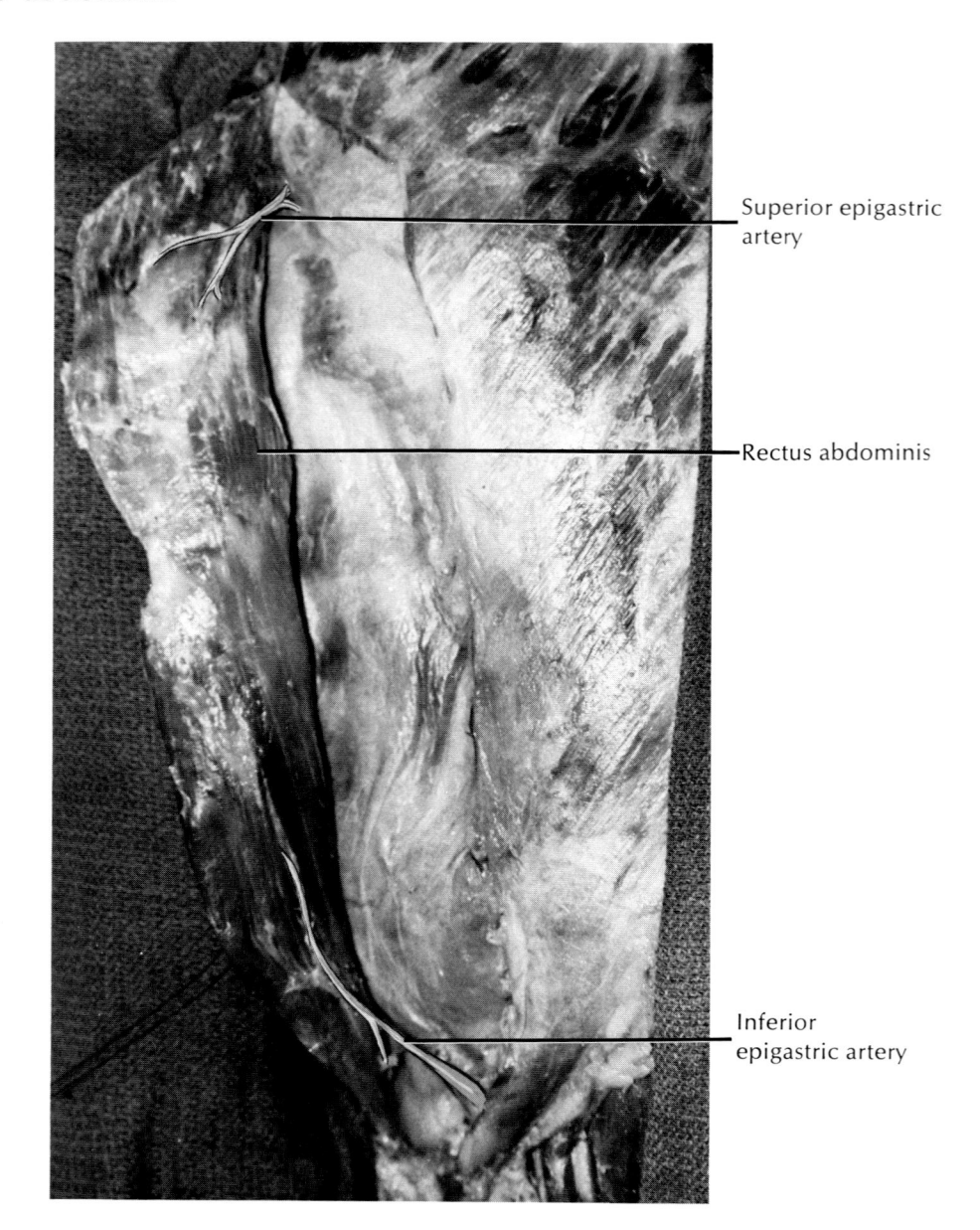

Blood supply

The rectus abdominis has two major pedicles, one entering close to the origin and the other close to the insertion. Each pedicle will support just over one-half the length of the muscle.

The upper pedicle is a continuation of the internal mammary artery, and the lower pedicle is a branch of the external iliac artery.

TRUNK: ANTERIOR TRUNK

Superior epigastric artery

Posterior rectus sheath

Rectus abdominis

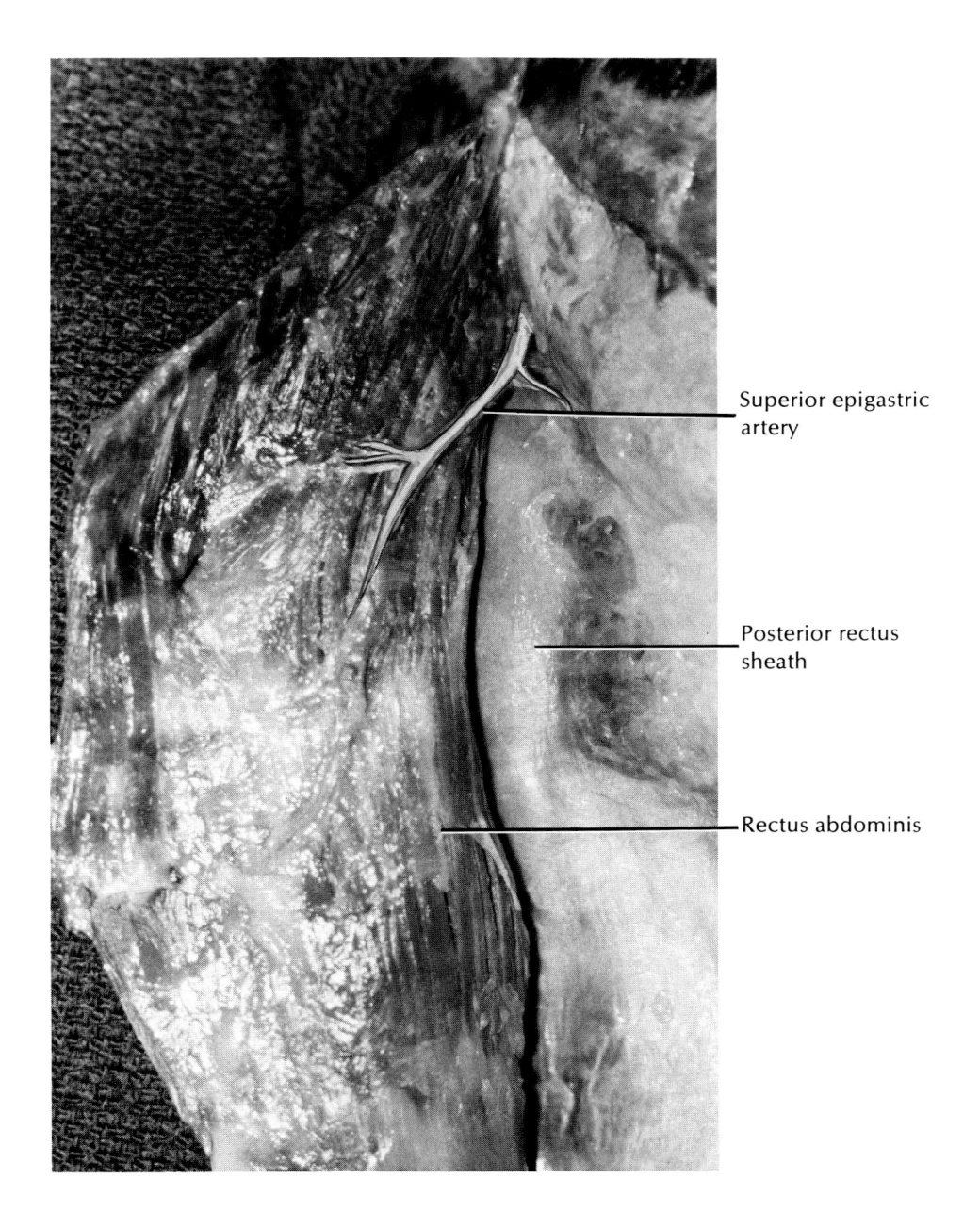

The internal mammary artery divides into the musculophrenic and superior epigastric arteries in the sixth intercostal space. The superior epigastric artery continues distally between the leaves of the diaphragm to enter the sheath of the rectus abdominis. The vessel first lies deep to the muscle between the muscle and the posterior sheath; it then enters the muscle and runs parallel longitudinally with the muscle fibers.

TRUNK: ANTERIOR TRUNK

Rectus abdominis

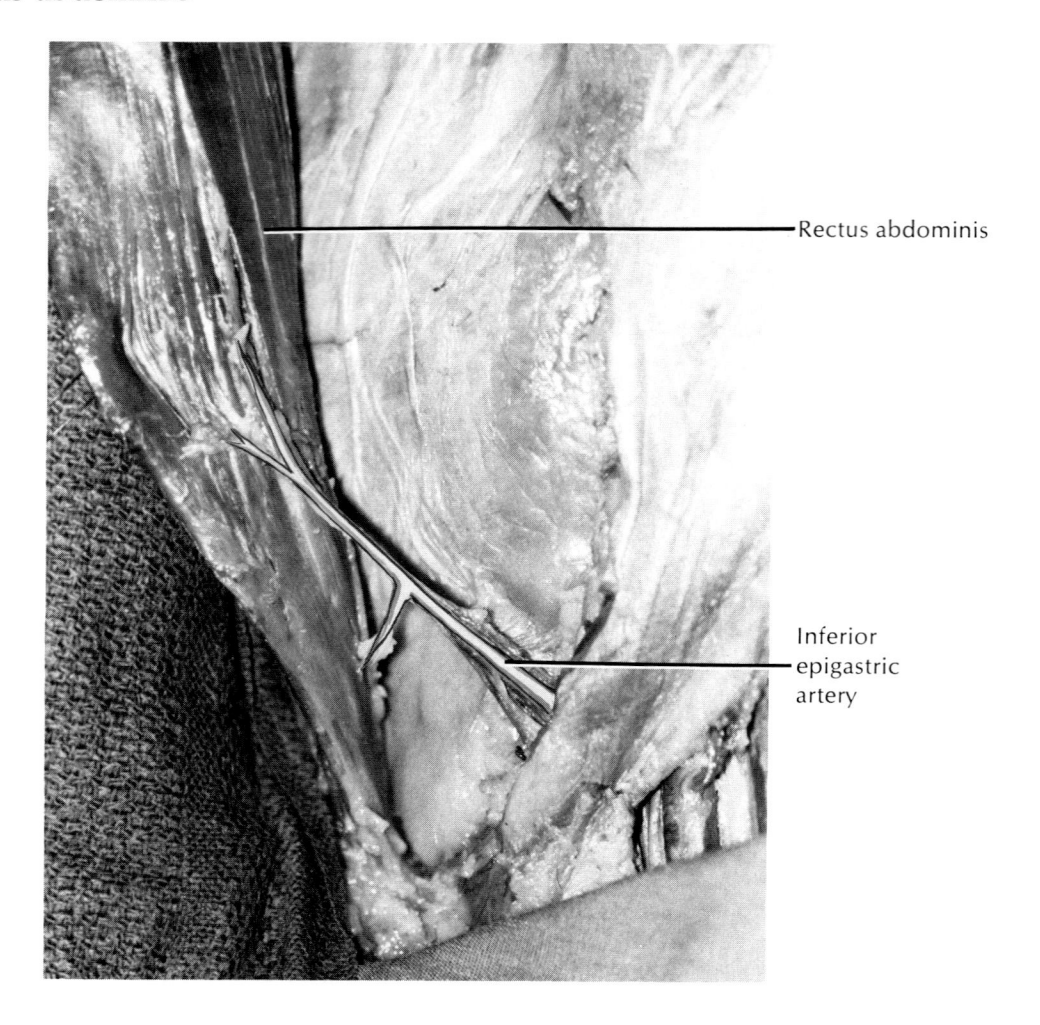

The inferior epigastric artery branches from the external iliac artery just proximal to the inguinal ligament and curves up along the medial wall of the internal inguinal ring, through the transversalis fascia into the posterior rectus sheath, and into the rectus abdominis muscle.

There are anastomotic connections between the terminal branches of the two major pedicles.

There are numerous musculocutaneous perforators through the muscle into the overlying skin. These perforating vessels not only supply the skin overlying the muscle, but will also support the anterolateral abdominal skin over the oblique muscles as far as the anterior axillary line. These perforators are the vascular bases for the transverse abdominal flap.

TRUNK: ANTERIOR TRUNK

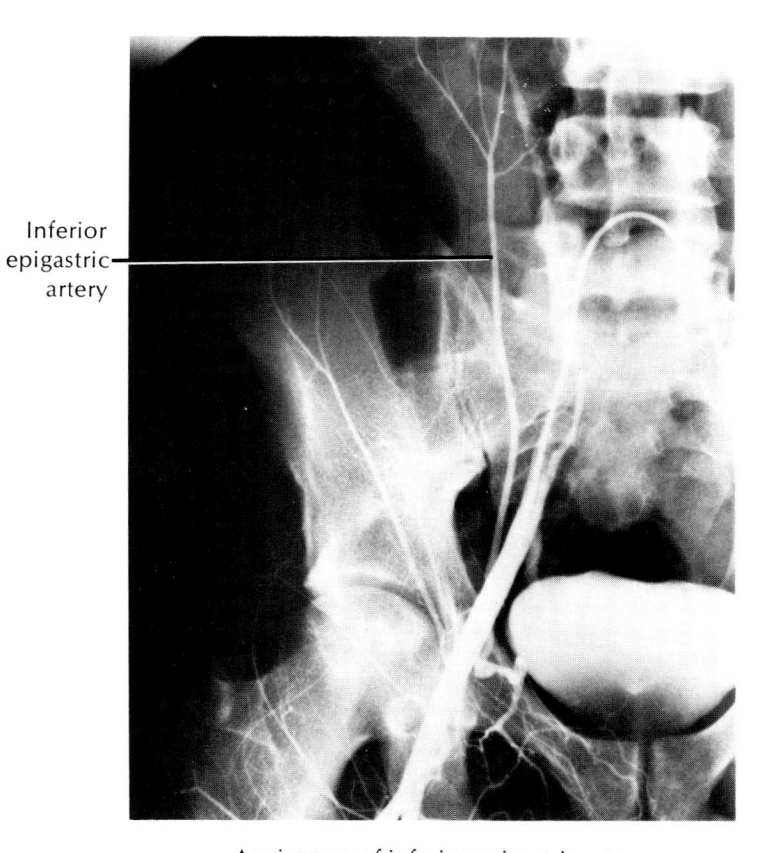

Angiogram of inferior epigastric artery

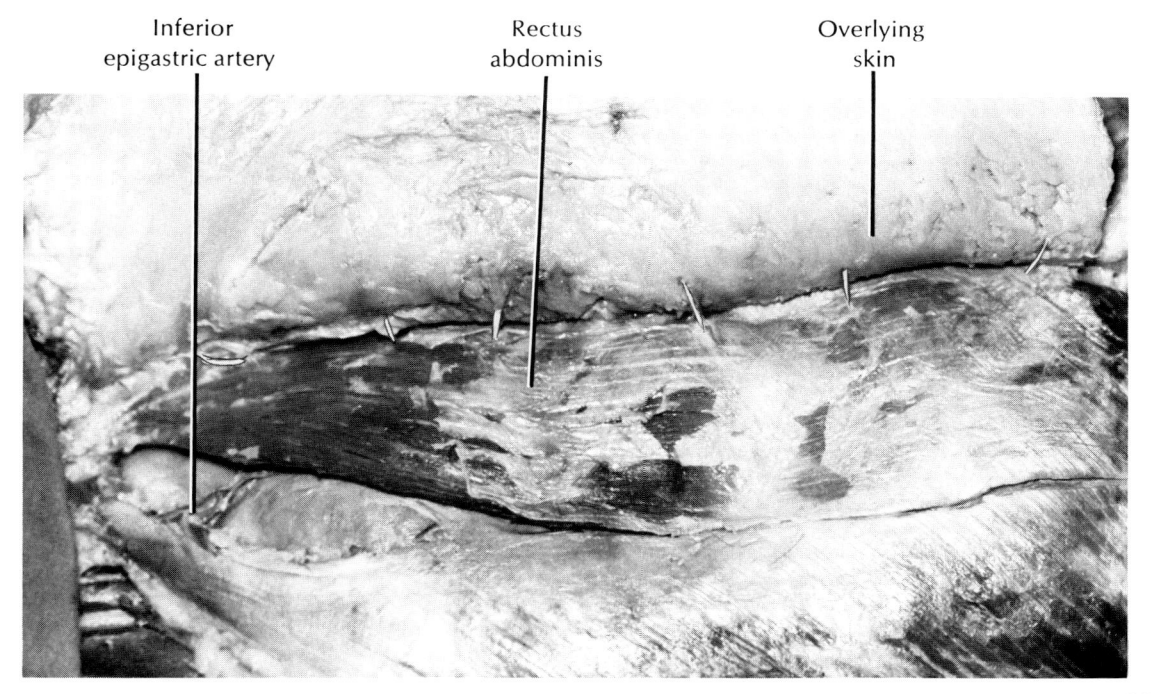

TRUNK: ANTERIOR TRUNK

Rectus abdominis

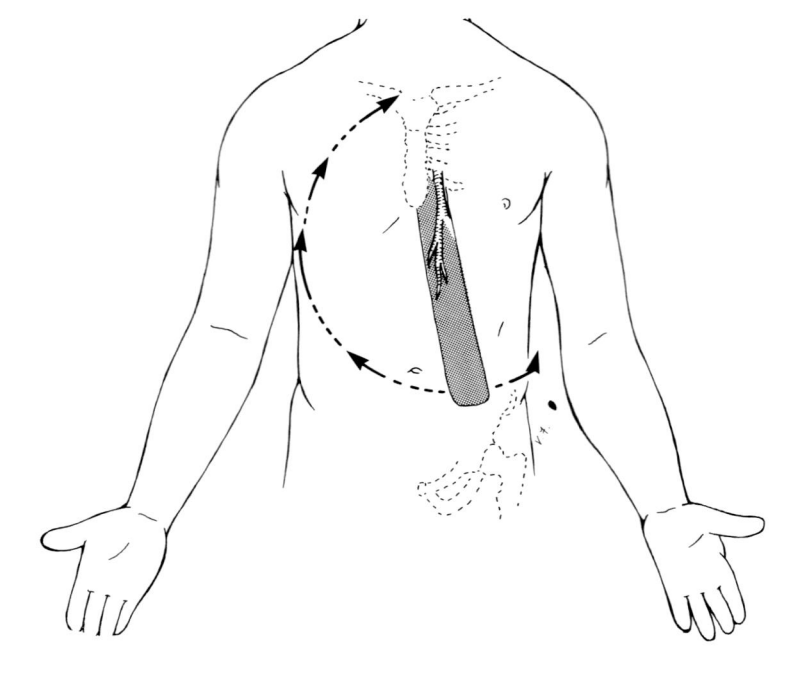

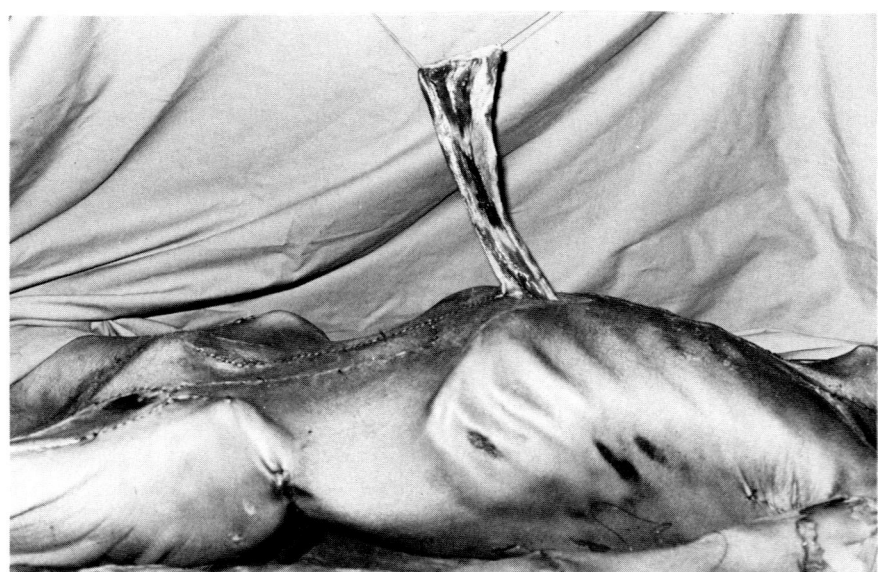

Superior arc

Arc of rotation

Just over half the unit may be elevated on either major pedicle. The superior arc based on the superior epigastric artery has a point of rotation just below the xiphisternum. This arc will cover the lateral abdomen, lower chest, and sternum.

TRUNK: ANTERIOR TRUNK

Inferior epigastric artery

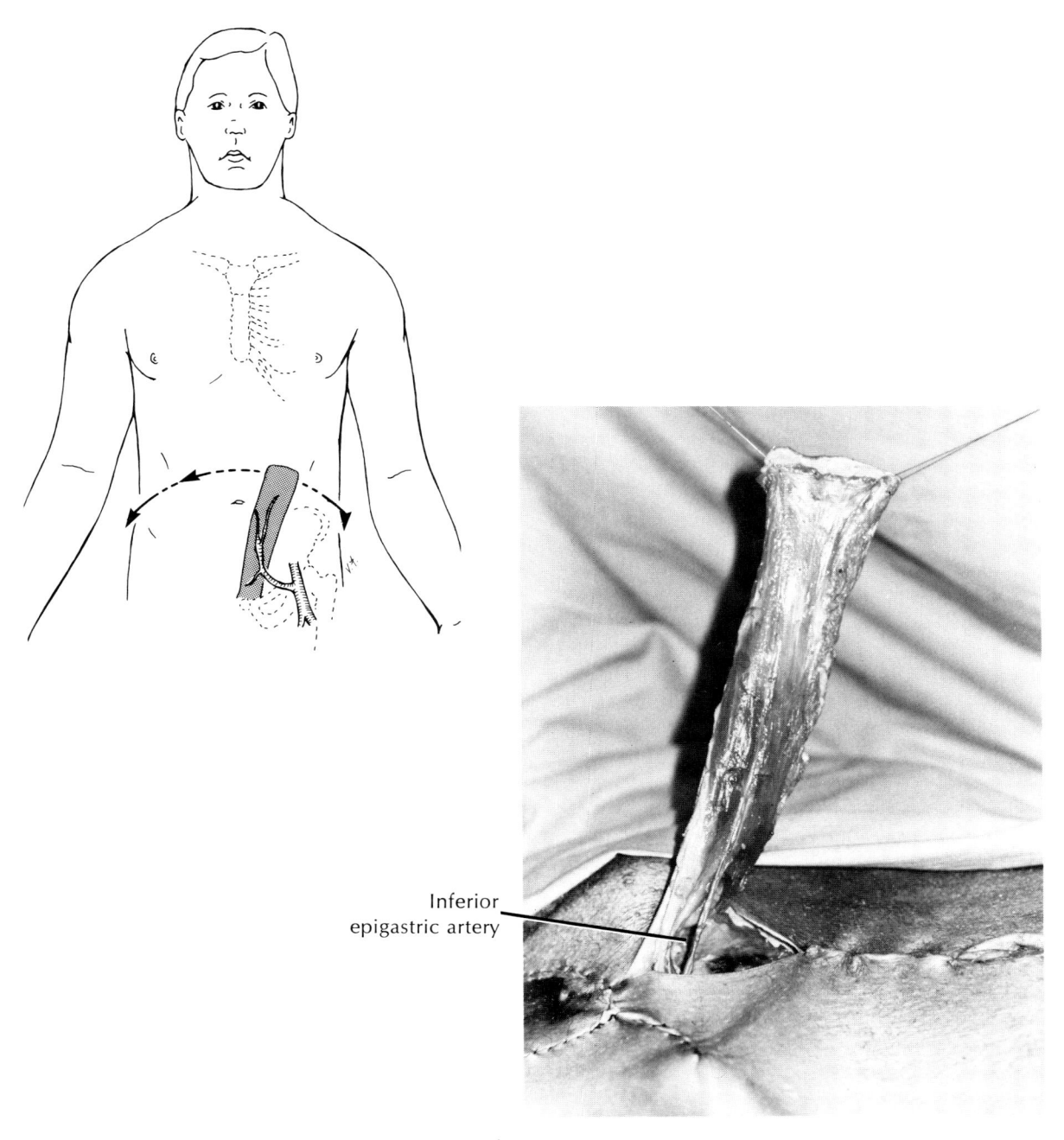

Inferior arc

The inferior arc based on the inferior epigastric artery will cover the lower abdomen and groin.

The transverse abdominal flap will cover the anterior chest wall, forearm, and hand.

The entire unit may be transferred as a free flap.

TRUNK: ANTERIOR TRUNK

Rectus abdominis

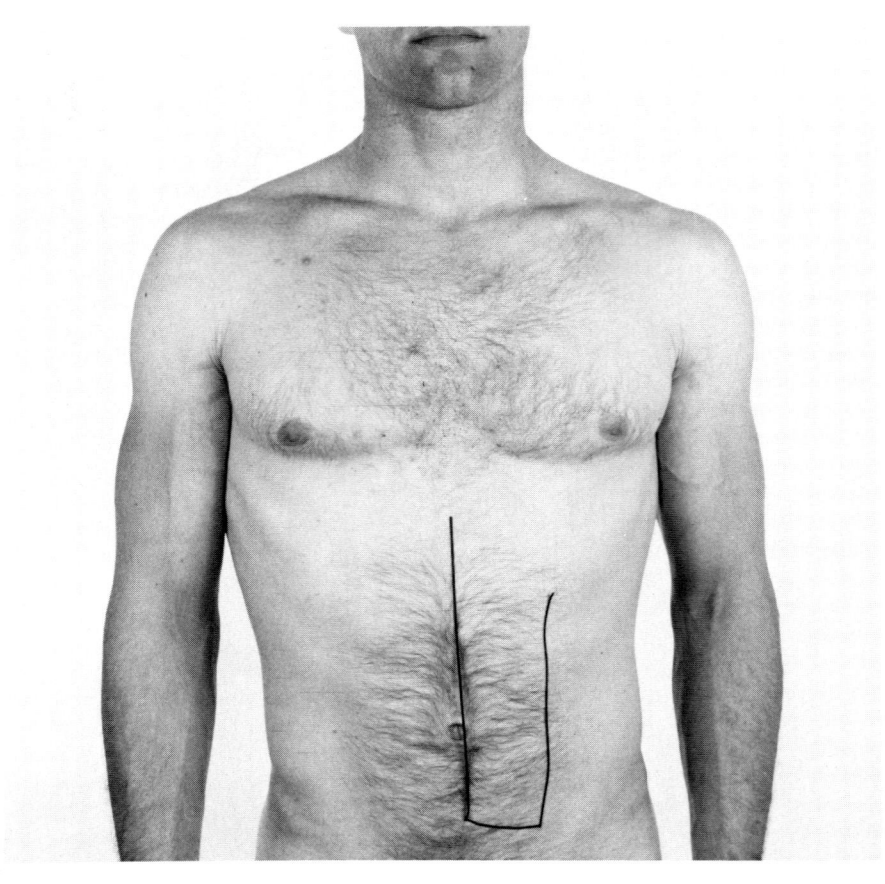

Superior flap

Elevation of flap
SUPERIOR ARC

The skin island is outlined, and the skin territory can be safely extended beyond the lateral margin of the muscle as far laterally as the anterior axillary line if desired. The donor defect would then require skin grafting. The skin island is then incised and the muscle is divided transversely just below the umbilicus. The posterior rectus sheath is not disturbed. The muscle is then elevated off the posterior sheath from distal to proximal. The superior epigastric artery is then identified below the costal margin. The muscle insertion may be divided off the fifth, sixth, and seventh costal cartilages. The resultant defect is closed by advancing the external oblique muscle medially and suturing it to the linea alba. The skin will close directly if only a small island over the muscle has been elevated.

TRUNK: ANTERIOR TRUNK

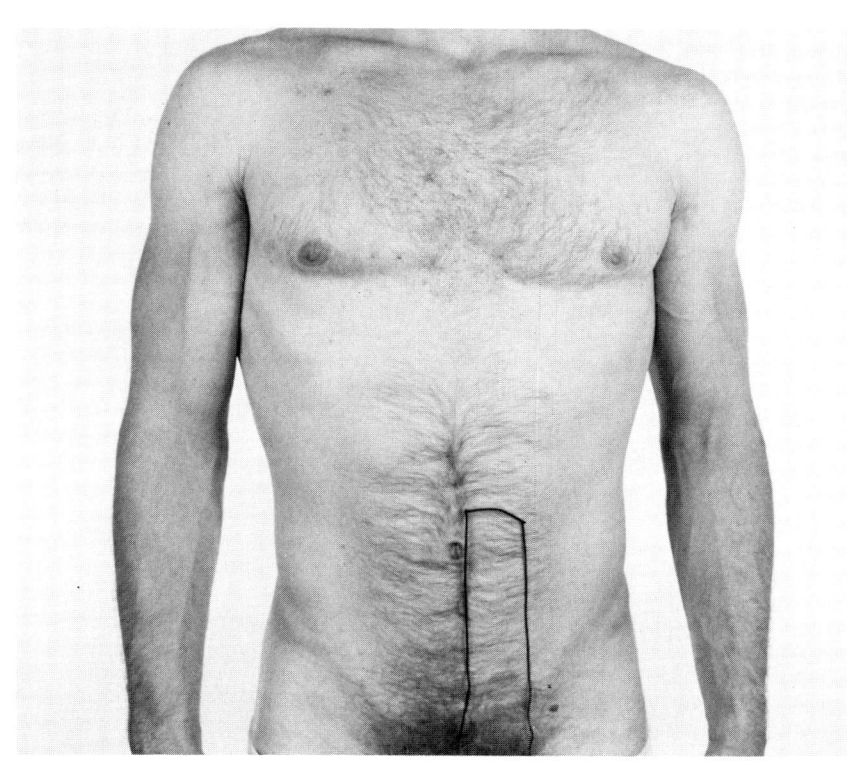

Inferior flap

INFERIOR ARC

The skin island is outlined and incised. The muscle is divided just above the umbilicus, elevated off the posterior sheath, and dissected distally toward its origin. The inferior epigastric artery is identified on the deep surface. The posterior rectus sheath in this area is less substantial. The tendon of origin may be divided for greater mobility. The donor defect is closed primarily.

Precautions

- The posterior rectus sheath should be preserved, as this facilitates donor site closure and prevents hernia formation.
- Below the semicircular line the posterior rectus sheath is less substantial, as in this location it is made up of only the transversalis fascia, while the fasciae of the internal and external oblique muscles contribute entirely to the anterior sheath. Inferior flap elevation in this area may lead to hernia formation and significant weakening of the abdominal wall.
- Elevation of an extended skin island will require skin grafting of the secondary defect over the oblique muscles.

TRUNK: ANTERIOR TRUNK

Rectus abdominis

Musculocutaneous island flap for abdominal wall reconstruction

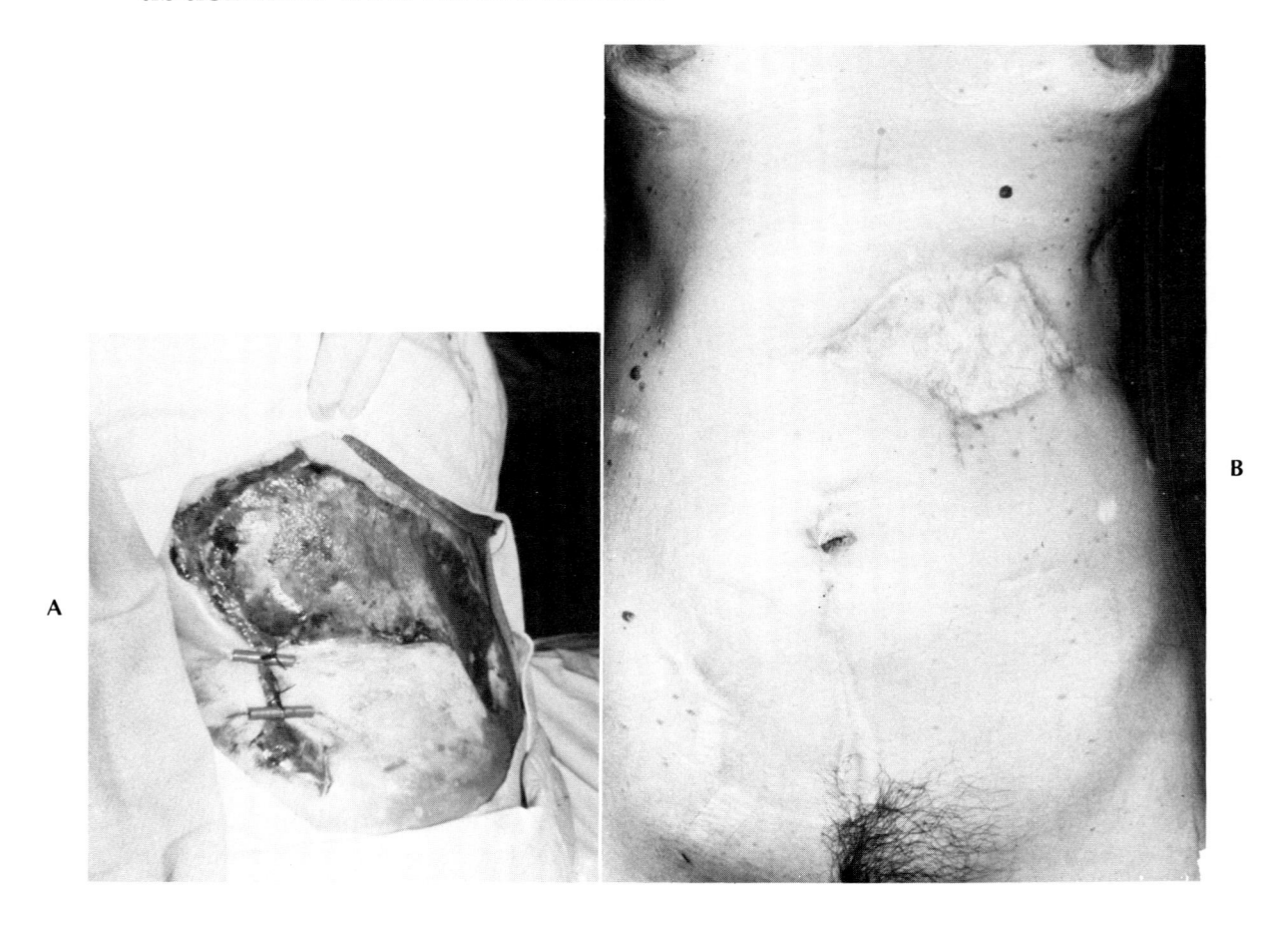

A   Shotgun wound of left upper quadrant reconstructed initially with Prolene mesh and split-skin graft.

B   Reconstructed abdominal wall with mesh covered by unstable skin.

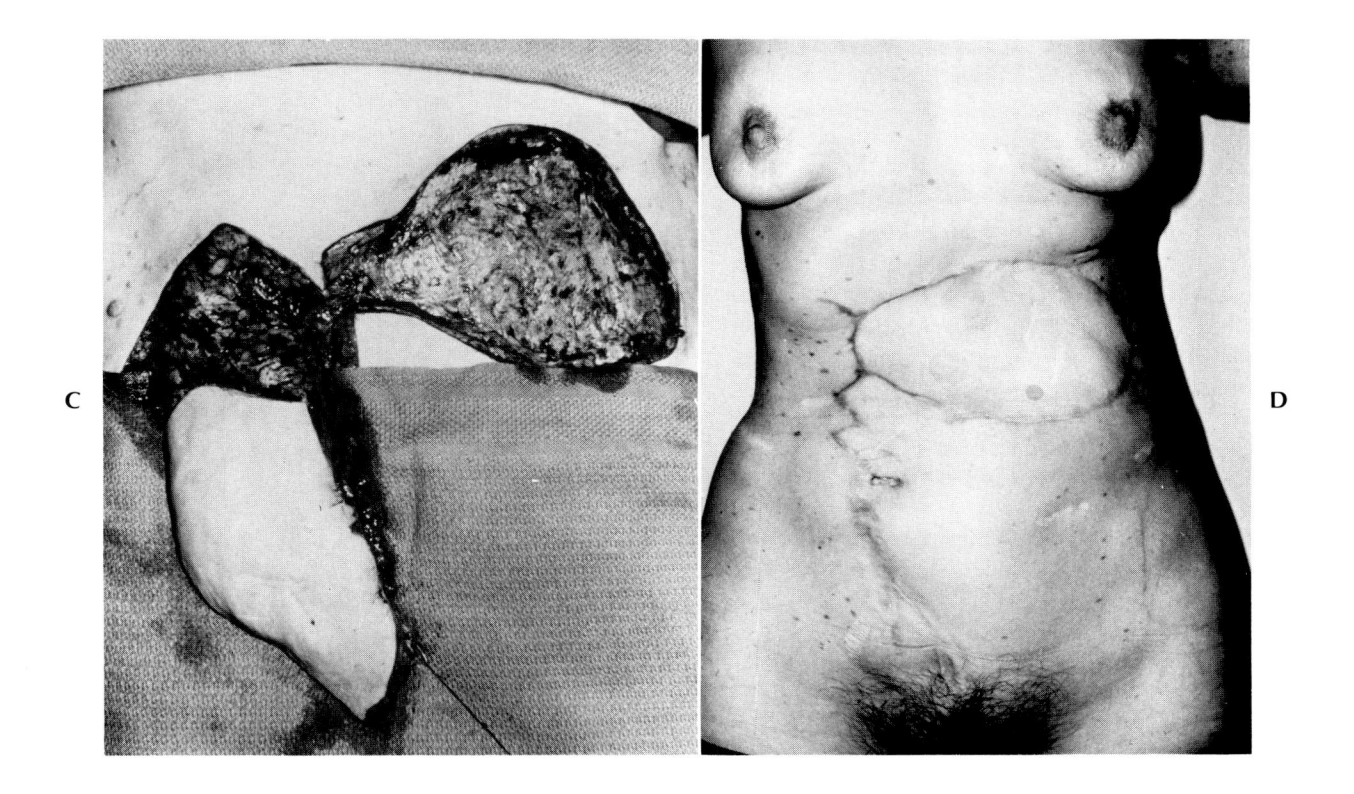

C Superior island rectus abdominis musculocutaneous flap elevated.

D Healed island flap. Secondary defect was closed directly. Single-stage reconstruction of abdominal wall. (From Mathes, S. J., and Bostwick, J., III: Br. J. Plast. Surg. 30:282, 1977.)

TRUNK: ANTERIOR TRUNK

Rectus abdominis

Transverse abdominal flap based on musculocutaneous perforating arteries of rectus abdominis

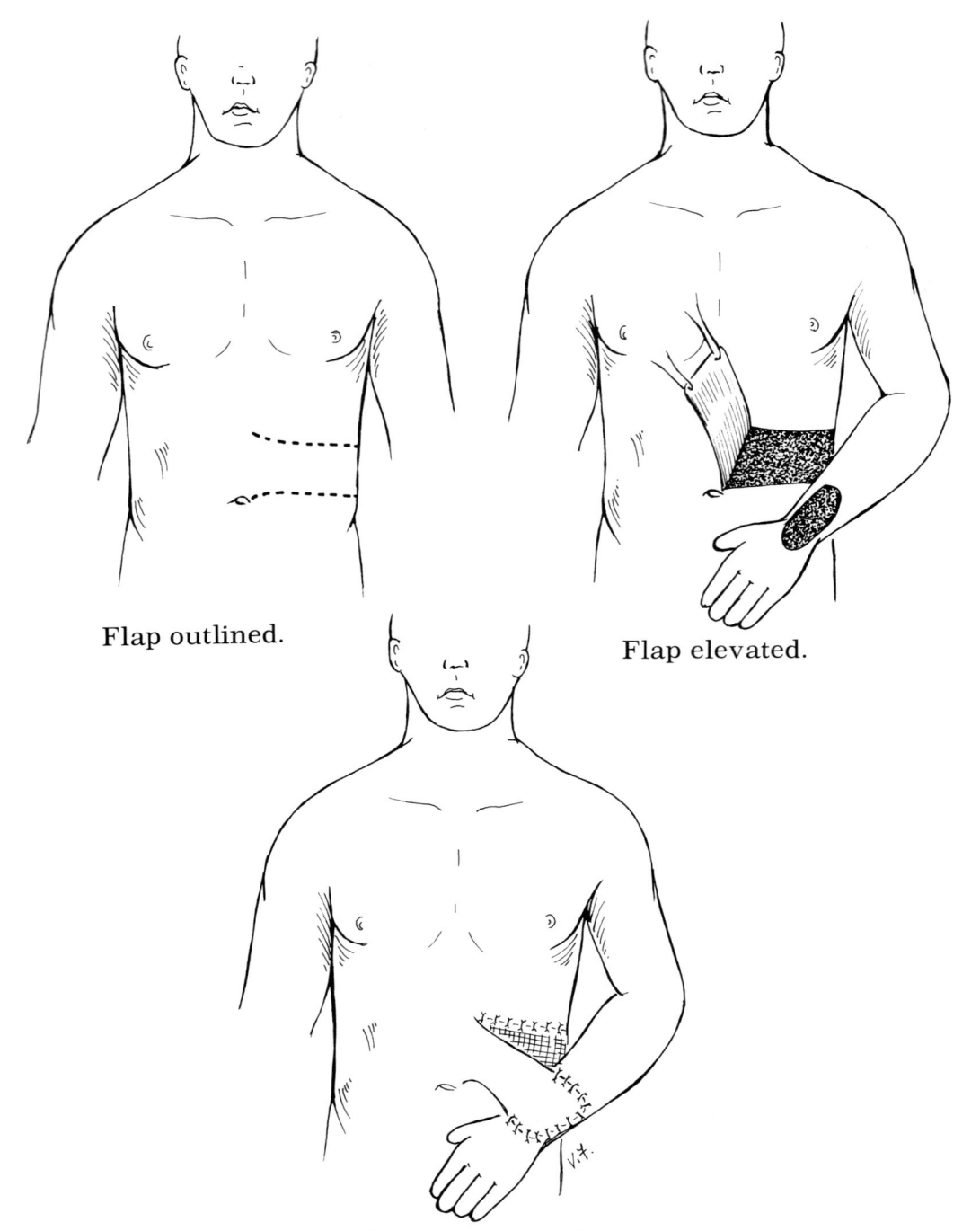

Flap outlined.

Flap elevated.

Flap for coverage of forearm.

Coverage of anterior chest

A Breast tumor with skin involvement (skin margins, outlined).
B Resultant defect following radical resection for breast tumor.
C Transverse abdominal flap provides stable wound prior to radiation.

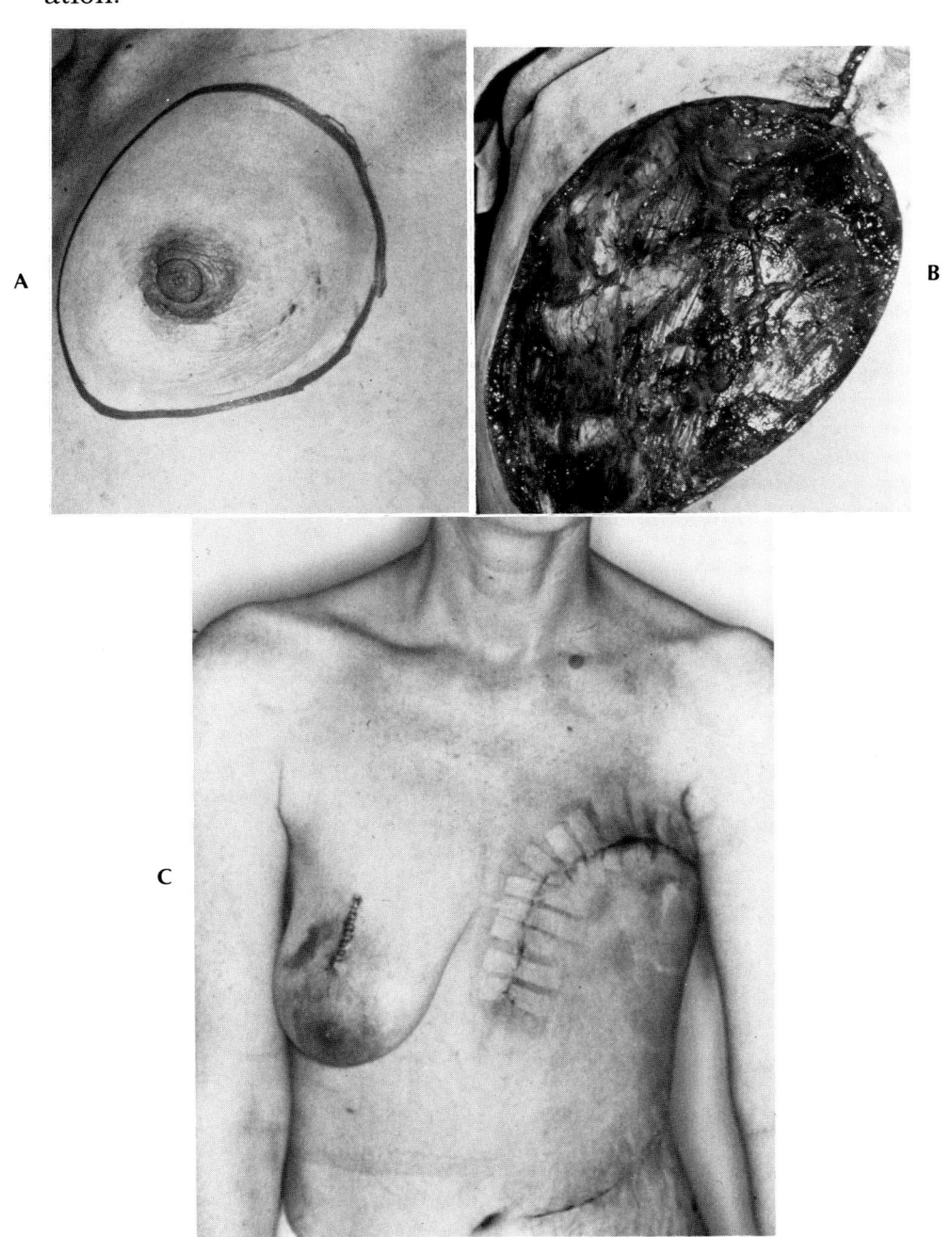

TRUNK

Posterior trunk
LATISSIMUS DORSI
TRAPEZIUS

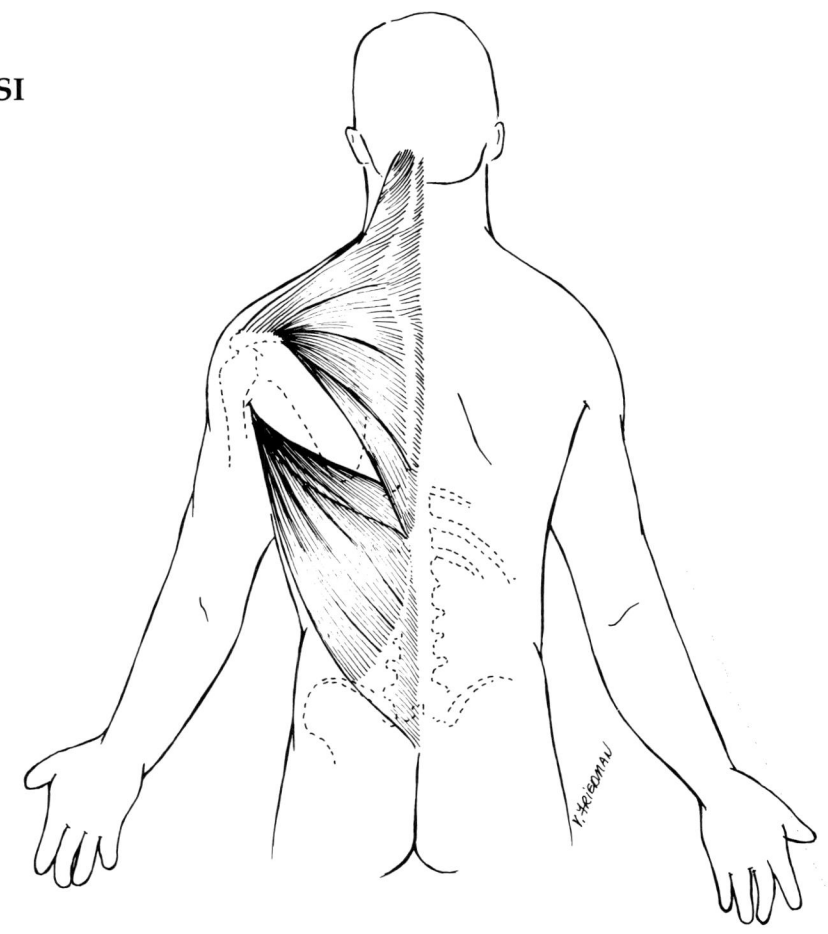

TRUNK: POSTERIOR TRUNK

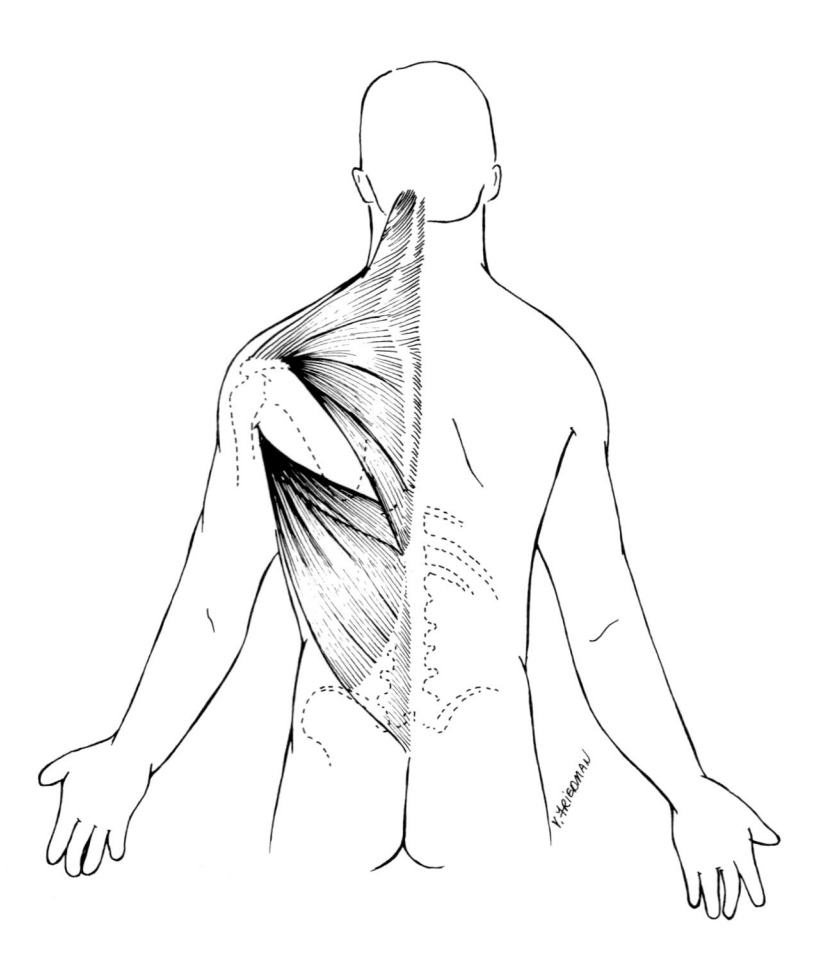

The two posterior trunk muscles are large, with proximal dominant vascular pedicles. Both have extensive anterior and posterior arcs of rotation and represent valuable units for reconstructive surgery. Both units have large cutaneous territories and are useful as musculocutaneous flaps. The perforating vessels of these units will carry skin extending into other muscle territories on the shoulder and back, further enhancing the usefulness of these muscles. These two units can be used in head and neck reconstruction and breast reconstruction. Both units will provide coverage of anterior and posterior trunk, neck, head, and upper extremity defects. Both units have potential for functional muscle transfers. Furthermore, the large, dominant vascular pedicles make these useful in free microvascular transfer procedures.

TRUNK: POSTERIOR TRUNK

- Trapezius
- Infraspinatus
- Latissimus dorsi
- External oblique

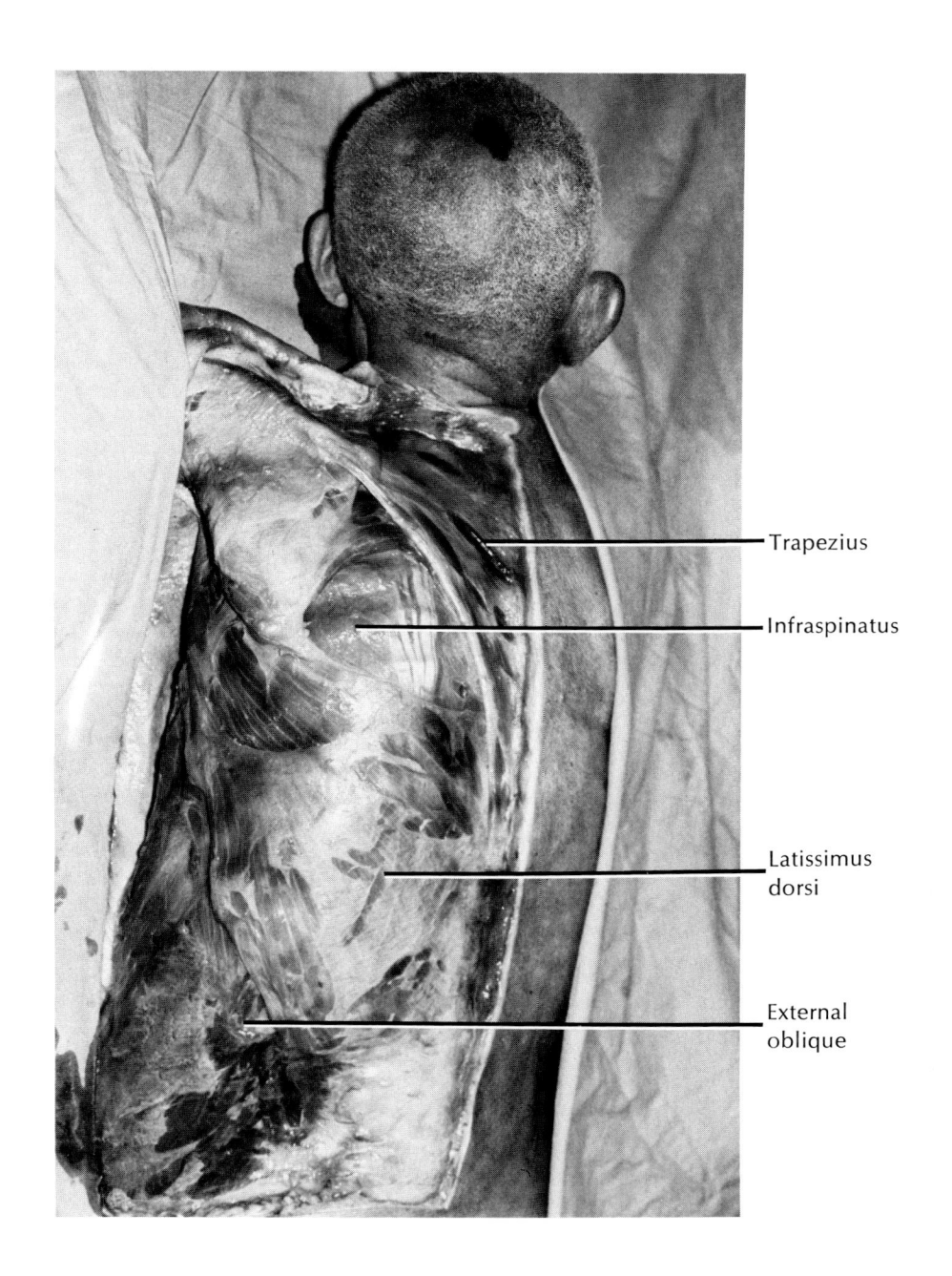

TRUNK: POSTERIOR TRUNK

Transverse cervical artery | Subscapular artery | Thoracodorsal artery | Circumflex scapular artery

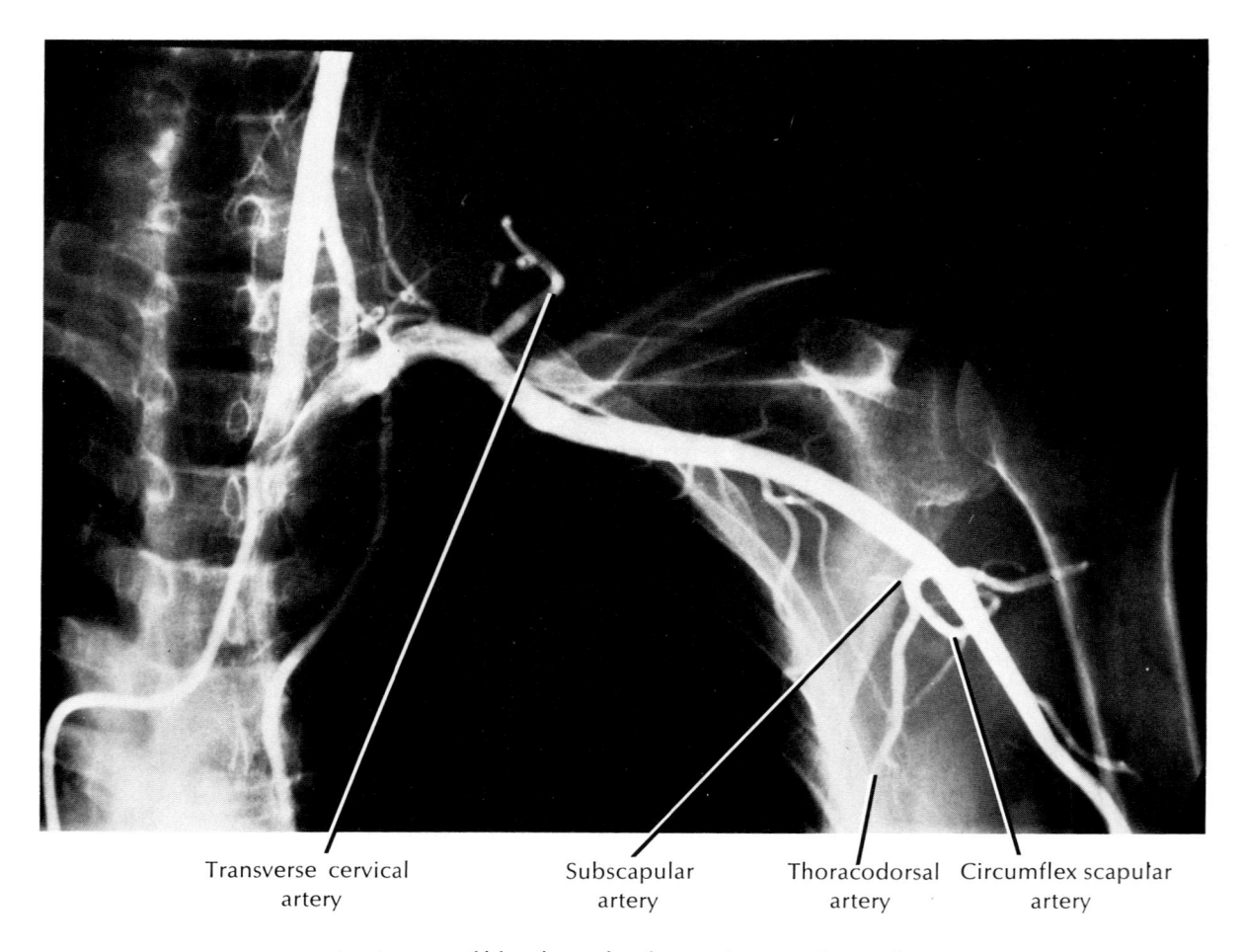

Angiogram of blood supply of posterior trunk muscles

Blood supply

Although these are posterior trunk muscles, both receive their dominant blood supply through pedicles originating from anterior arteries. The trapezius muscle receives the major vascular pedicle within the neck from the first and third portions of the subclavian artery via the transverse cervical artery. The latissimus dorsi muscle receives the major vascular pedicle from within the axilla from the third portion of the axillary artery via the subscapular artery.

TRUNK: POSTERIOR TRUNK

Transverse cervical artery
Posterior scapular artery
Thoracoacromial artery
Axillary artery
Subscapular artery
Circumflex scapular artery
Thoracodorsal artery
Lateral thoracic artery

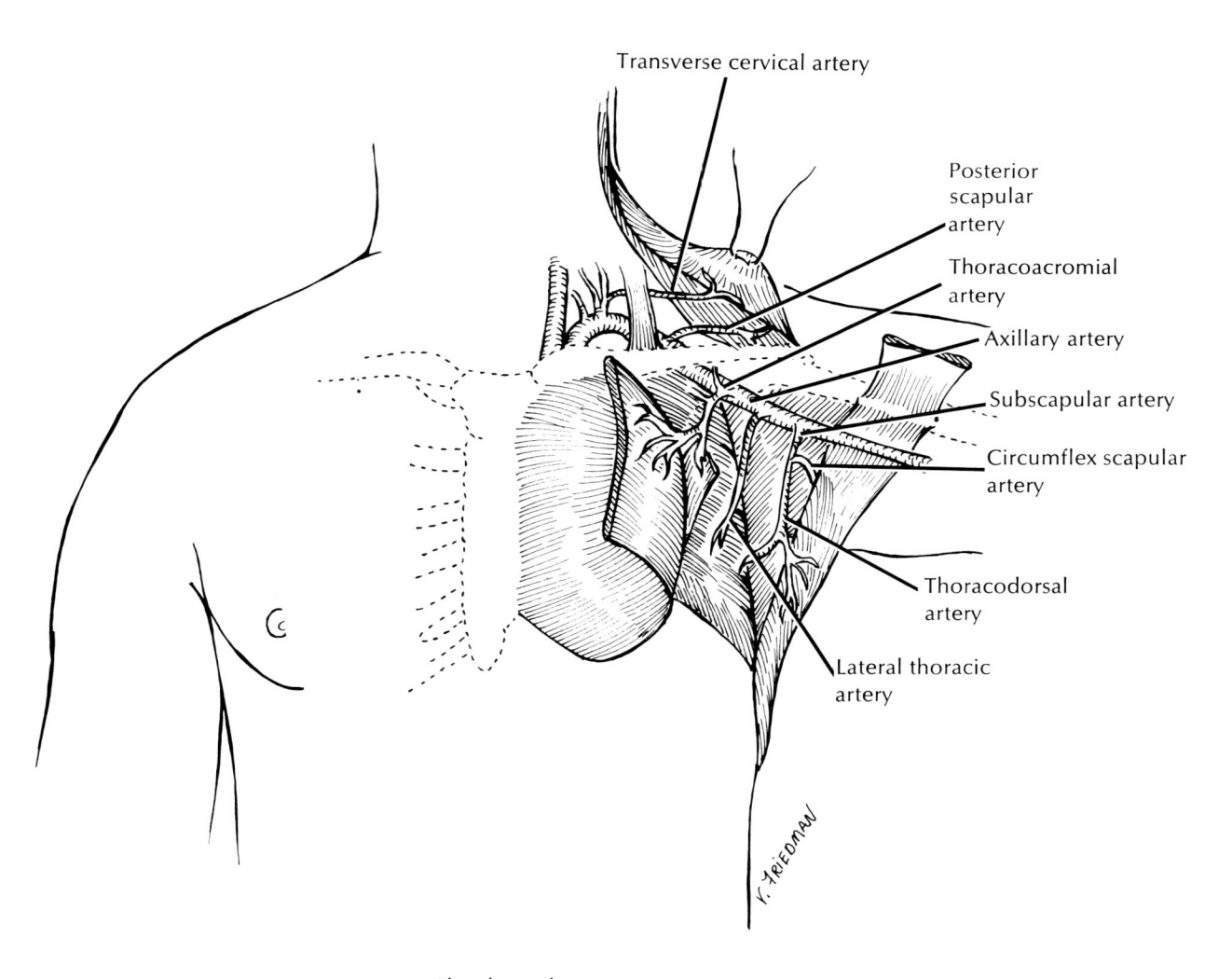

Blood supply

Latissimus dorsi

MUSCLE FLAP
MUSCULOCUTANEOUS FLAP
MUSCULOFASCIAL FLAP
FREE FLAP

Applications

Reconstruction of:

BREAST
ABDOMINAL WALL
MYELOMENINGOCELE

Coverage of:

CHEST WALL
HEAD AND NECK
LATERAL ABDOMEN
BACK
UPPER ARM

Functional muscle transfer

Free flap:

DISTANT COVERAGE
FUNCTIONAL MUSCLE

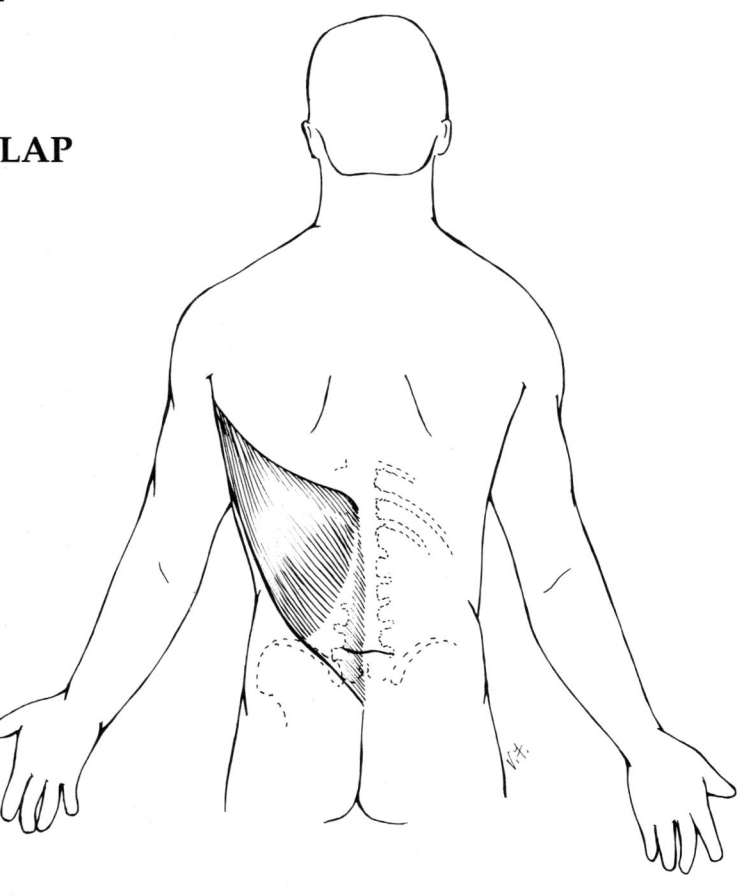

TRUNK: POSTERIOR TRUNK

Latissimus dorsi

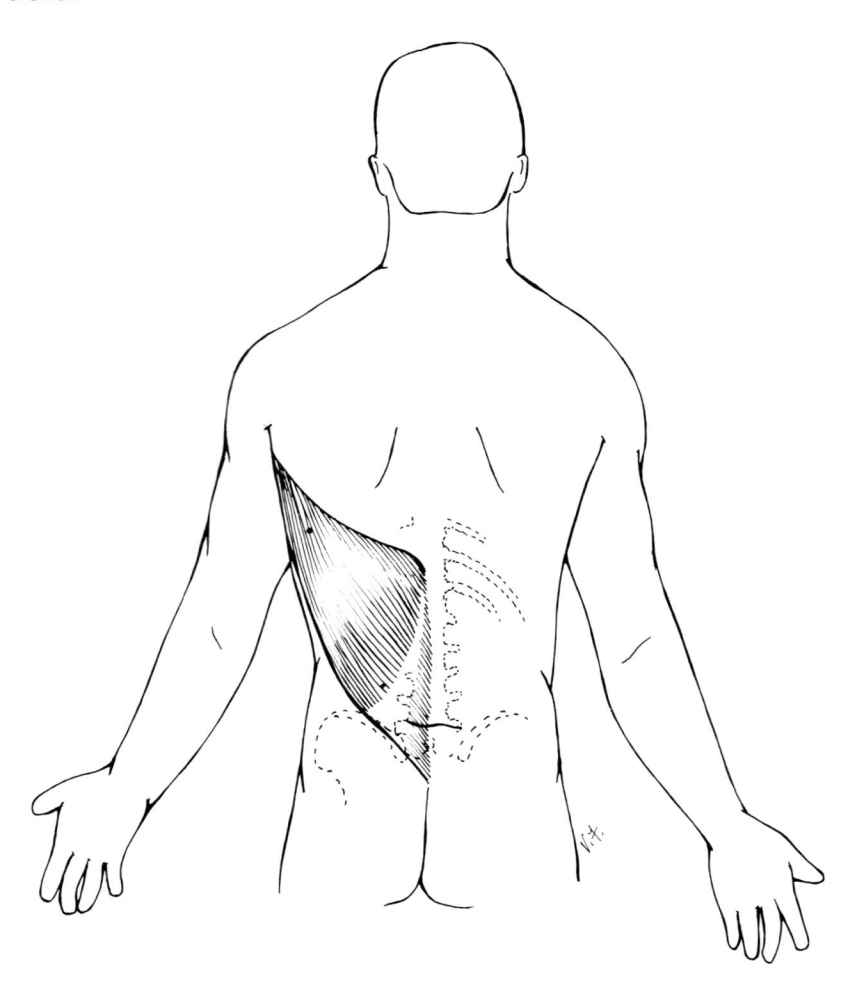

The latissimus dorsi is a flat, triangular muscle with a broad origin. It covers almost half the back. Superiorly it is related to the trapezius medially, and the teres major and minor muscles laterally. Deep to the latissimus dorsi lie the erector spinae, serratus posterior inferior, and the serratus anterior muscles. The tendon of the latissimus dorsi is lateral to the subscapularis muscle in the axilla.

Origin: The muscle arises from the spine of the lower six thoracic vertebrae, through the posterior layer of the thoracolumbar fascia, from the spine of the lumbar and sacral vertebrae, and the posterior crest of the ilium. The muscle also has some small muscular slips of origin from the lower four ribs, interdigitating with the slip of origin of the external oblique muscle of the abdomen. The upper and anterolateral borders are essentially free.

TRUNK: POSTERIOR TRUNK

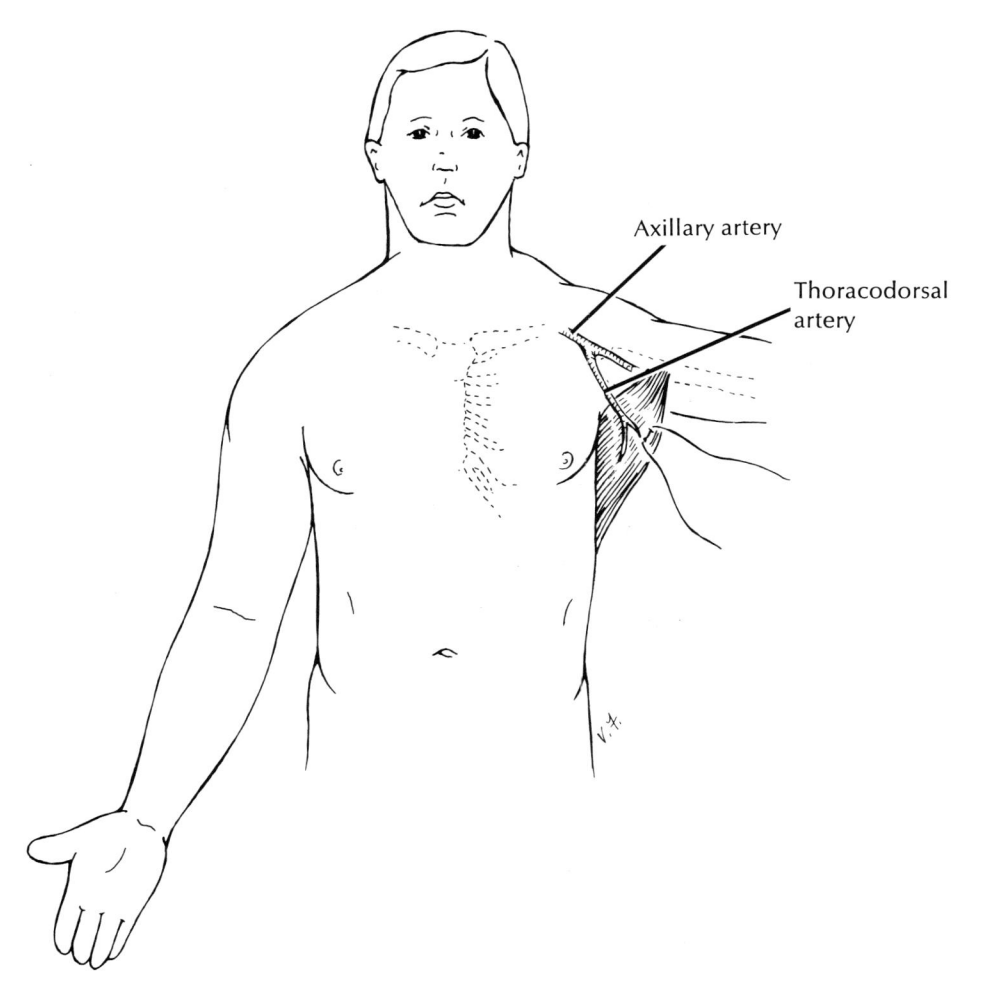

Blood supply

Insertion: From this broad origin, which is muscular above and fascial below, the fibers converge in spiral fashion into a tendon that wraps around the lower border of the teres major muscle to form the posterior fold of the axilla. This muscular tendon then inserts into the intertubercular groove of the humerus.

Nerve supply: The thoracodorsal nerve, a branch of the posterior cord of the brachial plexus, accompanies the subscapular artery to enter the muscle with the major pedicle approximately 10 cm from the insertion.

Function: The latissimus dorsi is an expendable muscle that is an adductor, extender, and medial rotator of the humerus.

TRUNK: POSTERIOR TRUNK

Latissimus dorsi

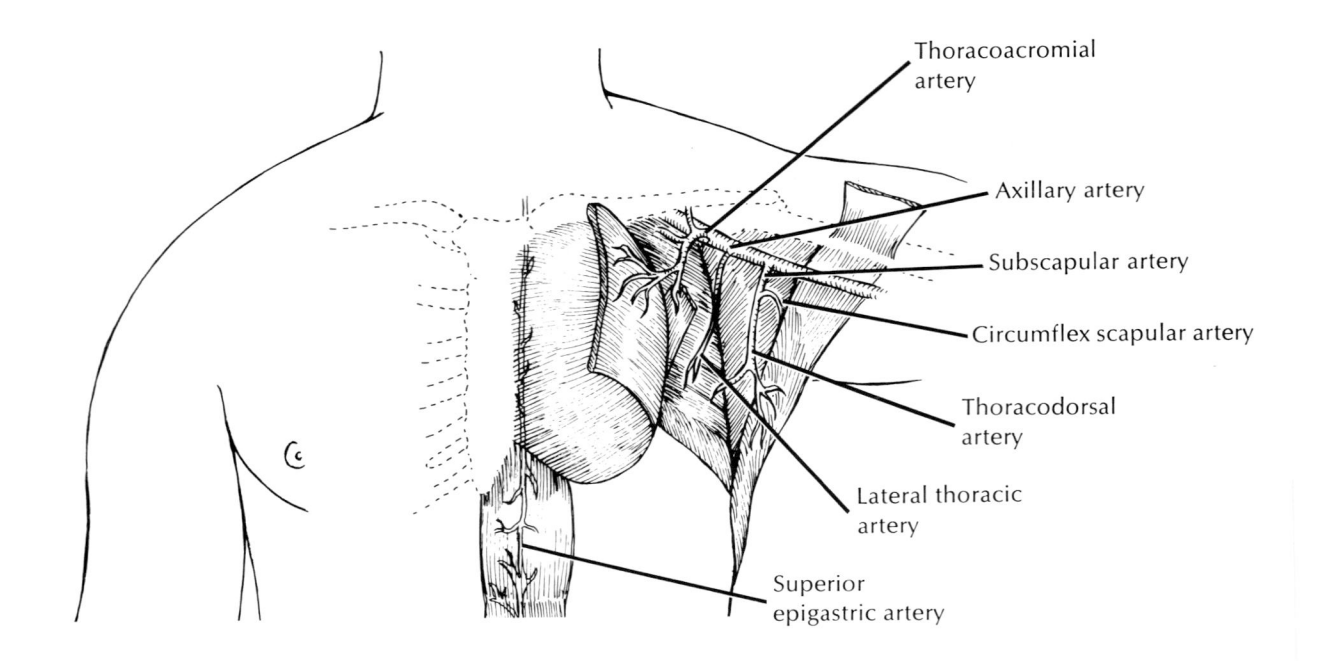

Blood supply

The thoracodorsal artery, a terminal branch of the subscapular artery, is the major pedicle to the latissimus dorsi muscle and will support the entire unit. It enters the muscle on the deep surface approximately 10 cm from the origin where the muscle forms the posterior axillary fold. The subscapular artery, the largest branch of the third part of the axillary artery, runs along the lower border of the subscapularis muscle and divides into the circumflex scapular and thoracodorsal arteries. The thoracodorsal artery continues distally along the border of the subscapularis muscle and sends a branch medially to the serratus anterior. The vessel then divides into two or three branches that enter the latissimus dorsi muscle.

TRUNK: POSTERIOR TRUNK

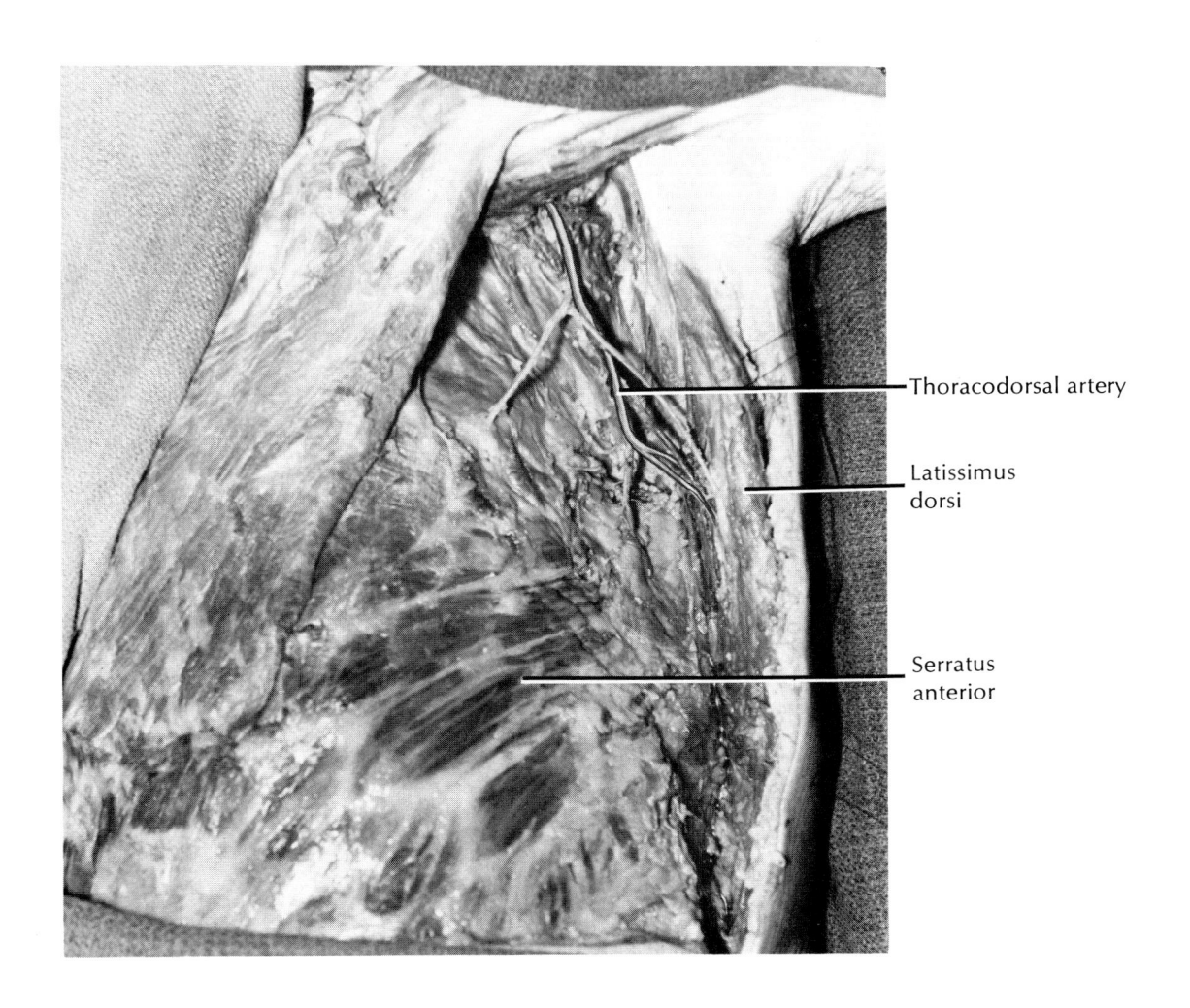

TRUNK: POSTERIOR TRUNK

Latissimus dorsi

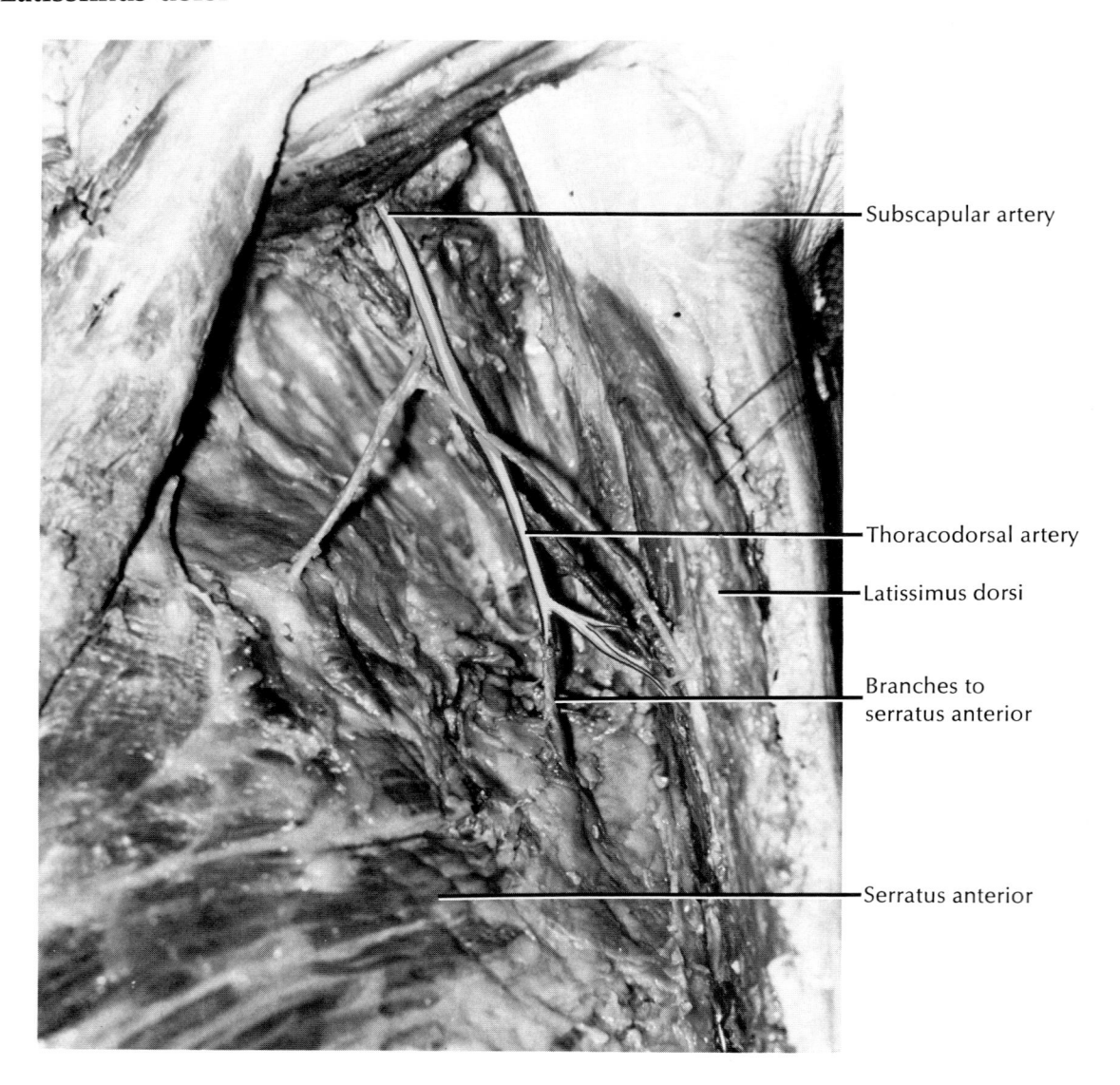

These vessels then divide into branches that run parallel to the muscle fibers, and through perforating vessels supply the entire skin overlying the muscle. A large set of perforating vessels is located along the posterior axillary fold, and smaller perforating vessels are located more distally.

TRUNK: POSTERIOR TRUNK

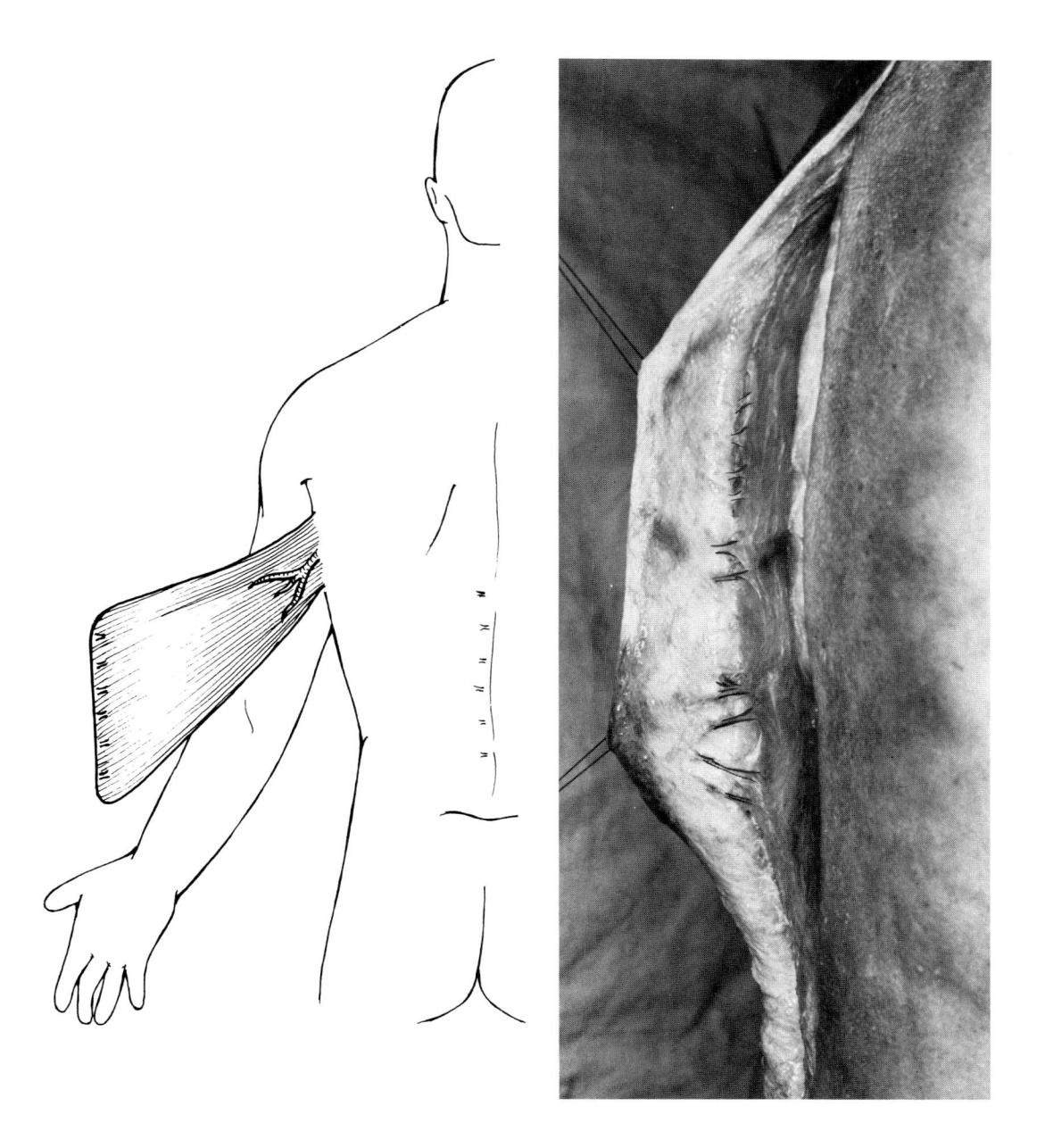

In addition to the major pedicle, segmental minor pedicles, perforating branches from the intercostal and lumbar arteries, pierce the latissimus dorsi to supply the muscle and the overlying skin. These vessels are the vascular basis of the transverse back flap and may support a medially based musculocutaneous flap.

TRUNK: POSTERIOR TRUNK

Latissimus dorsi

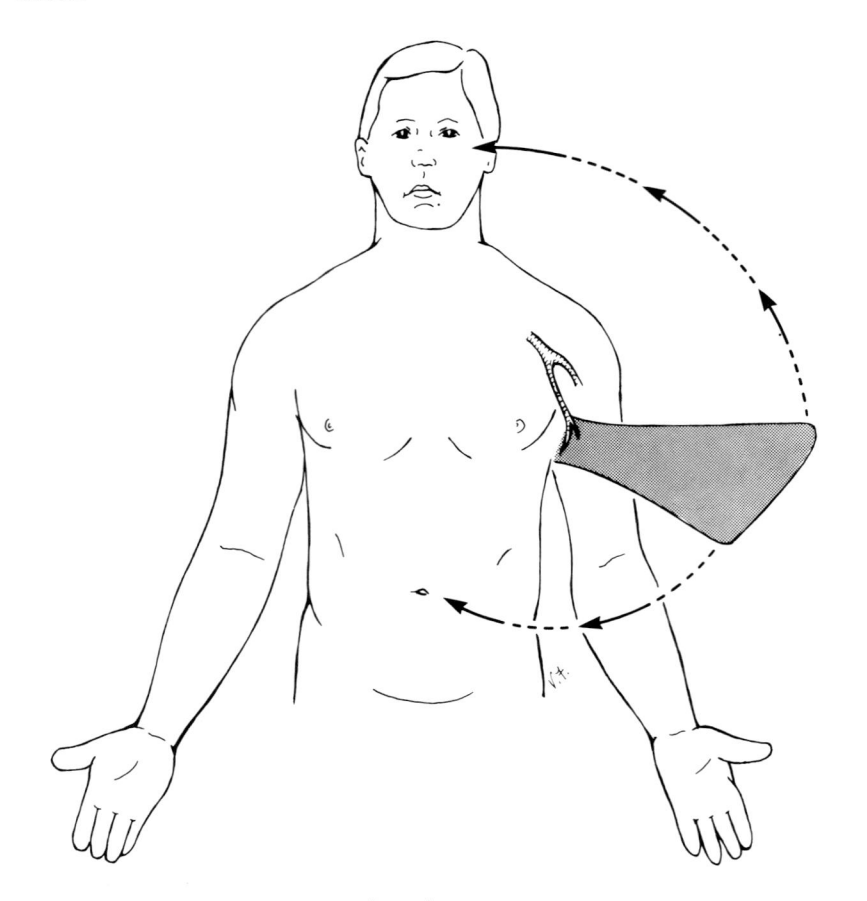

Anterior arc

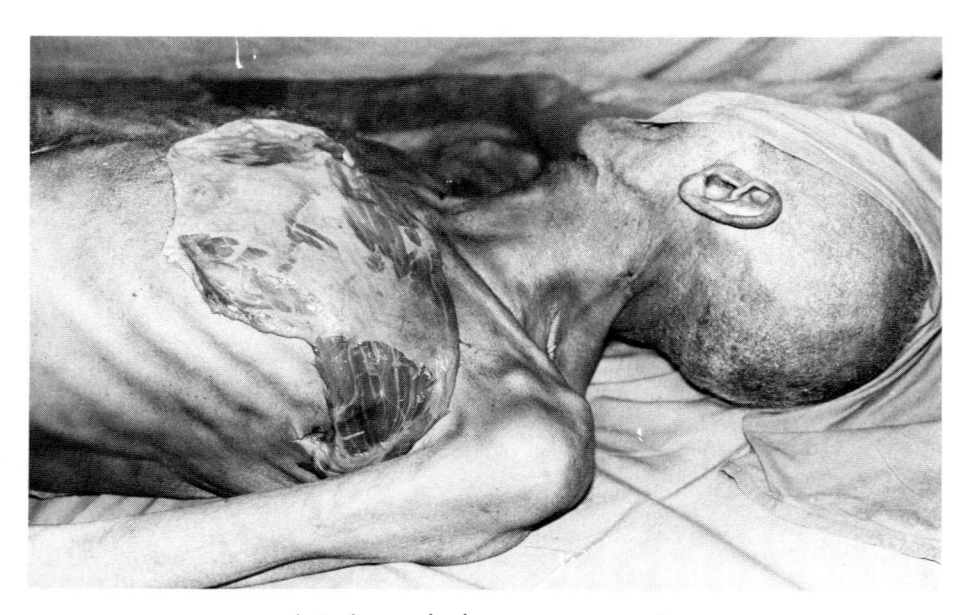

Anterior arc for breast reconstruction

TRUNK: POSTERIOR TRUNK

Anterior arc

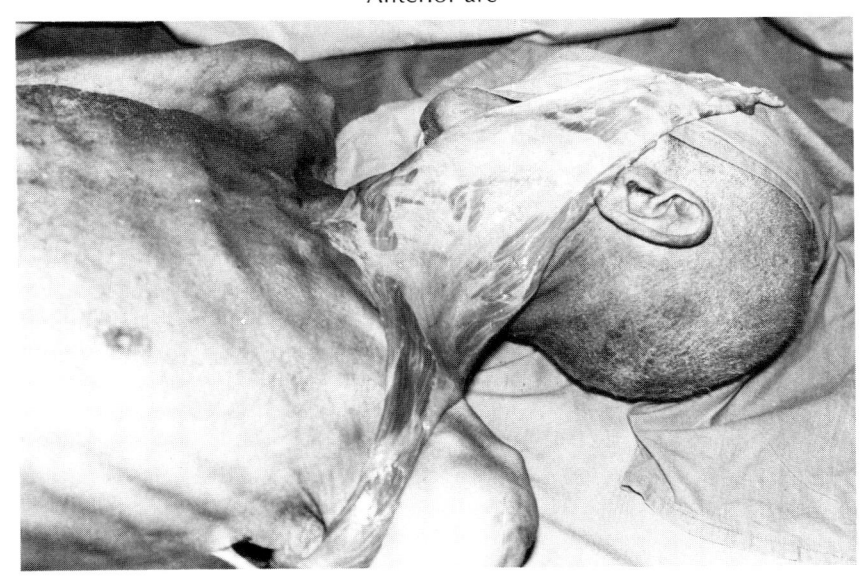

Anterior arc for head and neck reconstruction

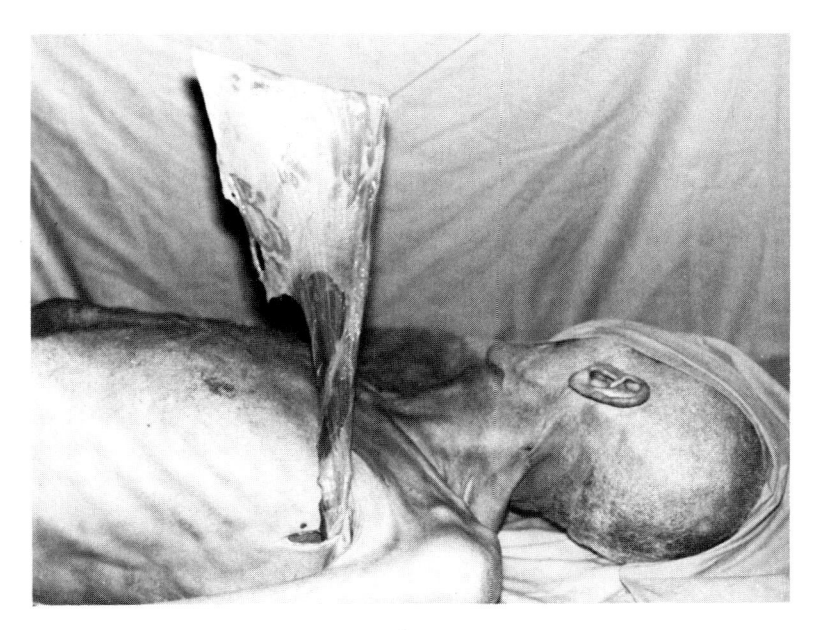

Arc of rotation

The latissimus dorsi, based on the major pedicle in the axilla, has a wide arc of rotation.

ANTERIOR ARC

The anterior arc will cover the lateral abdomen (musculofascial flap), chest wall, and head and neck region.

TRUNK: POSTERIOR TRUNK

Latissimus dorsi

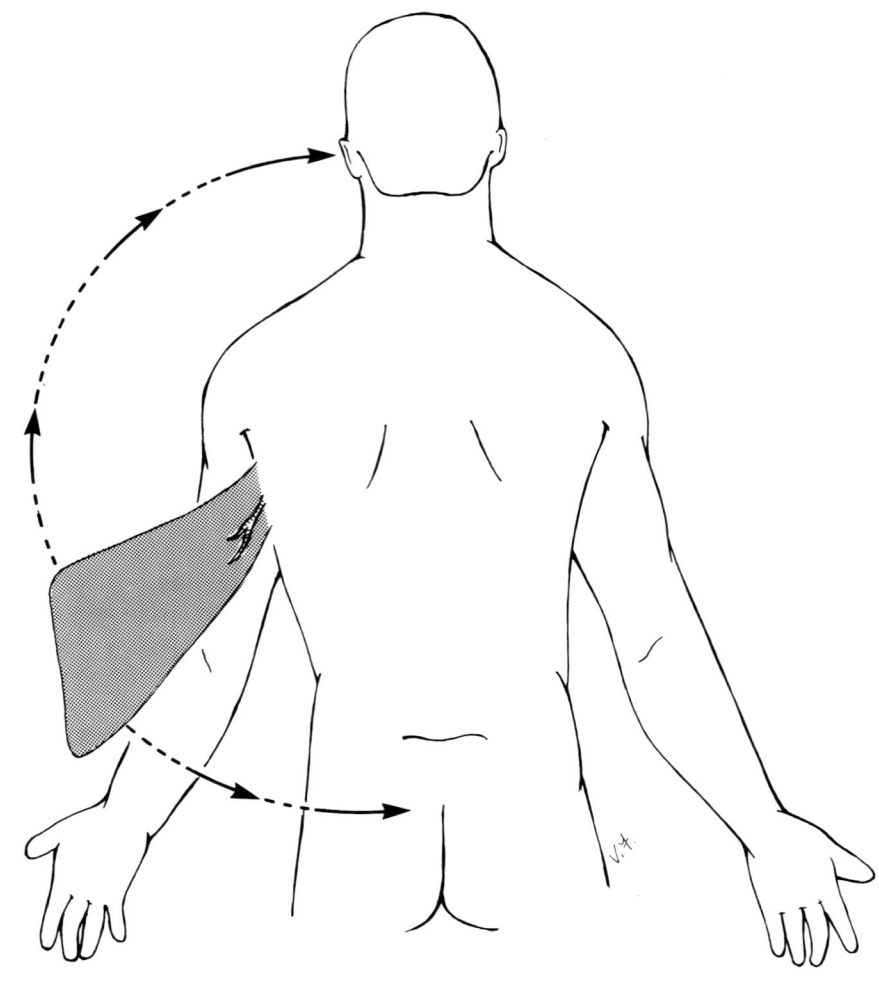

Posterior arc

POSTERIOR ARC

The posterior arc will cover the lumbar, thoracic, and cervical vertebrae and will reach the posterior neck.

TRUNK: POSTERIOR TRUNK

Posterior arc

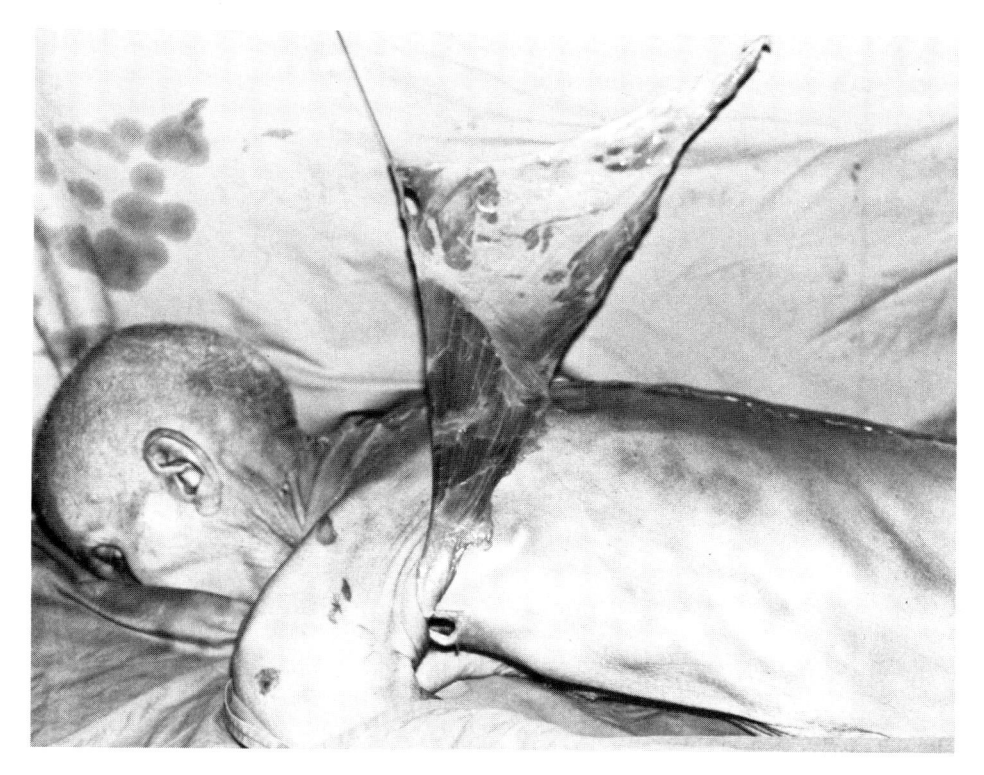

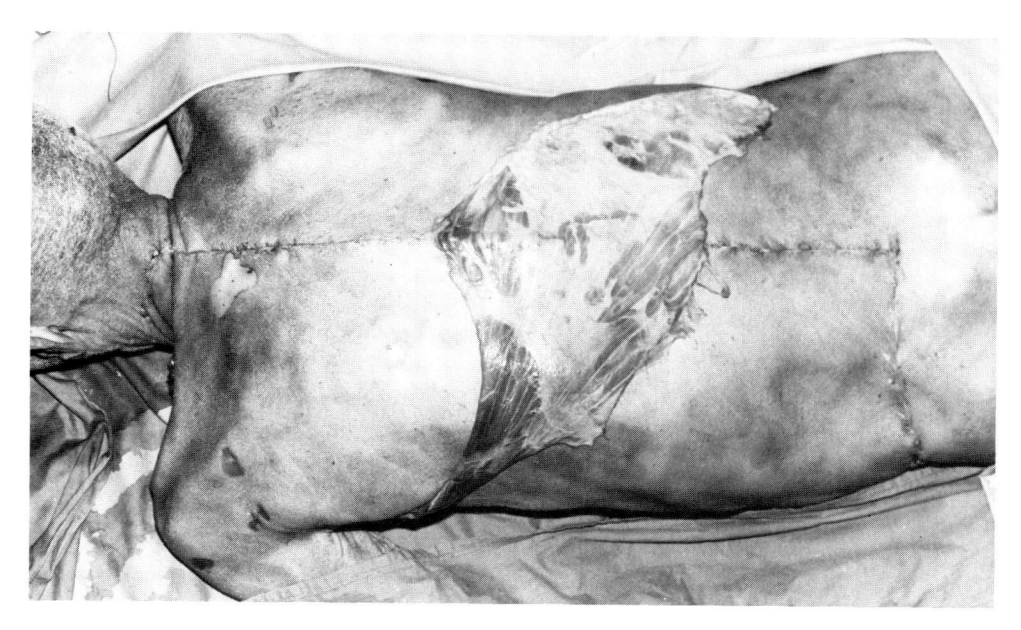

Posterior arc coverage of myelomeningocele

TRUNK: POSTERIOR TRUNK

Latissimus dorsi

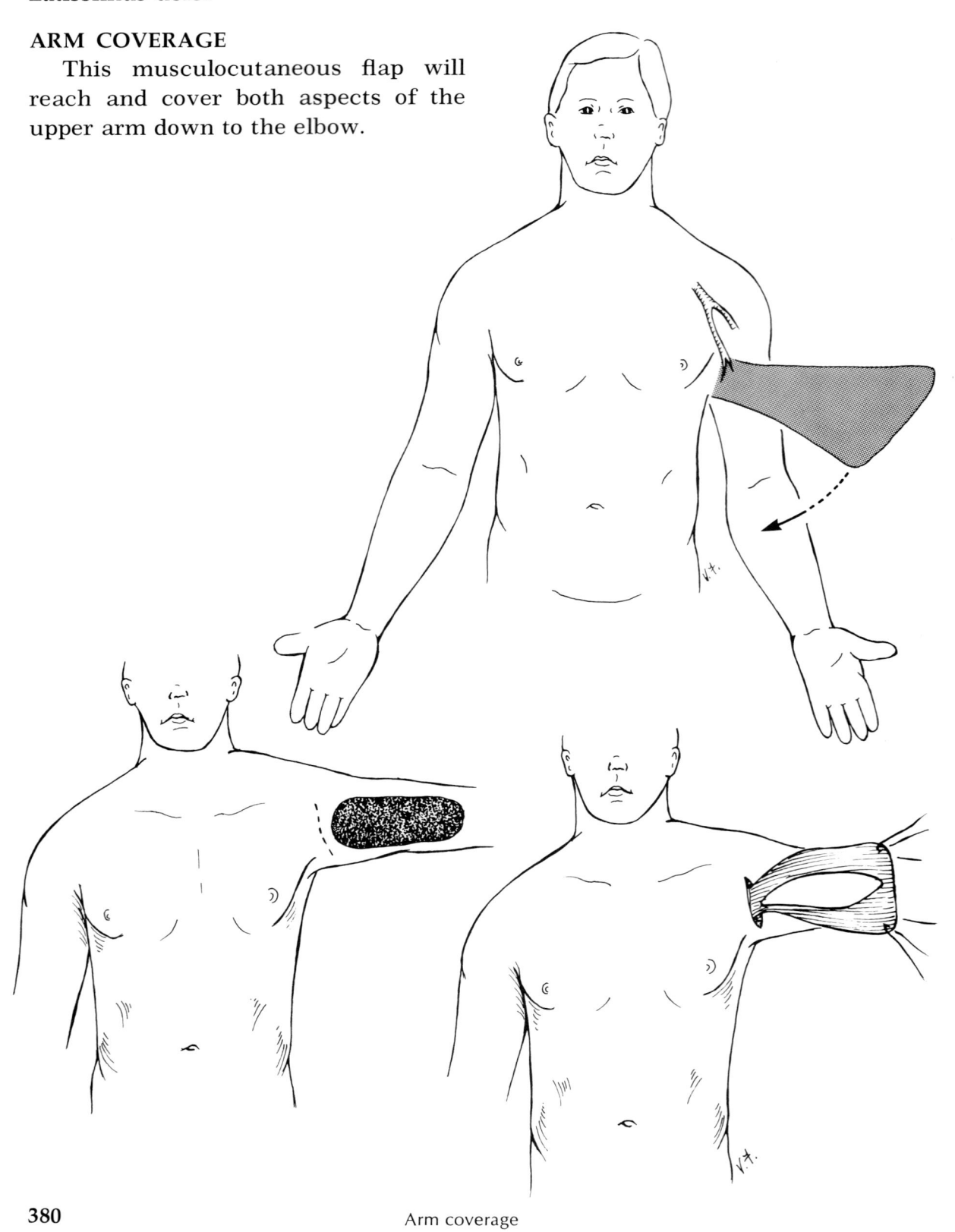

ARM COVERAGE

This musculocutaneous flap will reach and cover both aspects of the upper arm down to the elbow.

380   Arm coverage

TRUNK: POSTERIOR TRUNK

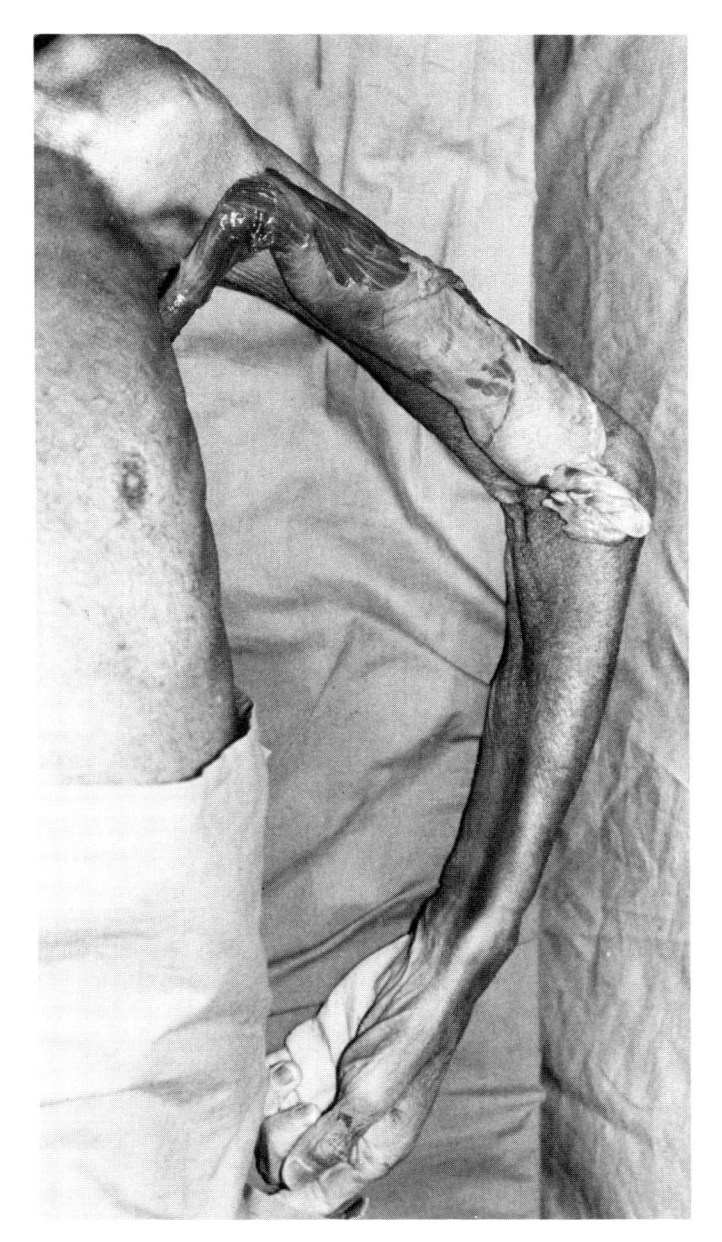

Arm coverage

TRUNK: POSTERIOR TRUNK

Latissimus dorsi

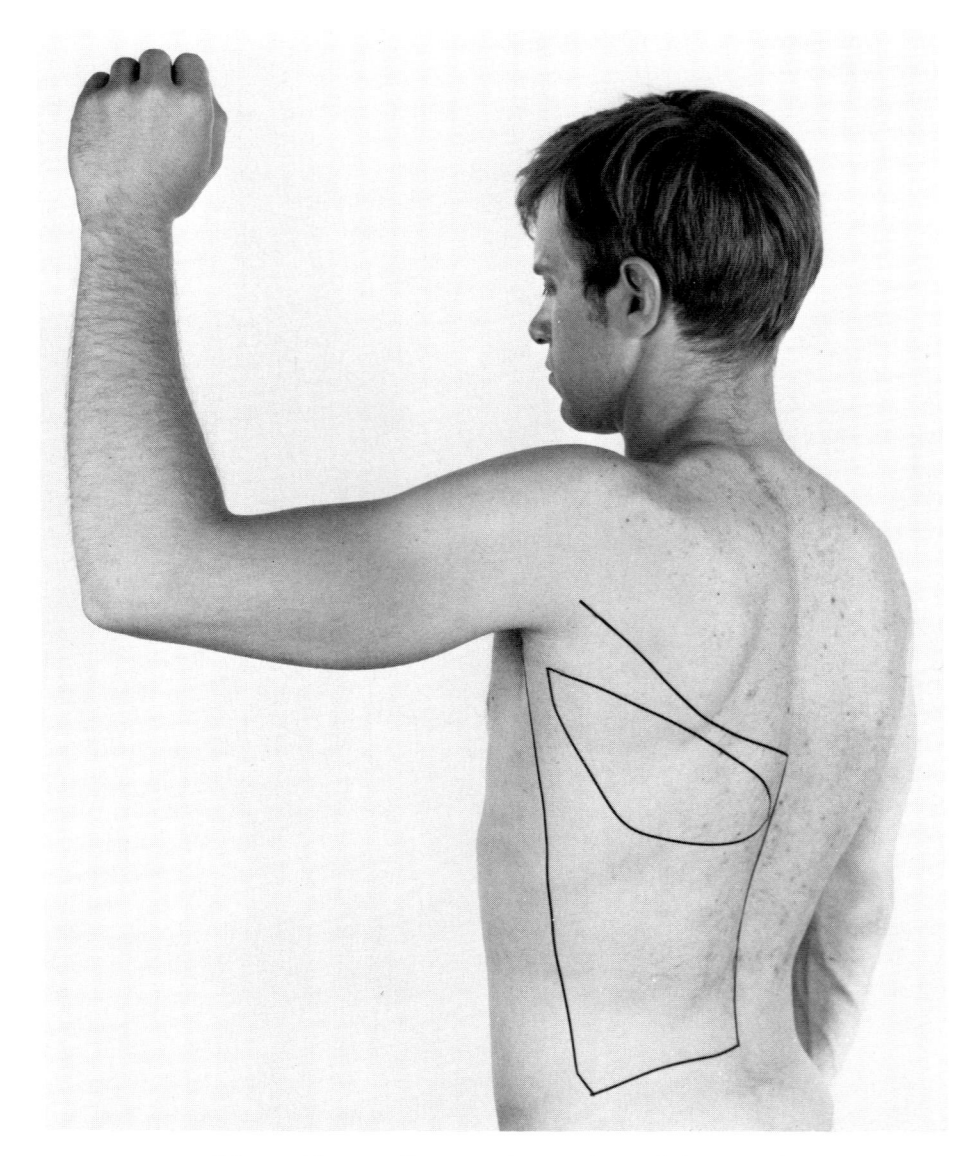

Skin markings outline muscle margins and skin island
along upper free border of muscle

Elevation of flap
ANTERIOR ARC

The flap may be elevated as a muscle or musculocutaneous island. The island of skin can be oriented along the anterolateral free border or over the upper free border of the muscle. For breast reconstruction, the island over the upper free muscle border is preferable, as the secondary defect, when closed, leaves a transverse scar that is easily concealed by a brassiere.

TRUNK: POSTERIOR TRUNK

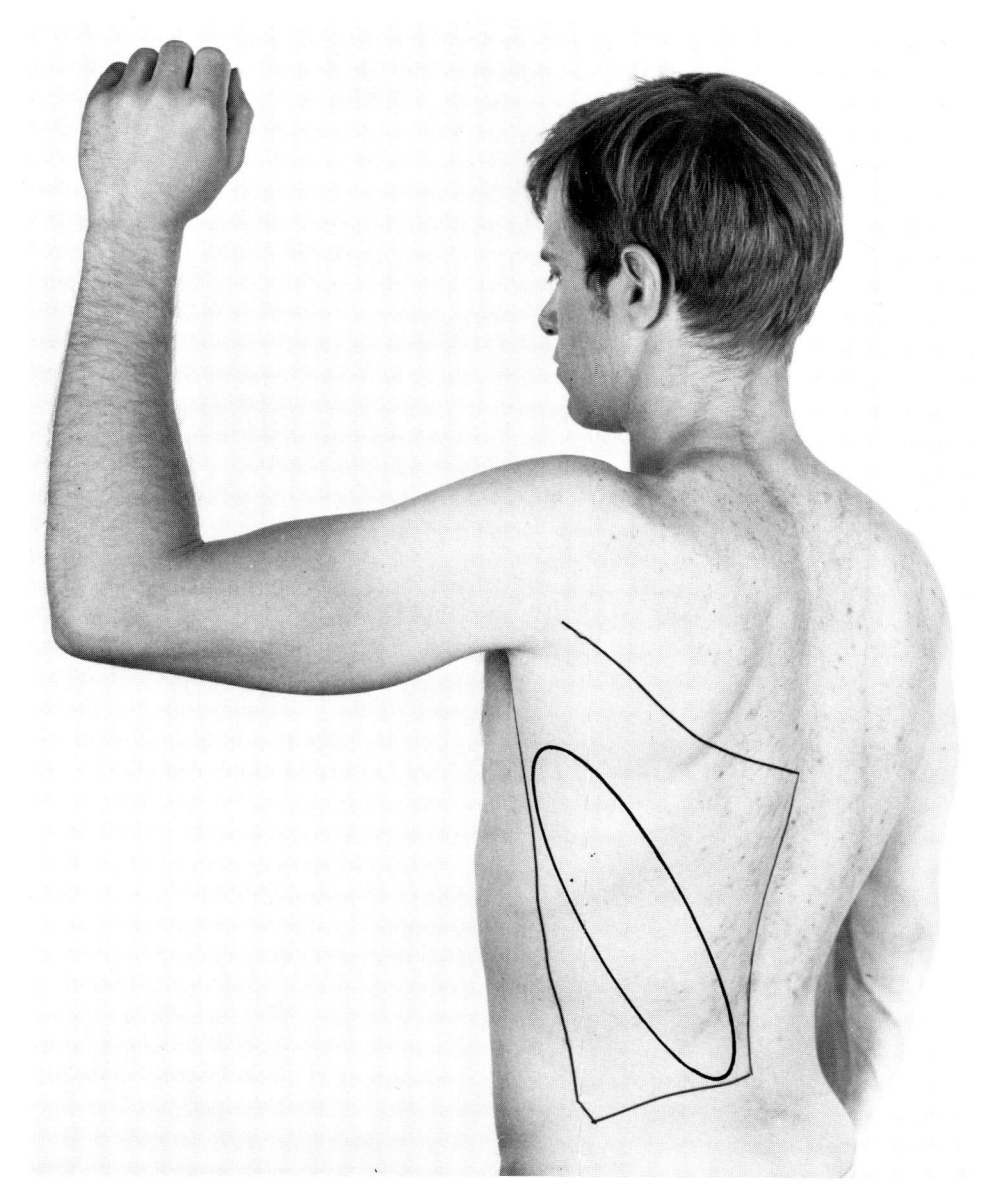

Skin markings outline muscle margins and skin island
along anterolateral free border

For head and neck reconstruction a skin island along the anterolateral margin of the muscle is necessary because the transverse island will not reach. Elevation of a muscle flap may be done through either incision.

TRUNK: POSTERIOR TRUNK

Latissimus dorsi

If there is any question concerning the presence or patency of the major vascular pedicle following mastectomy, then the thoracodorsal artery is identified prior to flap elevation. A small, longitudinal incision is made along the posterior axillary fold, the latissimus dorsi muscle is identified, the major pedicle is located on the deep surface of the muscle, and the thoracodorsal artery is traced proximally. At this stage a medial branch of the thoracodorsal artery to the serratus anterior is identified, and the vessel is divided and ligated to lengthen the vascular pedicle to the latissimus dorsi. For head and neck reconstruction or free flap transfer the pedicle may be further lengthened by dissecting proximally to identify the circumflex scapular artery. This is divided and the subscapular artery dissected up to the axillary artery. A pedicle 10 to 12 cm in length with an artery 2 to 3 mm in diameter is then developed.

After confirming the presence of the major pedicle, the desired skin island is incised and the skin edges are then sutured to the epimysium of the muscle. Skin flaps are then elevated above and below the island to expose the muscle. Either the entire muscle or part of it is then dissected off the origin. If fascia is required, then the thoracolumbar fascia, which forms a large part of the origin of the muscle, is included with the flap. The muscle with the overlying island of skin is then ready for transposition. The secondary defect can be closed directly for skin islands up to 20 cm in length and 10 to 12 cm in width. For larger skin islands, a skin graft may be required.

If the presence or patency of the thoracodorsal artery is not in doubt, the flap is elevated as outlined without prior identification of the pedicle. By dissecting the muscle from origin to insertion, the branches of the major pedicle are easily seen under the epimysium on the deep surface of the muscle in the posterior axillary fold. The branches are then traced back to the pedicle.

Elevation of flap
POSTERIOR ARC

For posterior transposition over the spine it is not necessary to identify the pedicle. A longitudinal incision is made, and the muscle or muscle and overlying skin is elevated and transposed.

The insertion of the muscle into the intertubercular groove of the humerus may be divided to extend the reach of the flap. For functional muscle transfer and in breast reconstruction when the muscle is used to simulate the missing pectoralis major muscle, the muscle is sutured under tension and the motor nerve preserved to prevent loss of muscle bulk. Electromyographic muscle reeducation may prove useful in preservation of muscle bulk.

ARM COVERAGE

For coverage of the arm, or functional biceps transfer, the muscle or musculocutaneous unit is elevated as described for the anterior arc and then transposed onto the arm through an axillary tunnel.

Precautions

- The motor nerve must be preserved and the muscle sutured under adequate tension to preserve muscle bulk in breast reconstruction, or functional muscle transfer.
- In some postmastectomy patients the thoracodorsal nerve may have been divided. In this situation the atrophied latissimus dorsi muscle may not have adequate bulk for pectoralis muscle simulation. An assessment of latissimus dorsi muscle bulk may be done by examining the posterior axillary fold while the humerus is adducted, extended, and medially rotated.
- When there is clinical evidence of muscle atrophy and hence suspected nerve damage in the postmastectomy patient, compromise of the vascular pedicle should be suspected.
- During flap elevation, care should be taken to avoid elevation of the serratus anterior with the latissimus dorsi along the anterolateral free border.

TRUNK: POSTERIOR TRUNK

Latissimus dorsi

Breast reconstruction with musculocutaneous island flap

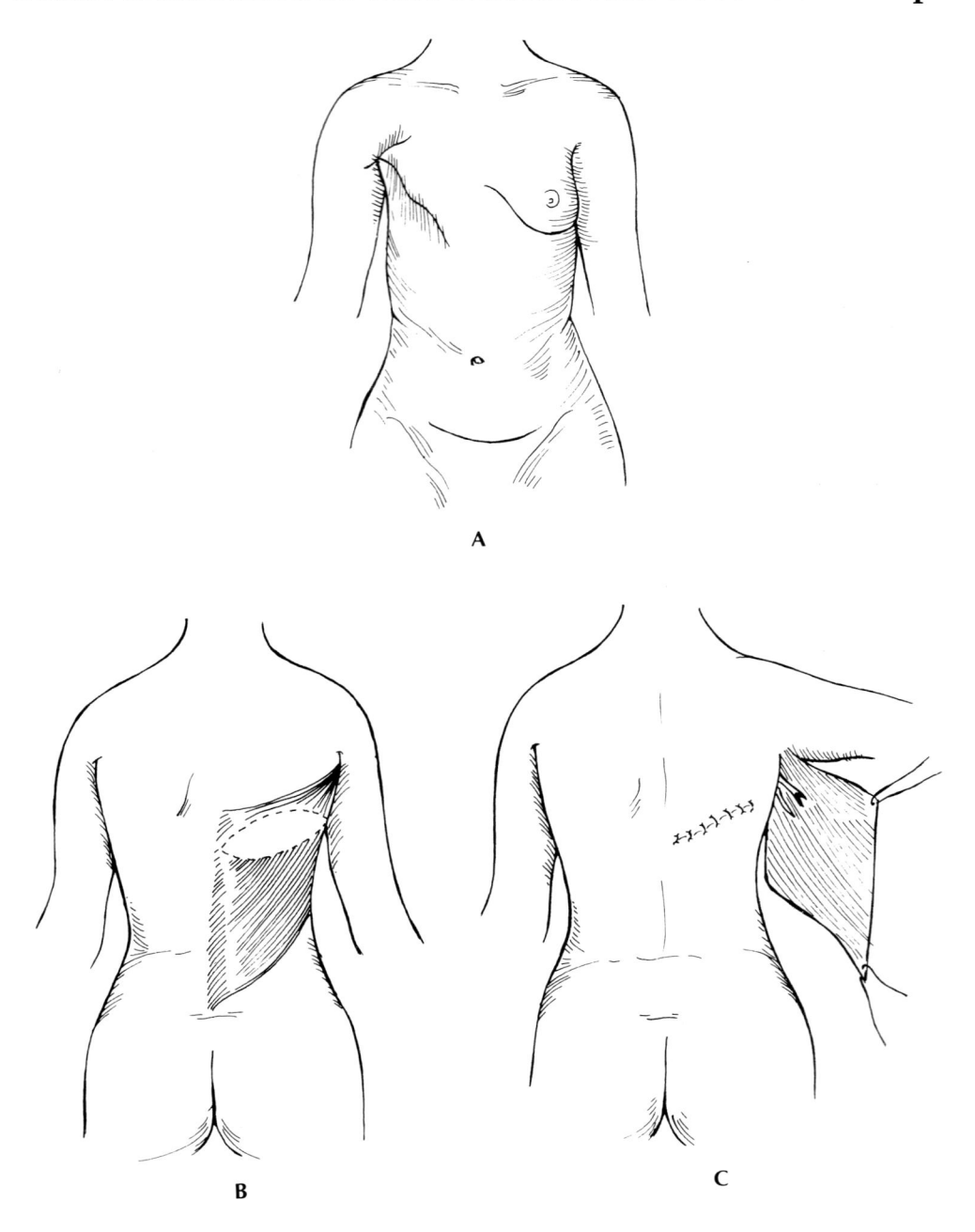

A   Postmastectomy defect, including missing pectoralis major.
B   Skin island outlined.
C   Donor defect closed directly.

TRUNK: POSTERIOR TRUNK

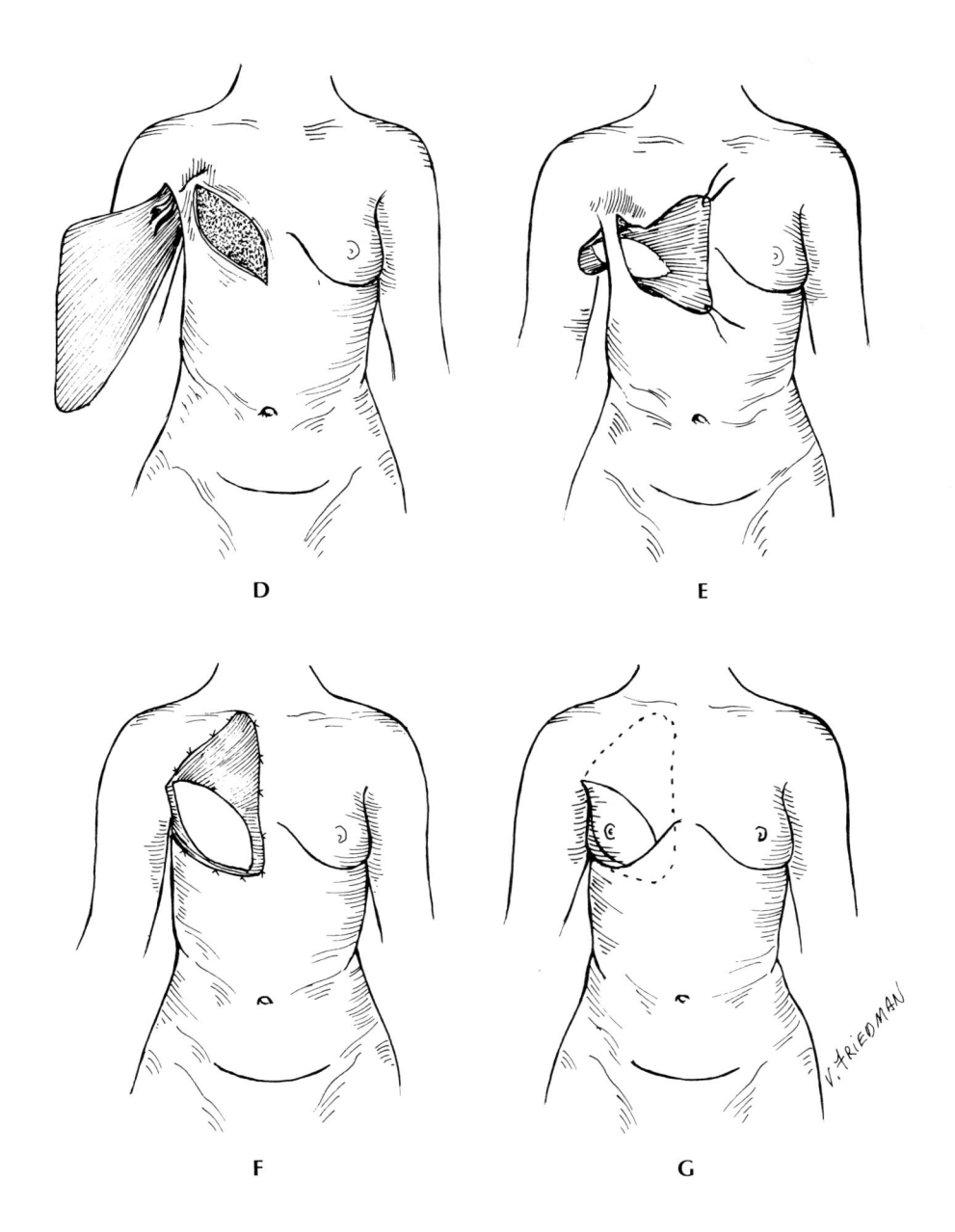

D and E Musculocutaneous unit is passed anteriorly.

F Muscle is sutured down to simulate missing pectoralis, and implant is placed under muscle.

G Single-stage reconstruction of breast with latissimus dorsi flap and implant.

TRUNK: POSTERIOR TRUNK

Latissimus dorsi

Breast reconstruction with musculocutaneous island flap

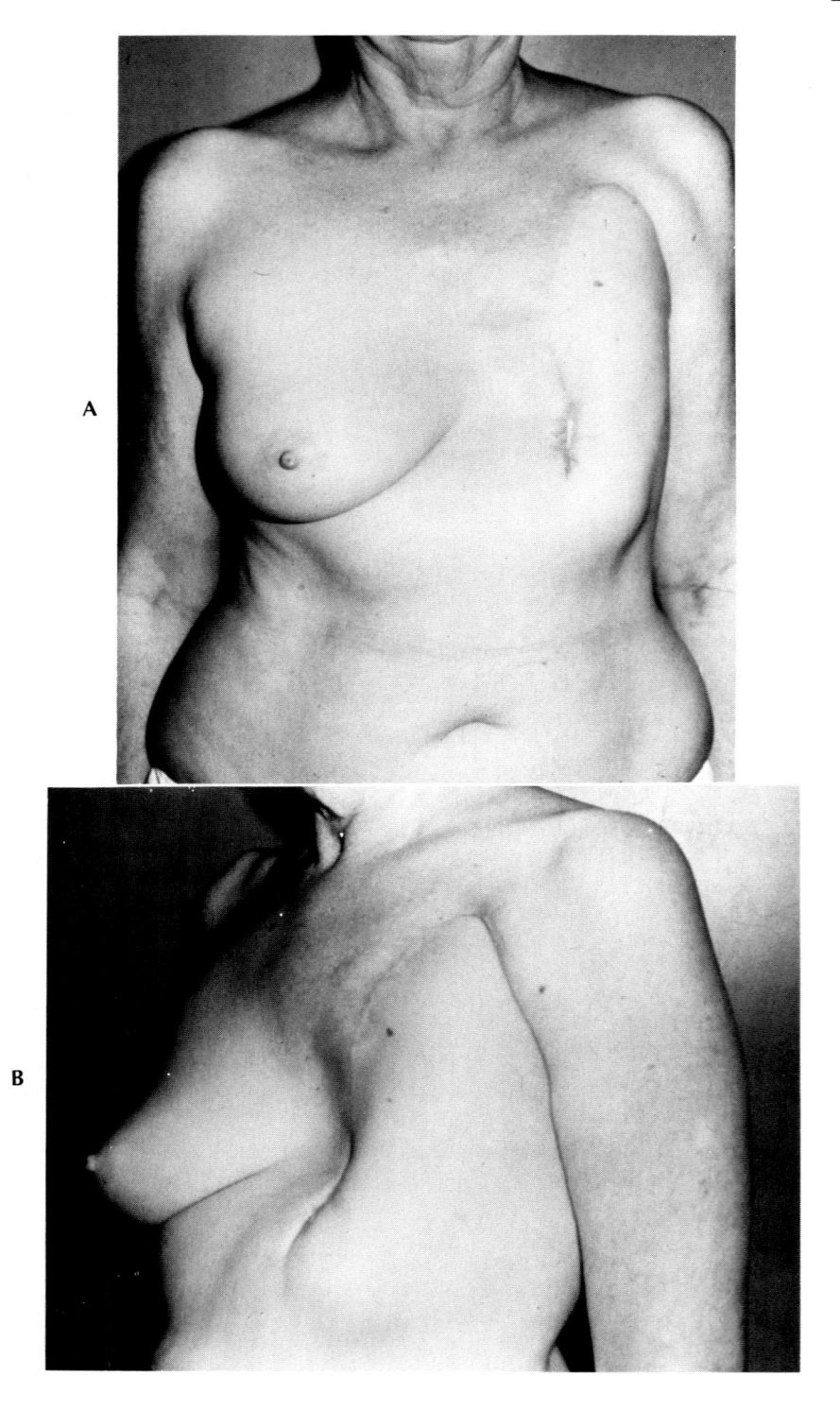

TRUNK: POSTERIOR TRUNK

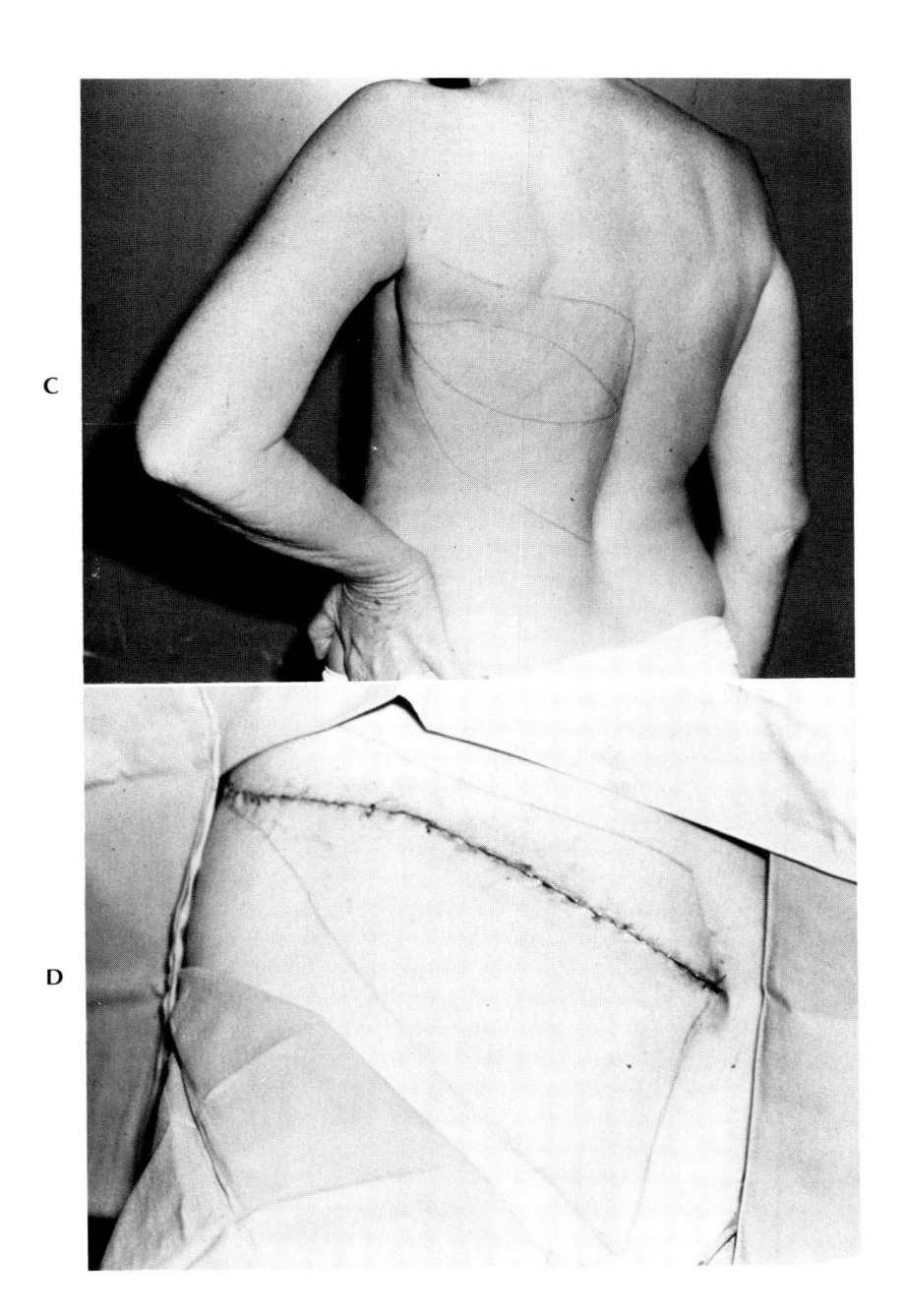

A   Postmastectomy patient requiring skin coverage, pectoralis simulation, breast mound, and nipple-areola complex.

B   Note missing pectoralis major muscle and severe pectus excavatum deformity.

C   Skin island and extent of muscle elevation is outlined.

D   Direct closure of secondary defect.

Continued.

TRUNK: POSTERIOR TRUNK

Latissimus dorsi

Breast reconstruction with musculocutaneous island flap—cont'd

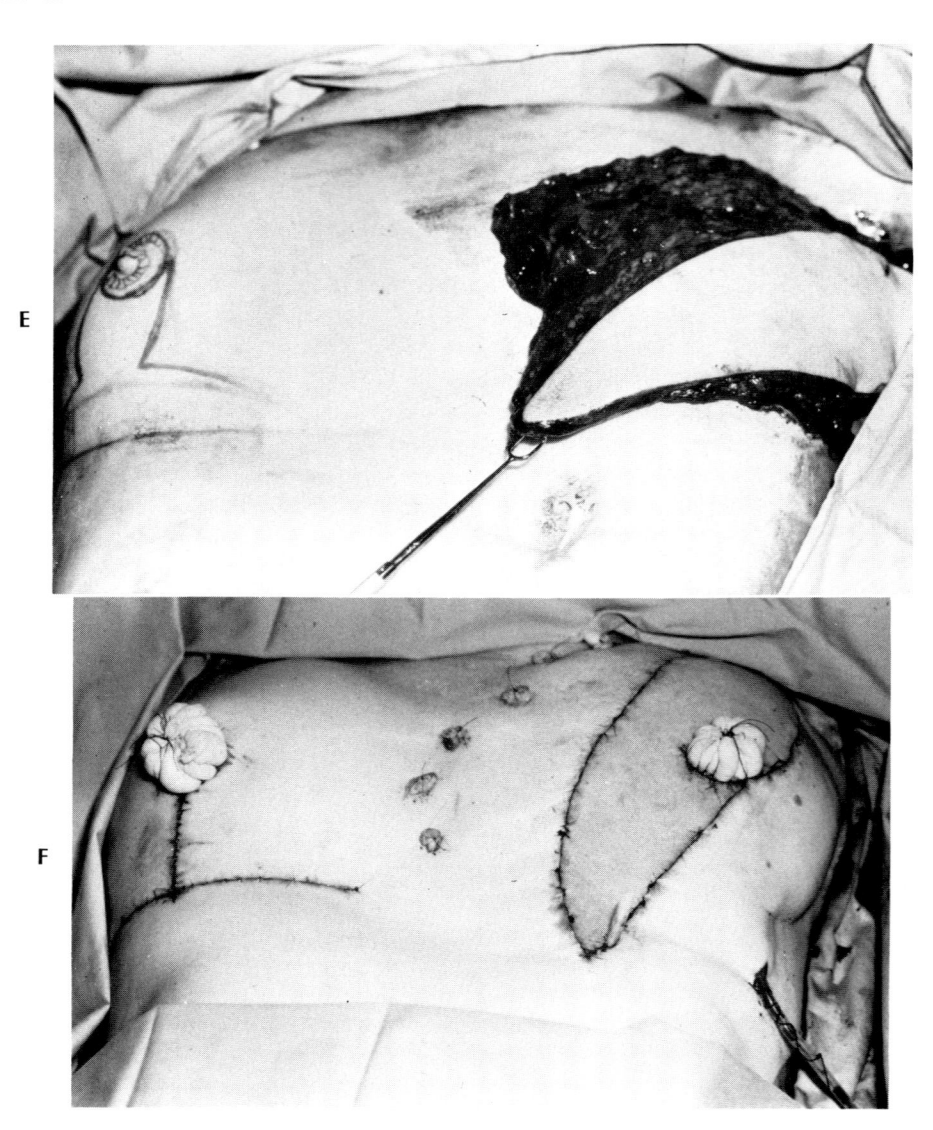

E Musculocutaneous unit is passed anteriorly onto chest wall.

F Muscle has been sutured down and implant positioned under flap. Total mastectomy with immediate reconstruction was performed on opposite breast.

TRUNK: POSTERIOR TRUNK

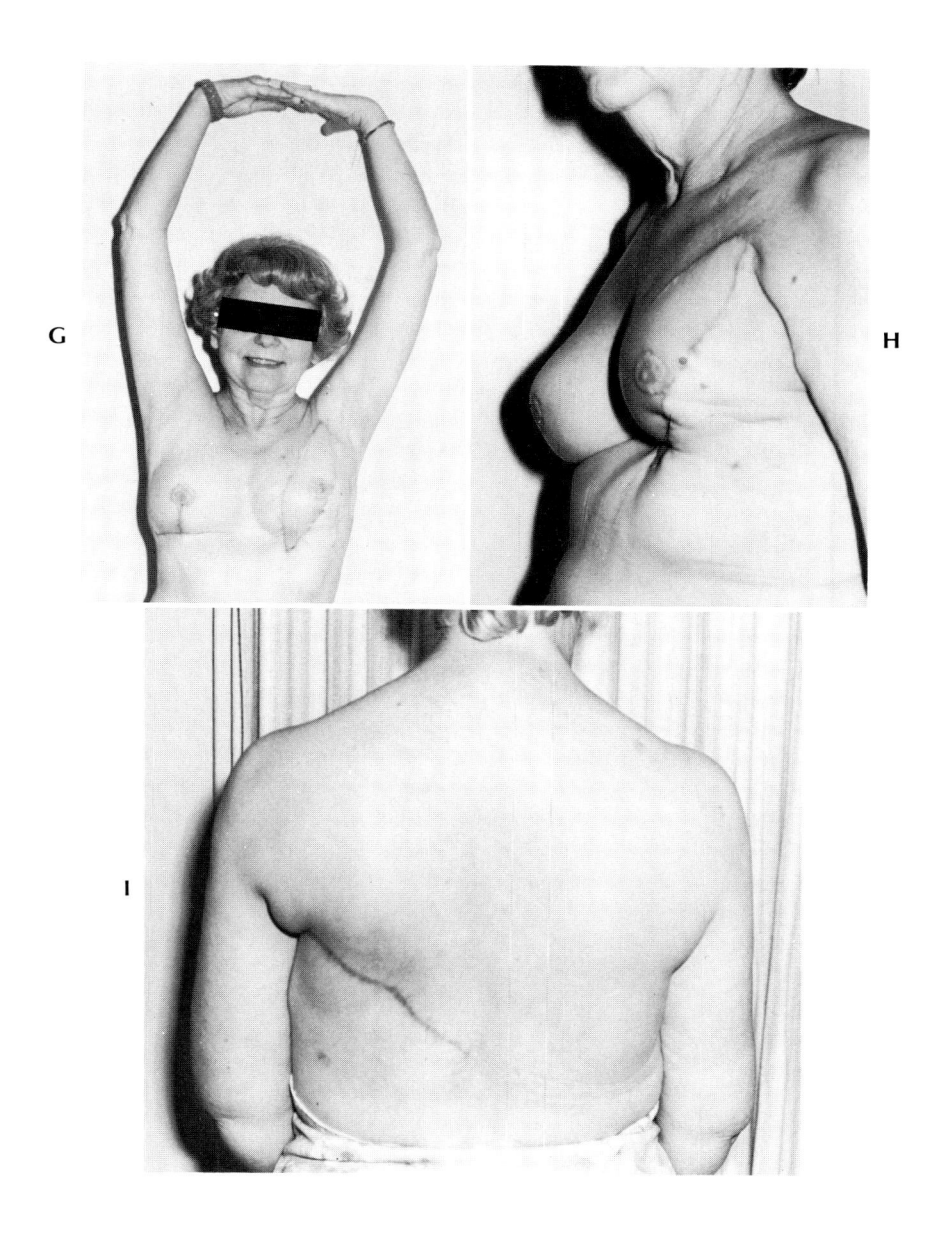

G Single-stage reconstruction of breast with latissimus dorsi island musculocutaneous flap, inflatable implant, and nipple sharing. No loss of upper extremity range of motion. Note correction of pectus deformity.

H Lateral view of reconstructed breast.

I Healed donor defect is easily concealed by brassiere.

Trapezius

MUSCLE FLAP
MUSCULOCUTANEOUS FLAP
FREE FLAP

Applications

Reconstruction of:
BREAST

Coverage of:
CHEST WALL
HEAD AND NECK
BACK
UPPER ARM

Functional muscle transfer

Free flap:
DISTANT COVERAGE

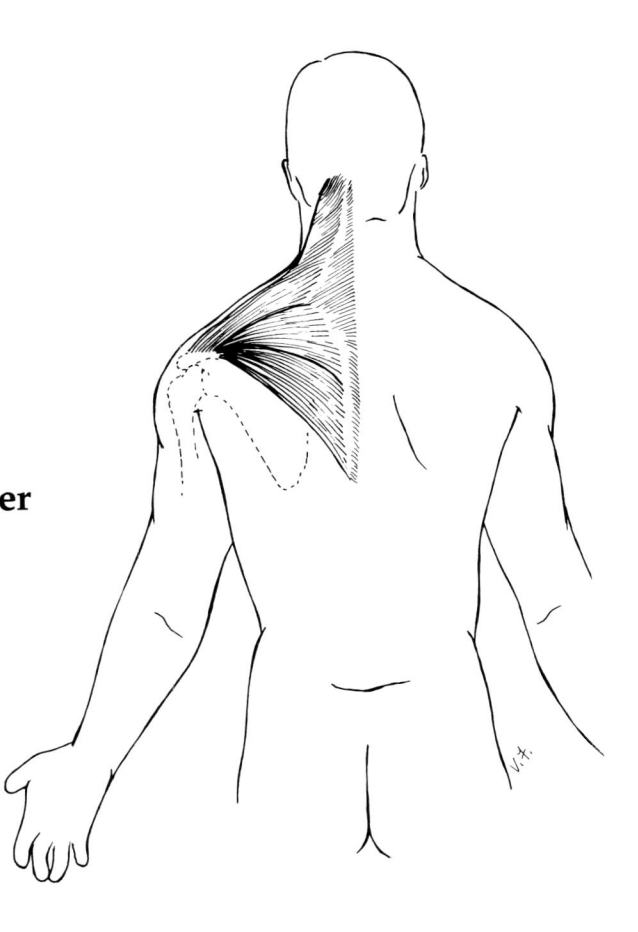

TRUNK: POSTERIOR TRUNK

Trapezius

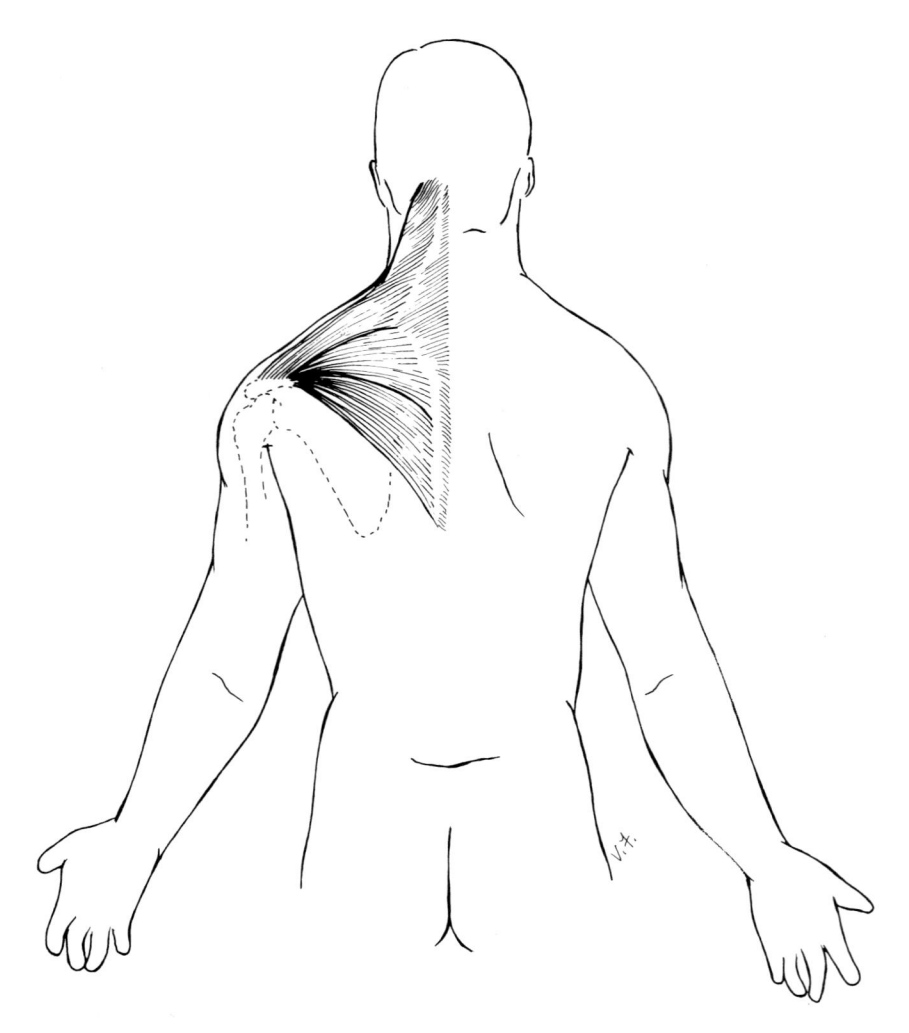

The trapezius is a large, flat muscle located in the posteroinferior neck and posterosuperior trunk. In the neck the muscle has a superficial relationship to the underlying splenius capitis muscle. In the back the muscle has a superficial relationship to the rhomboideus minor and major muscles. Inferiorly the muscle is superficial to the infraspinatus fascia and a portion of the superior latissimus dorsi muscle.

Origin: External occipital protuberance, superior nuchal line of the occipital bone, nuchal ligament, spinous processes of seventh cervical, and all thoracic vertebrae.

Insertion: Superior fibers—lateral third of clavicle; middle fibers—spine of scapula; inferior fibers—acromion.

TRUNK: POSTERIOR TRUNK

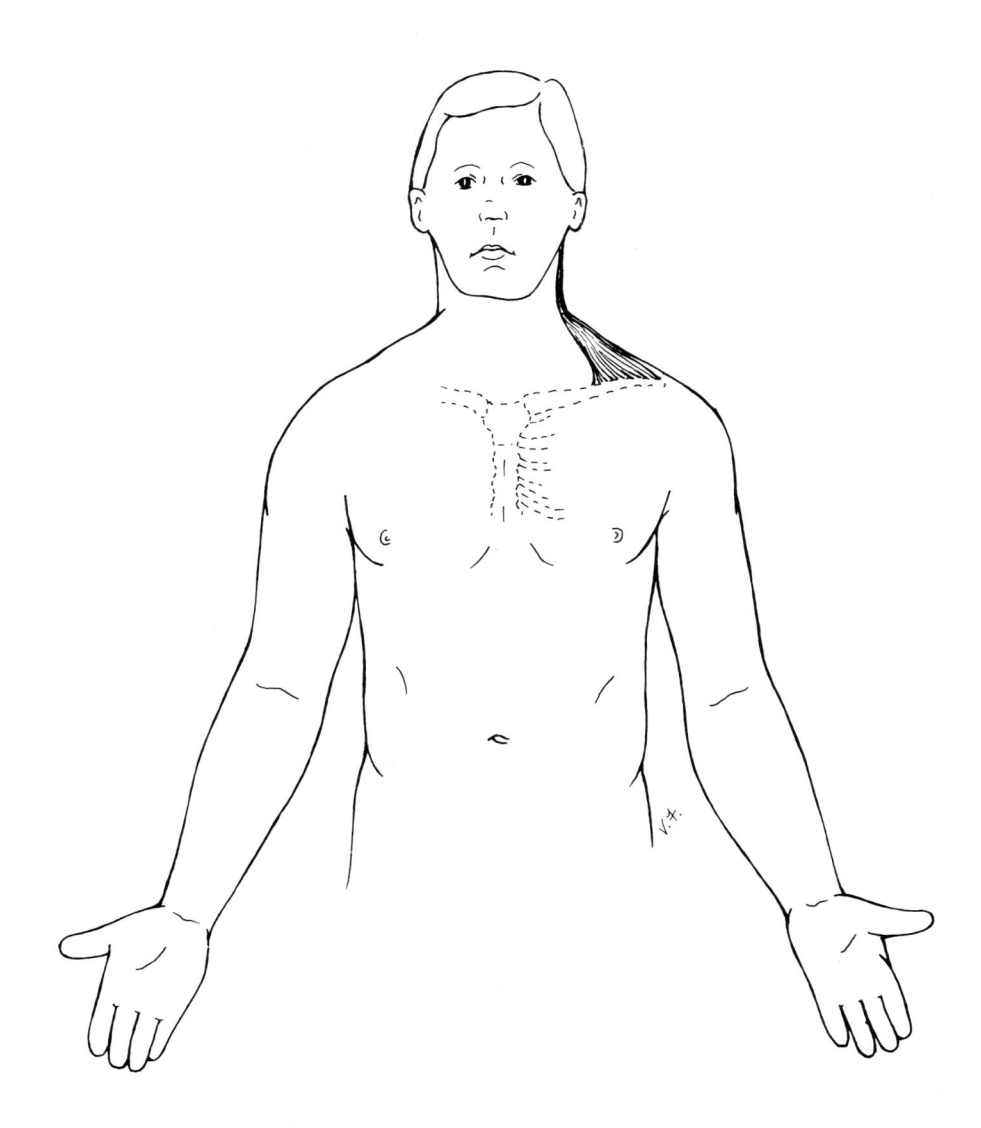

Nerve supply:

Motor — Spinal accessory, third and fourth cervical nerves. These motor nerves enter the proximal posterior aspect of the muscle adjacent to the descending pedicle from the transverse cervical artery.

Function: The muscle rotates the scapula and elevates the shoulder in full abduction and flexion of the arm, tilts the chin, draws back the acromion, and rotates the scapula. With loss of the entire muscle function, as in denervation injuries, the drooping shoulder deformity is present. However, the middle and inferior fibers, when used in transposition, are expendable.

TRUNK: POSTERIOR TRUNK

Trapezius

Transverse cervical artery

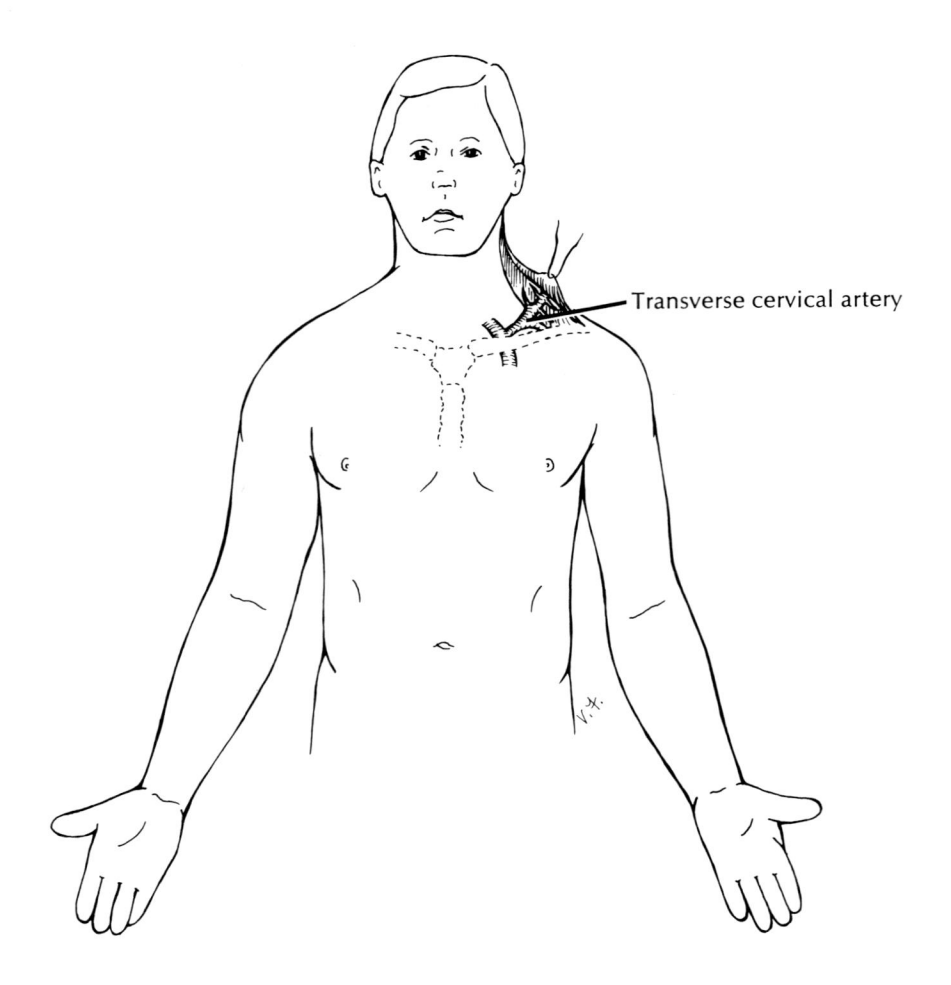

Blood supply

This posterior trunk muscle receives its dominant blood supply through pedicles originating from branches of the subclavian artery in the anterior neck. Although the nomenclature is variable, depending on the point of origin of the vascular pedicles from the subclavian artery, the dominant vascular pedicle to the trapezius consistently enters the muscle at the base of the neck with ascending and descending arterial branches.

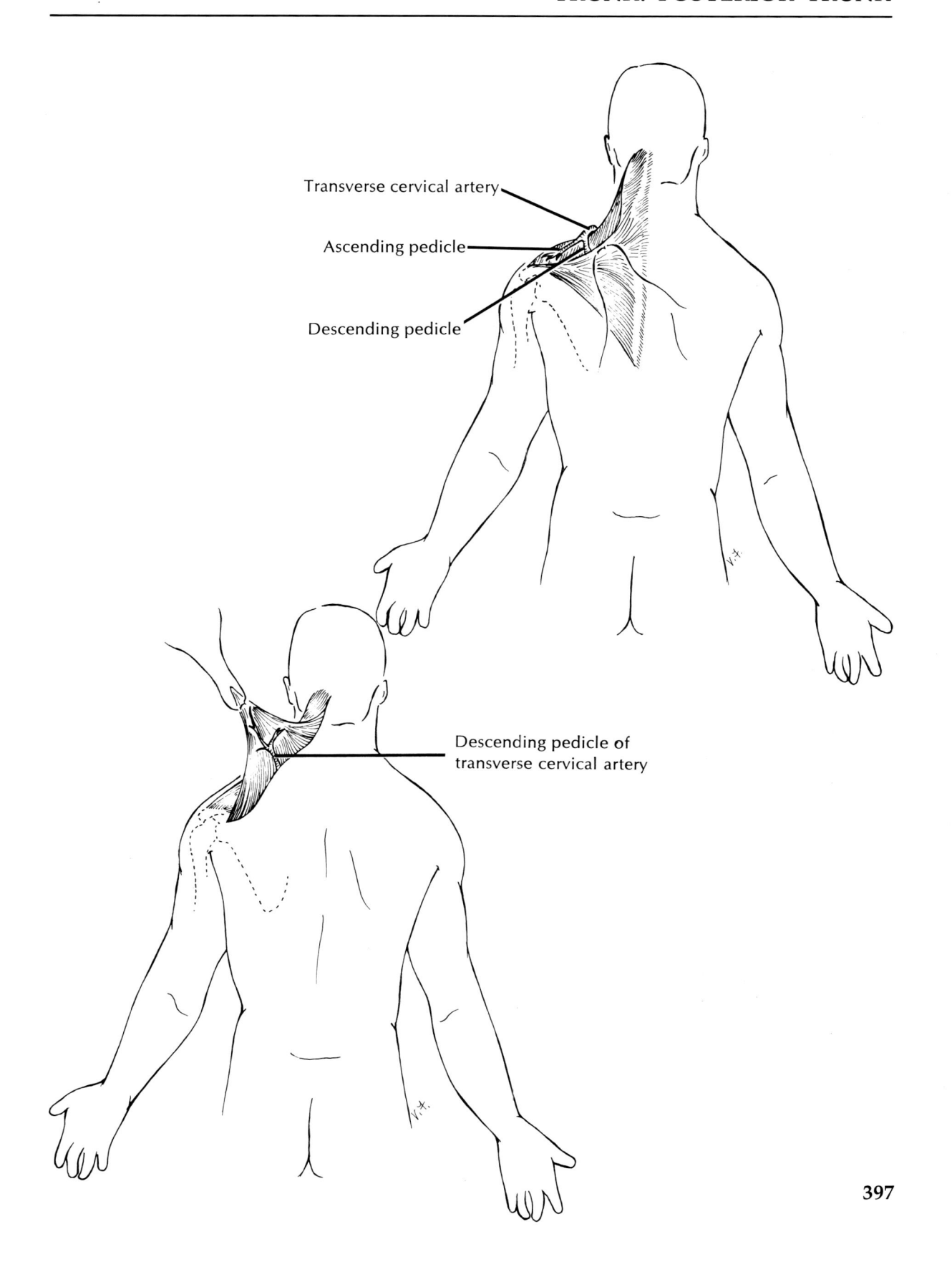

TRUNK: POSTERIOR TRUNK

Transverse cervical artery

Ascending pedicle

Descending pedicle

Descending pedicle of transverse cervical artery

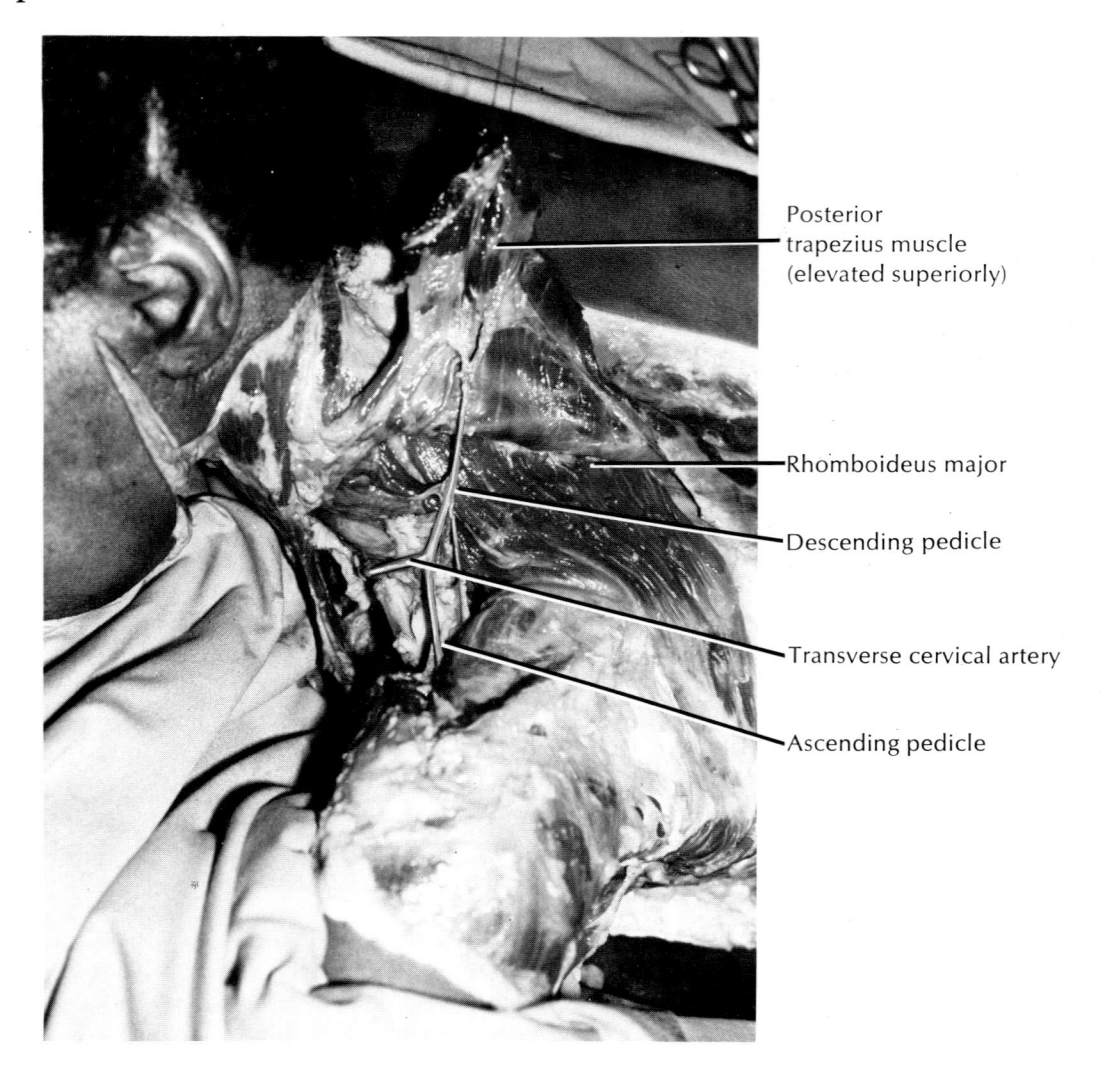

TRUNK: POSTERIOR TRUNK

Trapezius

- Posterior trapezius muscle (elevated superiorly)
- Rhomboideus major
- Descending pedicle
- Transverse cervical artery
- Ascending pedicle

The ascending artery, superficial cervical artery, generally arises as a branch of the transverse cervical artery. However, the superficial cervical artery may arise as a separate branch from the thyrocervical trunk. This artery courses between the sternocleidomastoid and scalenus muscles, entering the anterior margin of the trapezius muscle. Prominent perforating vessels from this artery extend onto the lateral shoulder and will support an extended cutaneous territory (cervical humeral flap).

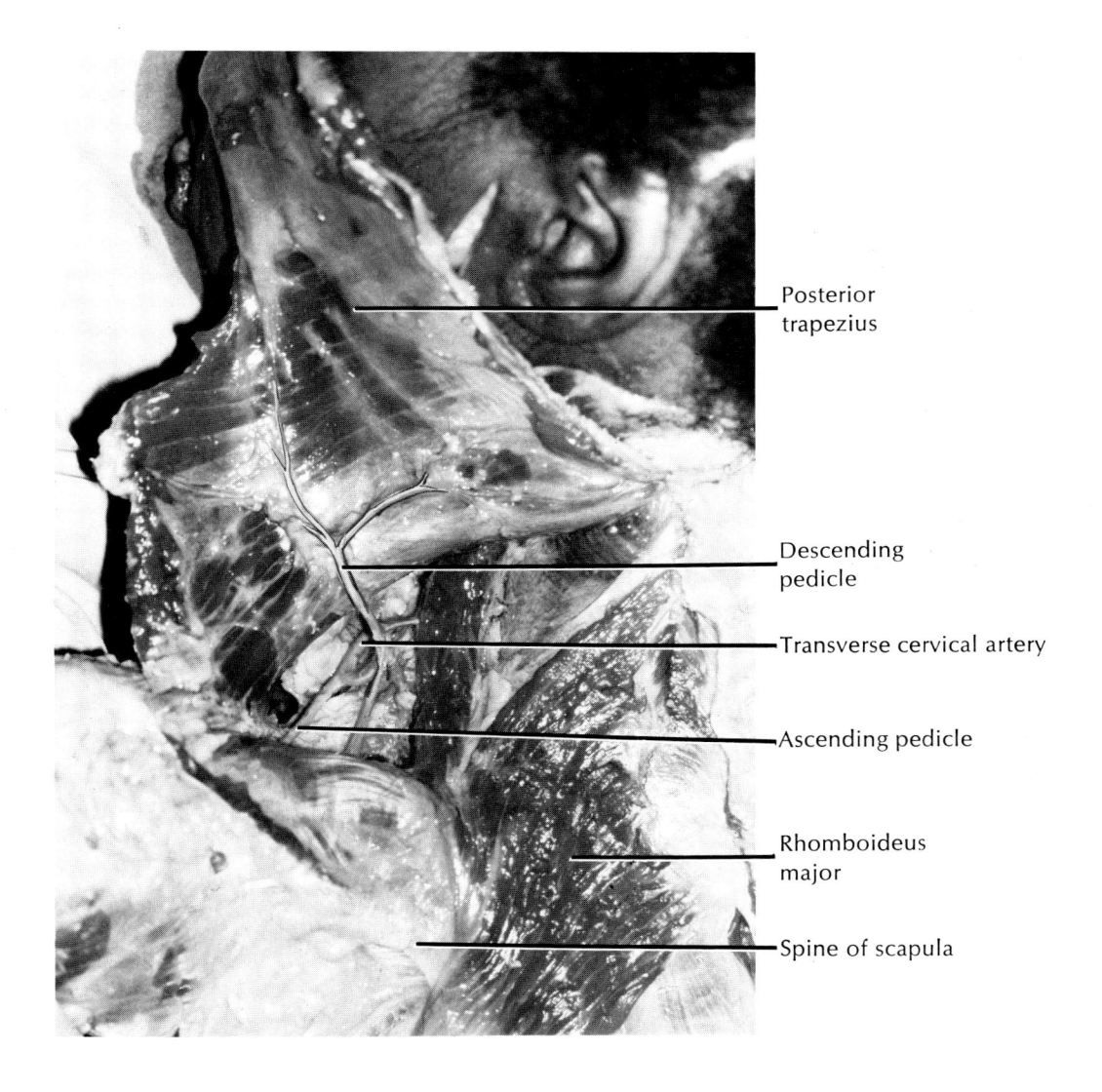

TRUNK: POSTERIOR TRUNK

- Posterior trapezius
- Descending pedicle
- Transverse cervical artery
- Ascending pedicle
- Rhomboideus major
- Spine of scapula

The descending artery, posterior scapular artery, originates generally as a branch of the transverse cervical artery and extends inferiorly along the deep surface of the trapezius muscle. The posterior scapular artery may arise as a separate branch from the subclavian artery. This artery will support the thoracic portion of the trapezius muscle. Its perforating musculocutaneous vessels extend into the skin located between the vertebral column and scapula. There are also segmental perforating vessels to this region through the paraspinous muscles via the intercostal arteries.

TRUNK: POSTERIOR TRUNK

Trapezius

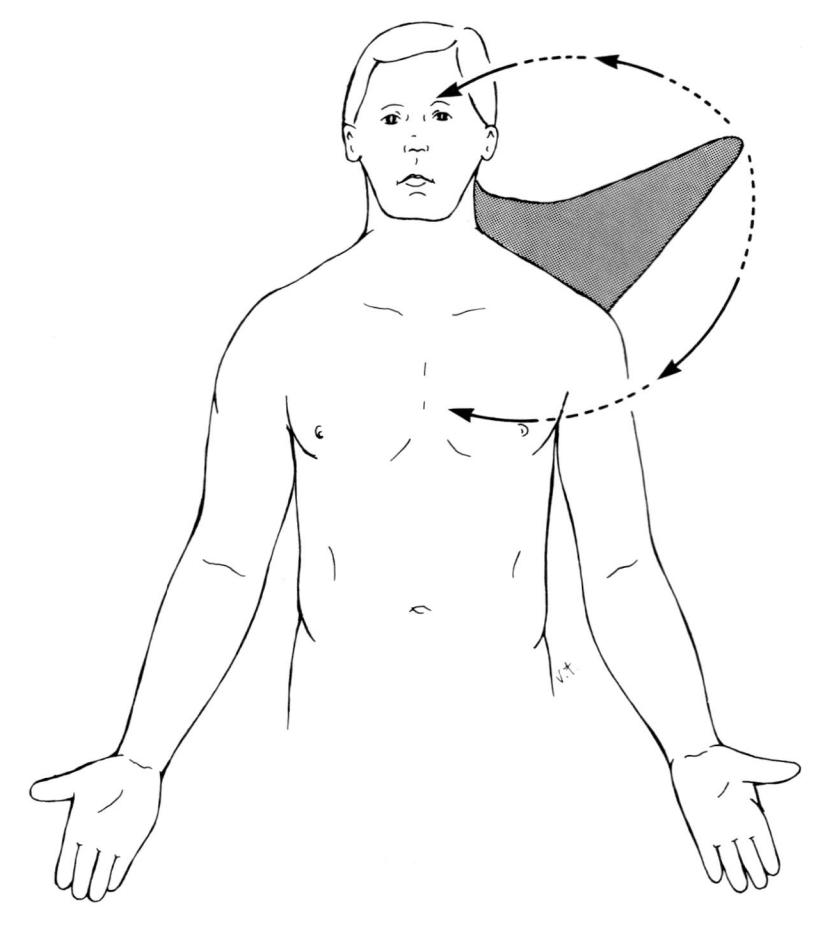

Anterior arc

Arc of rotation

This muscle has wide arcs of rotation on its dominant vascular pedicle in the anterior base of the neck. The muscle has both anterior and posterior arcs of rotation.

ANTERIOR ARC

The anterior arc of the muscle will reach the head, neck, and upper thoracic region. This arc is possible with the superior fibers left intact. If the insertion to the clavicle is detached, the muscle will have a wider arc in the head and neck region. If the origin from the superior nuchal line is detached, the muscle will have a wider arc in the superoanterior chest. With an extensive dissection, the unit can be mobilized as an island on the superficial portion of the transverse cervical artery with a greater arc of rotation.

TRUNK: POSTERIOR TRUNK

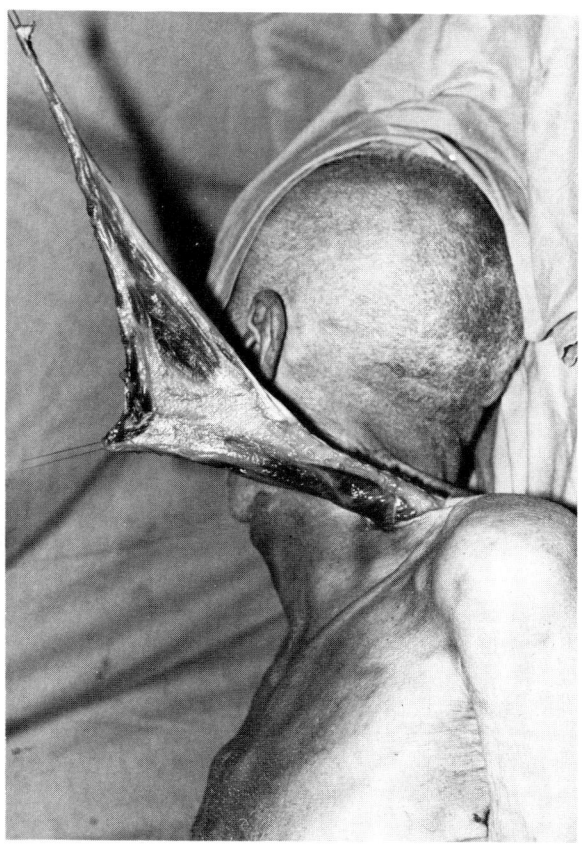

Anterior arc for reconstruction of head, neck, and upper thorax

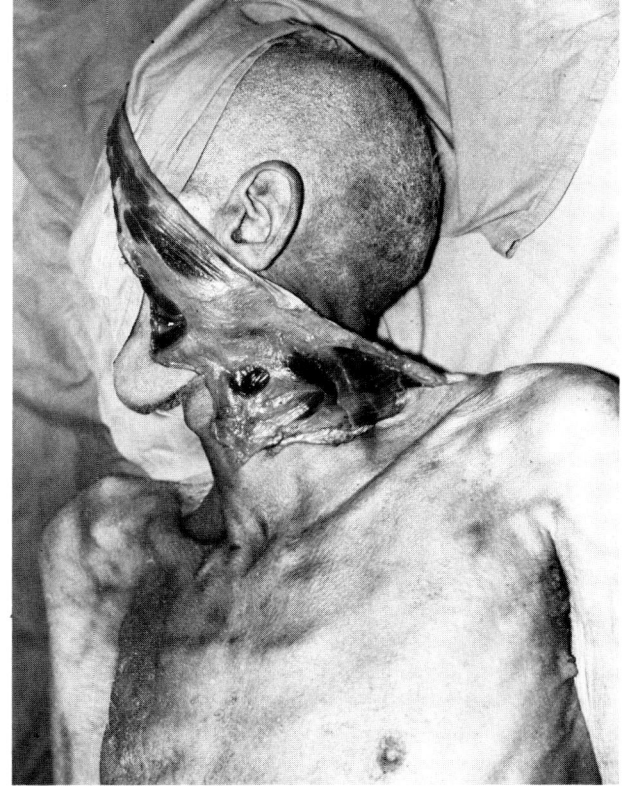

Anterior arc for head reconstruction

TRUNK: POSTERIOR TRUNK

Trapezius

Posteromedial arc (clavicular insertion detachment)

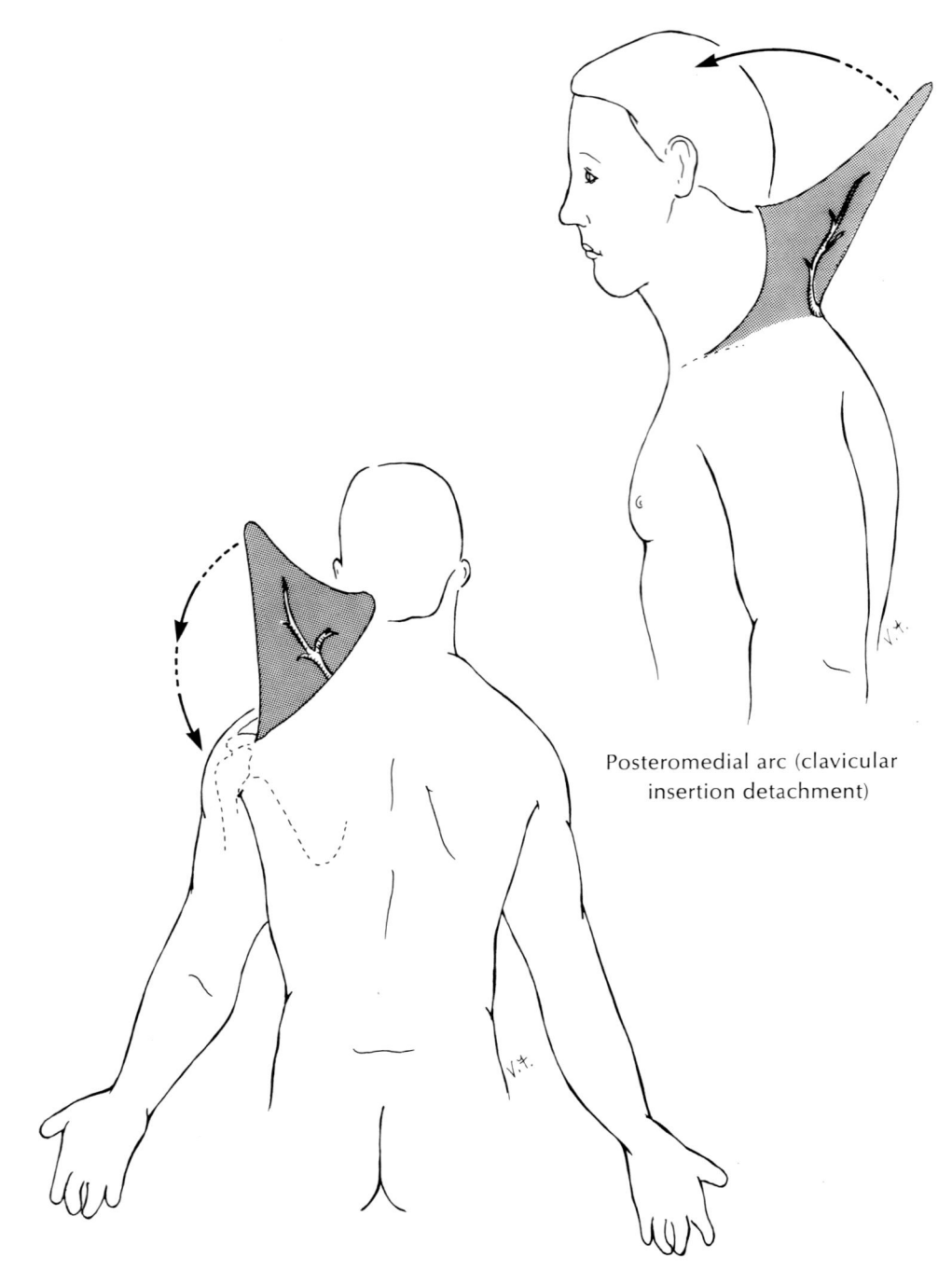

POSTERIOR ARC

This muscle has a posterior arc that will cover defects in the skull, posterior neck, and shoulder. This arc does not require release of the superior muscle origin in the neck and lateral clavicular insertion.

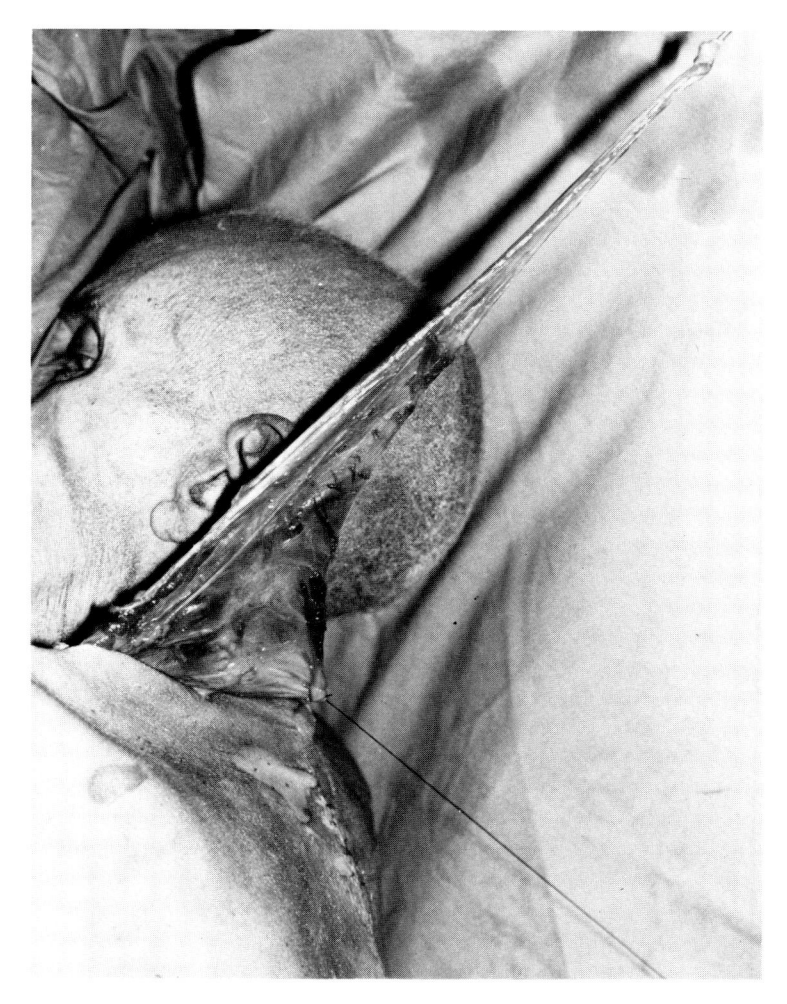

Posterior arc for neck and posterior skull reconstruction

TRUNK: POSTERIOR TRUNK

Trapezius

Elevation of flap

MUSCLE UNIT

The muscle can be located through a vertical back incision between the vertebral column and the vertebral border of the scapula. The muscle overlaps the latissimus dorsi at the level of the inferior angle of the scapula. The fascial attachments to the vertebral column and spine of the scapula are detached. The muscle is elevated to the level of the base of the posterior neck. At this level the descending artery on the posterior muscle can be visualized. The clavicular insertion of the muscle can be detached, depending on the required arc of rotation. The donor area can be closed primarily.

MUSCULOCUTANEOUS UNIT

Two musculocutaneous units can be designed, incorporating perforating vessels through the trapezius muscle. These vessels allow extension of the cutaneous territory of the flap into the territories of adjacent muscles.

A posterior musculocutaneous unit can be based on the descending branch of the superficial cervical artery or transverse cervical artery. The cutaneous territory is designed between the vertebral column and the lateral border of the muscle. This flap may extend 3 to 5 cm below the inferior angle of the scapula. The cutaneous flap is elevated over the latissimus dorsi muscle, and at the inferior margin of the trapezius muscle the muscle is elevated with the skin. The muscle fibers are divided laterally to the superior scapula and lateral clavicle and medially to the vertebral origin as needed for flap length. This musculocutaneous flap has a wide arc to the posterior neck, skull, anterior neck, and face.

TRUNK: POSTERIOR TRUNK

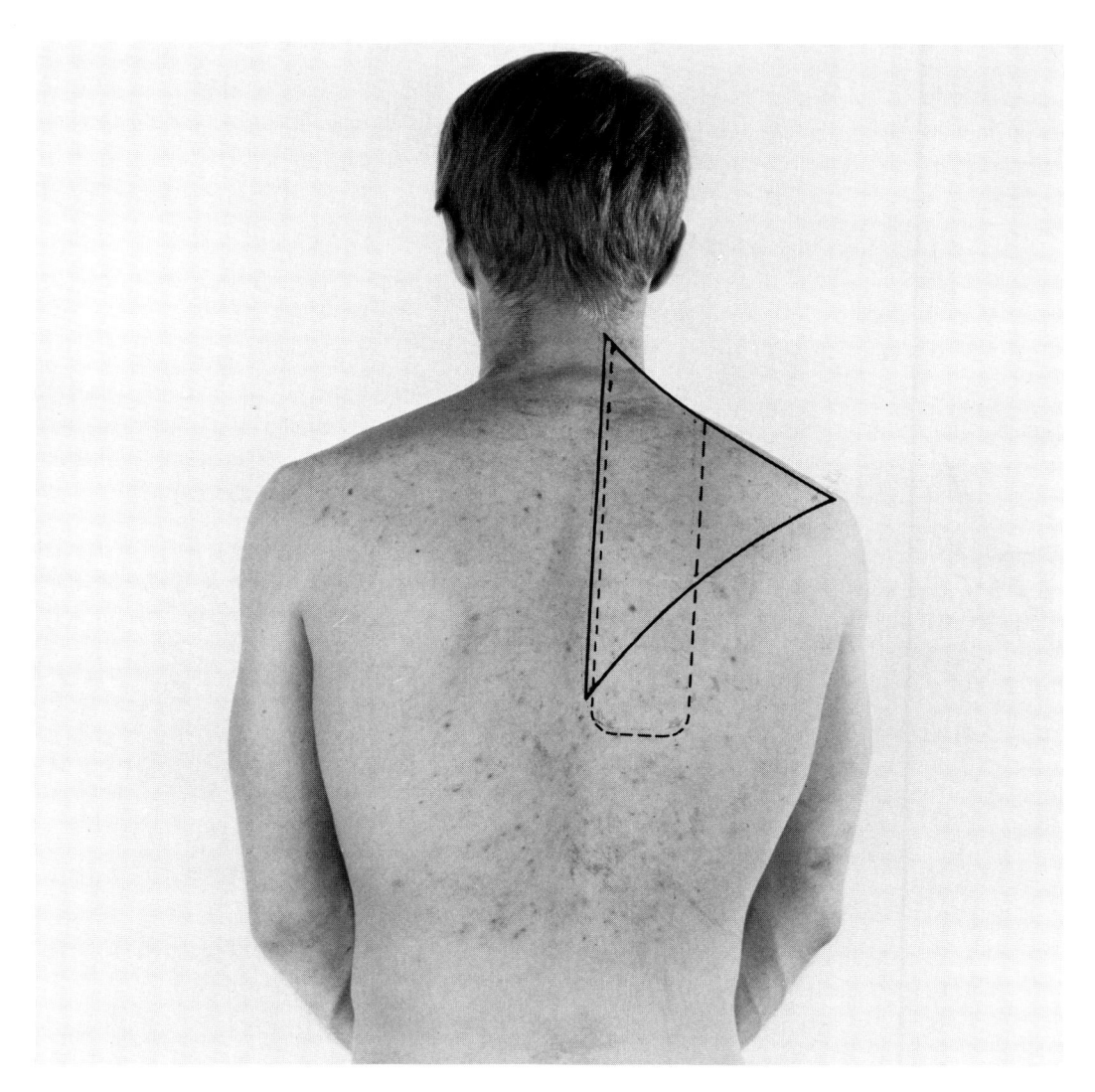

Skin markings for posterior musculocutaneous flap
(note: flap can be extended below inferior muscle margin)

TRUNK: POSTERIOR TRUNK

Trapezius

The ascending branch of the transverse cervical artery supplies the superior trapezius muscle fibers. The perforating vessels supply the overlying cutaneous territory in the nape of the neck. A musculocutaneous unit may be elevated between the posterior clavicle and superior border of the scapula, extending over the shoulder and inferiorly along the anterior upper arm. The unit may extend to the junction of the middle and lower thirds of the upper arm. The flap is elevated superiorly, including the fascia of the underlying deltoid, biceps, and triceps muscles. At the level of the acromioclavicular joint, the trapezius fibers are elevated with the muscle to the level of the superficial branch of the transverse cervical artery. This flap should have its point of rotation in the neck several centimeters superior to the acromioclavicular joint to prevent kinking of the flaps. For this reason the flap, which has a wide arc of rotation, should be elevated from the arm only the required length. Also, if a short flap is required from the neck, the trapezius musculocutaneous flap may be elevated starting at the acromioclavicular joint, dividing the superficial branch of the transverse cervical artery. This flap is then based on the minor pedicle to the superior trapezius muscle from the occipital artery. This flap will reach the lower third of the face. The extended cervical humeral flap will reach the head, neck, and anterior chest. With both flaps the donor area may not be closed primarily and will require skin grafting.

Precautions

- The spinal accessory nerve should be preserved to prevent denervation of trapezius fibers left intact. It is rare to completely mobilize the superior, middle, and inferior fibers of the muscle so some function is usually preserved.
- The cervical humeral flap must be centered over the acromioclavicular joint extending inferiorly along the anterolateral arm.
- The cervical humeral flap must be rotated in the base of the neck to prevent severe flap kinking during transposition.
- The superficial branch of the transverse cervical artery must be intact for use of the cervical humeral flap. This vessel may have been removed during a standard radical neck dissection or may have been subject to radiation injury following radiotherapy to the anterior neck region.

TRUNK: POSTERIOR TRUNK

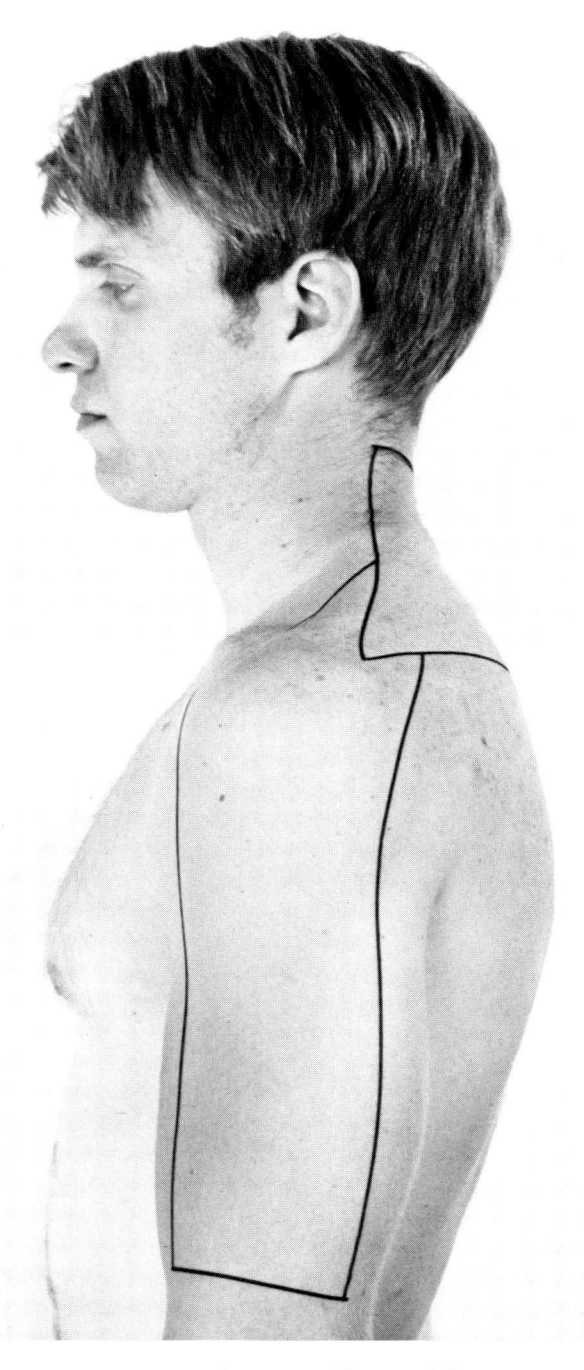

Skin markings for cervical humeral flap

TRUNK: POSTERIOR TRUNK

Trapezius

Head and neck reconstruction

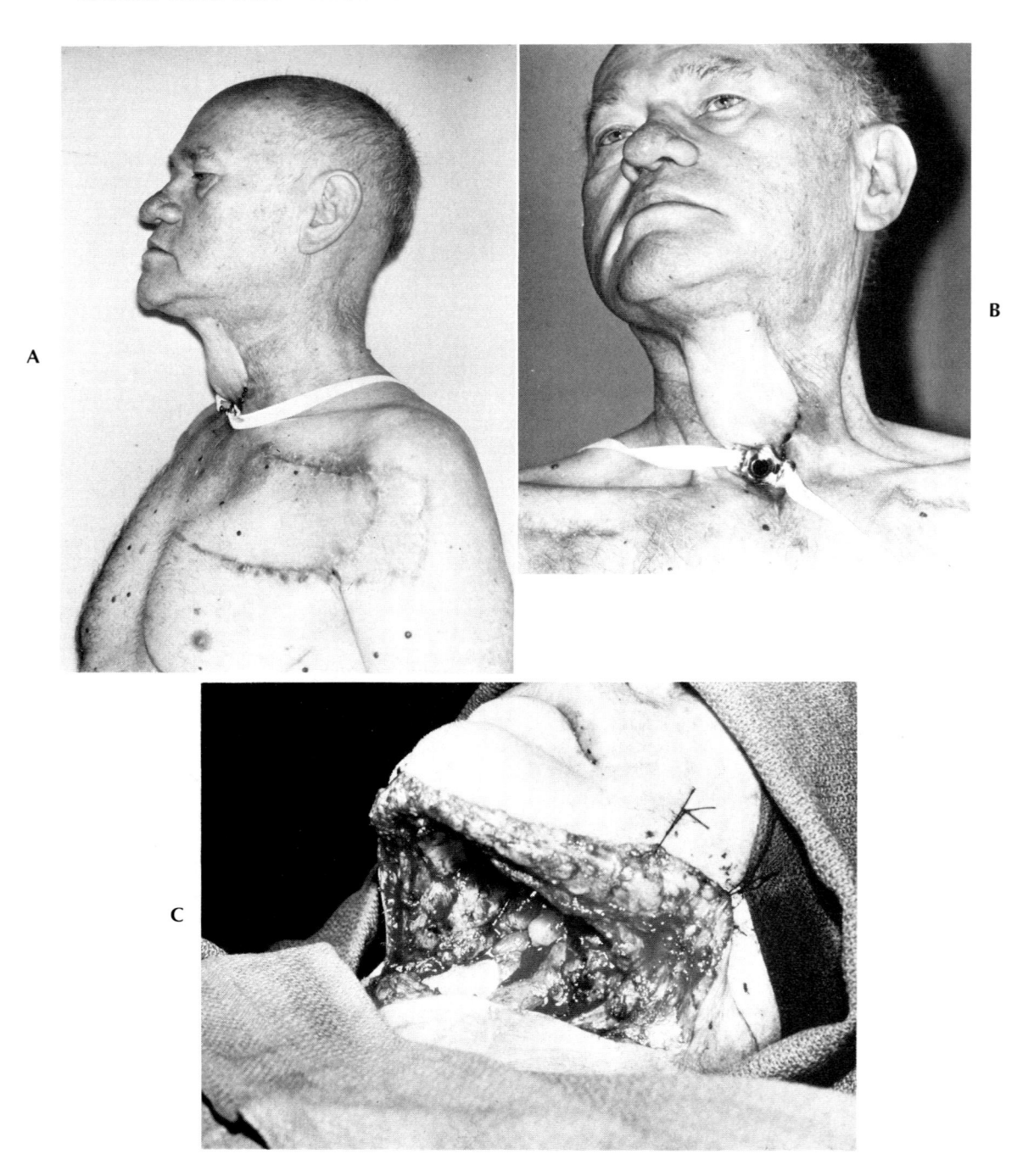

408

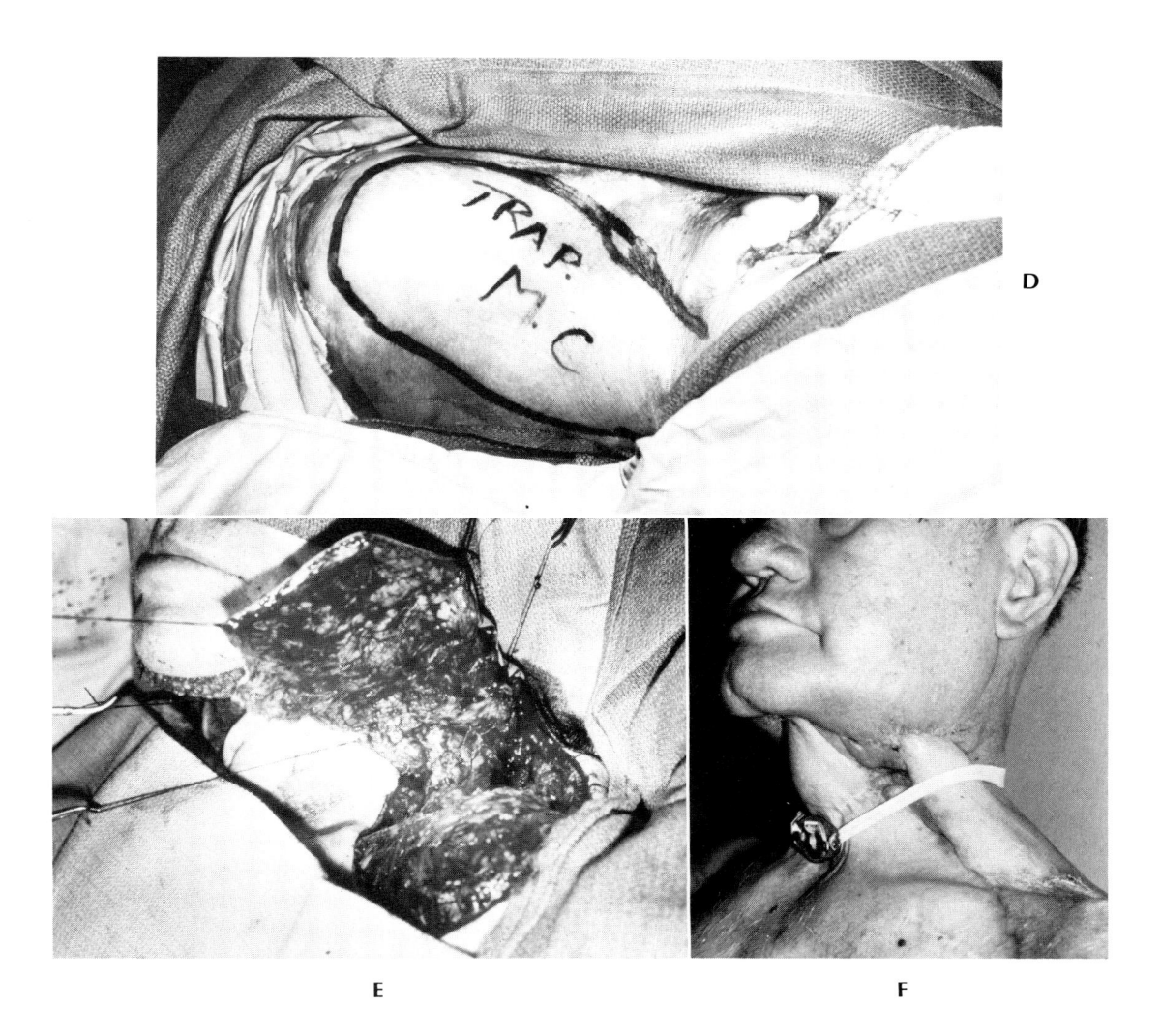

TRUNK: POSTERIOR TRUNK

A   Recurrent left neck carcinoma following extirpative surgery and radiotherapy.

B   Deltopectoral flap used for prior neck reconstruction.

C   Surgical defect following resection of local recurrence in anterior neck.

D   Trapezius musculocutaneous flap skin markings. Flap is based on occipital artery pedicle to muscle.

E   Trapezius musculocutaneous flap elevation.

F   Postoperative single-stage reconstruction of anterolateral neck defect with musculocutaneous flap transposition.

TRUNK: POSTERIOR TRUNK

Trapezius

Cervicohumeral flap for head and neck reconstruction

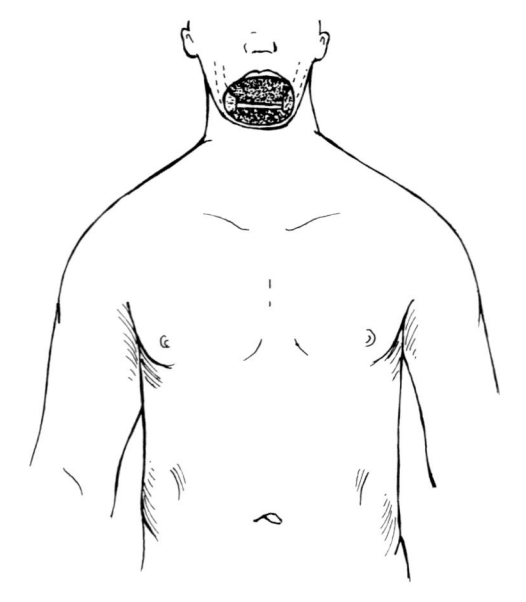

Traumatic defect with loss of central mandible and floor of mouth.

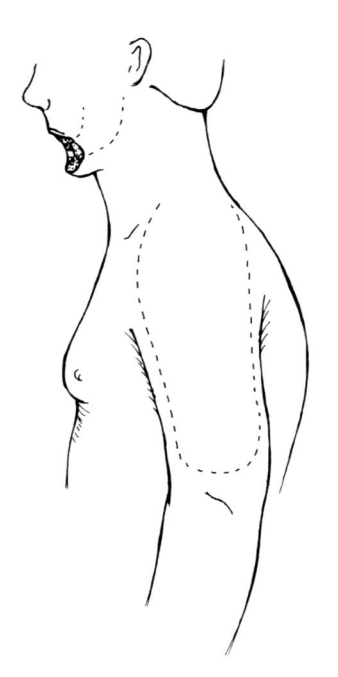

Skin outline of cervicohumeral flap.

TRUNK: POSTERIOR TRUNK

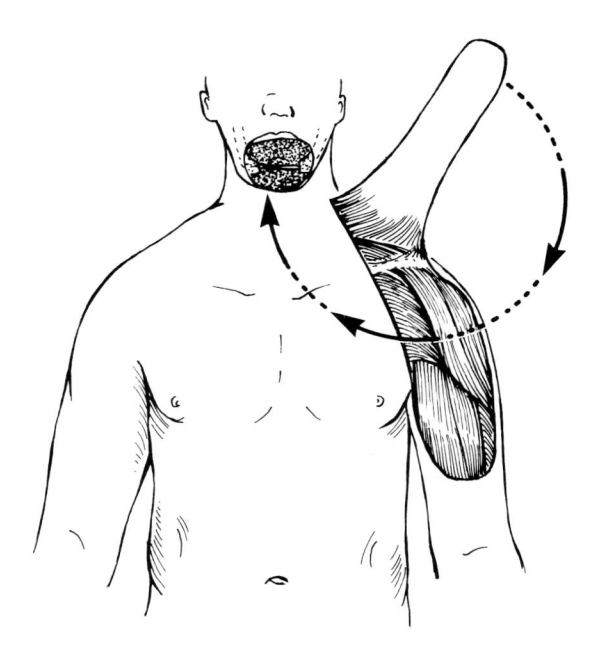

Flap elevation including cutaneous extension on arm and trapezius muscle at clavicular insertion. Donor defect requires skin grafts for closure.

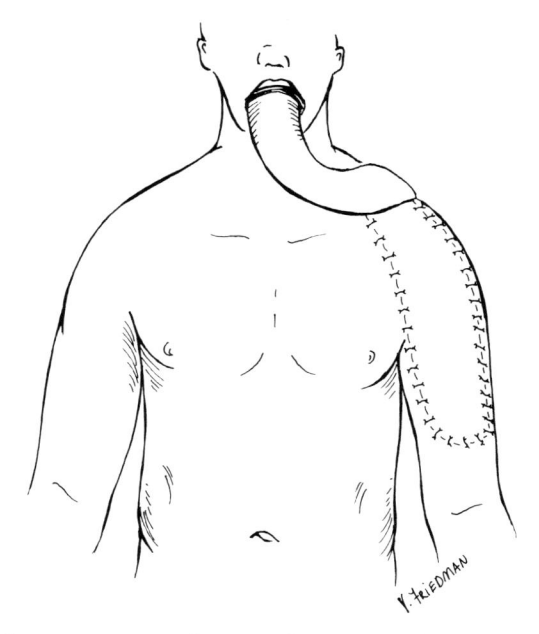

Flap point of rotation in base of neck. Flap inset in oral cavity.

TRUNK: POSTERIOR TRUNK

Trapezius

Cervicohumeral flap for head and neck reconstruction

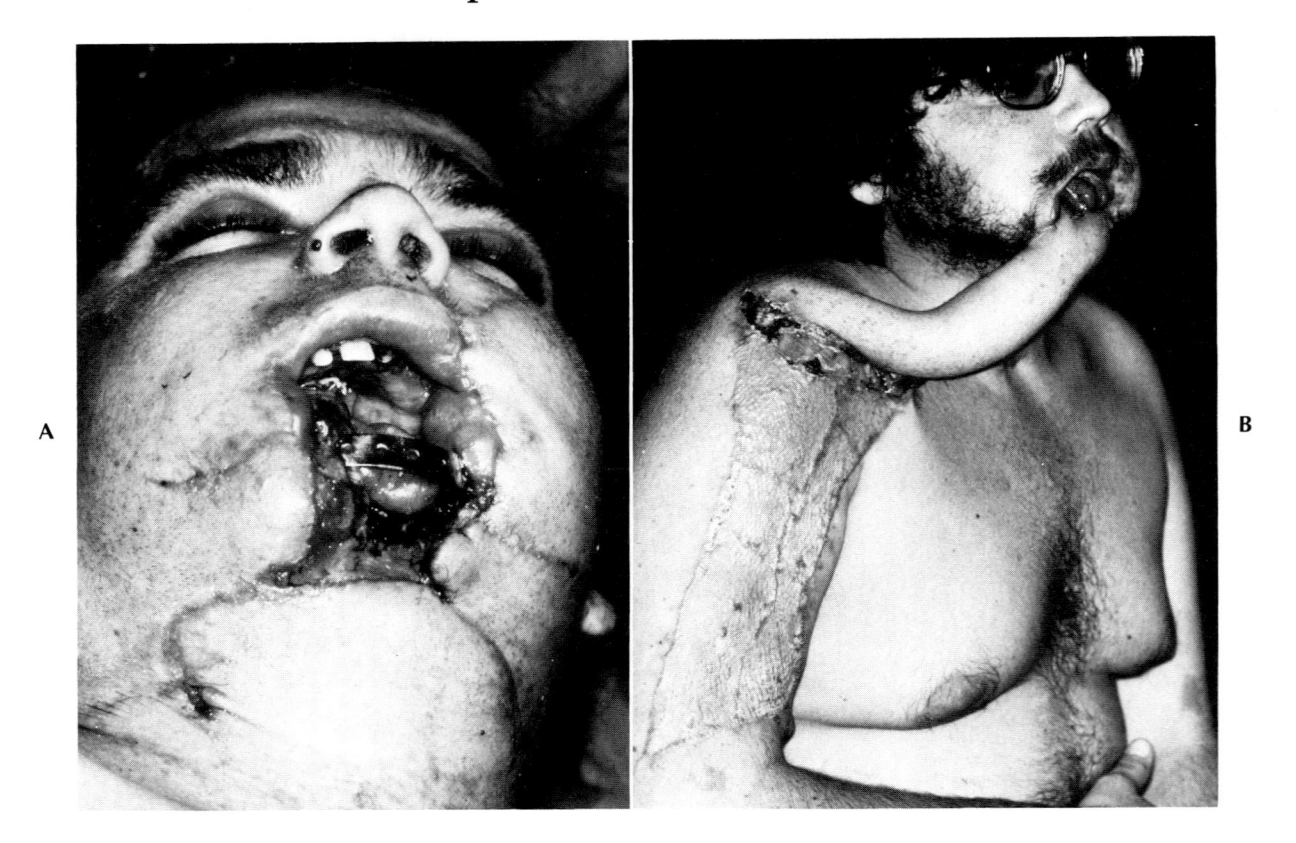

A to C Traumatic defect with loss of central mandible and floor of mouth.

D Postoperative reconstruction of oral cavity and chin with cervicohumeral flap.

TRUNK: POSTERIOR TRUNK

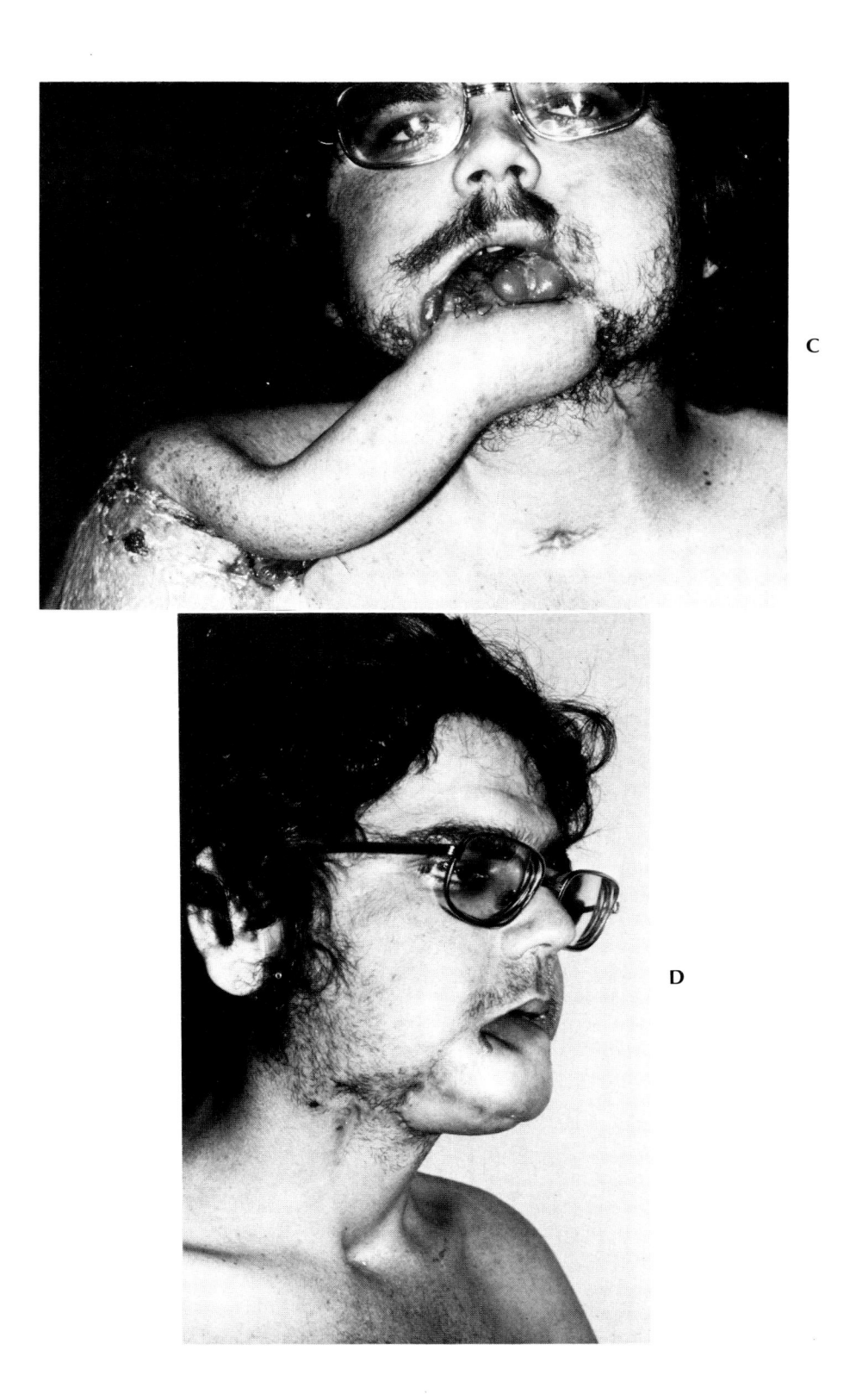

C

D

413

TRUNK: POSTERIOR TRUNK

Trapezius

Posterior musculocutaneous flap for neck reconstruction

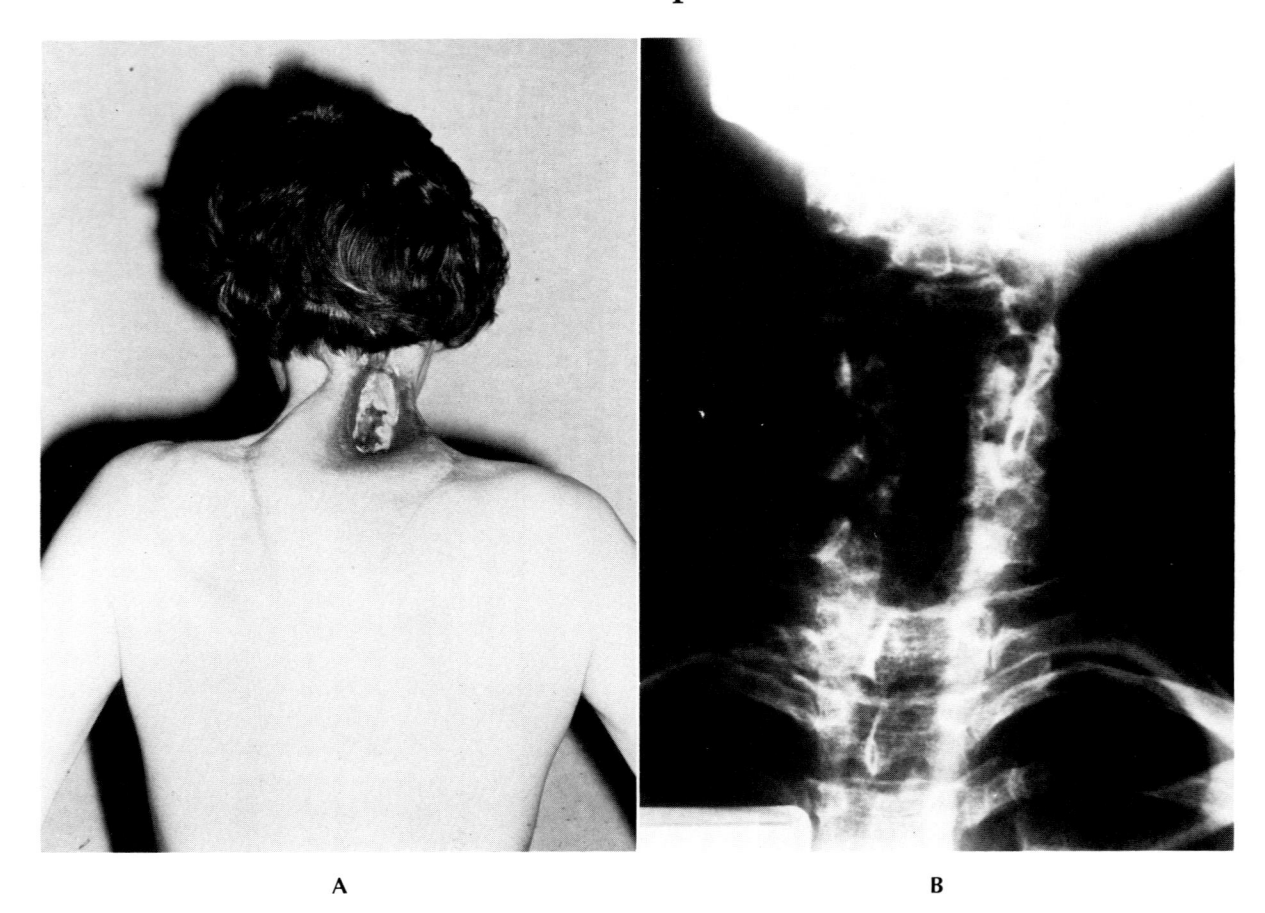

A   B

A Radiation necrosis of posterior neck. Local flaps have not provided defect coverage.

B Patient has undergone posterior laminectomy and neck radiation therapy.

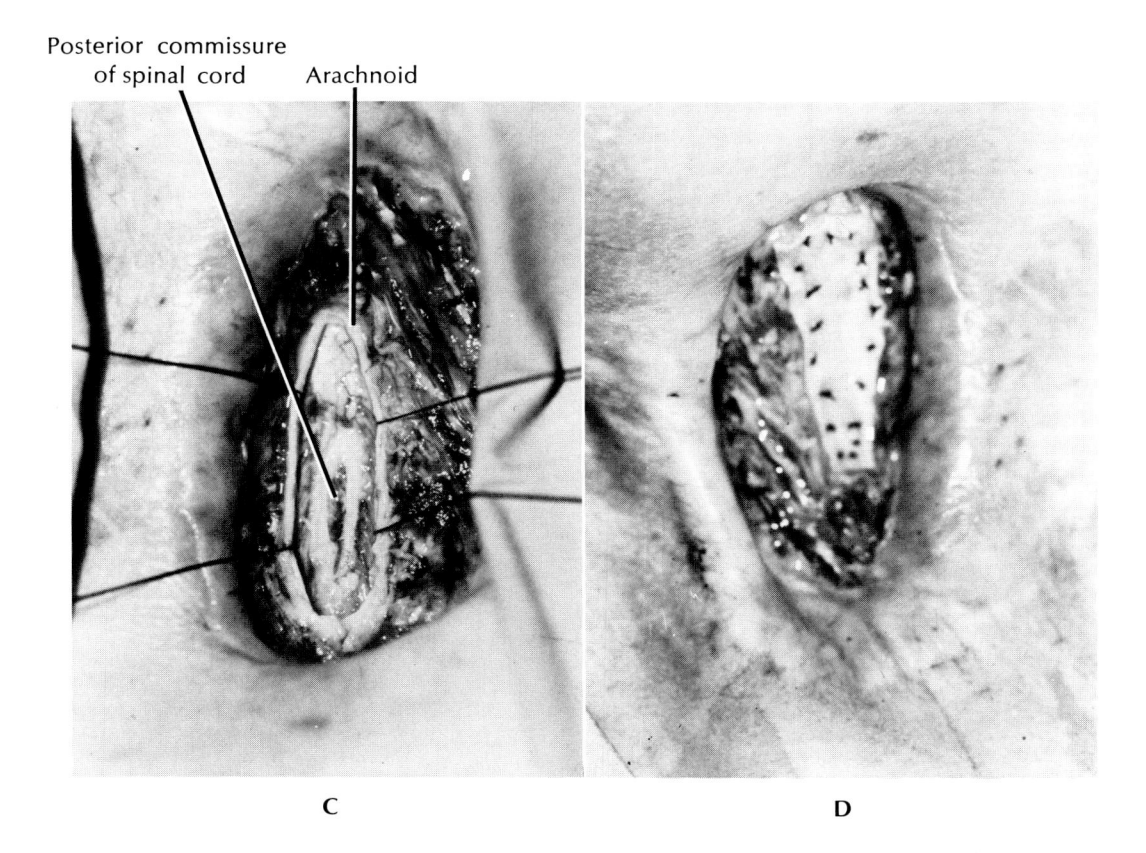

C Debridement of neck wound reveals necrosis of posterior arachnoid and posterior commissure of spinal cord.

D Arachnoid defect repaired with fascia lata.

Continued.

TRUNK: POSTERIOR TRUNK

Trapezius

Posterior musculocutaneous flap for neck reconstruction — cont'd

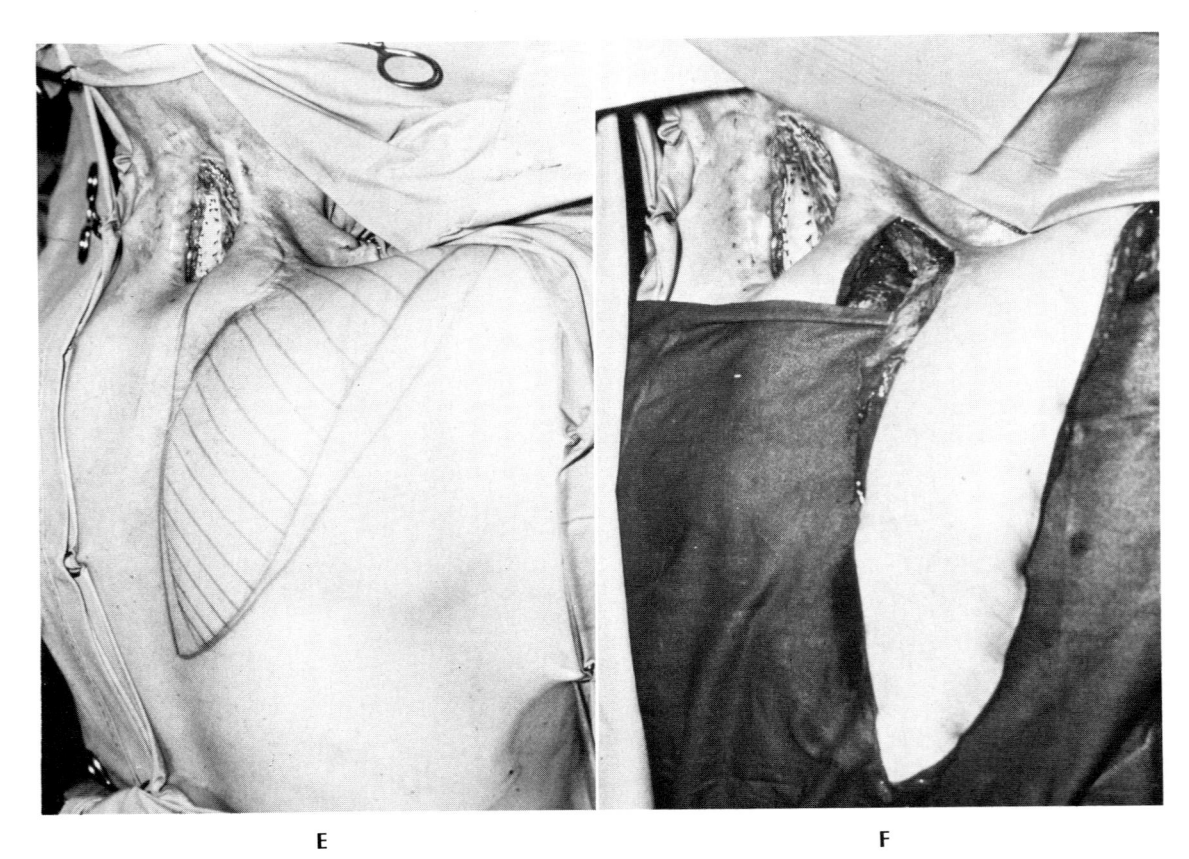

E

F

- E Design of posterior trapezius musculocutaneous flap.
- F Trapezius musculocutaneous flap elevated. Flap circulation via transverse cervical flap not affected by previous radiation therapy.
- G Posterior trapezius muscle is based on dominant vascular pedicle from transverse cervical artery.
- H Single-stage reconstruction of posterior cervical radionecrotic ulceration.
- I Long-term result.

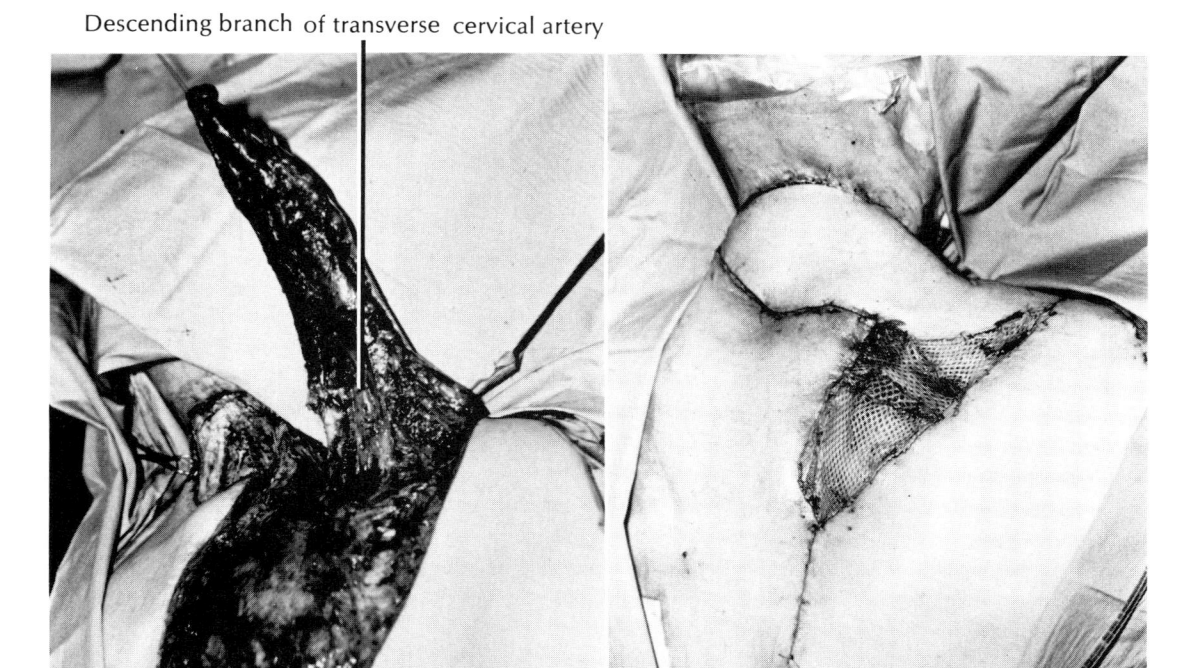

TRUNK: POSTERIOR TRUNK

Descending branch of transverse cervical artery

G

H

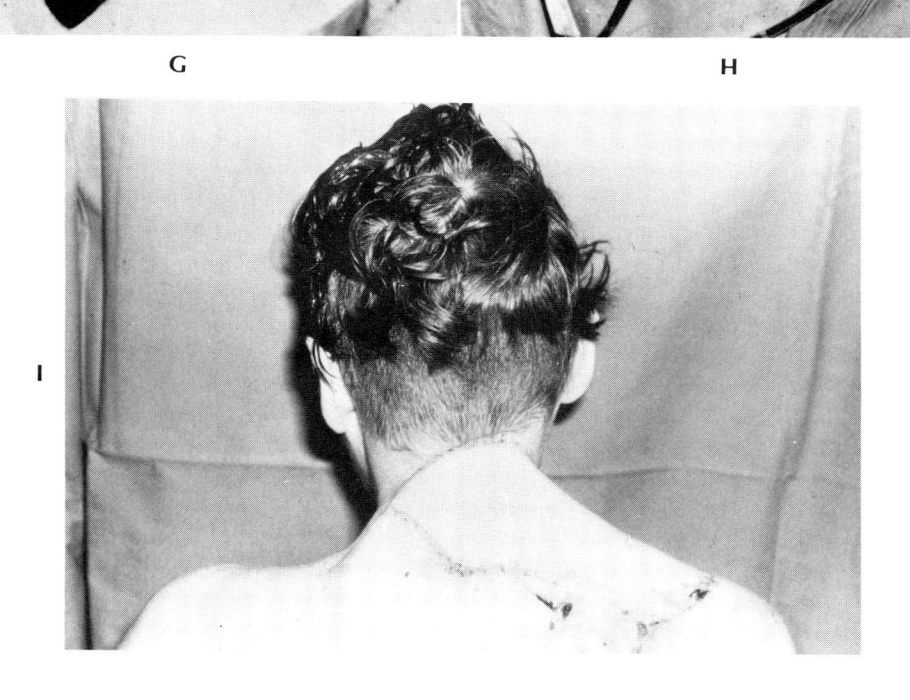

I

417

TRUNK: POSTERIOR TRUNK

Trapezius

Posterior musculocutaneous flap for head and neck reconstruction

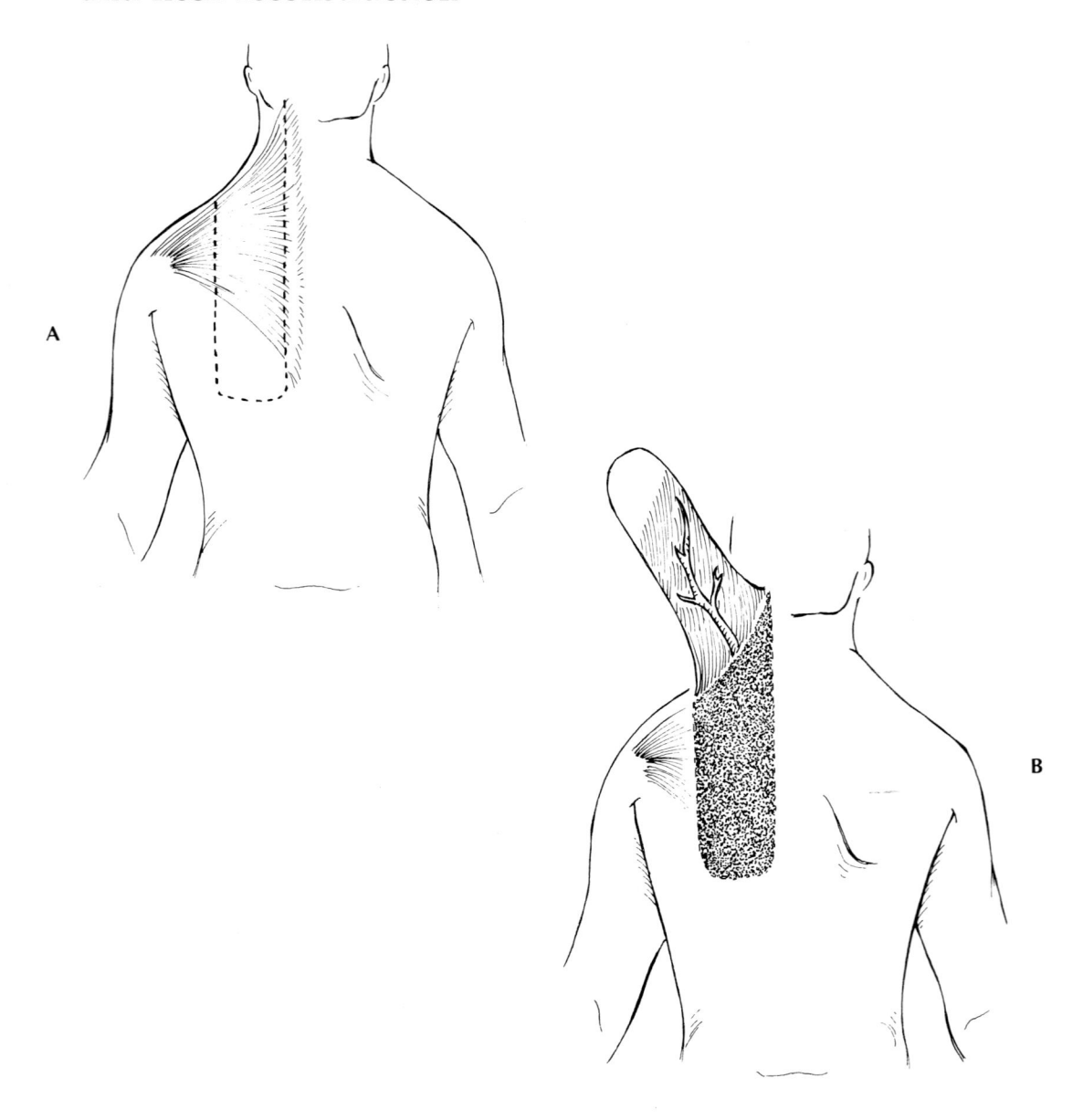

A Skin outline for flap elevation.

B Flap elevation. Lateral trapezius fibers to scapula are divided with descending pedicle under visualization. Donor defect is skin grafted.

TRUNK: POSTERIOR TRUNK

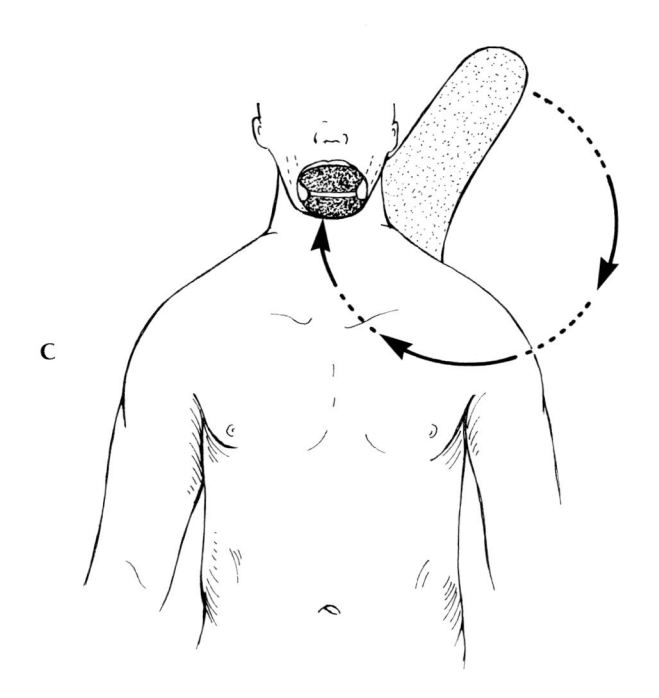

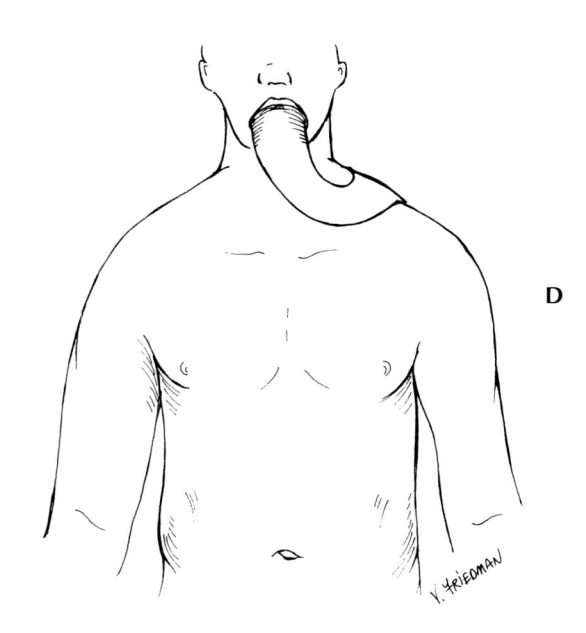

C Anterior arc of rotation.

D Cutaneous extension inset into floor of mouth. Musculocutaneous portion inset over chin defect.

SECTION SEVEN
UPPER EXTREMITY

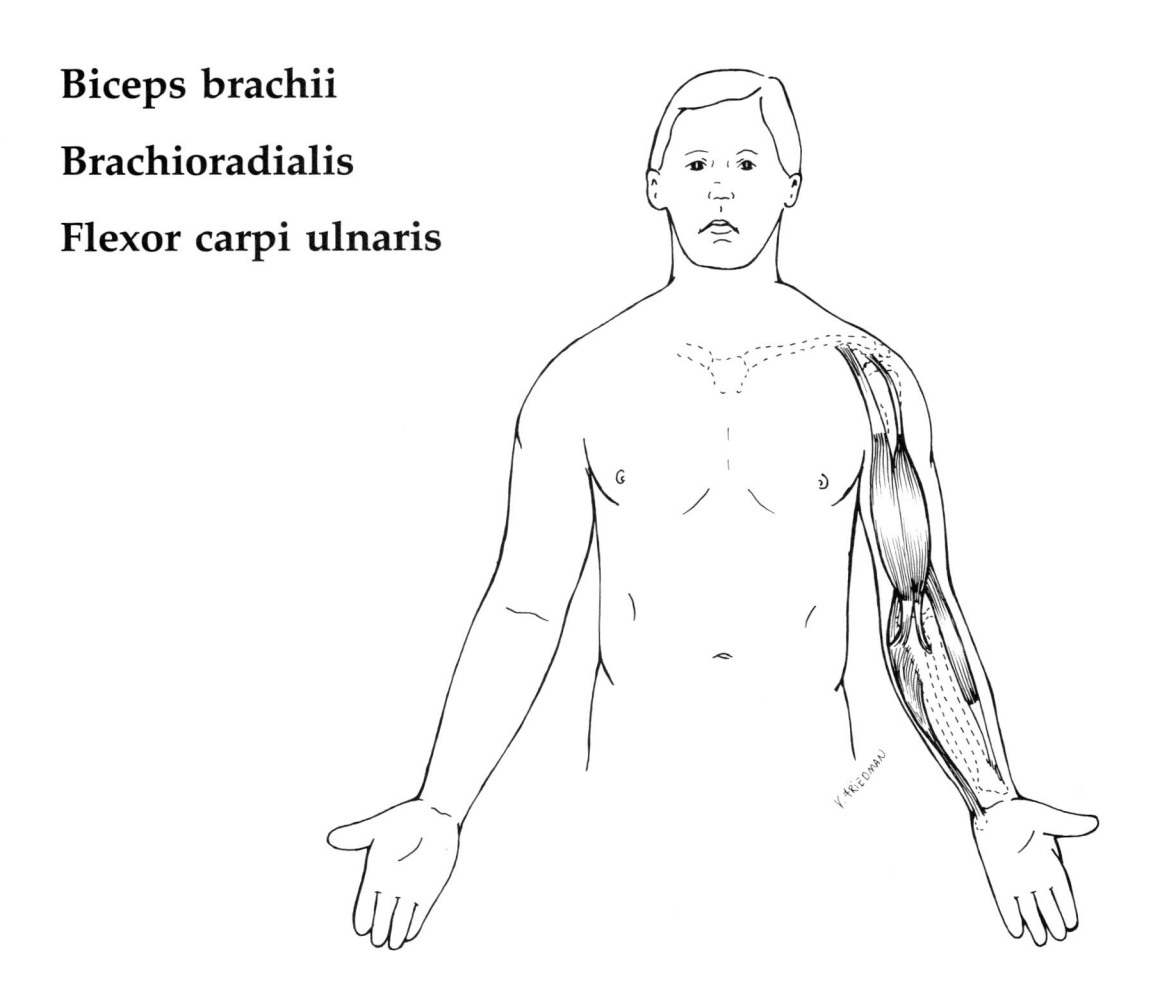

Biceps brachii

Brachioradialis

Flexor carpi ulnaris

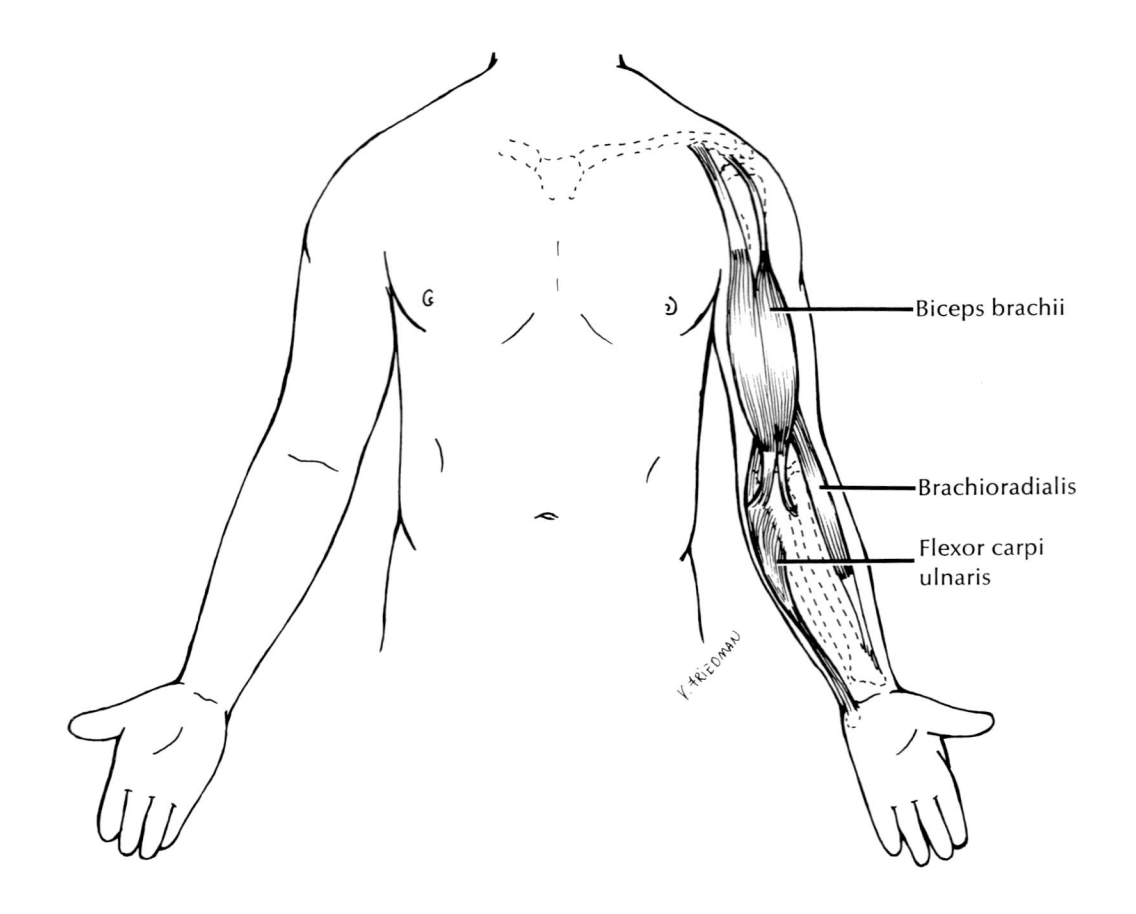

UPPER EXTREMITY

- Biceps brachii
- Brachioradialis
- Flexor carpi ulnaris

The application of muscle and musculocutaneous flaps for reconstruction of the upper extremity has not been as extensive as the application of such flaps in the lower extremity. Two major reasons account for this: the proximity of the upper extremity to the trunk and the relative inexpendability of the muscles of the upper extremity. Abdominal and groin flaps therefore remain the first choice for coverage of the forearm and hand. However, under certain circumstances, if trunk flaps are not available and free flap transfer is not possible, then muscle flap transposition may be indicated for upper extremity coverage.

It is with the realization that the muscles of the upper extremity are not expendable and should only be considered for muscle flap transposition when other methods are not suitable that this section on the upper extremity is included. The applications of the latissimus dorsi and pectoralis major muscles for axillary and upper arm coverage have already been described. The latissimus dorsi musculocutaneous flap is preferable to thoracic flaps for upper arm coverage because coverage is achieved in one operation and the donor defect is minimal.

UPPER EXTREMITY

- Triceps
- Biceps brachii
- Brachialis
- Brachioradialis

Upper arm

- Flexor carpi ulnaris
- Flexor carpi radialis
- Brachioradialis

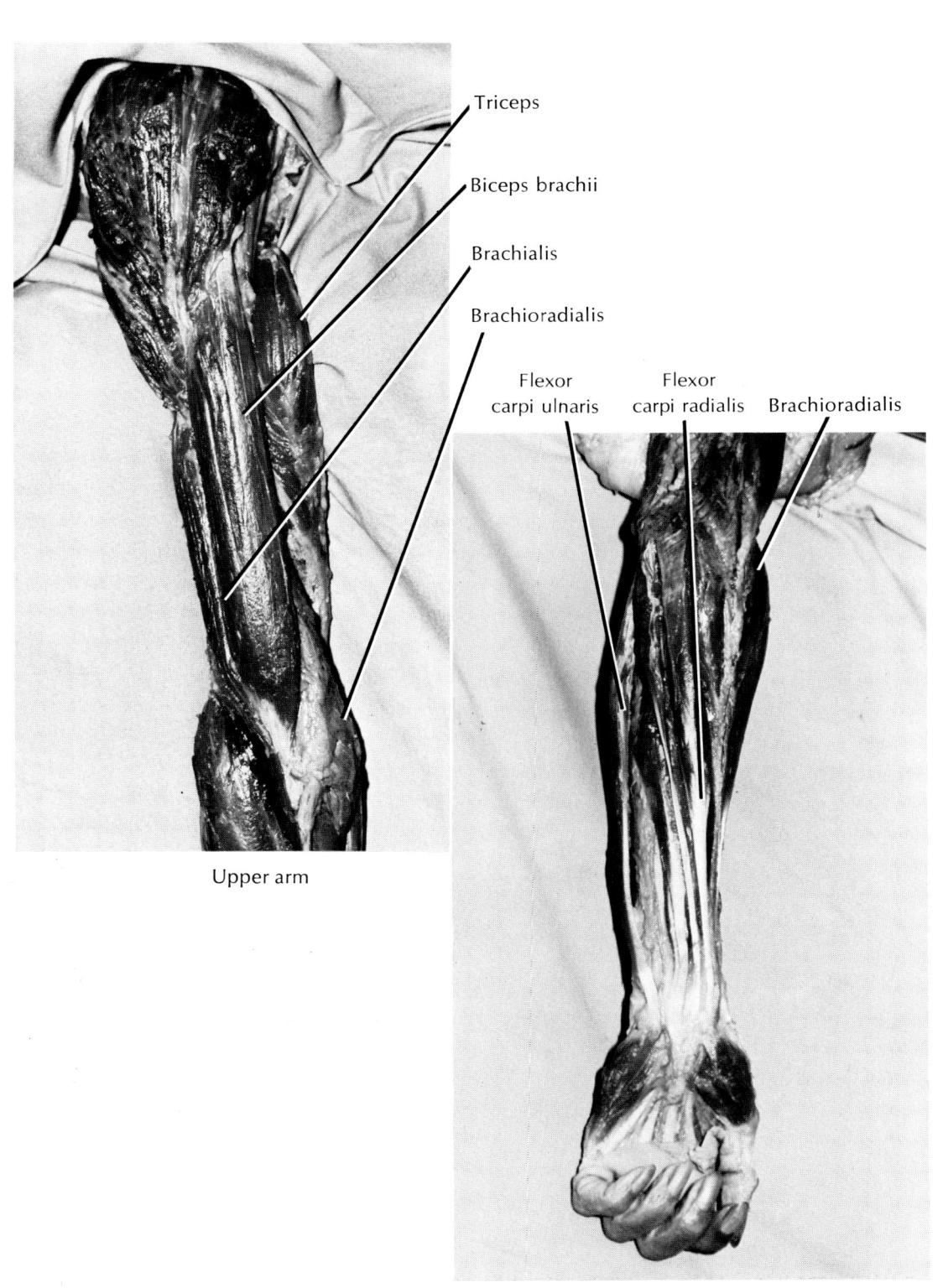

Forearm

423

UPPER EXTREMITY

Blood supply

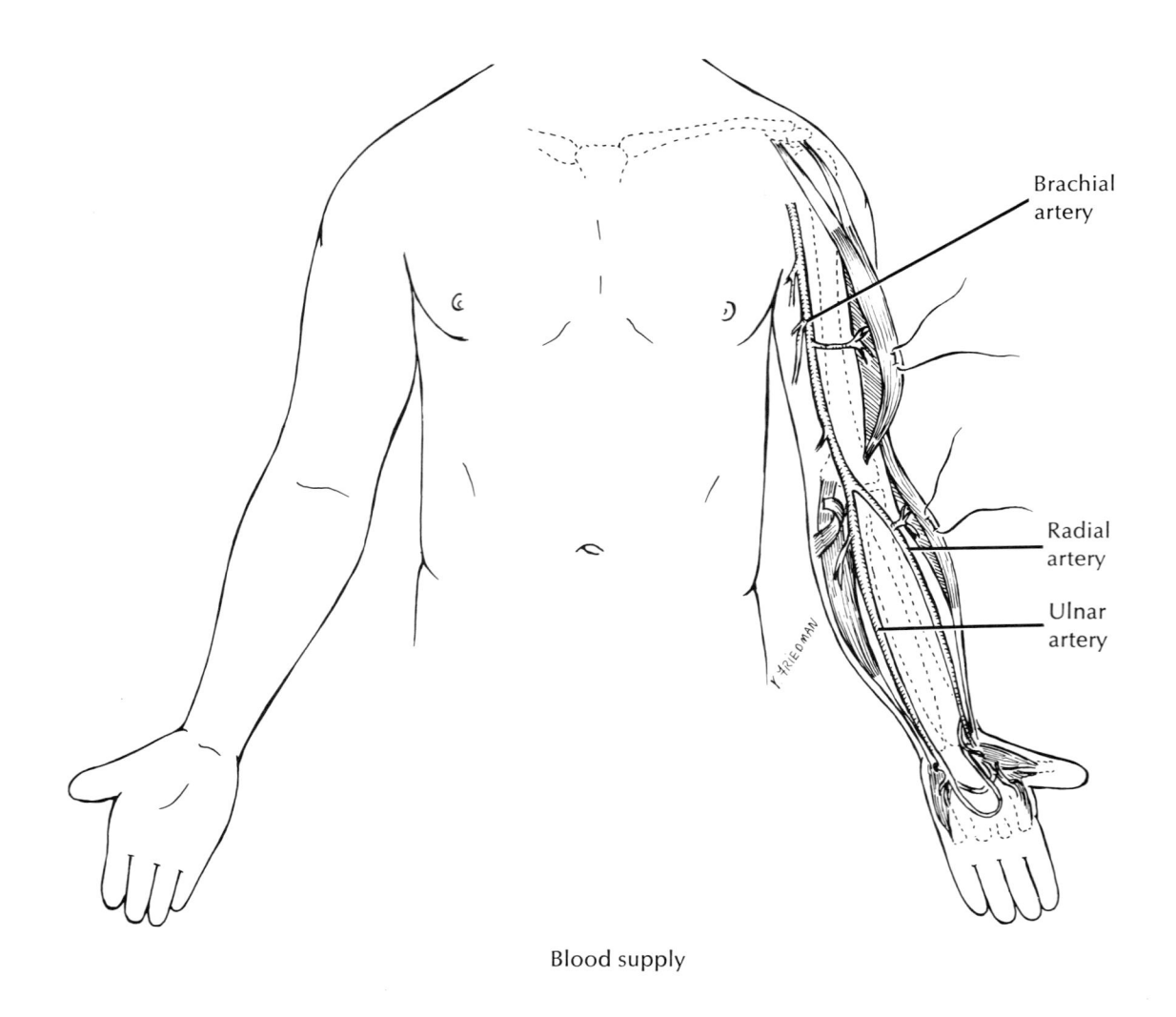

Blood supply

The blood supply to the upper arm muscles is based on the brachial artery. The forearm muscles are supplied by the ulnar and radial arteries. In general the major vascular pedicle enters the muscle proximally, with minor pedicles entering more distally. For muscle flap transposition, the major pedicle will carry the muscle. Although only the brachioradialis and flexor carpi ulnaris are described in detail, this basic general pattern represents the blood supply of all the anterior forearm muscles. The palmaris longus has been omitted because the small belly of this muscle limits its usefulness. The flexors of the fingers are not included as they are not expendable and should be preserved for functional tendon transfer.

UPPER EXTREMITY

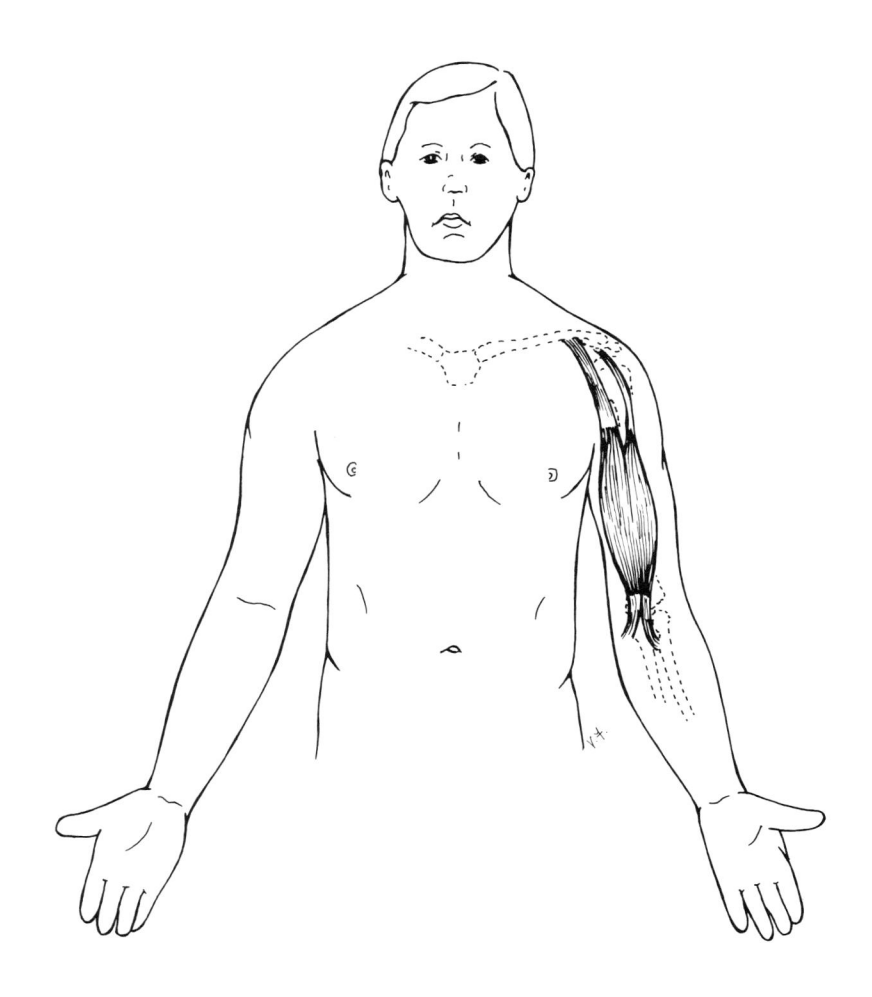

Biceps brachii

MUSCLE FLAP

Applications

Coverage of:
AXILLA
UPPER ARM

UPPER EXTREMITY

Biceps brachii

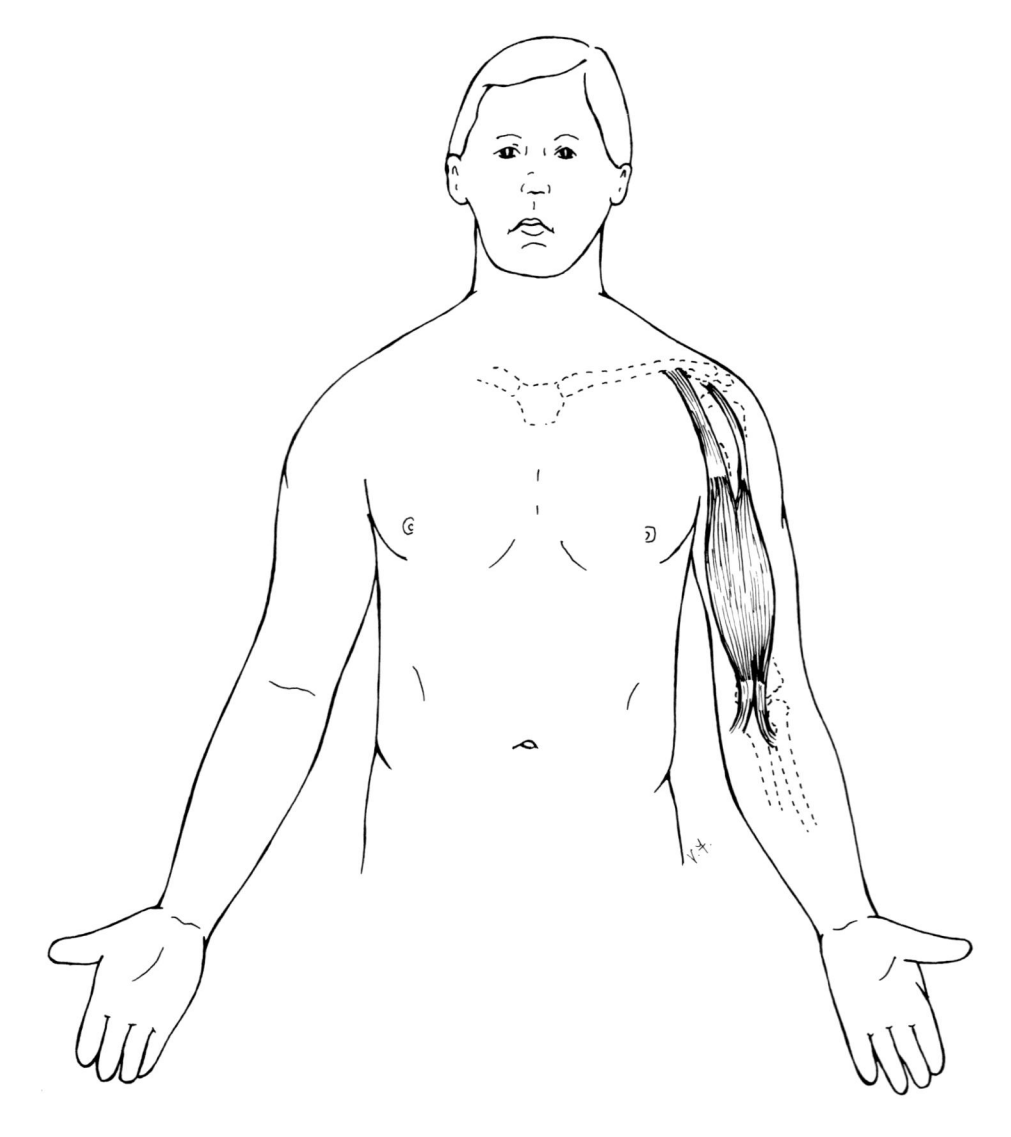

The biceps brachii lies superficially on the anterior aspect of the upper arm. Deep to the biceps brachii lies the brachialis muscle. Medially the brachial artery and median nerve run in a groove between these two muscles.

Origin: Short head-coracoid process of scapula; long head—supraglenoid tuberosity of the scapula.
Insertion: Radial tuberosity and the lacertus fibrosus onto the fascia covering the common flexor origin.

UPPER EXTREMITY

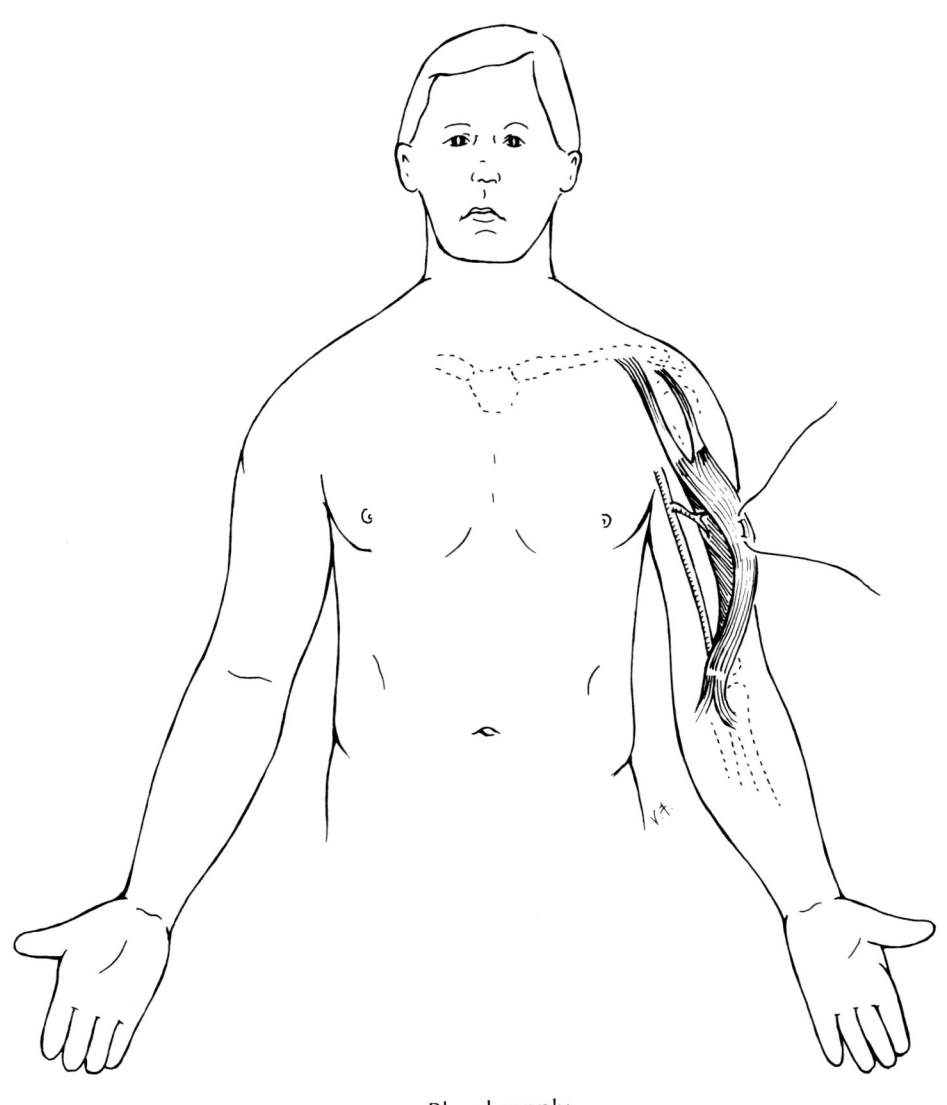

Blood supply

Nerve supply: Branch of musculocutaneous nerve. The musculocutaneous nerve crosses the arm from medial to lateral, lying between the biceps brachii and brachialis muscles. It innervates the biceps brachii on its deep surface.

Function: The biceps brachii is not expendable and is a flexor and supinator of the forearm.

UPPER EXTREMITY

Biceps brachii

- Biceps brachii
- Vascular pedicle
- Brachial artery
- Brachialis

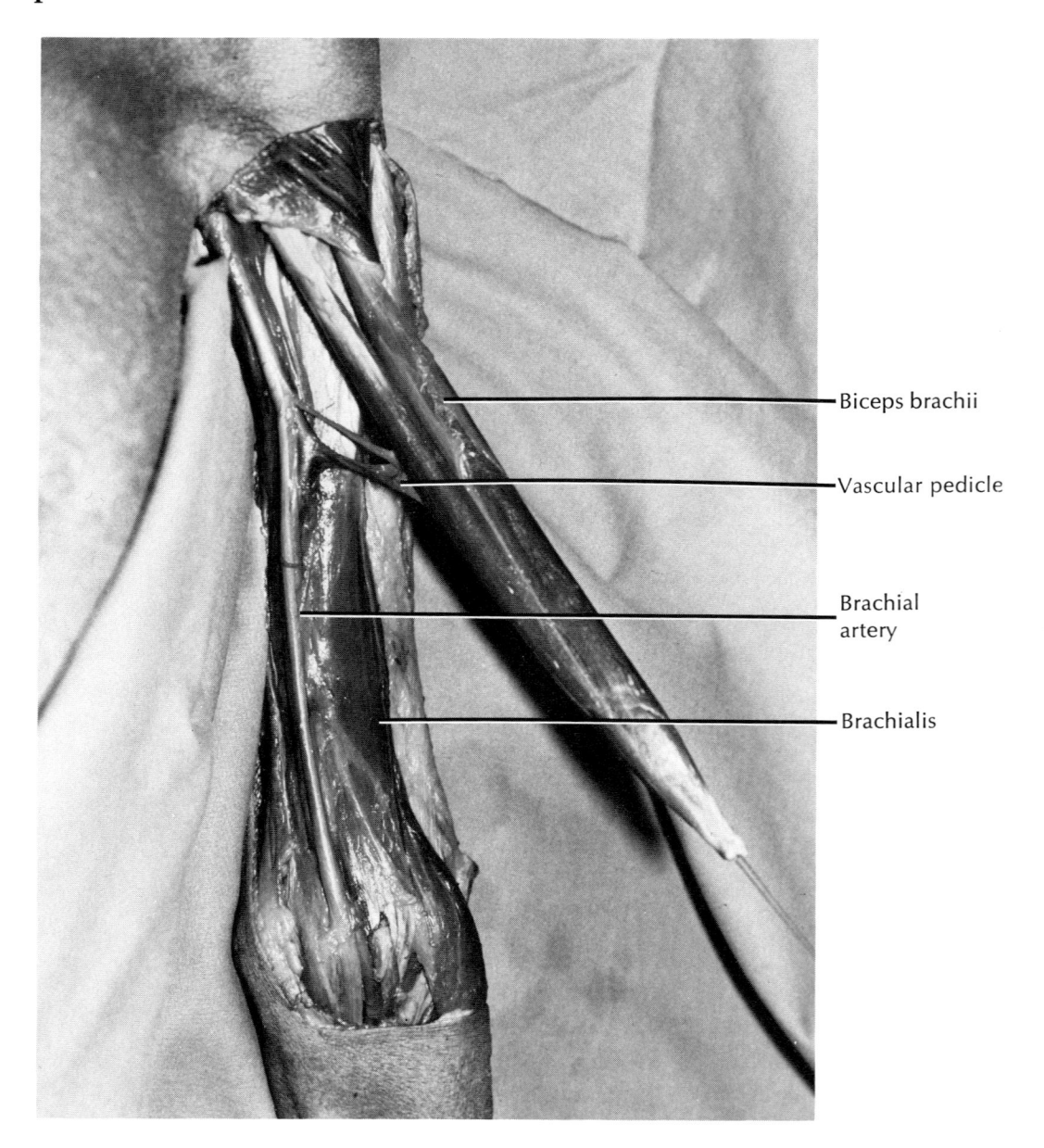

Blood supply

The brachial artery runs along the medial aspect of the biceps brachii and, through one or two proximal branches, supplies the biceps brachii. The branches, one to each head, are the dominant pedicles to the muscle. These enter the muscle approximately 5 to 8 cm below the anterior axillary fold.

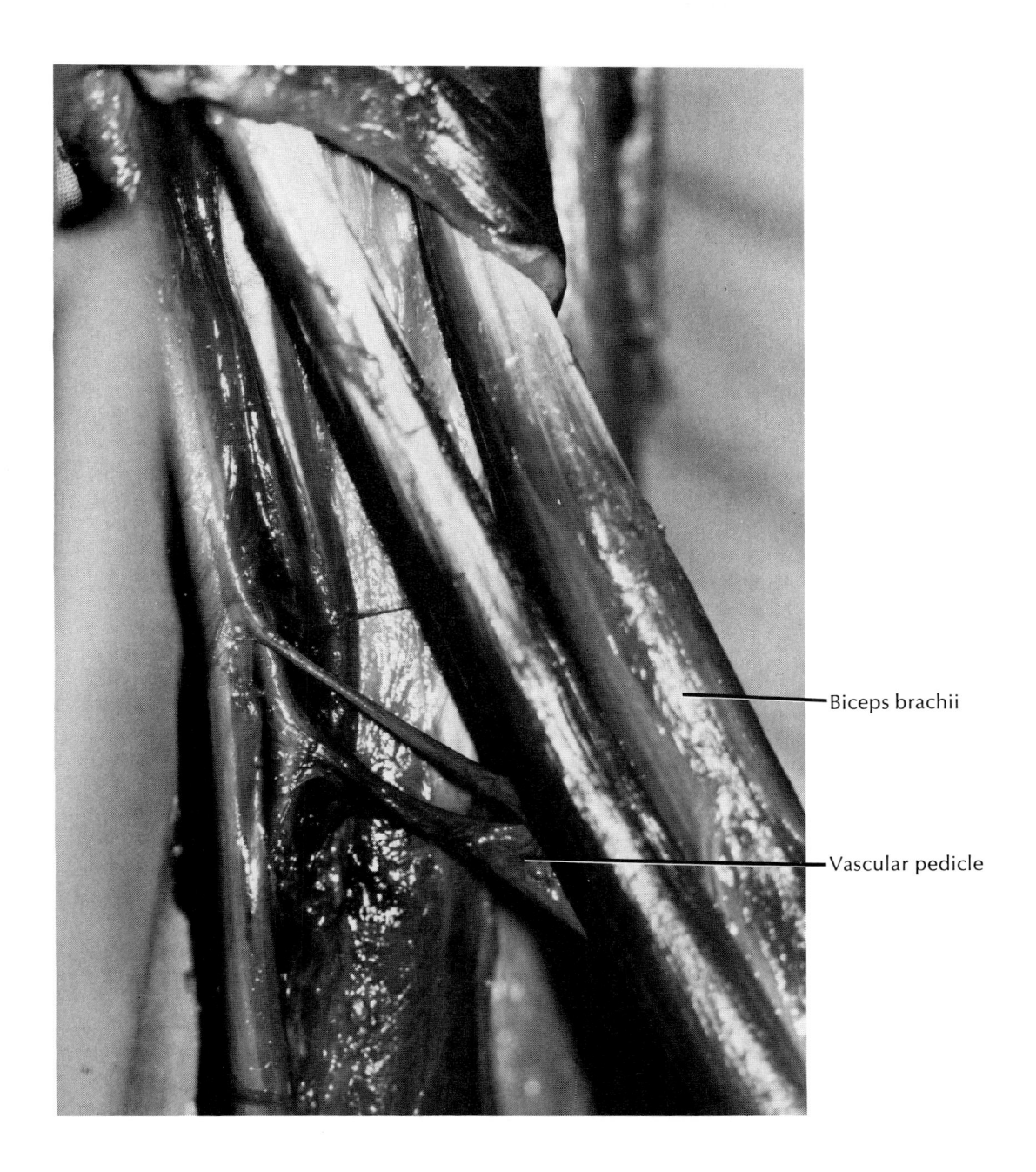

— Biceps brachii

— Vascular pedicle

UPPER EXTREMITY

Biceps brachii

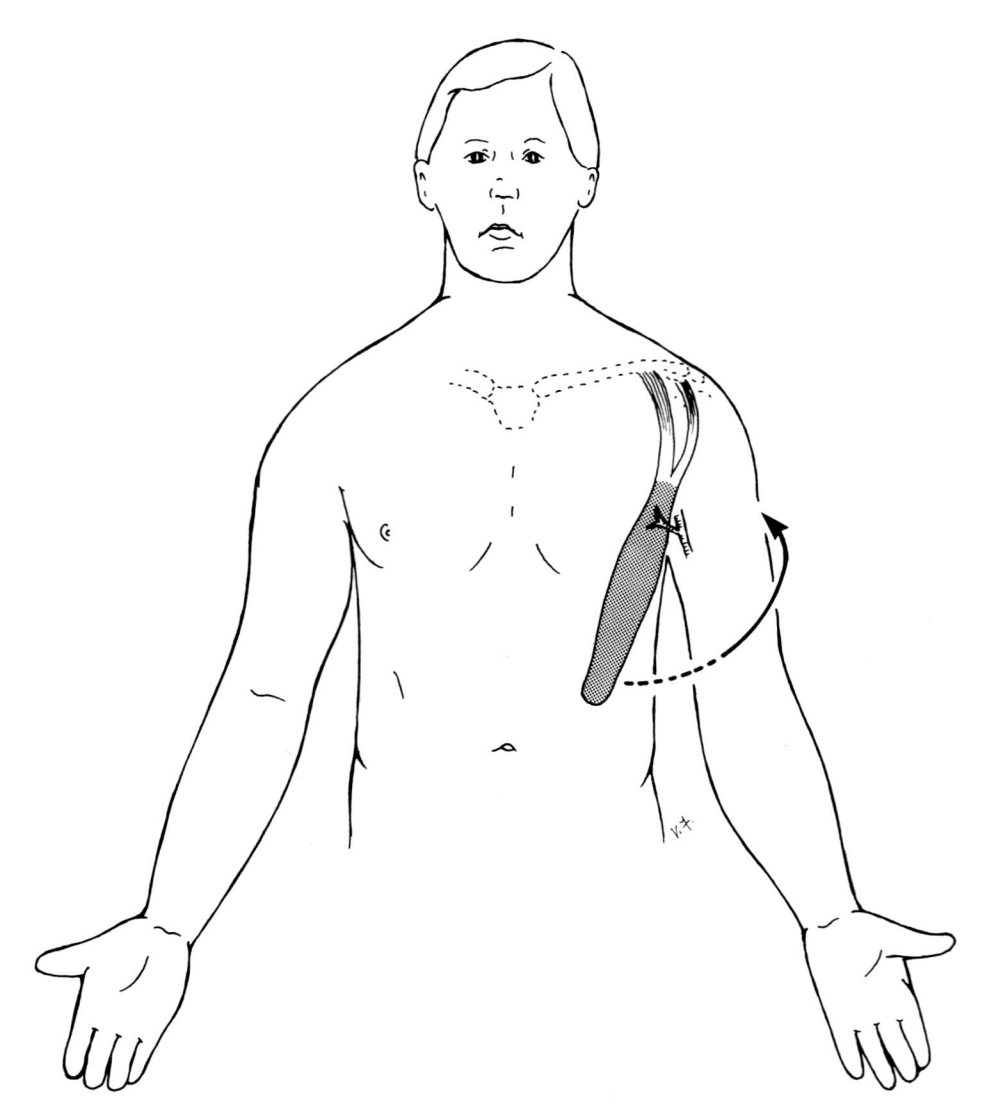

Arc of rotation

The biceps brachii, based on its major pedicle, will rotate upward to cover the axilla and parts of the upper arm. Part of the biceps muscle could be transposed medially by separating muscle fibers from the tendon (thus preserving the tendon) to cover exposed shunts for dialysis access.

UPPER EXTREMITY

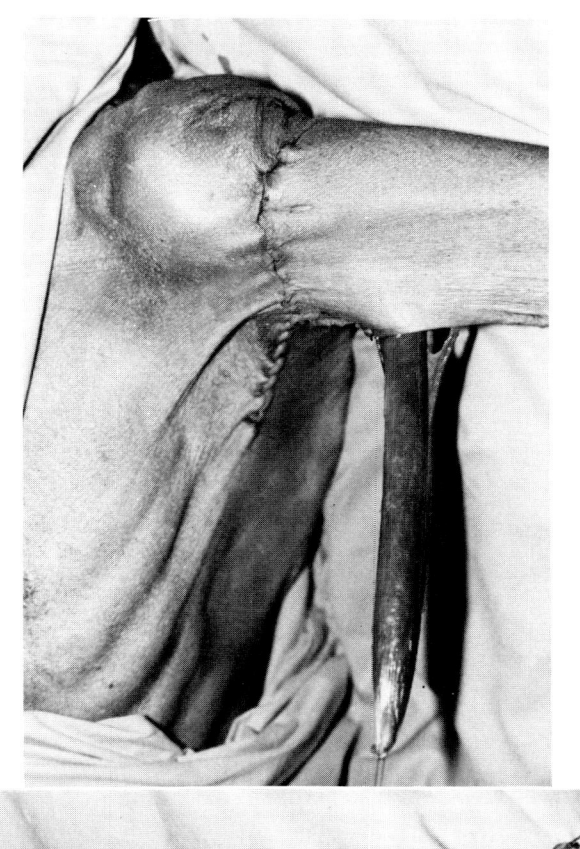

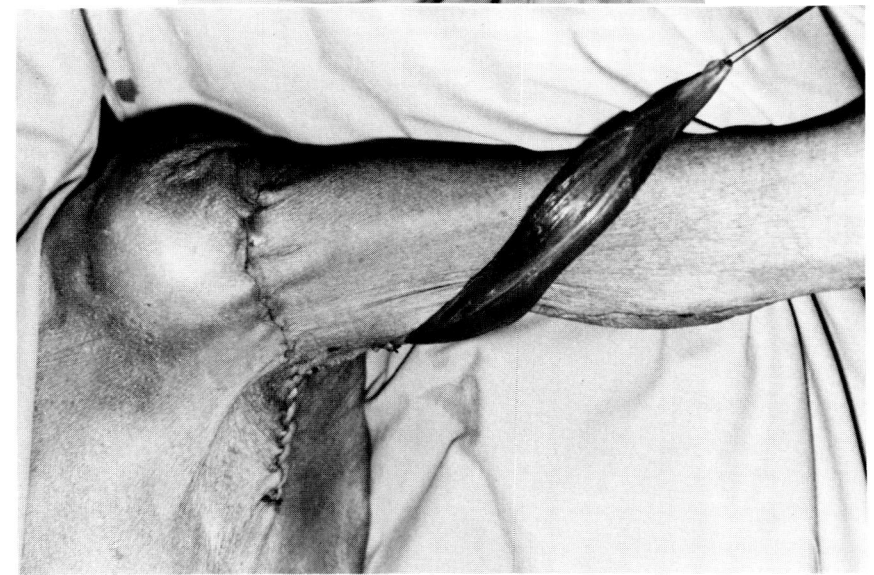

Coverage of arm

UPPER EXTREMITY

Biceps brachii

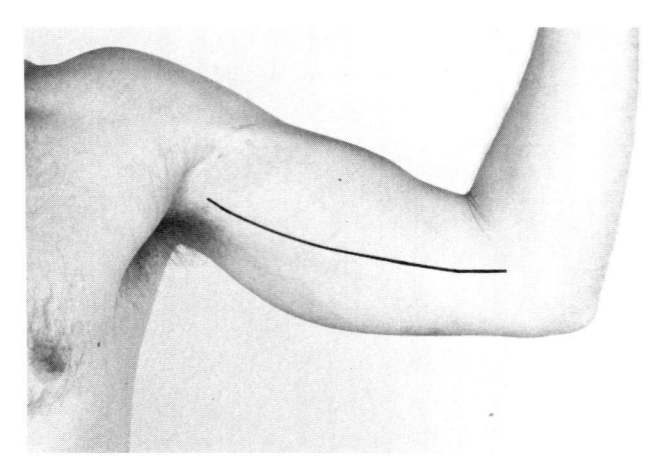

Skin incision

Elevation of flap

Although the biceps brachii may be elevated as a musculocutaneous unit, this is not recommended, as the resulting defect may leave the brachial artery and median nerve exposed.

The muscle is approached through a medial incision in the upper arm. The median nerve and brachial artery are identified, and the biceps brachii is elevated off the brachialis muscle. The musculocutaneous nerve lies between these two muscles. The tendon is divided and the muscle elevated from insertion toward the origin. The vascular pedicle is identified approximately 5 to 8 cm below the anterior axillary fold.

For coverage of exposed dialysis access shunts in the arm, part of the biceps brachii can be transposed medially. The medial fibers of the muscle are separated from the tendon, which remains functionally intact. These fibers are then transposed medially to cover the exposed shunt.

Precautions

- The biceps brachii is not expendable and should only be used if alternate flaps are not available. The latissimus dorsi is the first choice for coverage of upper arm and axilla.
- The brachial artery and median, musculocutaneous, and ulnar nerves are closely related to the biceps brachii and should be preserved during flap elevation.
- The musculocutaneous flap is not recommended, as the resulting defect may leave the brachial artery and median nerve exposed.

UPPER EXTREMITY

Brachioradialis

MUSCLE FLAP
MUSCULOCUTANEOUS FLAP
FREE FLAP

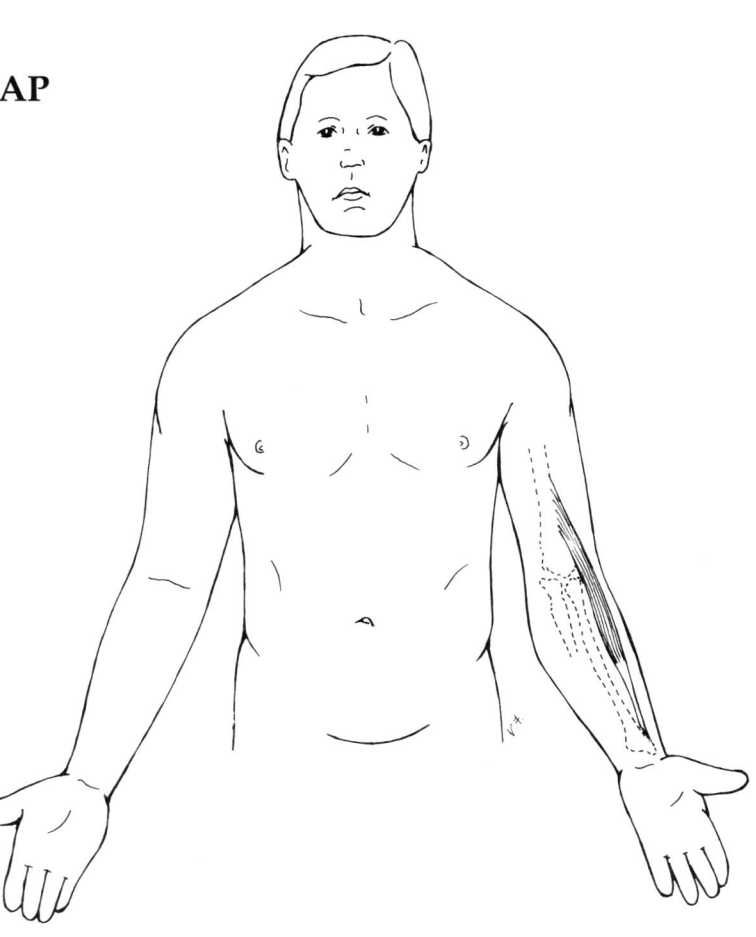

Applications

Coverage of:
ANTECUBITAL FOSSA
UPPER FOREARM
LOWER ARM

Free flap:
FUNCTIONAL MUSCLE

UPPER EXTREMITY

Brachioradialis

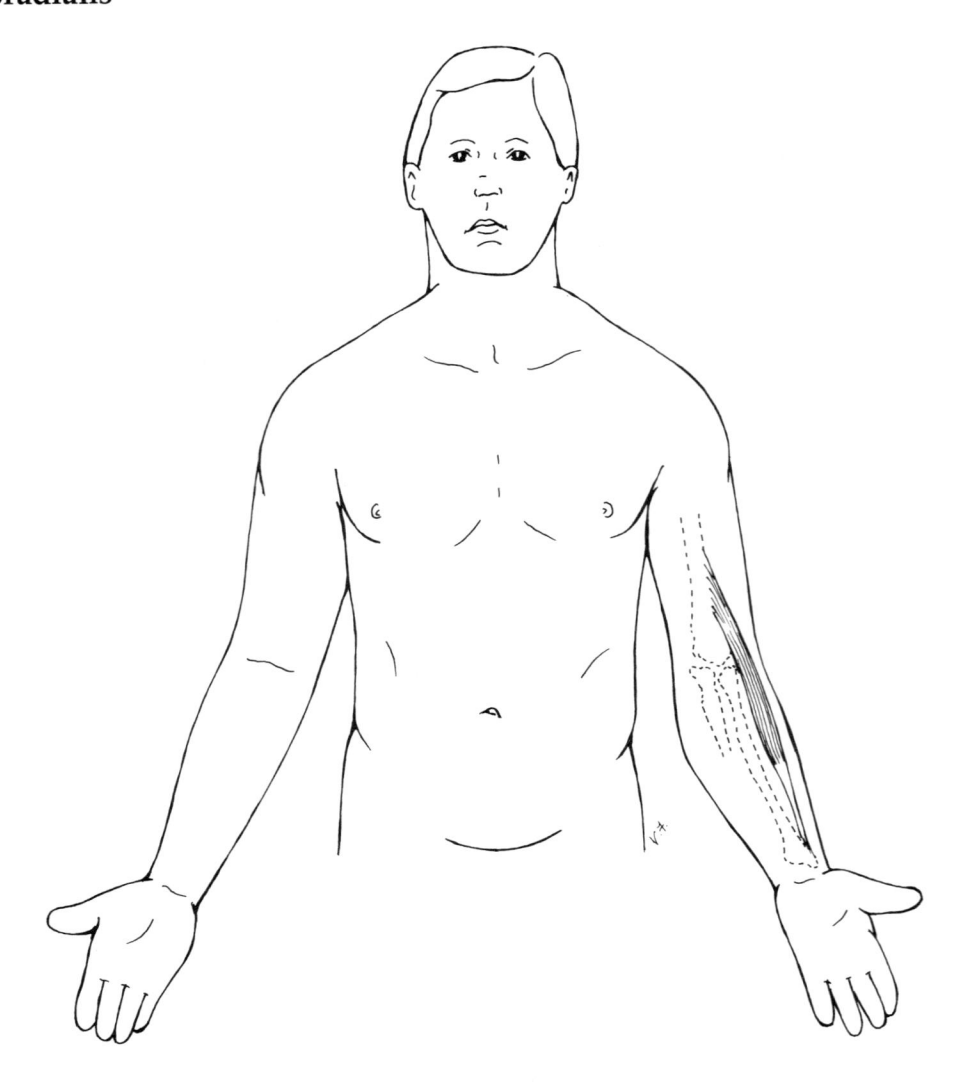

The brachioradialis is the most superficial muscle on the radial border of the forearm. Deep to this muscle lie the radial nerve artery and the extensor group of muscles.

Origin: Lateral supracondylar ridge of the humerus.
Insertion: Styloid process of the radius.

UPPER EXTREMITY

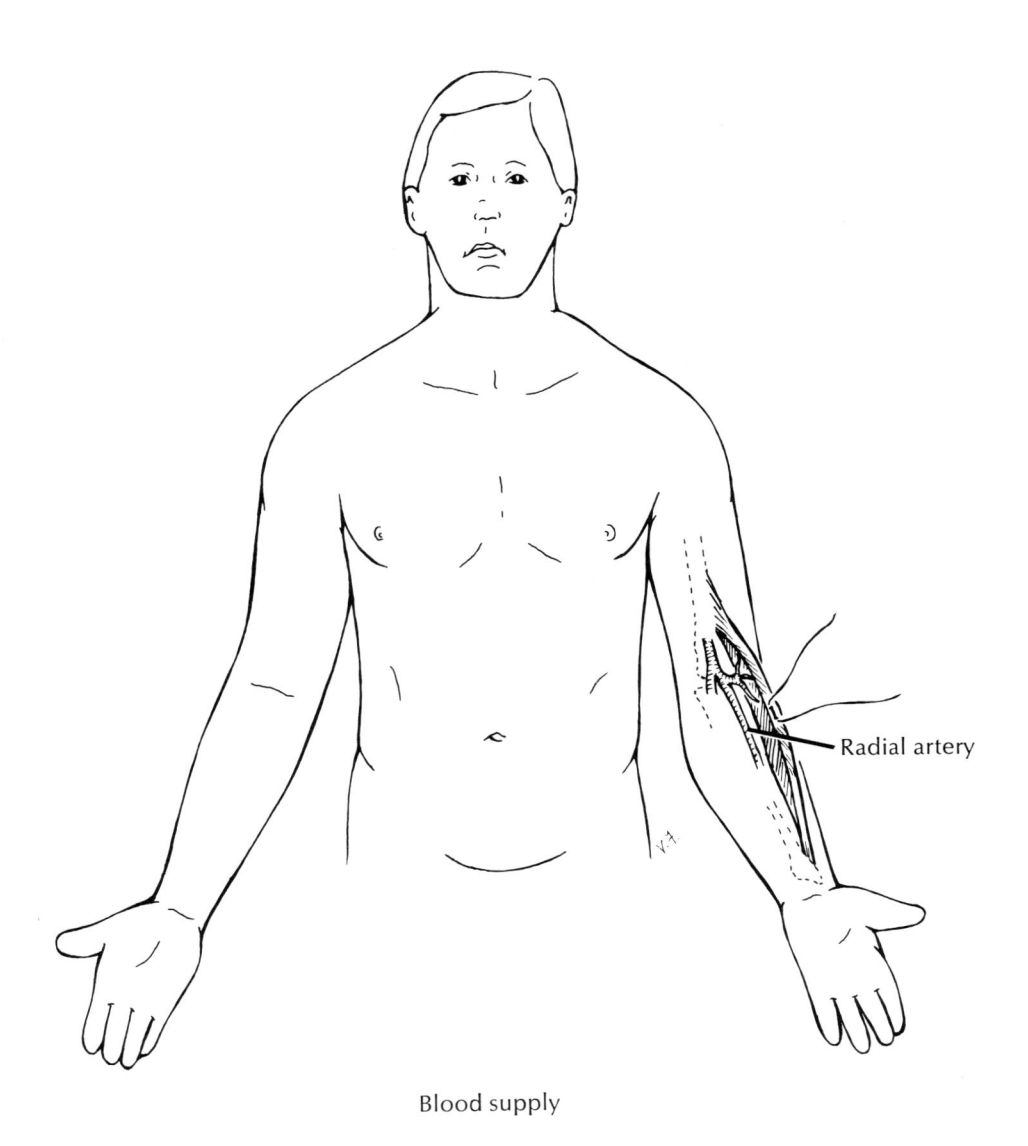

Blood supply

Nerve supply:
    Motor — Branch of radial nerve. This muscle is intimately related to the radial nerve along its entire course and is innervated by a branch of the radial nerve just below the elbow on its deep surface.
    Sensory — The skin overlying the muscle is supplied by the lateral antebrachial cutaneous nerve.
Function: The brachioradialis is a flexor of the forearm and is relatively expendable.

UPPER EXTREMITY

Brachioradialis

Vascular branches to radial nerve

Major pedicle to brachioradialis

Superficial radial nerve

Brachioradialis

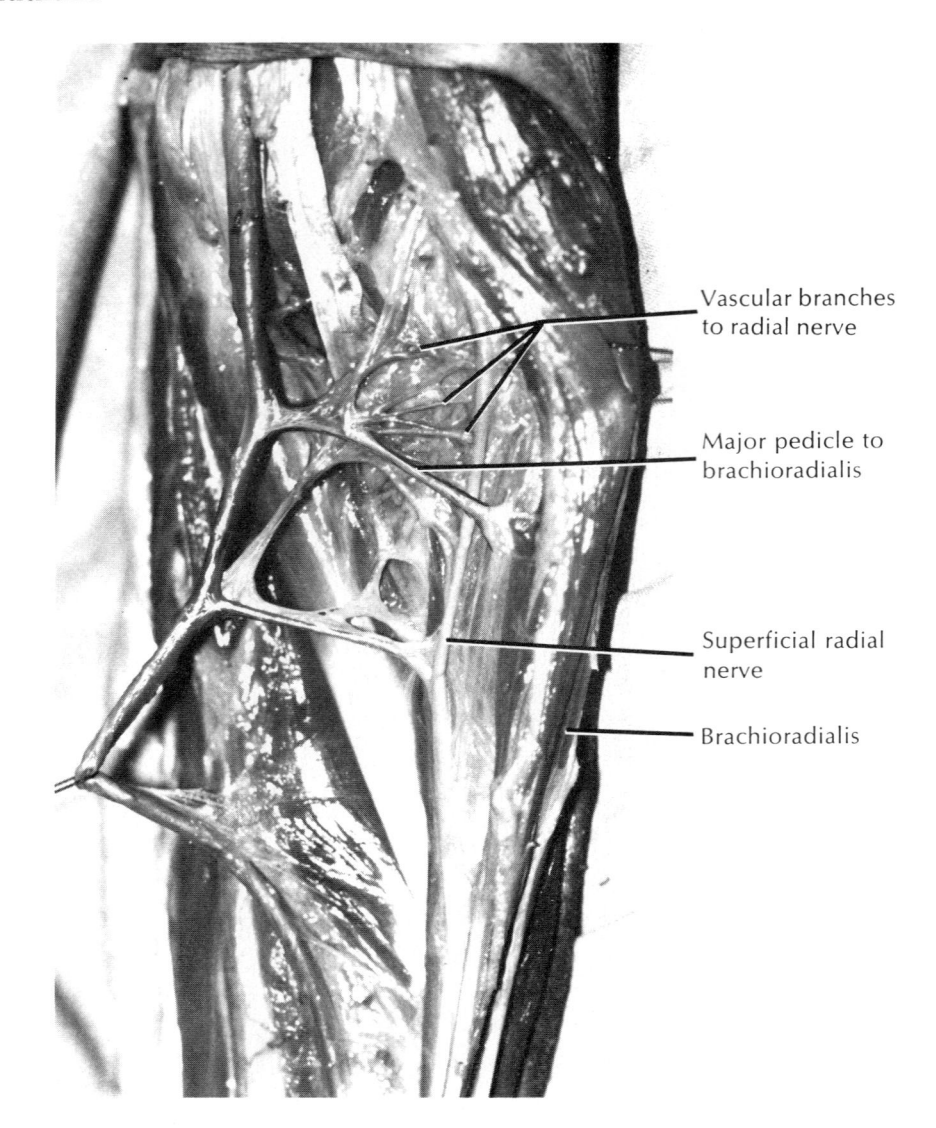

Blood supply

The radial artery through its radial recurrent branch supplies the brachioradialis. The major pedicle of the muscle is located in the antecubital fossa on the deep aspect of the muscle.

The radial artery lies deep to the brachioradialis throughout its length, as does the superficial radial nerve. In the antecubital fossa the radial recurrent artery sends several small branches to the radial nerve.

UPPER EXTREMITY

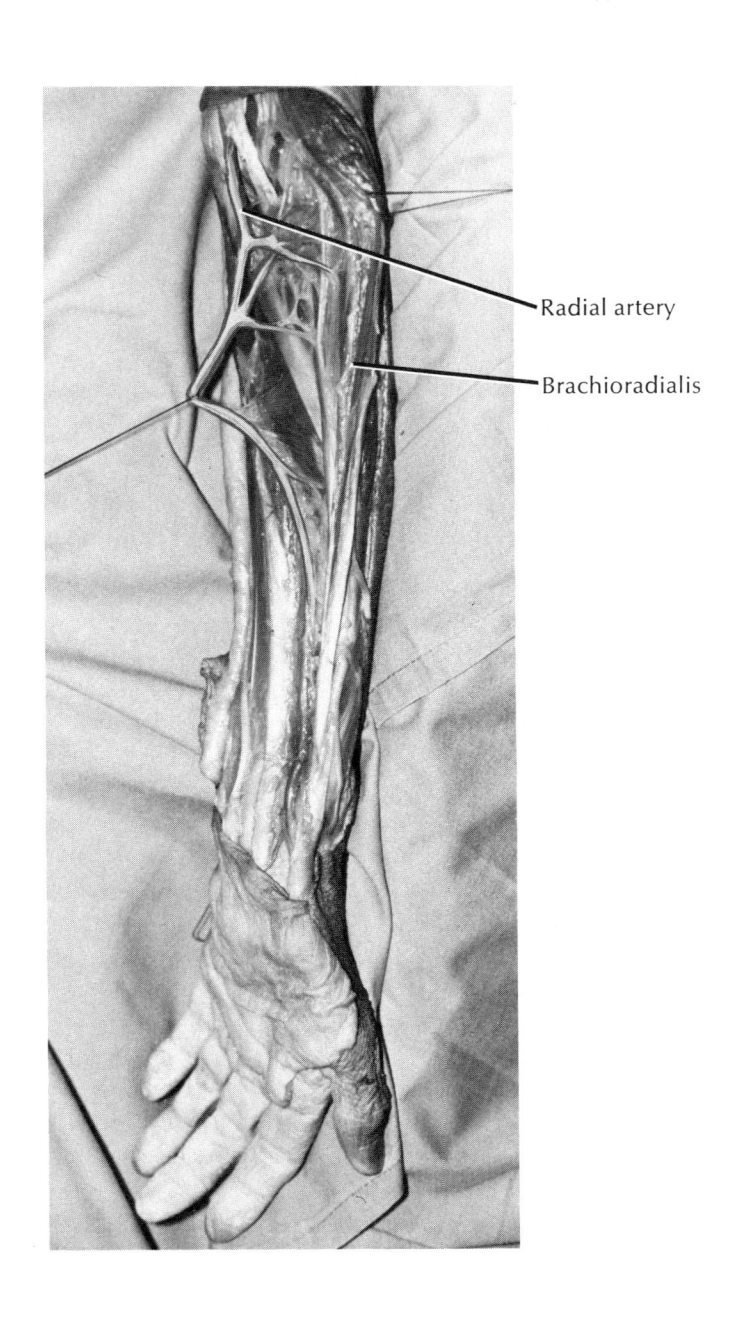

Radial artery
Brachioradialis

UPPER EXTREMITY

Brachioradialis

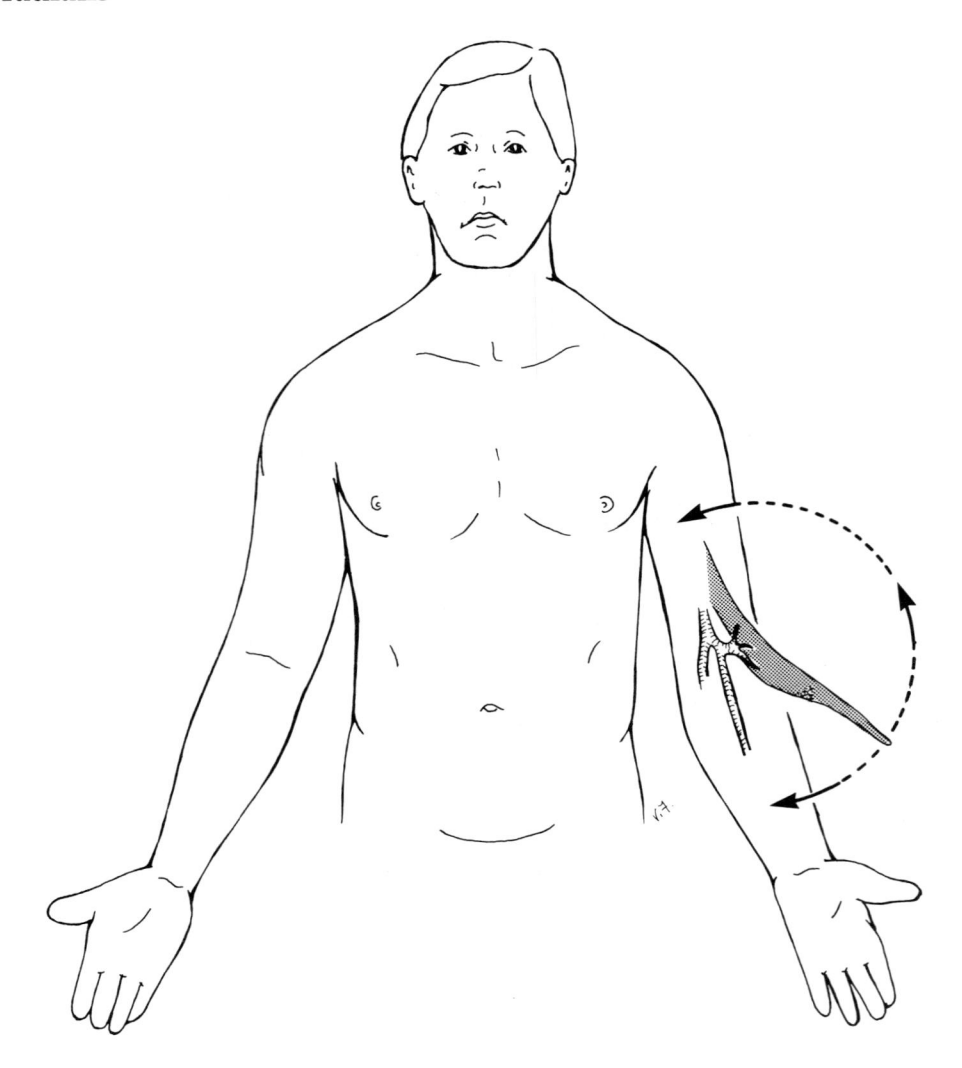

Arc of rotation

Based on the major pedicle in the antecubital fossa, the muscle has an arc of rotation that will reach the lower portion of the upper arm and the upper portion of the forearm.

It may prove to be an excellent unit for free transfer as a functional muscle. The radial artery and superficial radial nerve may be included with the muscle for free flap transfer.

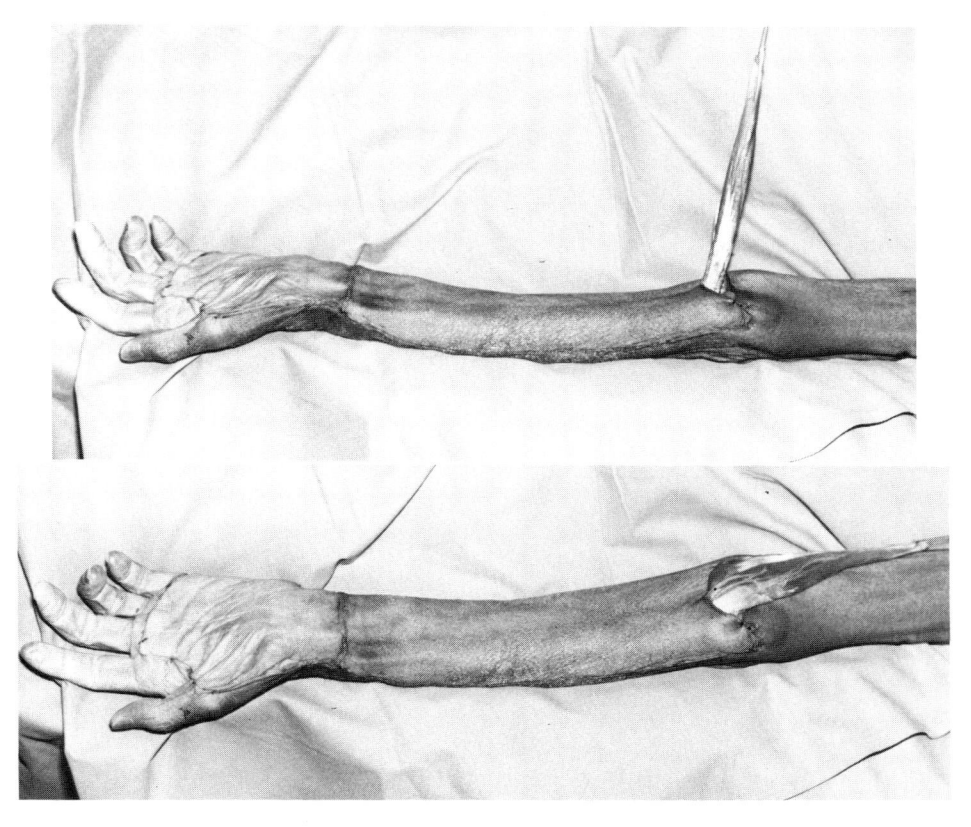

Coverage of antecubital fossa

Elevation of flap

Although the skin overlying the proximal part of the muscle can be elevated with the muscle as a musculocutaneous flap, the resultant defect may leave the radial artery and superficial radial nerve exposed.

An incision along the radial border of the arm is made, and the muscle is retracted radially to expose the radial artery and superficial radial nerve. The tendon of the brachioradialis is identified over the distal radius lying deep to the tendon of the abductor pollicis longus and extensor pollicis brevis. The tendon is divided, and by dissecting from insertion to origin, the major pedicle is identified in the antecubital fossa.

Precaution

- The muscle is closely related to the radial nerve and artery, which must be preserved during flap elevation.

UPPER EXTREMITY

Flexor carpi ulnaris

MUSCLE FLAP
MUSCULOCUTANEOUS FLAP

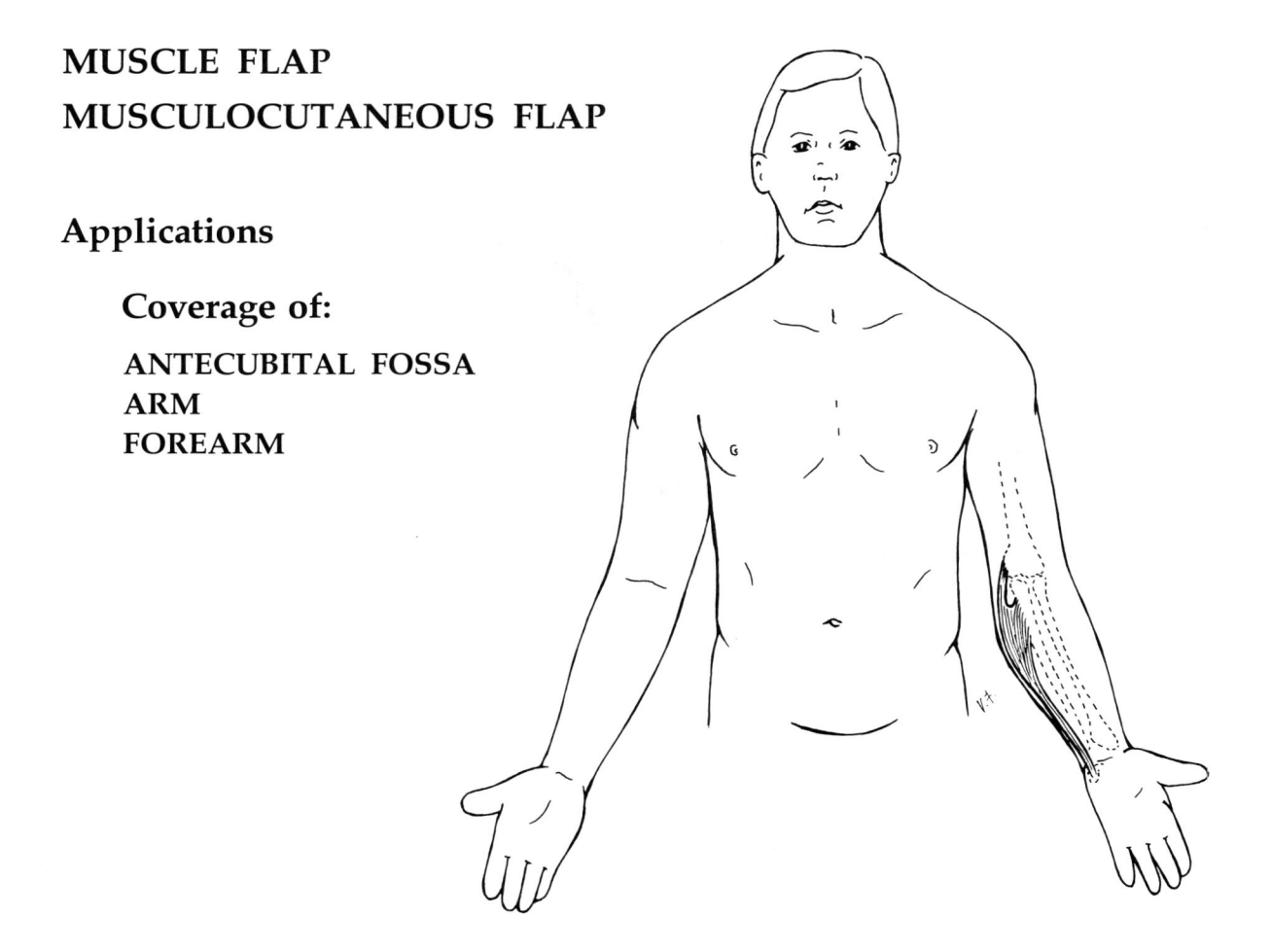

Applications

Coverage of:

ANTECUBITAL FOSSA
ARM
FOREARM

UPPER EXTREMITY

Flexor carpi ulnaris

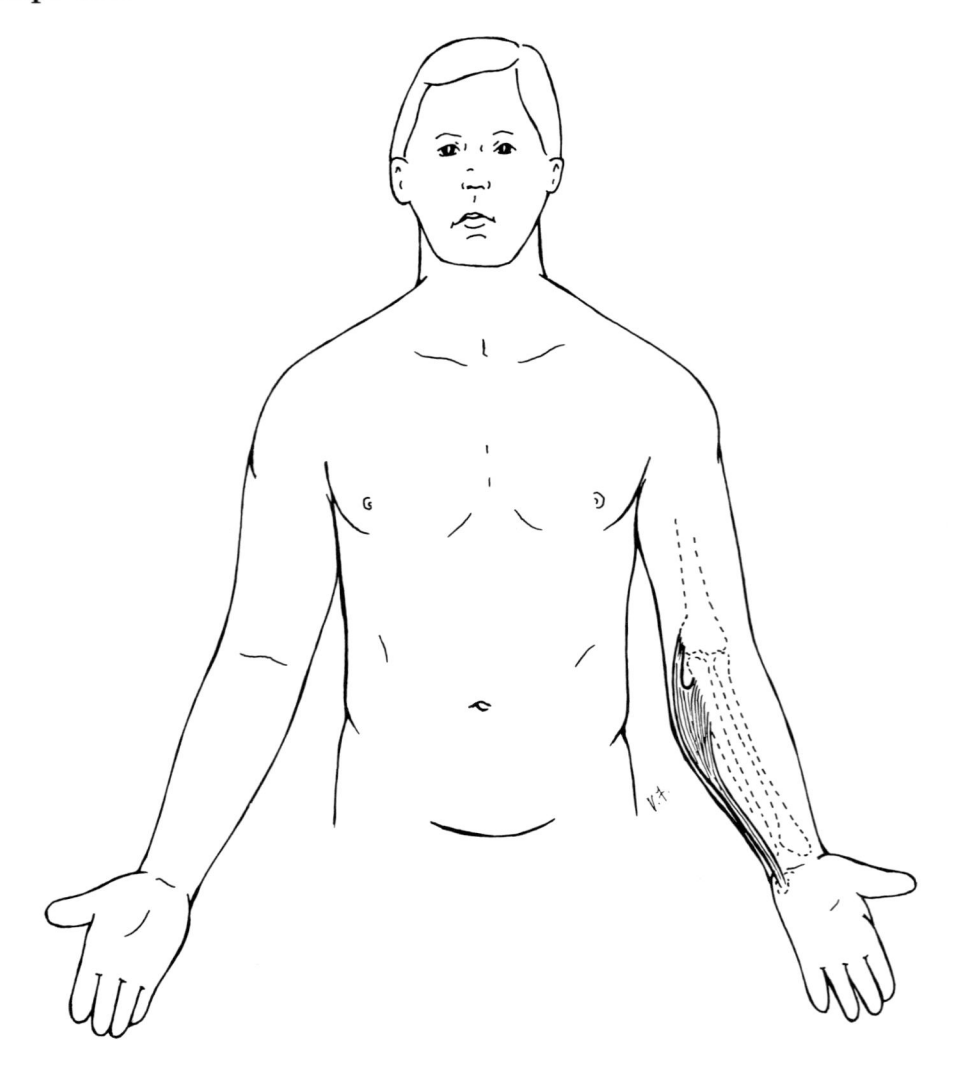

The flexor carpi ulnaris is the most superficial muscle on the ulnar aspect of the forearm. Deep to this muscle lies the flexor muscle group, the ulnar nerve, and artery.

Origin: Humeral head—common flexor tendon from medial epicondyle; ulnar head—posterior border of ulna. The ulnar nerve passes from the arm into the forearm through the two heads of origin of this muscle.

Insertion: Pisiform, hamate, and fifth metacarpal.

UPPER EXTREMITY

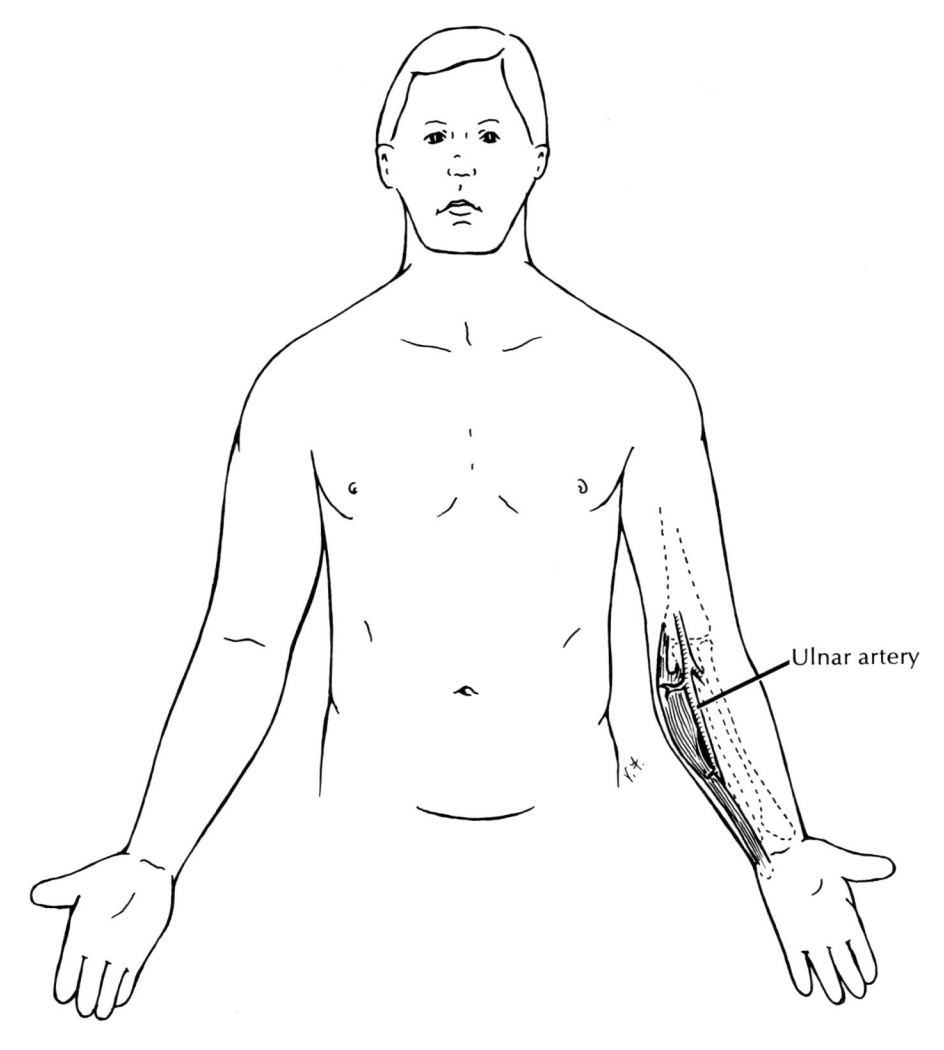

Blood supply

Nerve supply:
  Motor — Branch of ulnar nerve. The ulnar nerve innervates the muscle proximally as it passes between its two heads of origin.
  Sensory — Medial antebrachial cutaneous nerve. The skin overlying the muscle is innervated by the medial antebrachial cutaneous nerve.
Function: This is a relatively expendable muscle, which is a wrist flexor and adductor.

UPPER EXTREMITY

Flexor carpi ulnaris

Flexor digitorum superficialis

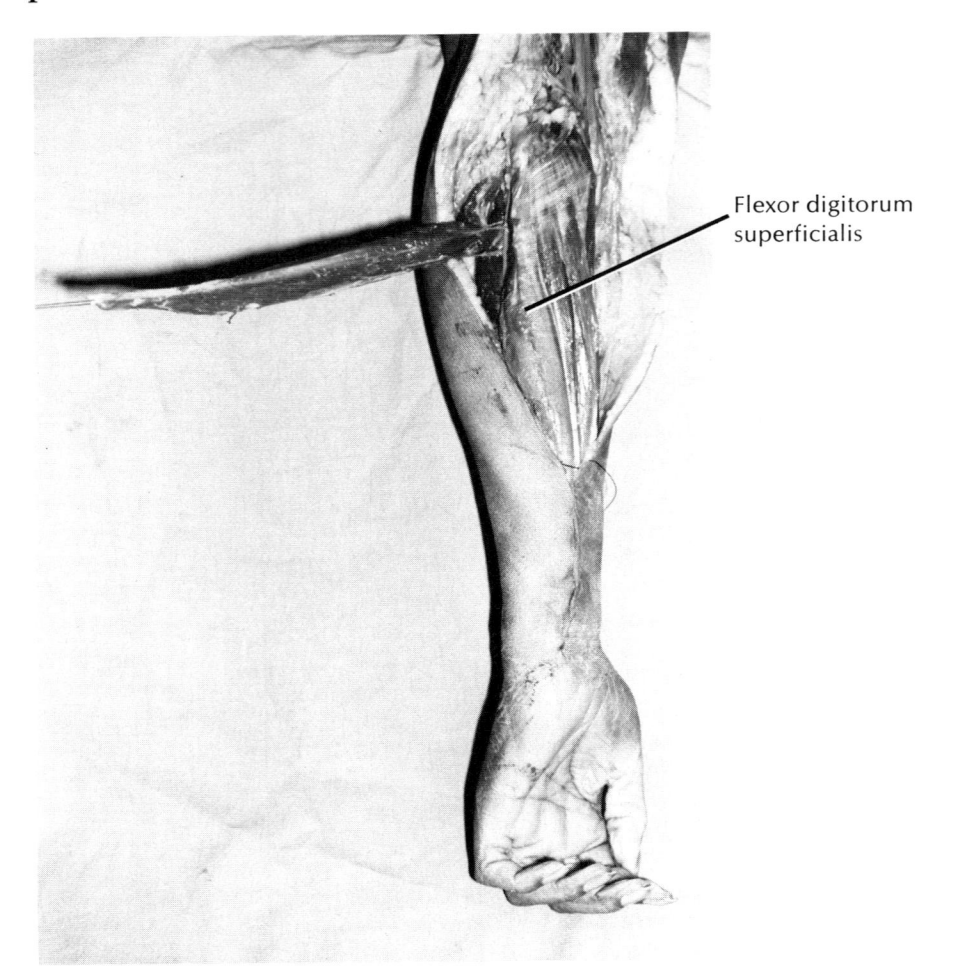

Blood supply

The ulnar artery, the larger branch of the brachial artery, courses to the ulnar side deep to the pronator teres, flexor carpi radialis, and the flexor digitorum sublimis, lying between these muscles and the flexor digitorum profundus. The vessel then runs distally, lying deep to the flexor carpi ulnaris and the ulnar nerve.

The major pedicle of the flexor carpi ulnaris is based on the posterior ulnar recurrent branch of the ulnar artery. It enters the proximal part of the muscle close to the origin on the deep aspect. The muscle has one or two distal minor pedicles that are direct branches of the ulnar artery.

UPPER EXTREMITY

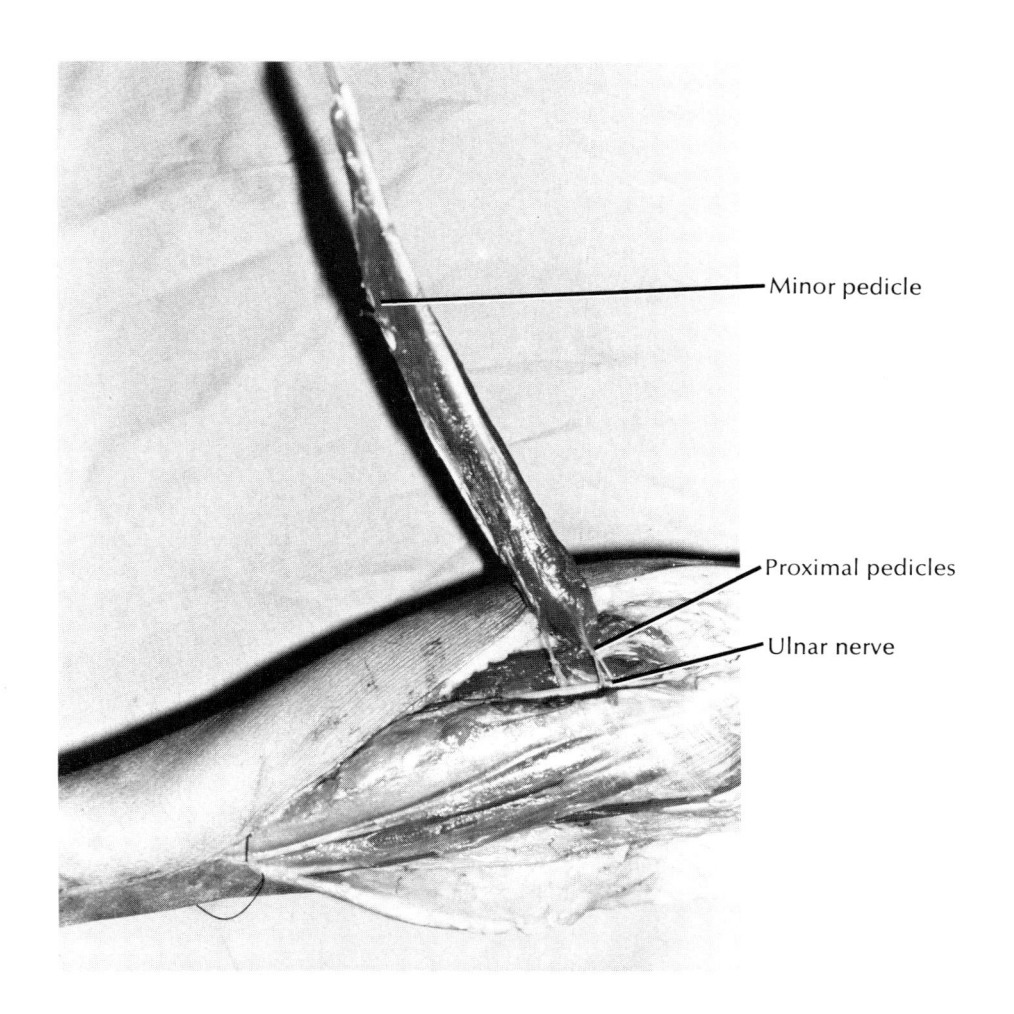

Minor pedicle

Proximal pedicles

Ulnar nerve

UPPER EXTREMITY

Flexor carpi ulnaris

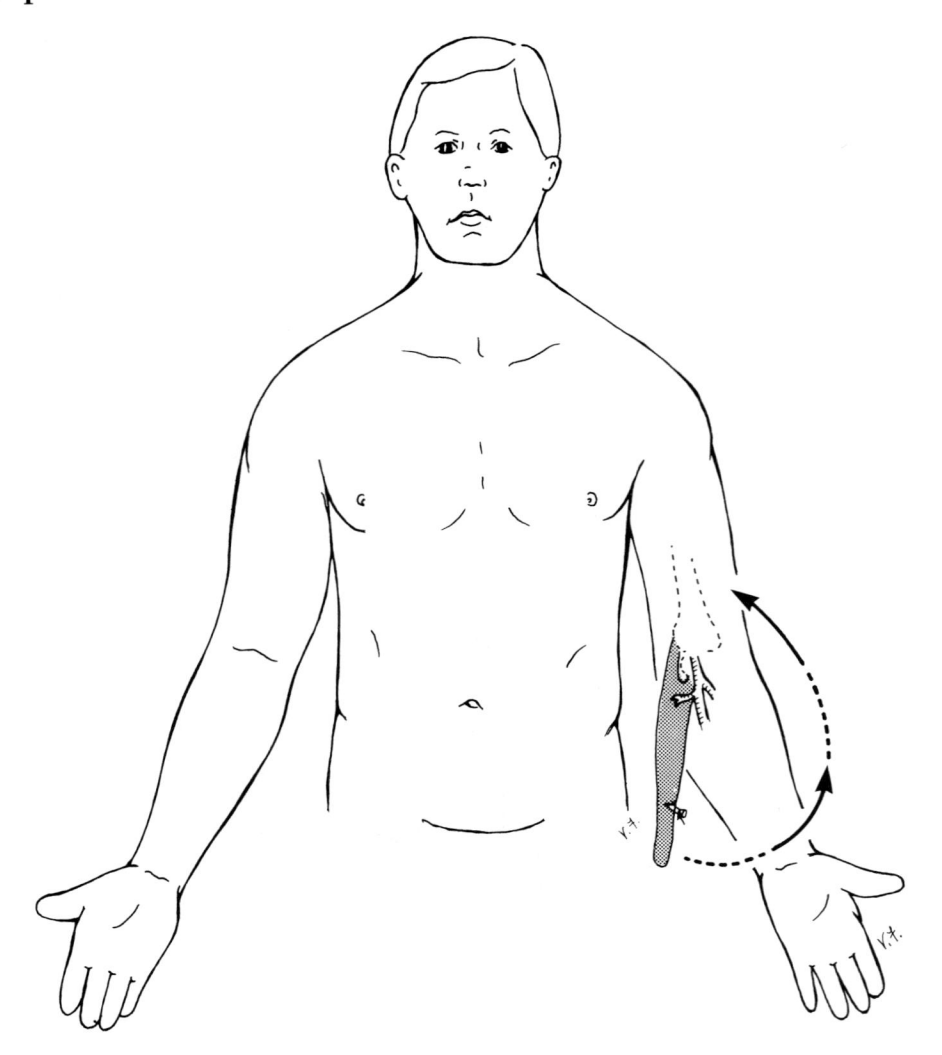

Arc of rotation

Based on the major pedicle in the antecubital fossa, the unit has an arc of rotation that will cover the lower part of the upper arm, the antecubital fossa, and upper half of the forearm.

UPPER EXTREMITY

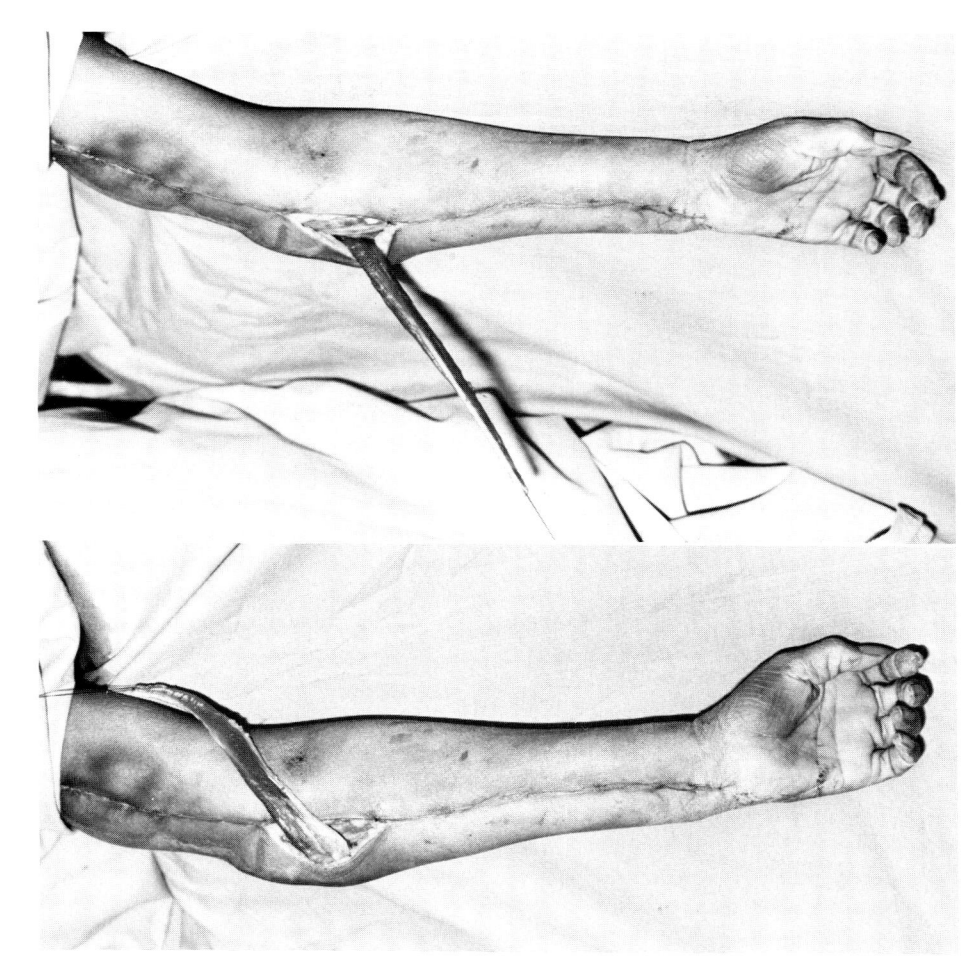

Coverage of antecubital fossa

Elevation of flap

An incision is made along the ulnar aspect of the forearm. The distal part of the muscle is identified. The distal muscle is retracted medially to locate the ulnar artery and nerve. The tendon is then divided above the wrist and the muscle elevated from insertion to origin. The distal minor pedicles are divided and ligated. The major pedicle is identified approximately 5 to 8 cm from the elbow.

Precautions

- The ulnar artery and nerve are closely related to the deep surface of the muscle and must be protected during flap elevation.
- Elevation as a musculocutaneous flap may leave the ulnar artery and nerve exposed.

SECTION EIGHT
HAND

First dorsal interosseus

Abductor pollicis brevis

Abductor digiti minimi

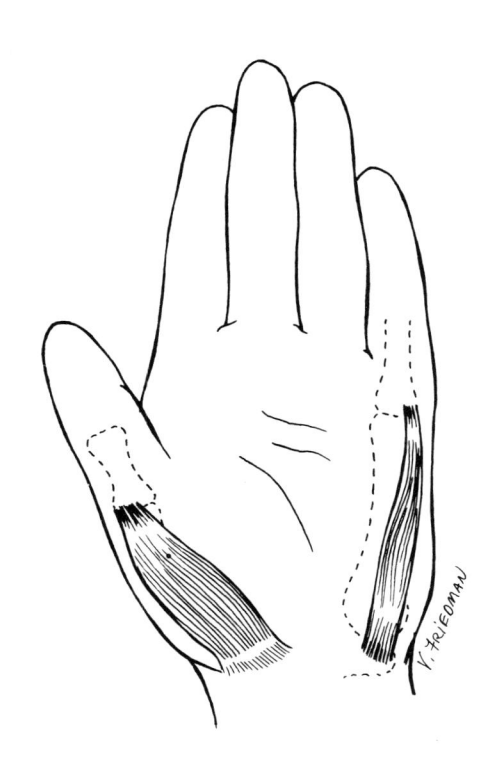

HAND

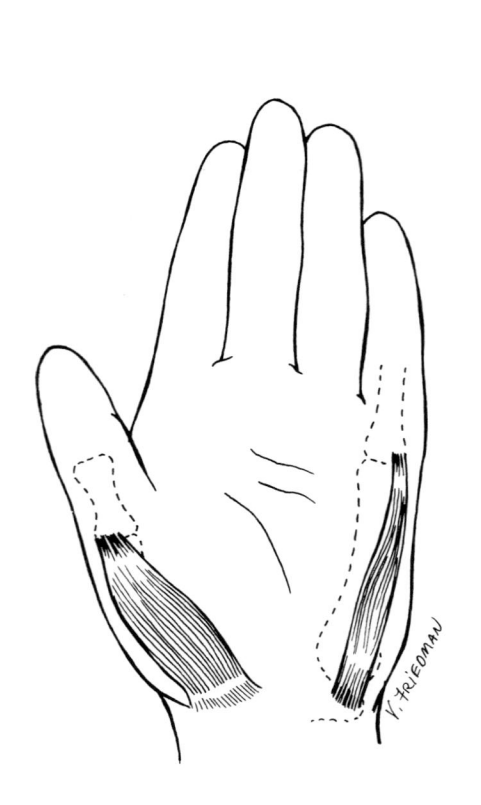

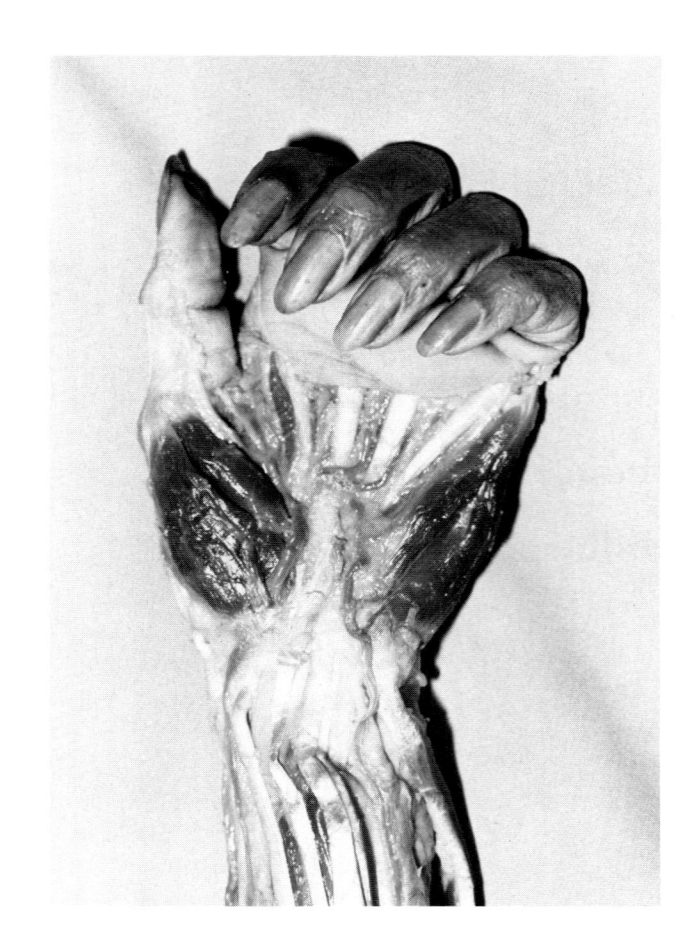

As with the forearm and upper arm muscles, the small muscles of the hand are not expendable, and their use as muscle flaps has been limited. For coverage of the hand, pedicle groin flaps and free flaps are preferable to muscle flaps. However, in particular instances when the defects are small and alternatives are not available, the small muscles of the hand may serve as suitable muscle flaps. Only three of these small muscles are described.

Labels on figure: Superficial palmar arch; Deep palmar arch; Ulnar artery; Radial artery

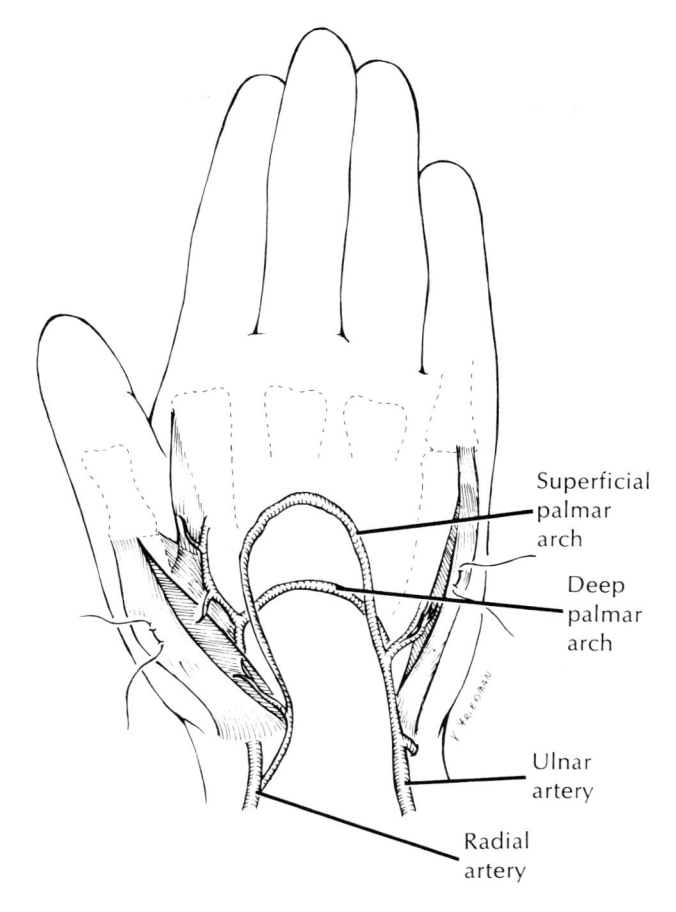

Blood supply

The ulnar and radial arteries are the basis for the blood supply to hand musculature. The radial artery at the wrist has a small branch on the volar aspect, the superficial palmar artery, which runs into the palm and, by anastomoses with the terminal part of the ulnar artery, completes the superficial palmar arch. The radial artery courses dorsally deep to the tendons of the abductor pollicis longus and extensor pollicis brevis, through the anatomic "snuffbox," deep to the extensor pollicis longus tendon, between the two heads of the first dorsal interosseus, then transversely across the palm where, by anastomoses with the deep palmar branch of the ulnar artery, it completes the deep palmar arch. Before the artery courses deep between the heads of the first dorsal interosseous, the princeps pollicis artery branches off.

The ulnar artery enters the hand deep to the palmar carpal ligament, with the pisiform bone on its ulnar side. It then divides into branches that form the deep and superficial palmar arches.

First dorsal interosseous

MUSCLE FLAP

Application

Coverage of:
FIRST AND SECOND METACARPALS

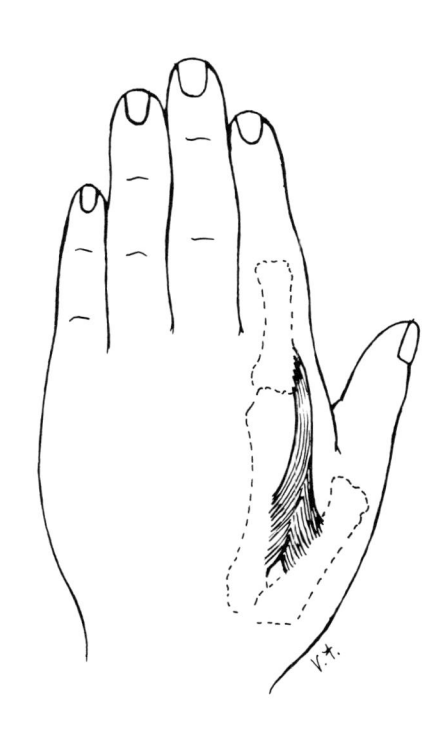

HAND

First dorsal interosseous

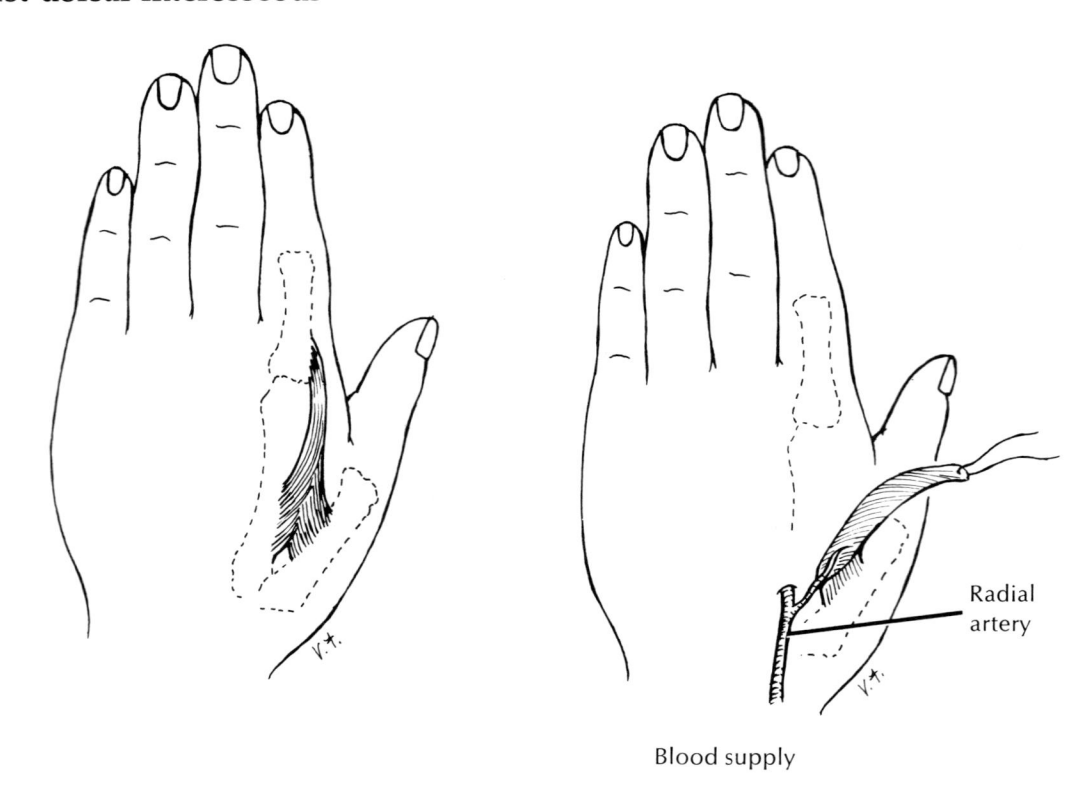

Blood supply

The first dorsal interosseus is a small bipennate muscle that lies in the thumb web space. The first dorsal interosseus is the largest dorsal interosseous. Although only the first dorsal interosseus is described, all four dorsal interossei have a proximal major pedicle and, based on the pedicle, may be transposed to cover exposed metacarpals.

These muscles are not expendable, but under certain circumstances their use for muscle flap transposition may be justified.

Origin: Two heads of origin: proximal half of the ulnar aspect of the first metacarpal, and radial aspect of the second metacarpal.

Insertion: The two heads converge into a tendon, which is inserted into the radial side of the base of the proximal phalanx of the index finger.

Nerve supply: Deep palmar branch of the ulnar nerve.

Function: The muscle is not expendable, and abducts the index finger, flexes it at the metacarpophalangeal joint, and extends it at the interphalangeal joints.

HAND

First dorsal interosseous

Radial artery

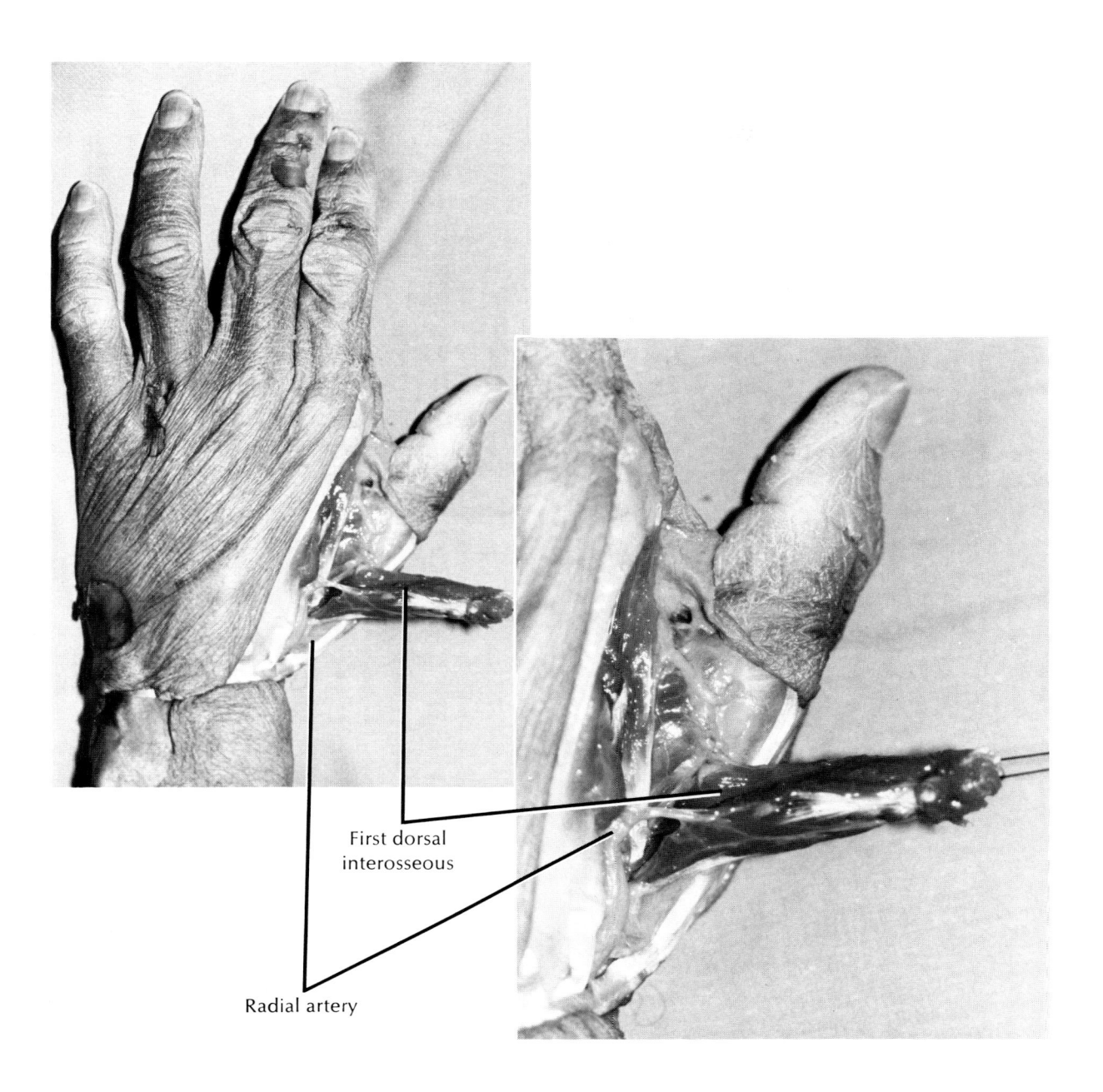

Blood supply

The radial artery passes between the two heads of origin of the muscle into the palm, where it forms the deep palmar arch. As it passes through, a small branch enters the muscle on its deep surface close to the origin. This is the major pedicle of the muscle.

HAND

First dorsal interosseous

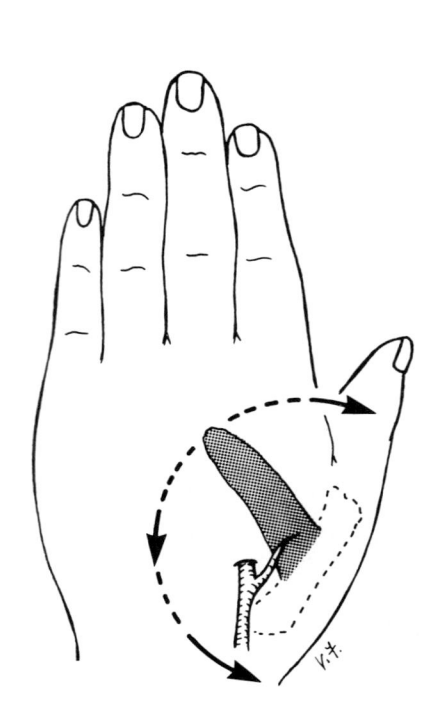

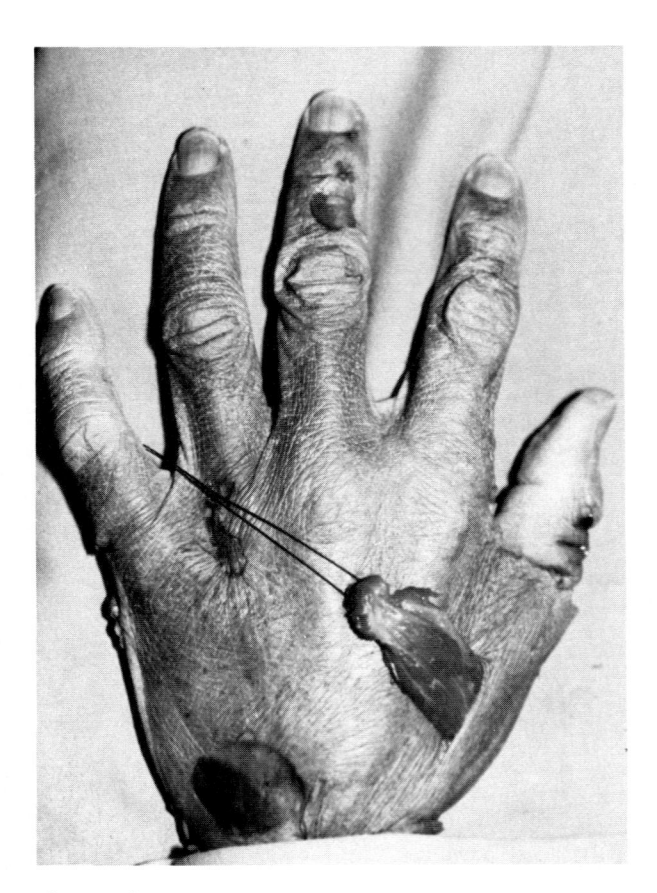

Coverage of second metacarpal

Arc of rotation

Based on the proximal major pedicle, the muscle may be transposed to cover the first and second metacarpal bones.

456

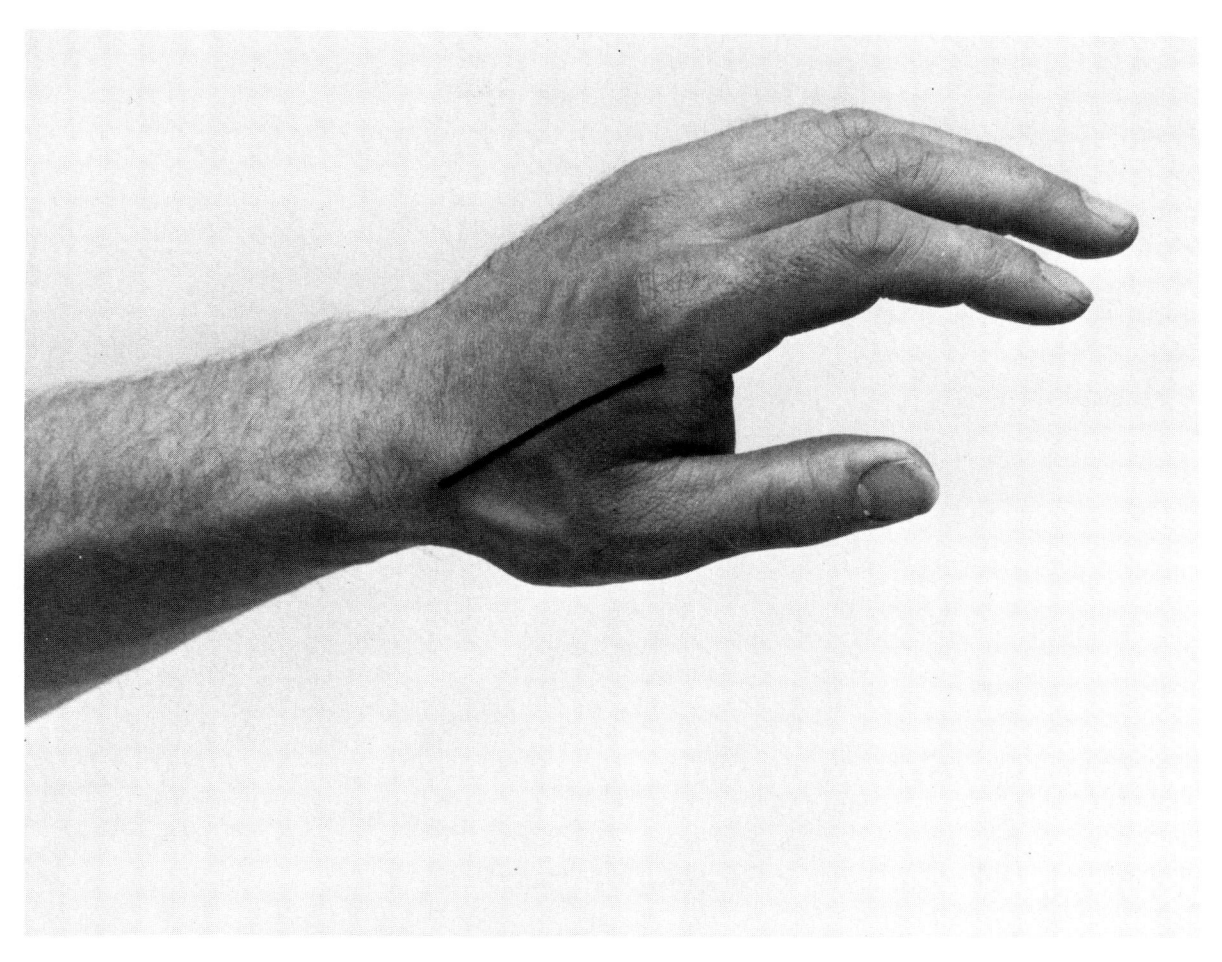

Skin incision

Elevation of flap

The muscle is approached through a longitudinal incision along the radial border of the first metacarpal. The tendon is divided and the muscular origin of the muscle dissected off the first and second metacarpal bones. Proximally the radial artery and the major pedicle are identified.

Precautions

- Sensory branches of the radial nerve to the index finger must be preserved during this dissection.
- The radial artery as it courses deep between the two heads of origin of the muscle must be preserved.

Abductor pollicis brevis

MUSCLE FLAP

Application

Coverage of:

SMALL DEFECTS OF WRIST

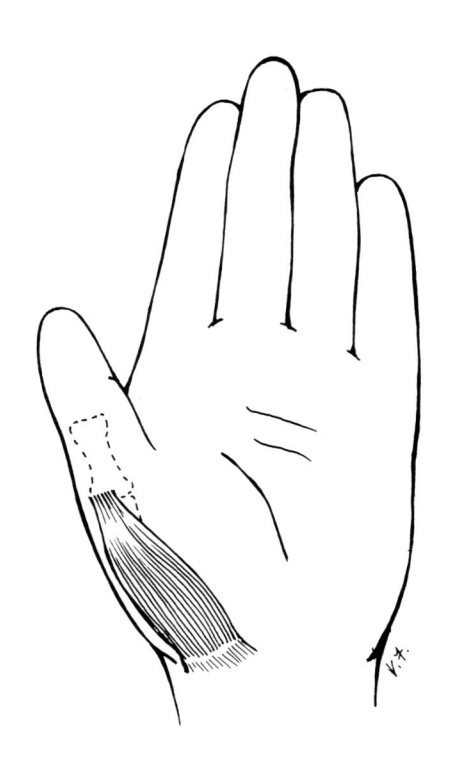

HAND

Abductor pollicis brevis

Radial artery

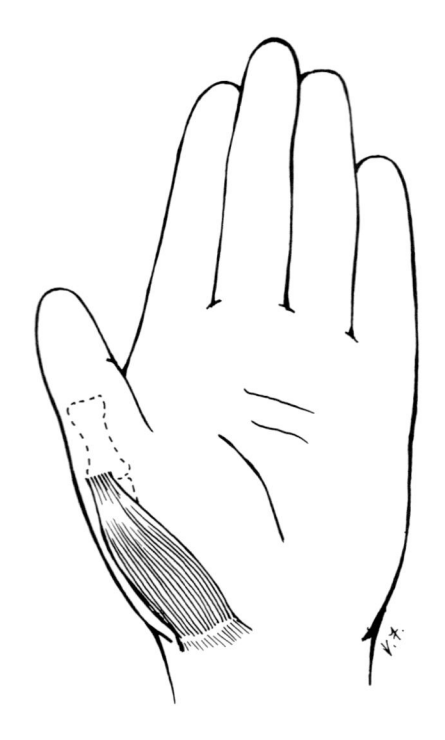

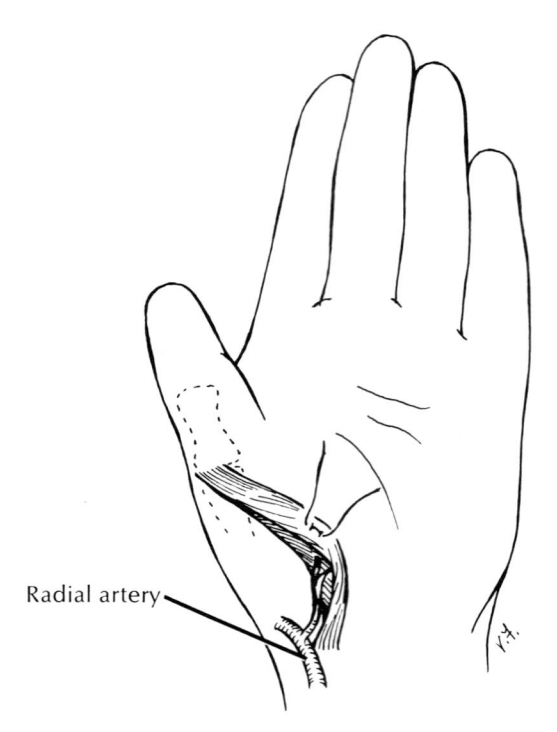

Blood supply

The abductor pollicis brevis is the most superficial of the thenar muscles. This is a thin, flat muscle that lies over the opponens pollicis muscle.

Origin: Navicular trapezium and the transverse carpal ligament.
Insertion: The tendon inserts on the radial side of the base of the first phalanx of the thumb.
Nerve supply: Branch of median nerve. The motor branch enters the muscle on its deep surface.
Function: This muscle is an abductor of the thumb.

HAND

Abductor pollicis brevis

Vascular pedicles

Radial artery

Blood supply

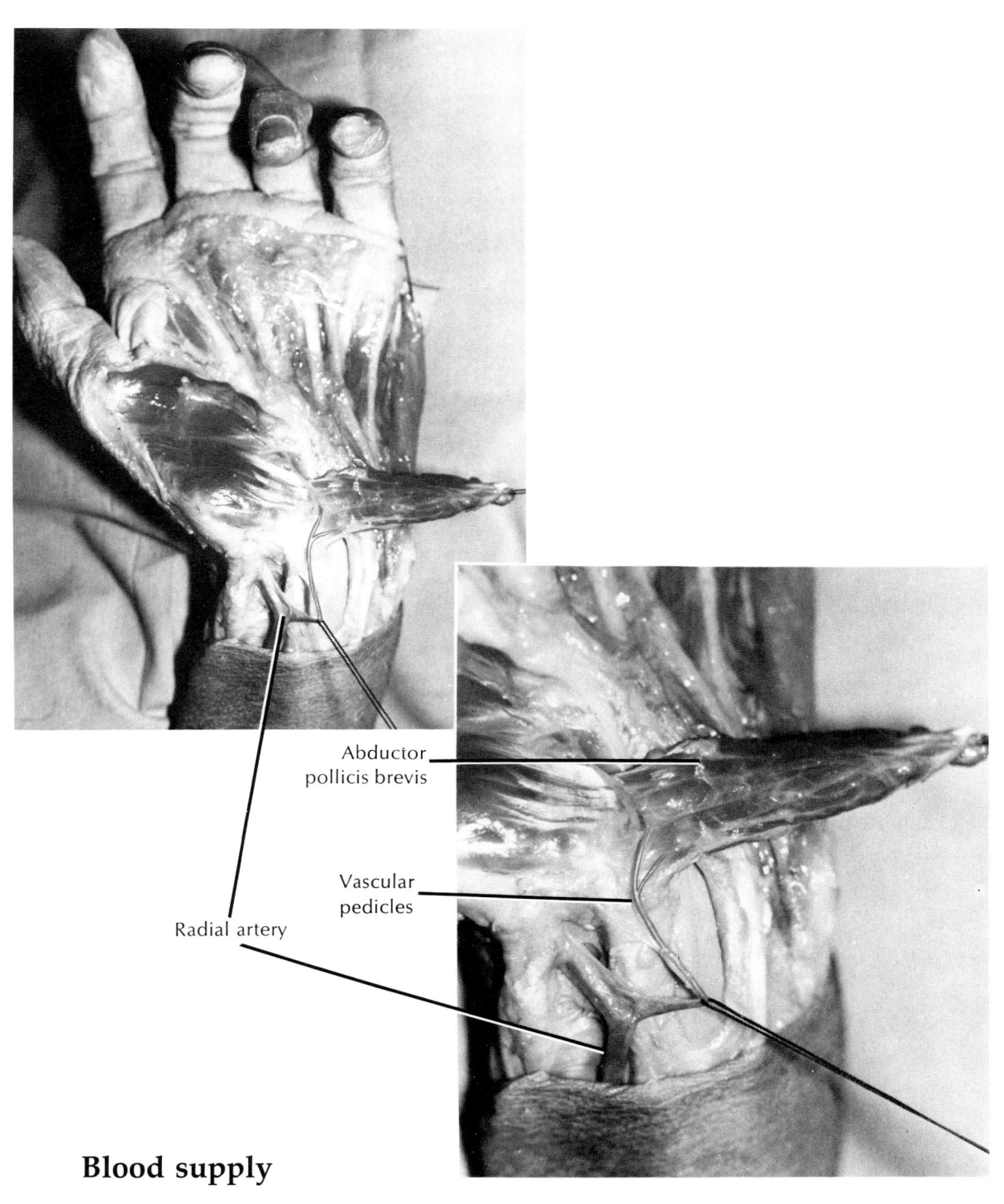

The major pedicle enters the deep surface of the muscle close to the origin. The pedicle is a branch of the superficial palmar branch of the radial artery.

HAND

Abductor pollicis brevis

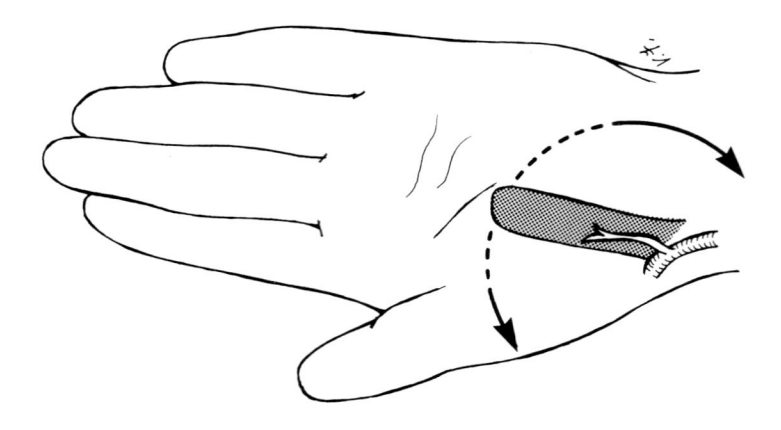

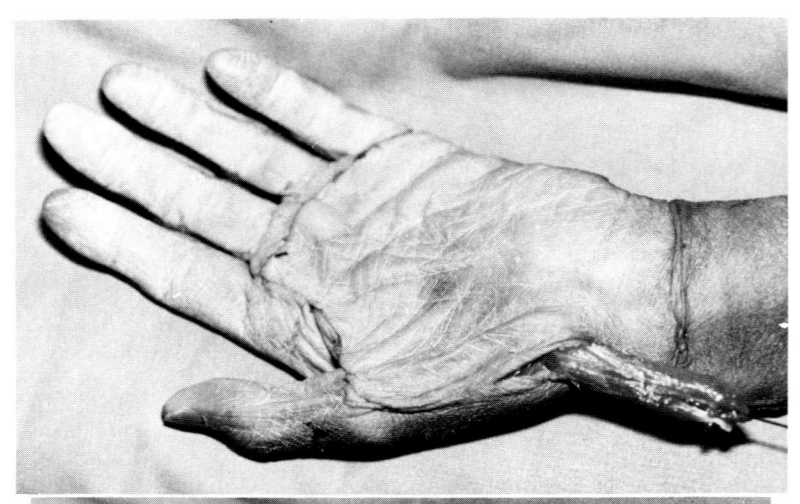

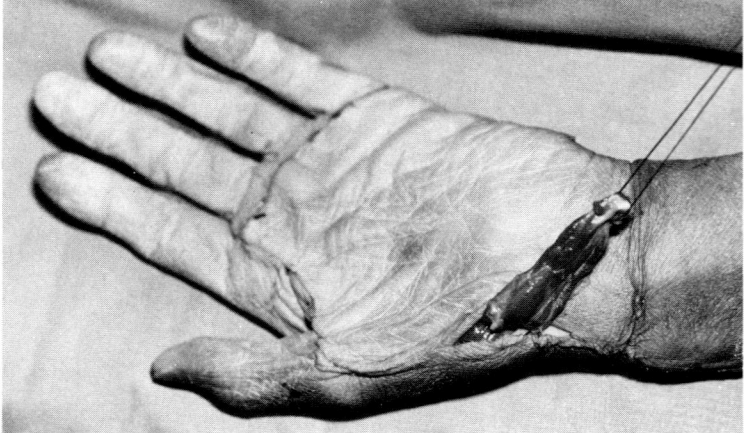

Arc of rotation

Based on the proximal major pedicle, the muscle will reach the volar aspect of the radial side of the wrist.

HAND

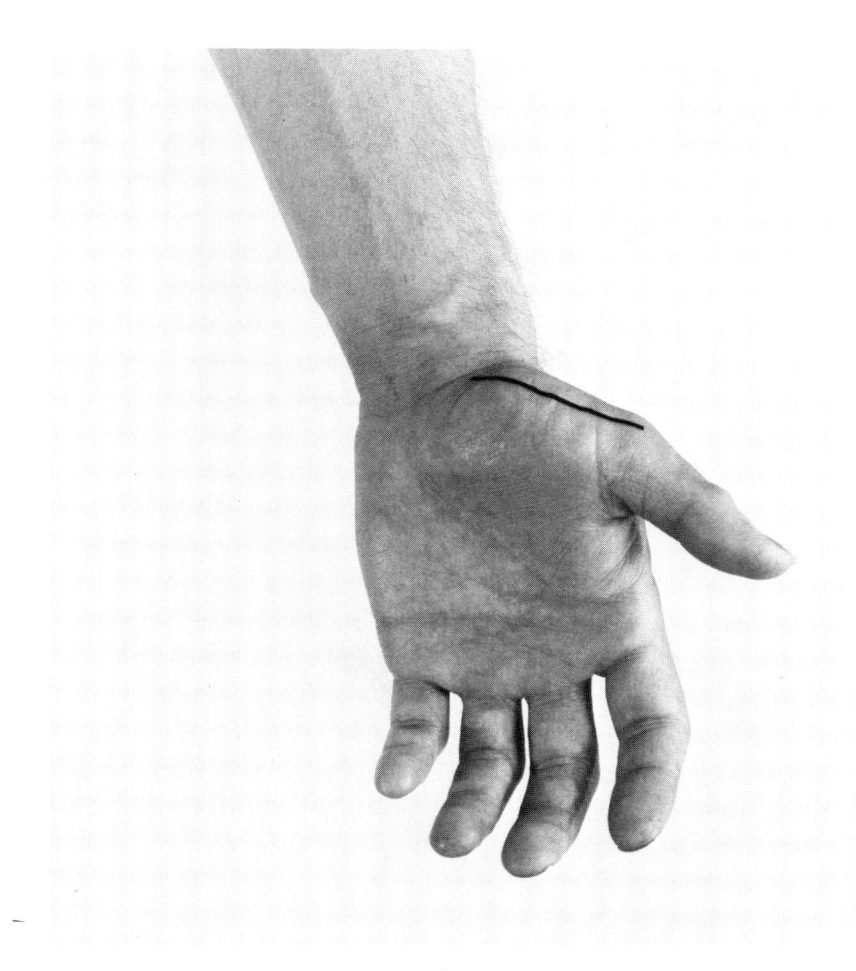

Skin incision

Elevation of flap

An incision along the upper border of the thenar eminence will expose the muscle. The tendon is divided on the radial side of the bone of the proximal phalanx. The muscle is then dissected from the insertion to origin. The major pedicle is identified on the deep surface, and the muscle is transposed.

Precautions

- This is a small muscle that will only cover small defects.
- The overlying skin can be elevated with the muscle as a musculocutaneous flap, but the secondary defect may not be acceptable.

Abductor digiti minimi

MUSCLE FLAP

Application

Coverage of:
SMALL DEFECTS OF WRIST

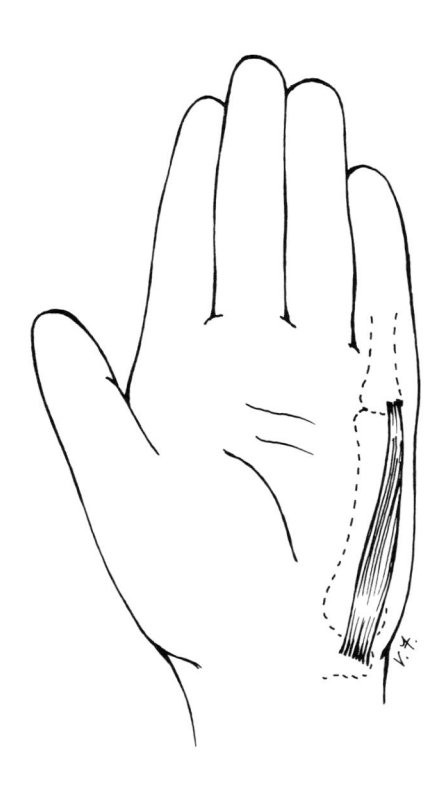

HAND

Abductor digiti minimi

Ulnar artery

Blood supply

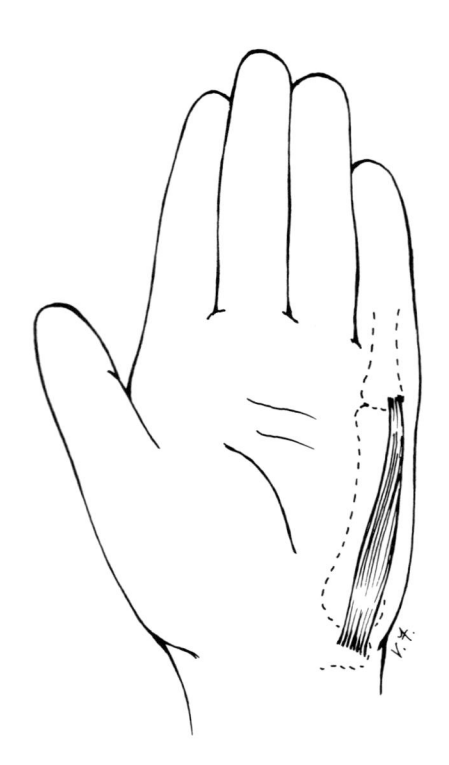

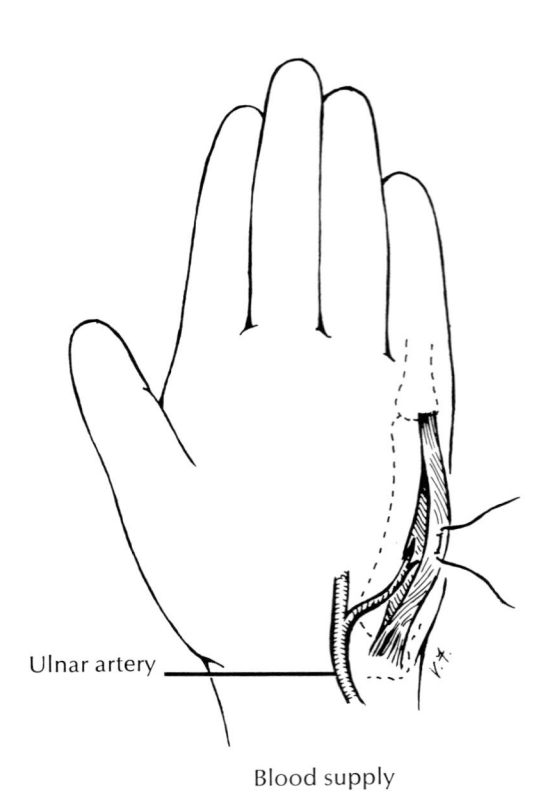

The abductor digiti minimi is a small, thin muscle that may be most useful for coverage of exposed structures over the volar aspect of the wrist. It is the most superficial of the hypothenar muscles.

Origin: Pisiform bone.
Insertion: Ulnar side of the first phalanx of the little finger.
Nerve supply: Branch of ulnar nerve. The motor branch enters the muscle on its deep proximal surface.
Function: The muscle is relatively expendable and is an abductor of the little finger.

HAND

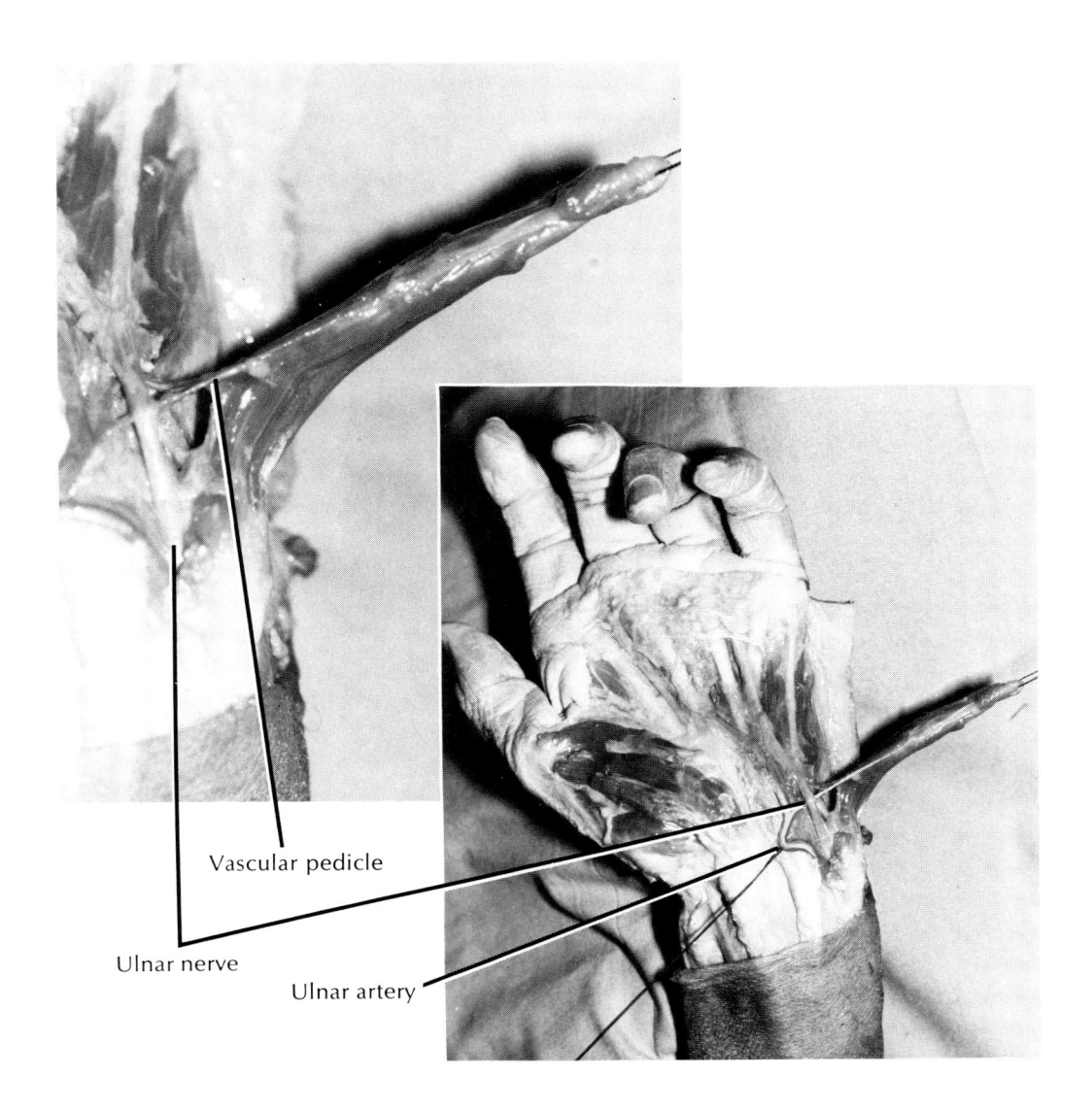

Vascular pedicle
Ulnar nerve
Ulnar artery

Blood supply

The major pedicle of this muscle enters the muscle deep on the proximal surface. This major pedicle is a direct branch of the ulnar artery and passes behind the ulnar nerve just distal to Guyon's canal.

HAND

Abductor digiti minimi

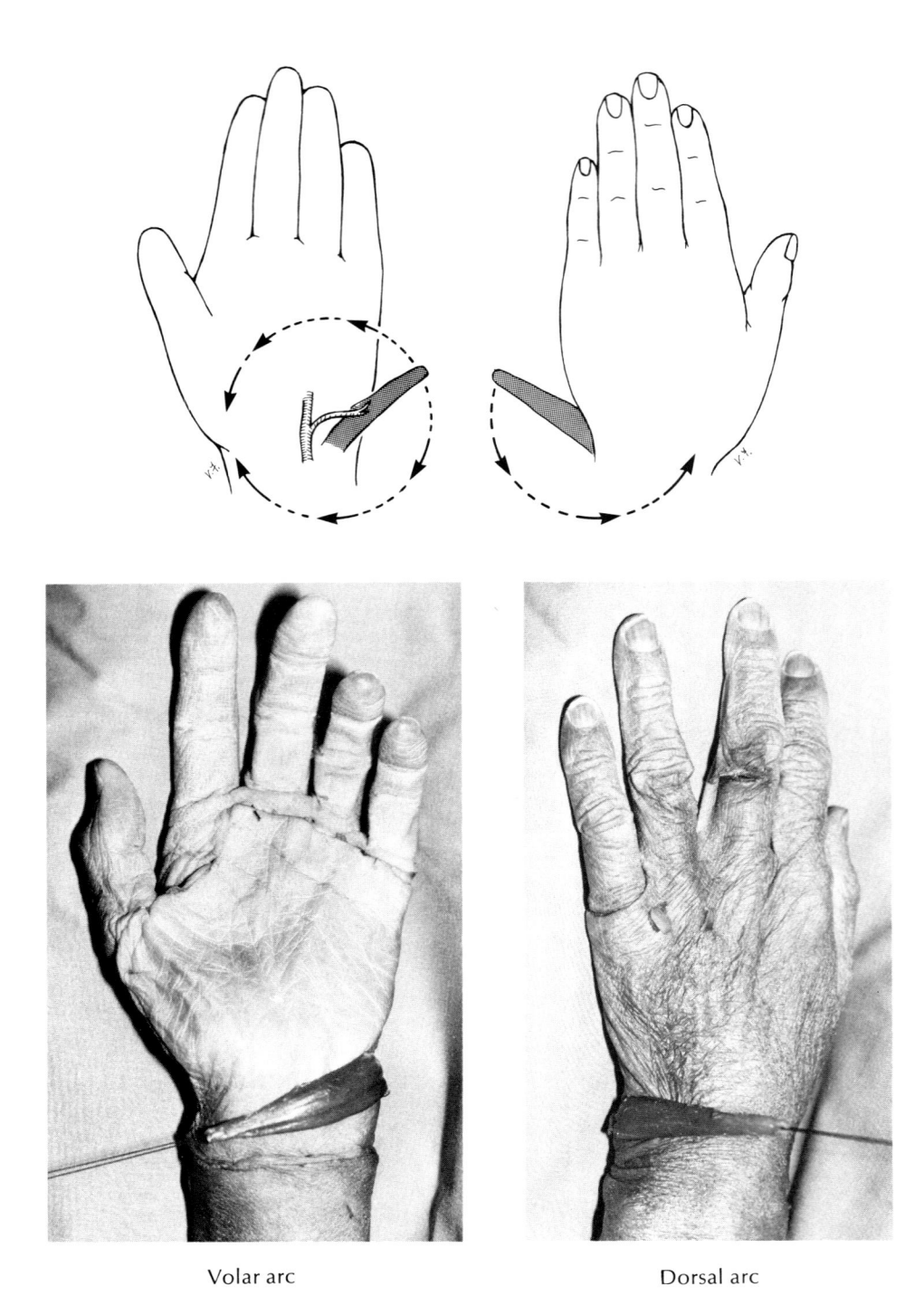

Volar arc

Dorsal arc

HAND

Arc of rotation

Based on the major pedicle at the level of the pisiform, the muscle may be transposed to cover the volar or dorsal aspect of the wrist.

Elevation of flap

Through an incision on the ulnar border of the hand, this superficial muscle is identified. The tendon is divided, and by elevating the muscle from insertion to origin, the major pedicle is identified. The major pedicle runs deep to the ulnar nerve.

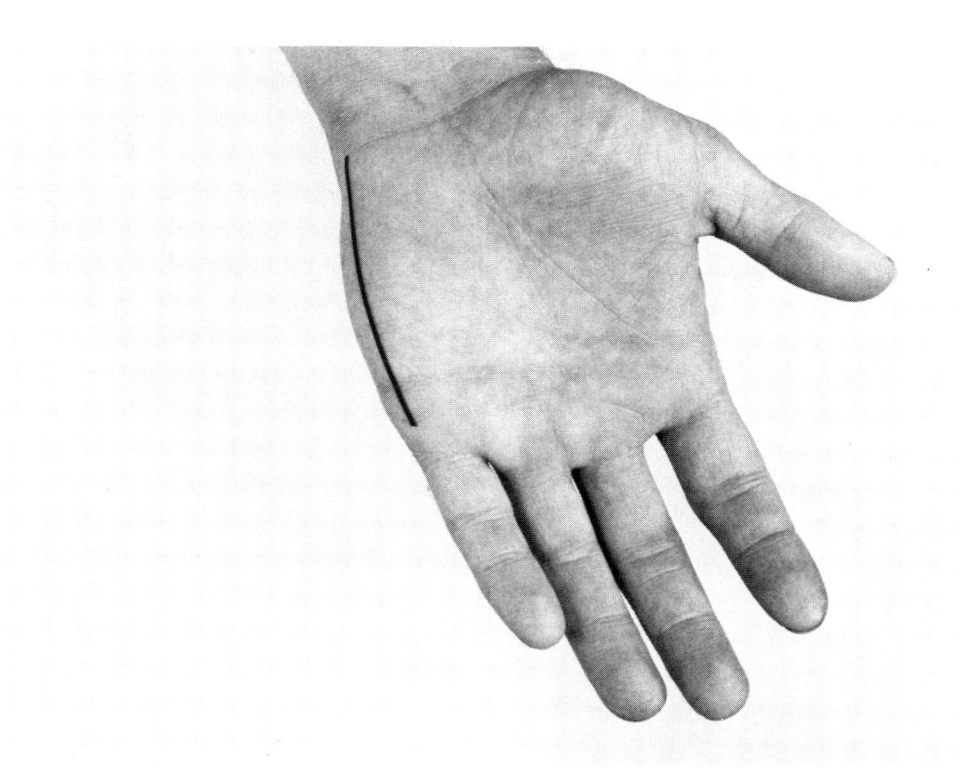

Skin incision

Precautions

- Near the origin of the muscle the ulnar nerve lies on its radial side.
- Although overlying skin may be elevated with the muscle as a musculocutaneous flap, the secondary defect may be unacceptable.

SECTION NINE
HEAD AND NECK

Sternocleidomastoid

Temporalis

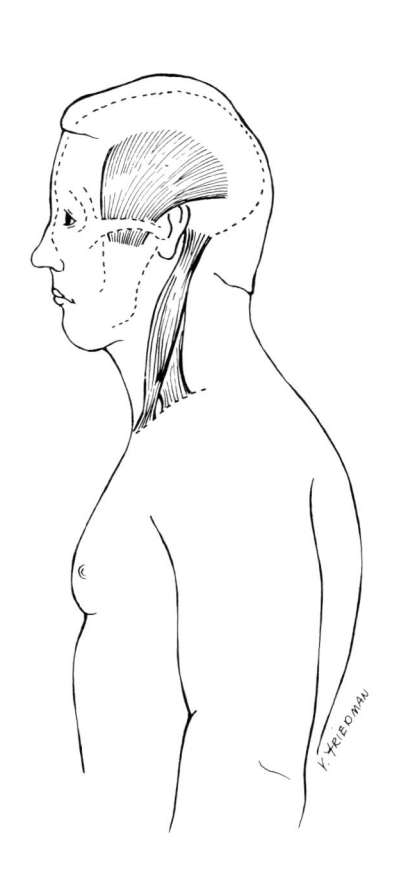

HEAD AND NECK

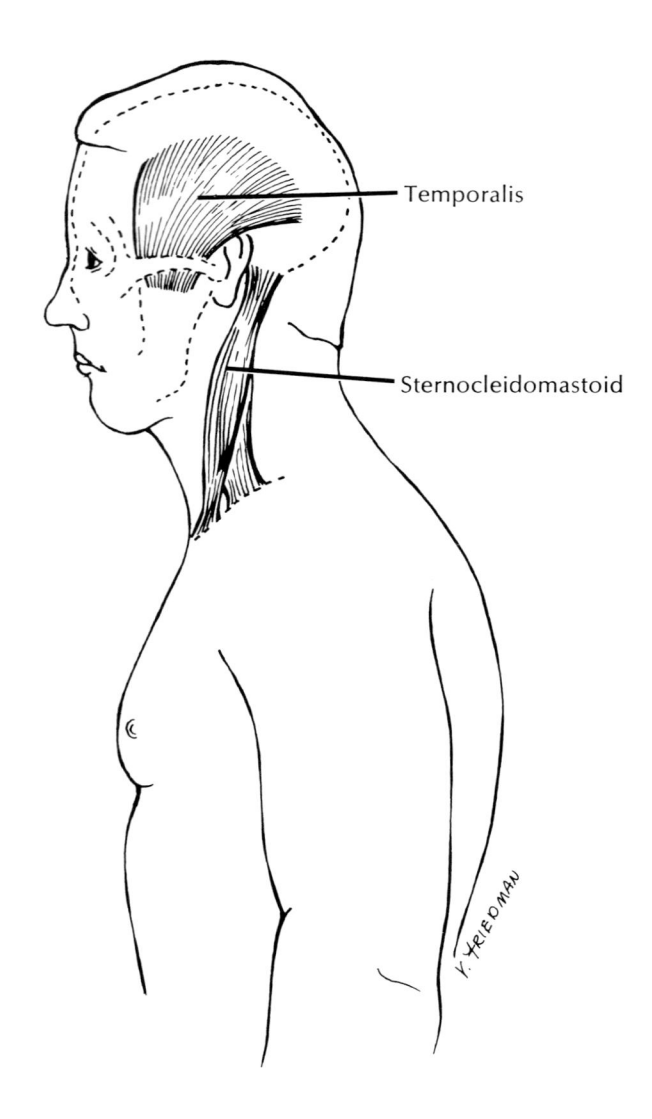

The sternocleidomastoid and temporalis muscles are suitable for transposition for reconstruction and coverage of the face and skull. However, the pectoralis major, trapezius, and latissimus dorsi are trunk muscles with a wide arc of rotation that easily reaches the head and neck region. The pectoralis major and latissimus dorsi muscles have major vascular pedicles from the axillary artery. This is an advantage for head and neck reconstruction, as local vascular pedicles may be compromised by the resection or radiation. These distant muscles useful in head and neck reconstruction are discussed in the trunk section.

HEAD AND NECK

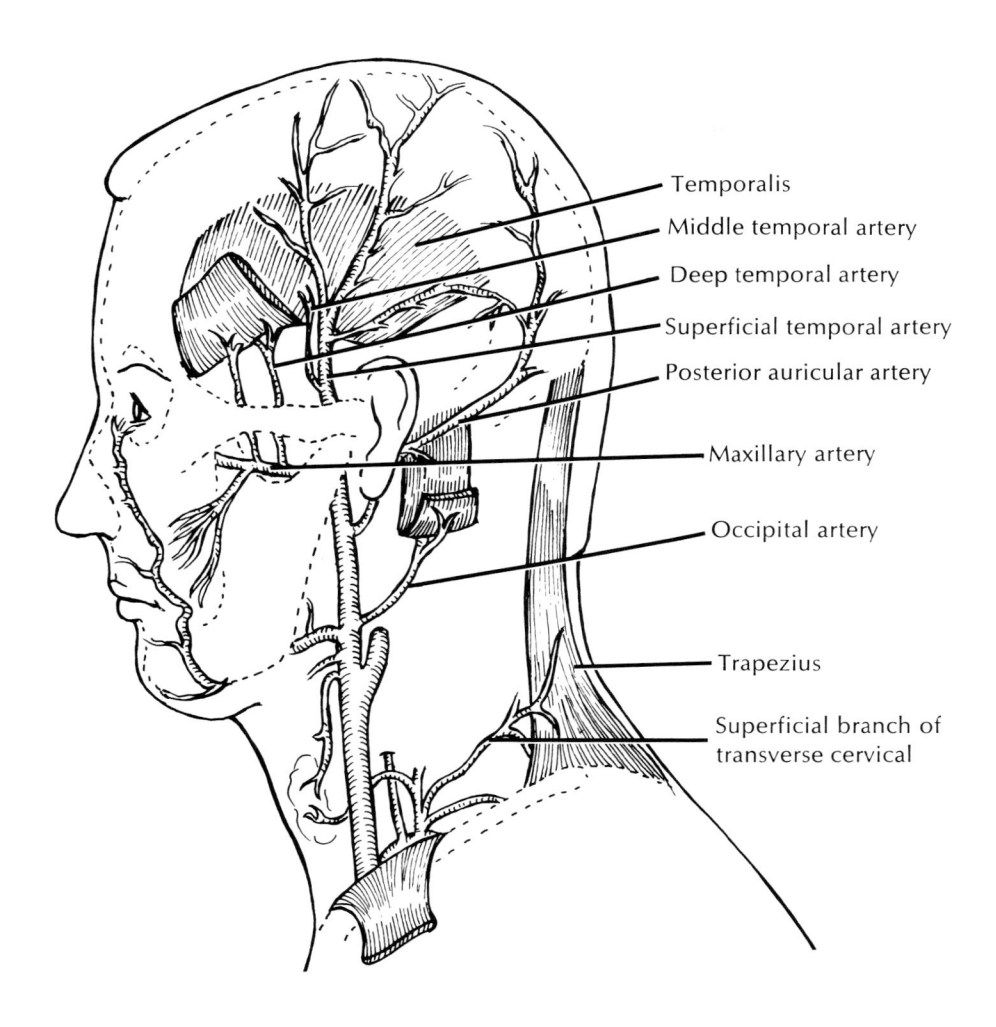

Blood supply

The posterior neck muscles receive vascular pedicles originating from the subclavian artery via the thyrocervical trunk. The trapezius muscle is included in the posterior trunk section, although this muscle is vascularized via the subclavian artery. The carotid artery bifurcates into the external and internal carotid arteries at the level of the superior border of the thyroid cartilage. This occurs in the superior carotid triangle bordered by the sternocleidomastoid, omohyoid, and digastric muscle bellies. At this level the occipital artery contributes pedicles to the sternocleidomastoid muscle. The terminal branches of the external carotid artery and the superficial temporal and maxillary arteries supply pedicles to the temporalis muscle.

ND AND NECK

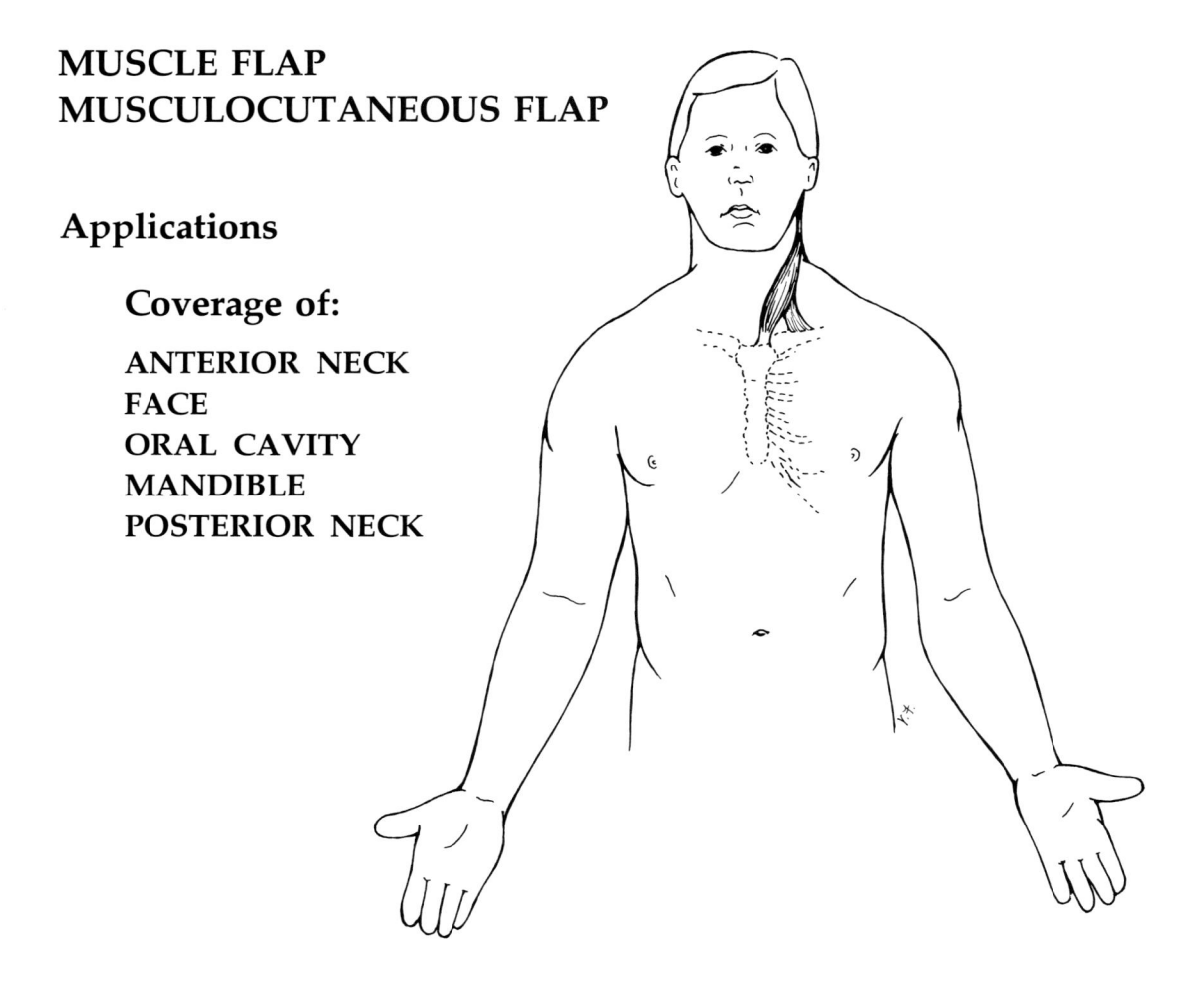

Sternocleidomastoid

MUSCLE FLAP
MUSCULOCUTANEOUS FLAP

Applications

   Coverage of:

      ANTERIOR NECK
      FACE
      ORAL CAVITY
      MANDIBLE
      POSTERIOR NECK

HEAD AND NECK

Sternocleidomastoid

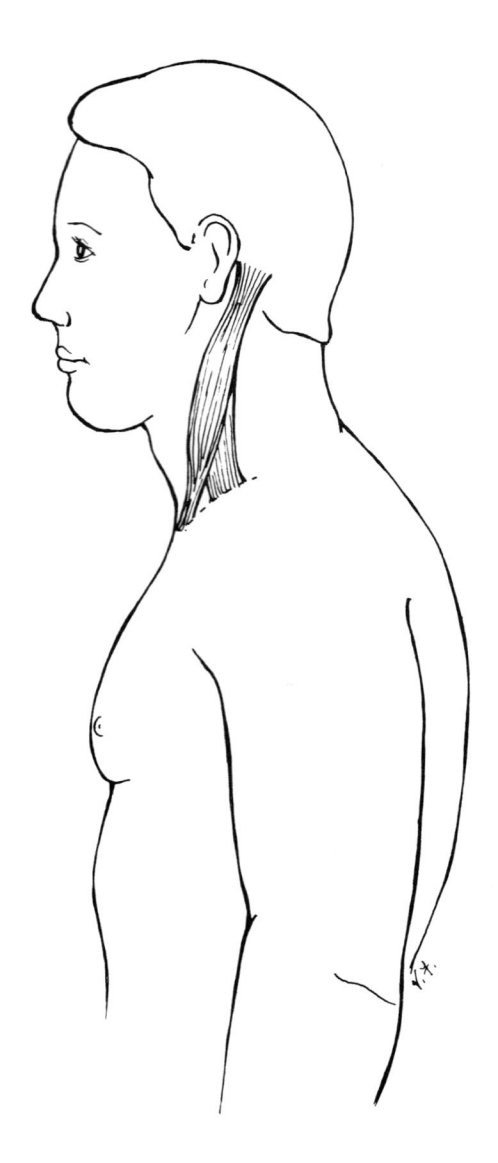

The sternocleidomastoid is a long, thick muscle located in the lateral neck coursing obliquely from the anteroinferior neck to the superolateral neck. In its inferior course this muscle has a close relationship posteriorly with the carotid sheath and sternohyoid muscles. In its mid muscle belly the muscle has a close relationship posteriorly with the omohyoid muscle, scalenus anterior muscle, and carotid bifurcation. The superior muscle has a close relationship posteriorly with the posterior belly of the digastric muscle, scalenus medius, levator scapula, and splenius capitis muscles.

HEAD AND NECK

Branch of posterior auricular artery

Occipital artery

Vascular pedicle

Lingual nerve

Blood supply

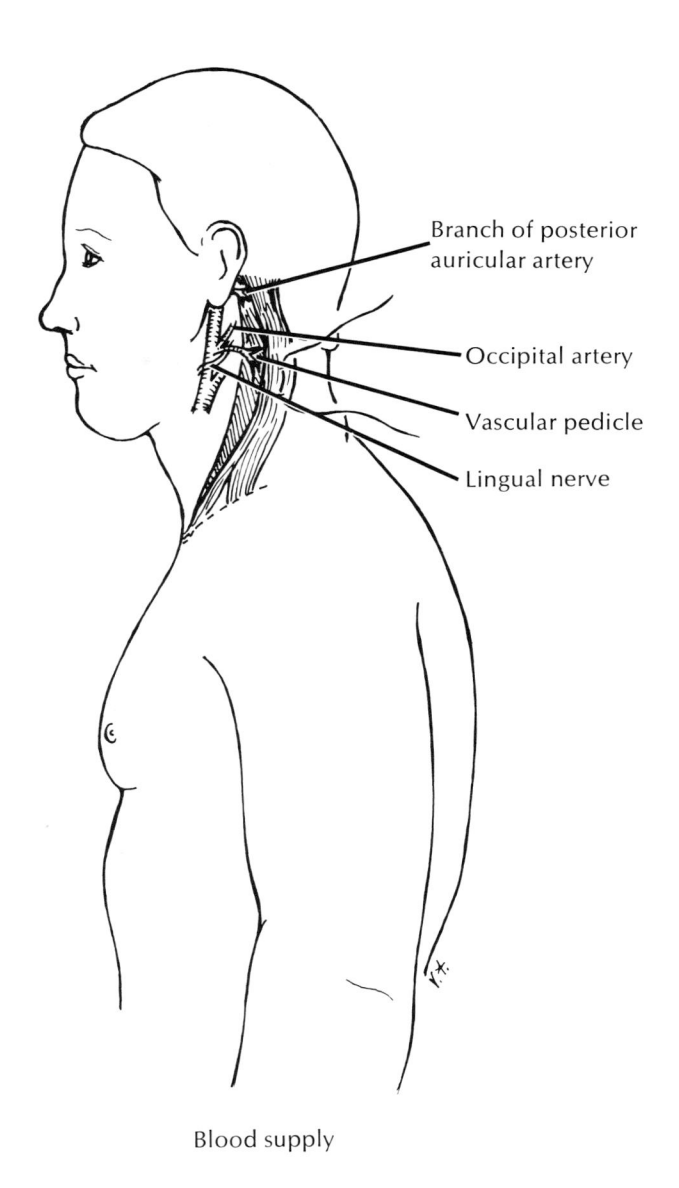

Origin: Sternal head—manubrium; clavicular head—medial third of clavicle.

Insertion: Mastoid process, superior nuchal line of occipital bone.

Nerve supply: Branches of second cervical and spinal portion of accessory nerve. These motor nerves enter the proximal portion of the posterior muscle belly.

Function: This is an expendable muscle. The function of drawing the head toward the shoulder and rotating the head will be preserved by the remaining lateral neck muscles.

HEAD AND NECK

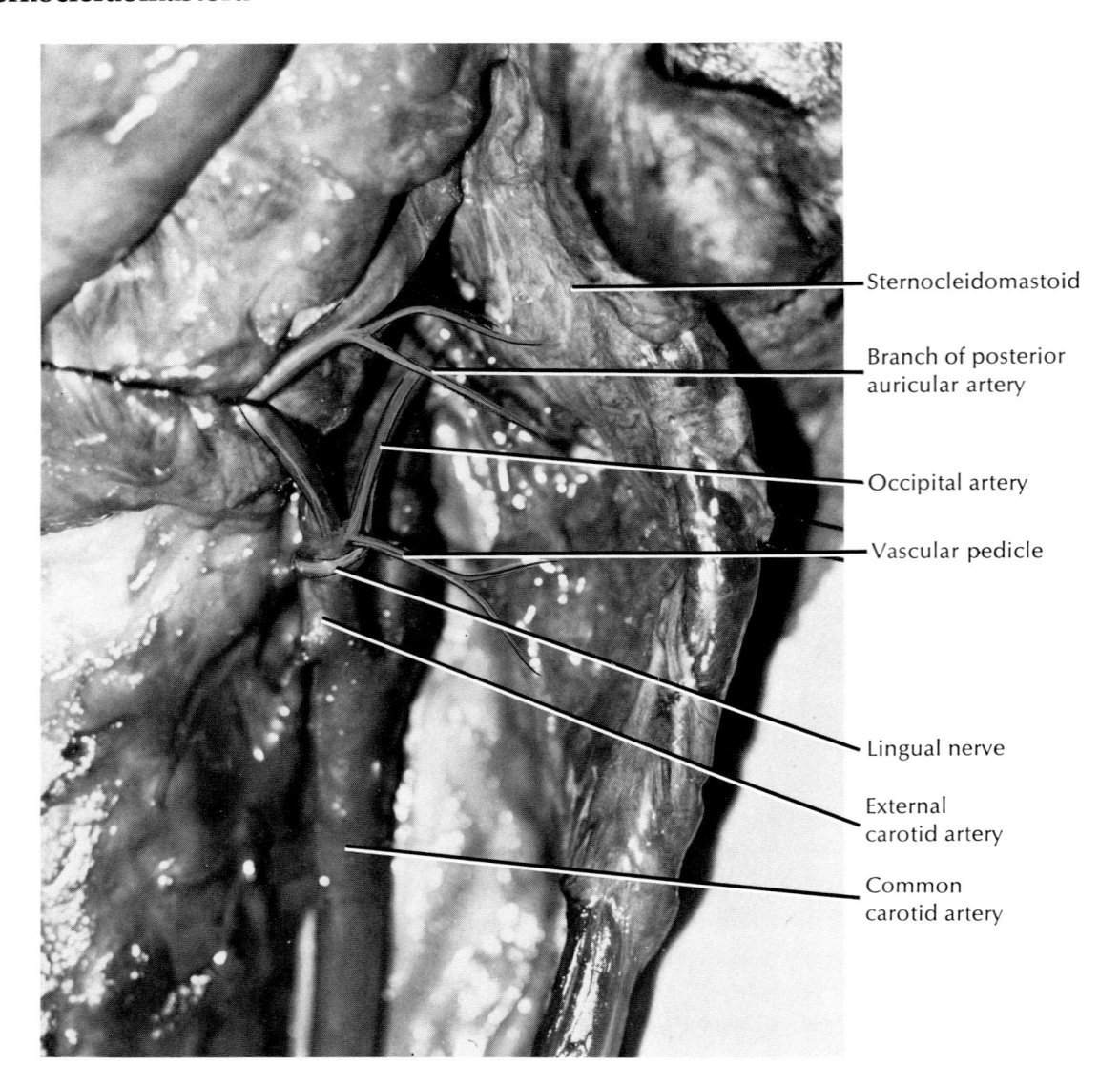

Sternocleidomastoid

- Sternocleidomastoid
- Branch of posterior auricular artery
- Occipital artery
- Vascular pedicle
- Lingual nerve
- External carotid artery
- Common carotid artery

Blood supply

The sternocleidomastoid muscle has a superior dominant vascular pedicle from the occipital artery entering posteriorly in the upper third of the muscle belly. The occipital artery courses beneath the posterior belly of the digastric muscle. This pedicle to the sternocleidomastoid muscle courses anteriorly to the lingual nerve into the posterior muscle belly. The muscle also receives a pedicle near its origin from the posterior auricular artery. This pedicle enters the anterior surface of the muscle.

HEAD AND NECK

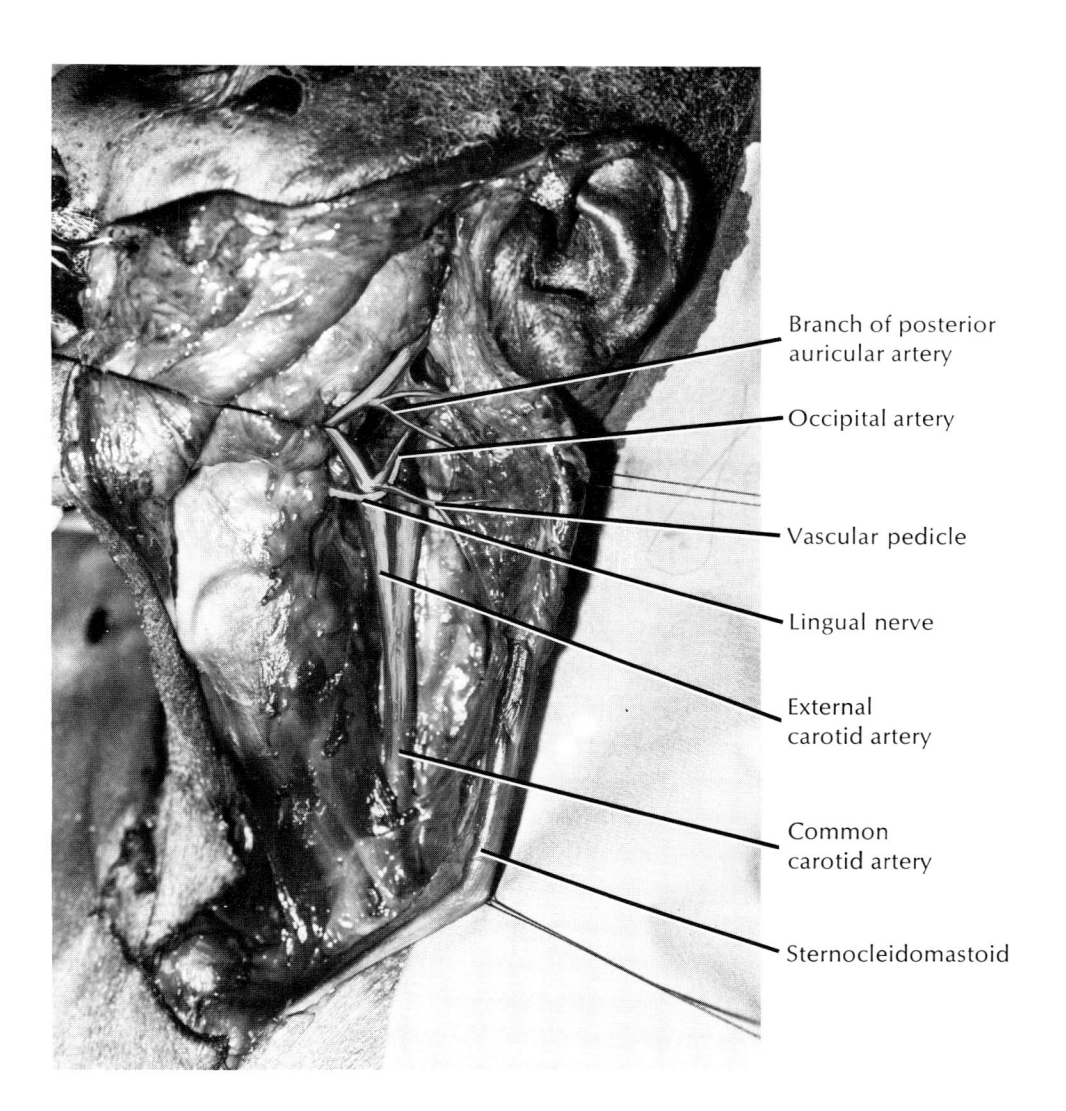

Branch of posterior auricular artery

Occipital artery

Vascular pedicle

Lingual nerve

External carotid artery

Common carotid artery

Sternocleidomastoid

HEAD AND NECK

Sternocleidomastoid

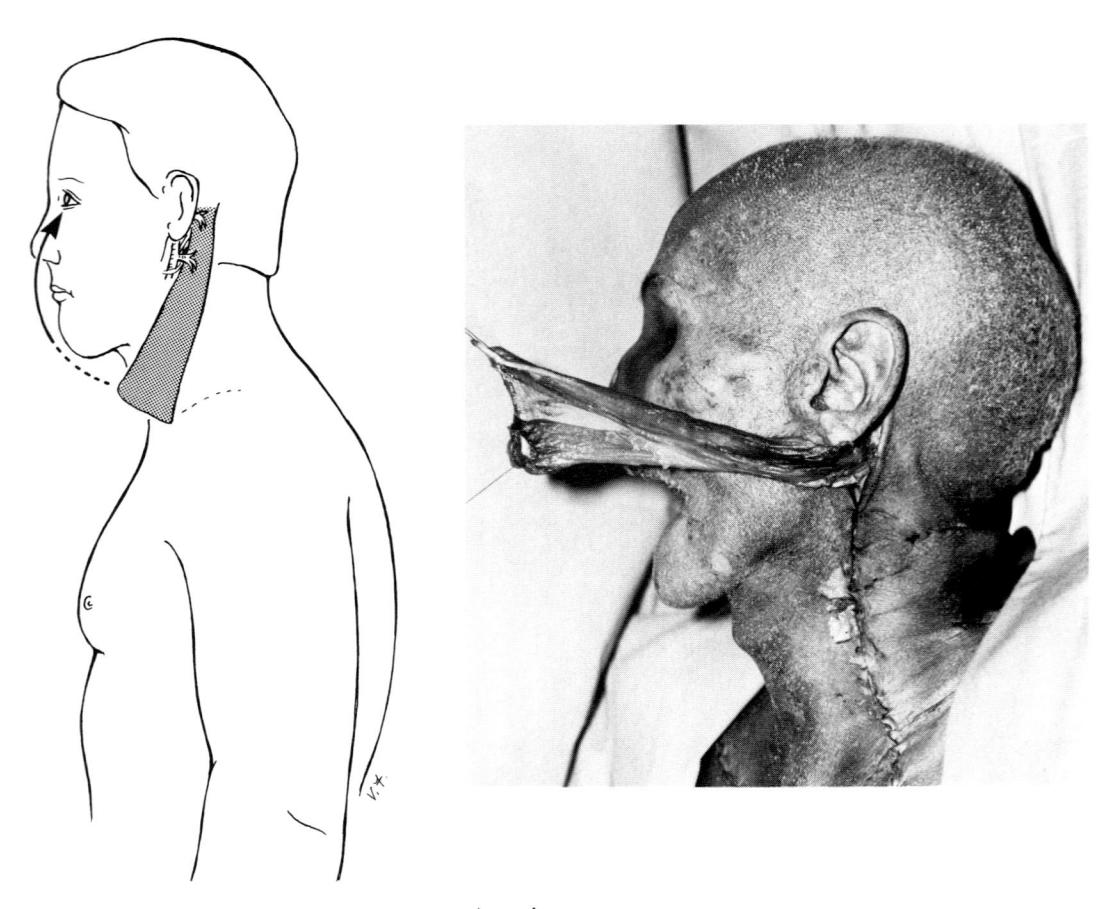

Anterior arc

Arc of rotation

This muscle has a point of rotation approximately 2 cm above the carotid bifurcation at the level of the dominant vascular pedicle to the muscle. The anterior arc of rotation will reach the anterior neck, face, and forehead. This muscle can be transposed posteriorly to cover the posterior neck, skull, and posterior mastoid regions.

HEAD AND NECK

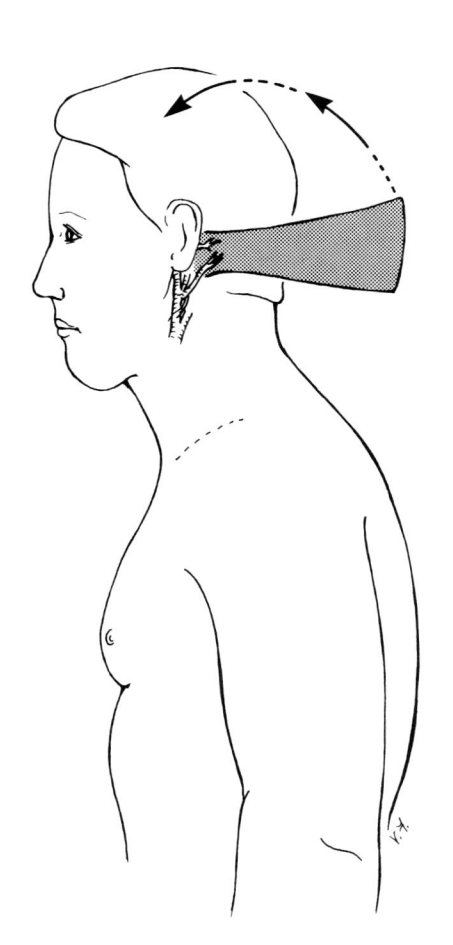

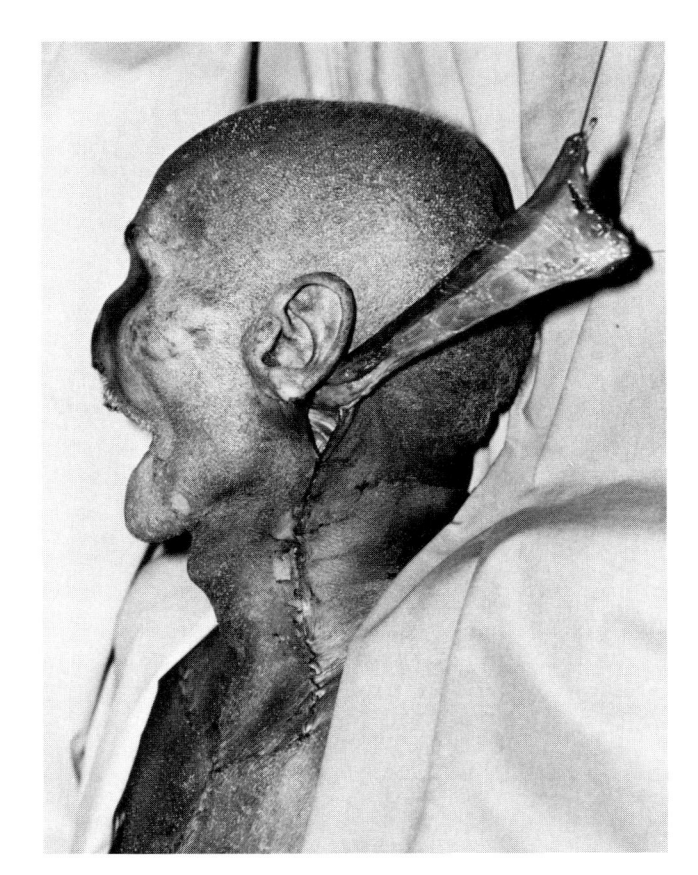

Posterior arc

481

HEAD AND NECK

Sternocleidomastoid

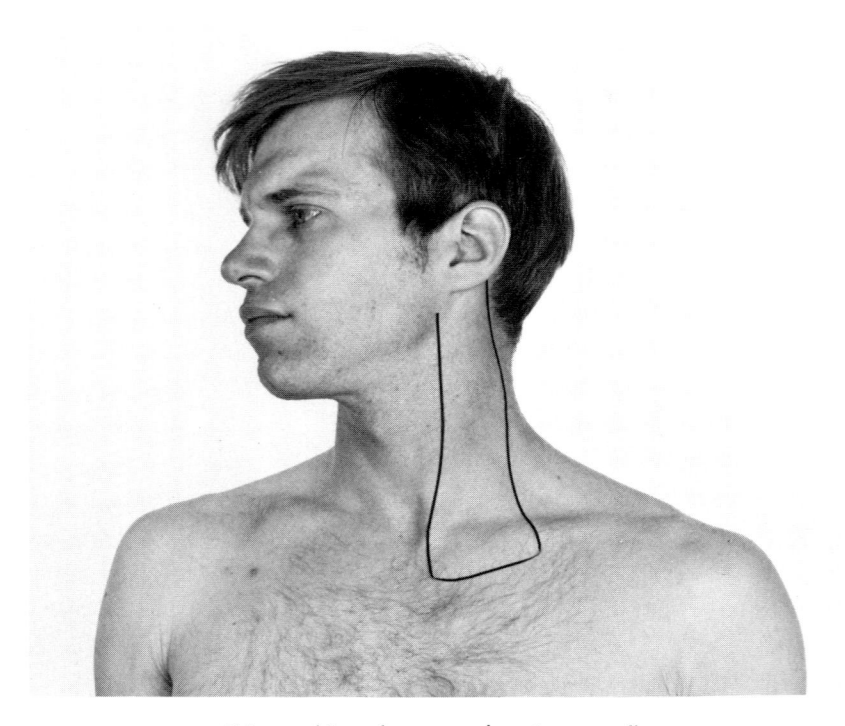

Skin markings for musculocutaneous flap

Elevation of flap

The muscle can be exposed through parallel transverse neck incisions or through a vertical neck incision overlying the muscle. The muscle is detached from its manubrium and clavicular origins and elevated superiorly. At the level of the carotid bifurcation, the posterior dissection must be performed with the occipital artery visualized. The pedicle from the occipital artery should be preserved as the transposition is performed. The donor defect can be closed primarily.

This unit can be elevated as a musculocutaneous flap. A 2 cm skin extension inferiorly over the clavicle may be elevated safely with the muscle in its cutaneous territory. The donor defect can generally be closed primarily. This flap may be elevated either with a proximal skin bridge or as a skin island with the muscle, depending on reconstructive requirements.

Anatomic dissections demonstrate vascular connections between the muscle circulation and bony attachments of muscle. Since the inferior muscle insertion to the clavicle is muscular, it is possible to carry a segment of clavicle with this muscle to incorporate into facial reconstruction.

Precautions

- Internal jugular vein has a close relationship with the clavicular head of the muscle.
- The greater auricular nerve has a close relationship to the anterosuperior muscle belly and should be preserved during flap elevation.

HEAD AND NECK

Sternocleidomastoid

Musculocutaneous flap reconstruction of mid face

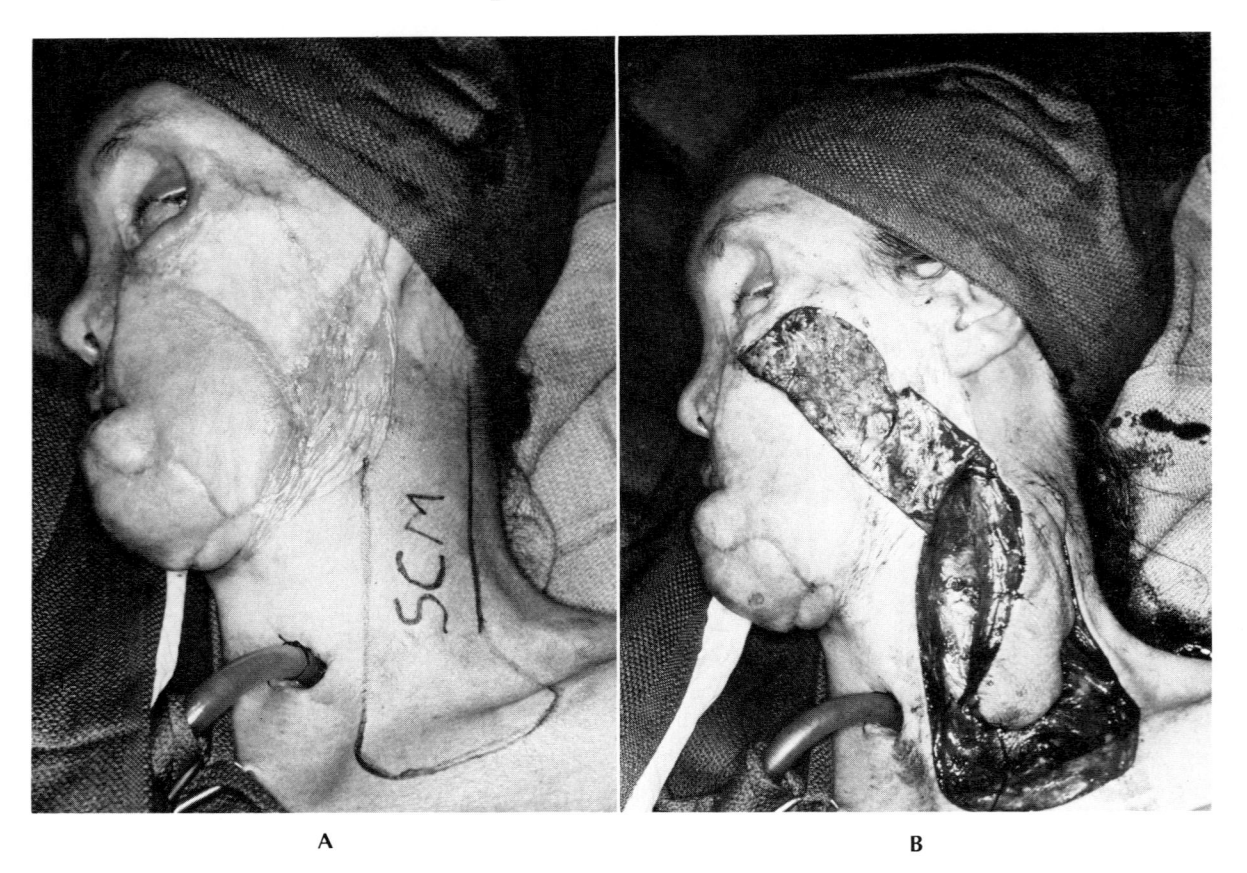

A

B

A   Traumatic defect of mid face. Skin markings for musculocutaneous flap.

B   Flap elevation with temporary suture of skin to muscle.

HEAD AND NECK

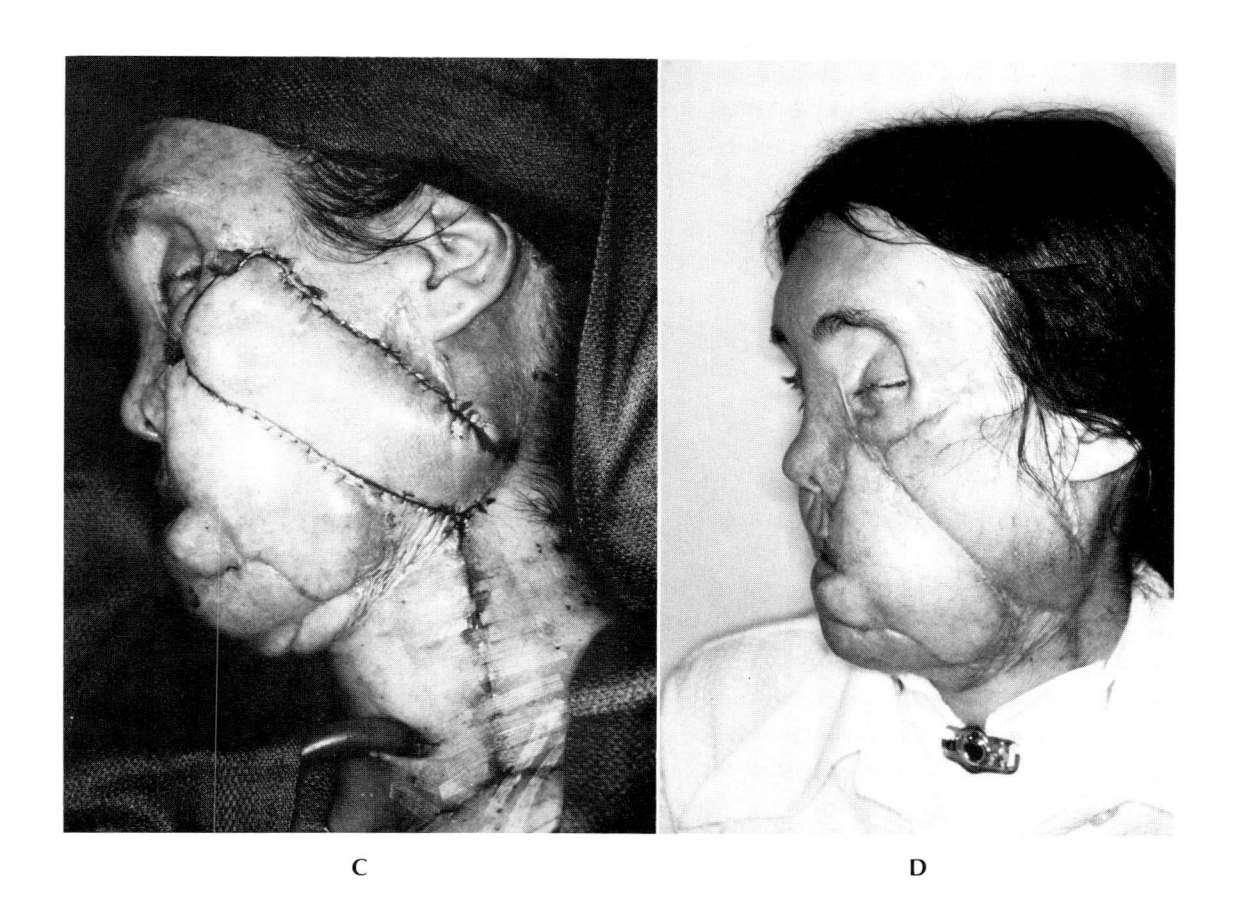

C Flap transposition with primary closure of donor region of neck.

D Reconstruction of mid-face defect with single-stage musculocutaneous flap transposition.

HEAD AND NECK

Temporalis

MUSCLE FLAP
MUSCULOFASCIAL FLAP

Applications

Coverage of:

ANTERIOR SKULL
FACE
FACIAL NERVE
ORBITAL CAVITY
MASTOID BONE

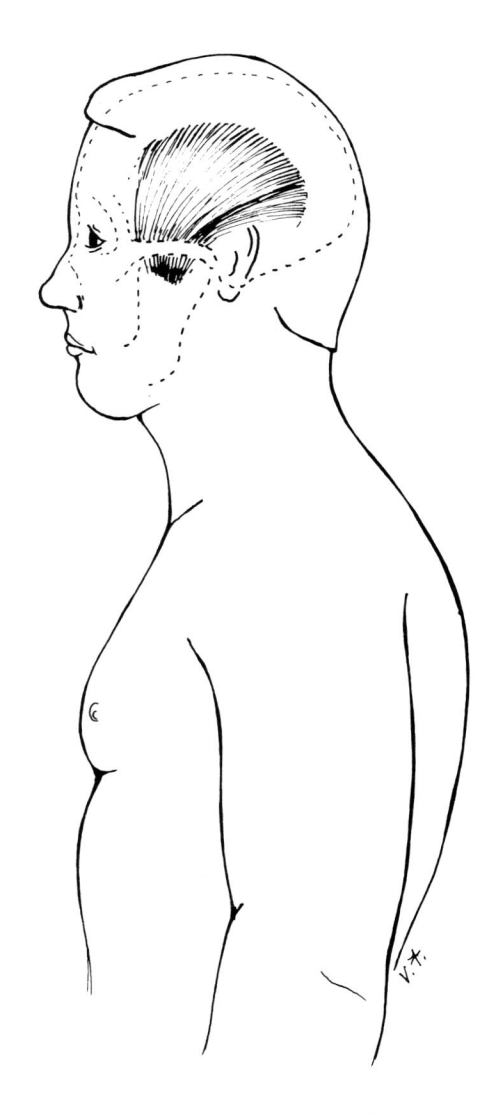

HEAD AND NECK

Temporalis

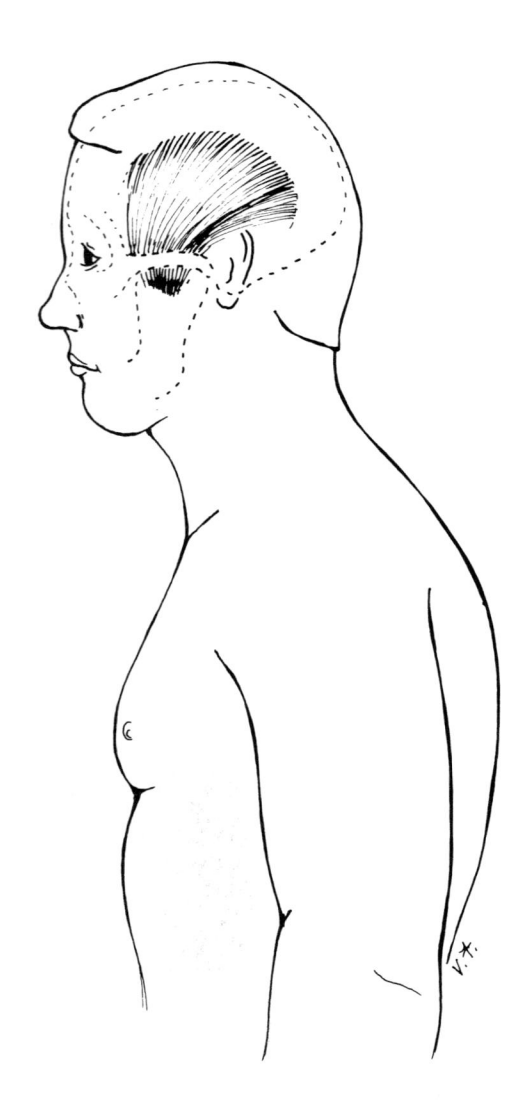

The temporalis, a fan-shaped muscle, is located on the lateral skull passing deep to the zygomatic arch to insert on the mandible. The muscle has a posterior relationship to the sphenoidal and temporal bones of the skull.

Origin: Temporal fossa and temporal fascia.
Insertion: Anterior coronoid process and anterior ramus of mandible.

HEAD AND NECK

- Middle temporal artery
- Superficial temporal artery
- Deep temporal artery
- Maxillary artery
- External carotid artery

Blood supply

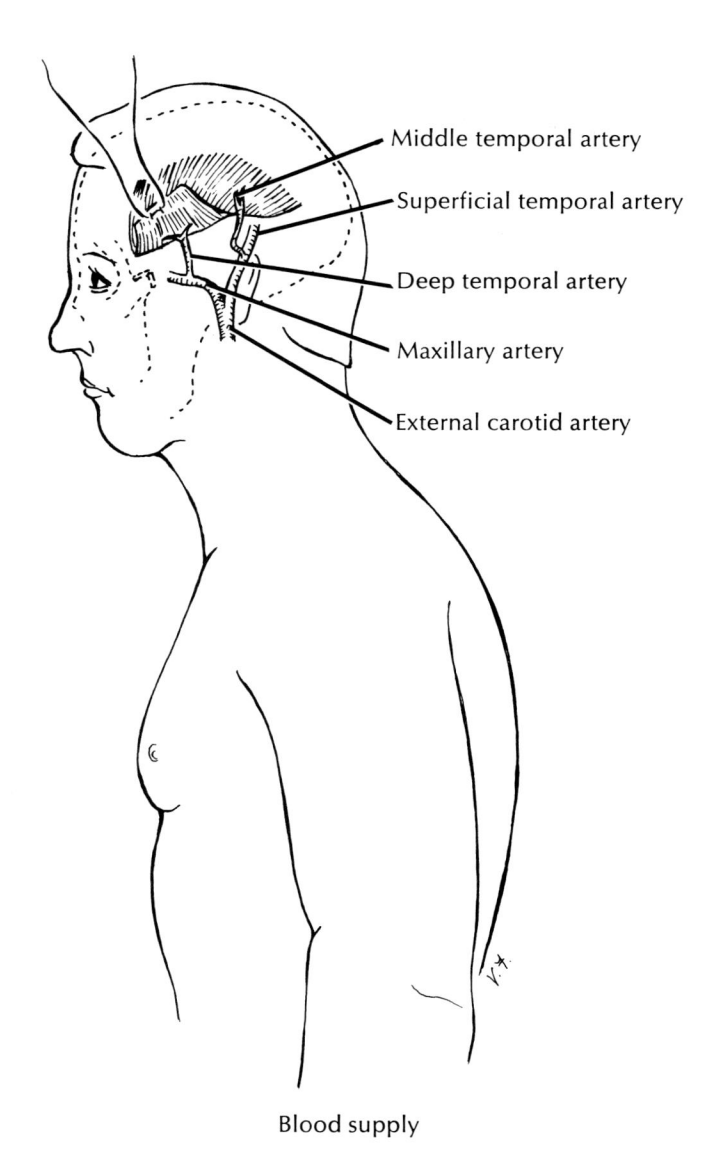

Nerve supply: Mandibular division of the trigeminal nerve. The deep temporal branches of the anterior trunk enter the posterosuperior muscle belly.

Function: This is an expendable muscle. The function of jaw elevation and mandibular retraction will be maintained by the remaining muscles of mastication.

HEAD AND NECK

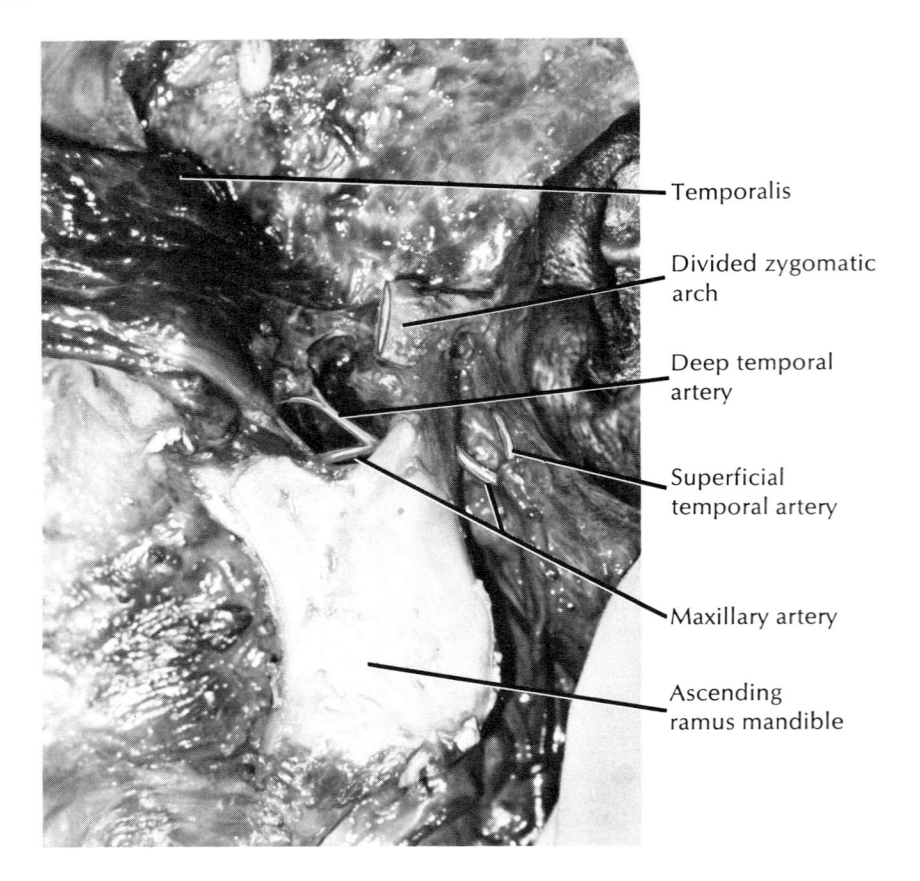

Temporalis

- Temporalis
- Divided zygomatic arch
- Deep temporal artery
- Superficial temporal artery
- Maxillary artery
- Ascending ramus mandible

Blood supply

This muscle receives its blood supply from the two terminal branches of the external carotid artery, the maxillary and superficial temporal arteries. The maxillary artery arises as an anterior terminal branch extending deep to the mandible in close relationship to the lateral pterygoid muscle. This artery then courses posterior to the insertion of the temporalis muscle to the ramus of the mandible. As the maxillary artery courses adjacent to the lateral pterygoid muscle, two pedicles course deep to the temporalis muscle and enter the muscle belly at the level of the zygomatic arch as its dominant vascular pedicles.

The superficial temporal artery courses in the preauricular region. Immediately superior to the zygomatic arch, the middle temporal artery branches from the superficial temporal artery. This artery penetrates the temporal fascia and is a posterior minor pedicle to the temporalis muscle.

HEAD AND NECK

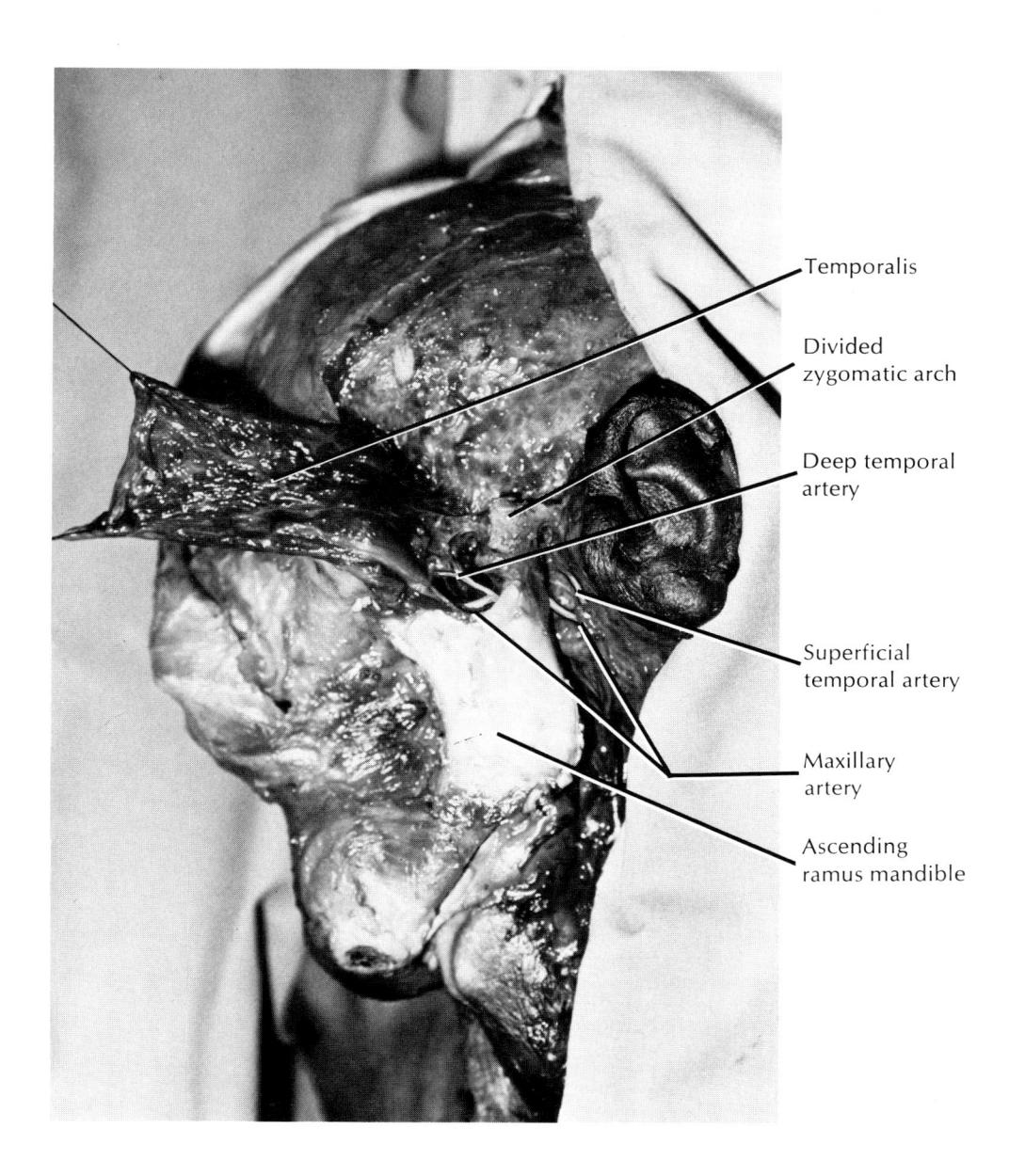

- Temporalis
- Divided zygomatic arch
- Deep temporal artery
- Superficial temporal artery
- Maxillary artery
- Ascending ramus mandible

491

HEAD AND NECK

Temporalis

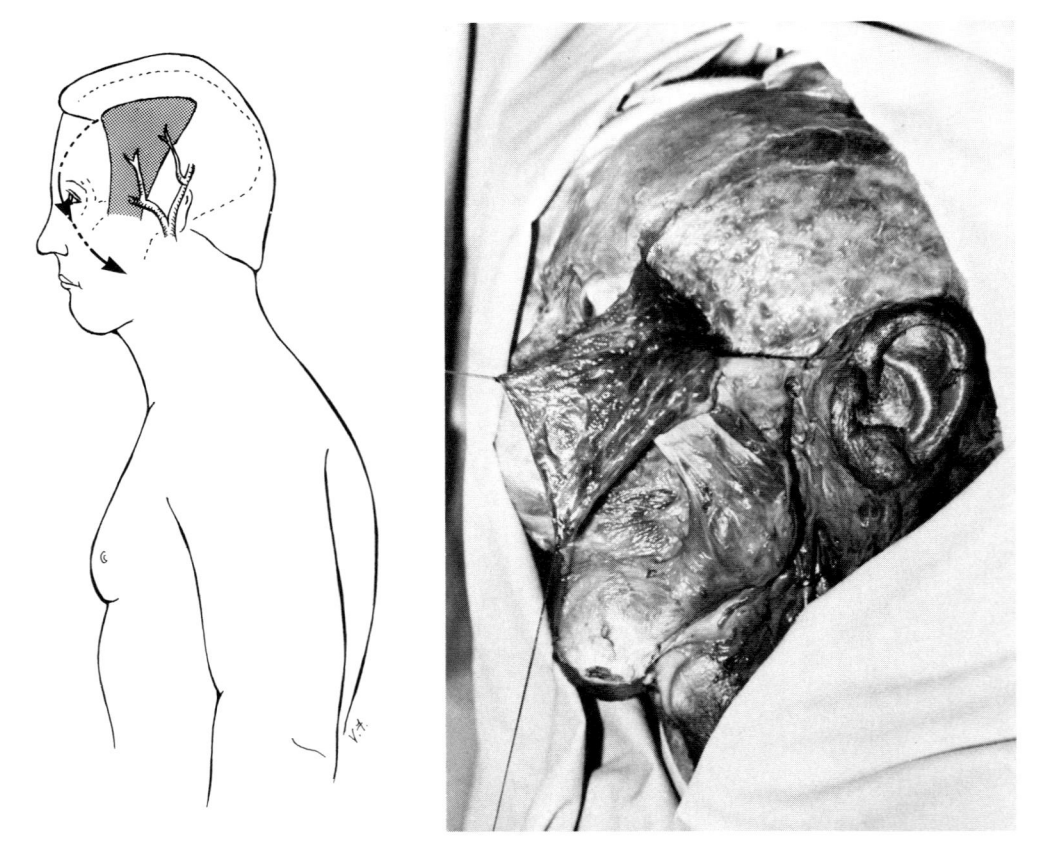

Anterior arc

Arc of rotation

This muscle has a point of rotation at the level of the zygomatic arch. The muscle will reach the anterior face or may be folded over the arch to cover the inferior lateral face. The muscle can be extended into the orbital cavity. With removal of the zygomatic arch, the arc of the flap can be extended. The flap also has a posterior arc to cover the mastoid region of the skull.

Elevation of flap

The temporalis muscle is exposed through a preauricular incision extending over the temporal scalp. The muscle is exposed at its insertion in the temporal fossa. It is elevated inferiorly after detaching the

HEAD AND NECK

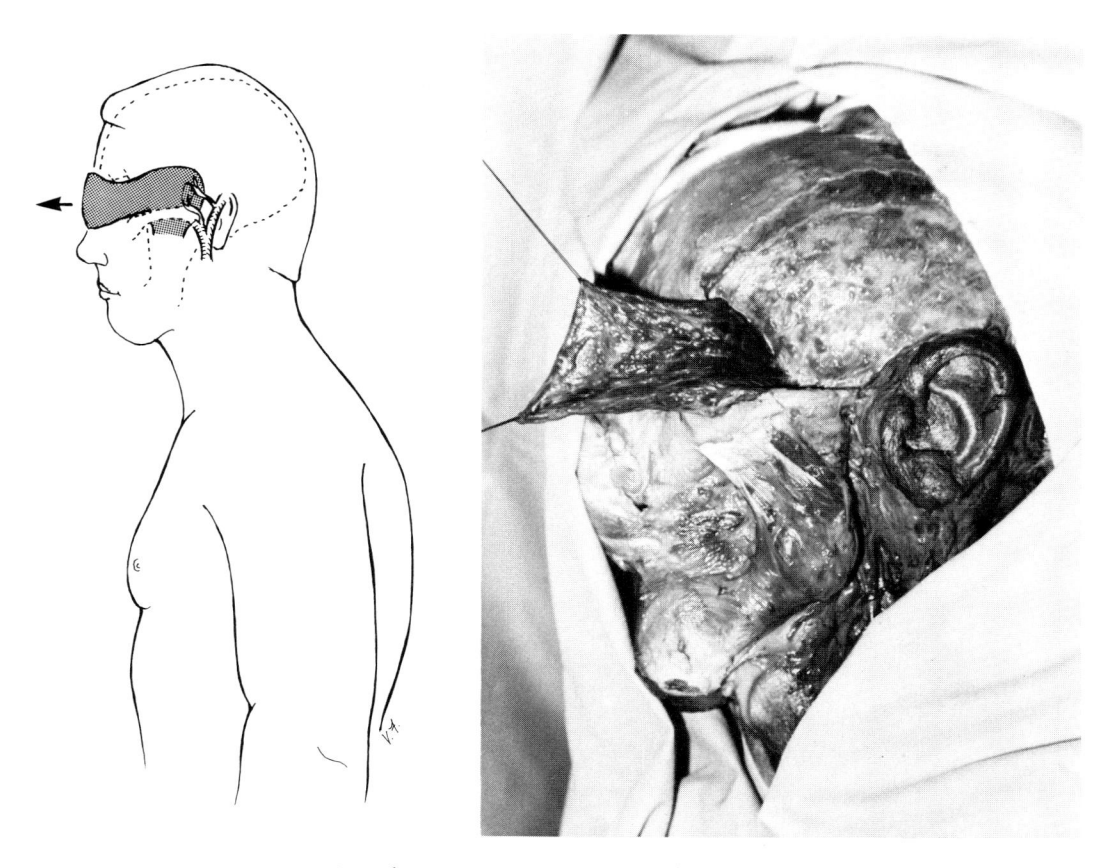

Arc of rotation (transposition into orbit)

origin. The muscle will fold to cover the zygomatic arch or exposed facial nerve grafts without the necessity of dividing the middle temporal arterial pedicle. However, if a wide anterior arc is required in facial reconstruction, the middle temporal artery is divided. The dominant vascular pedicles via the deep temporal branches of the maxillary artery are safely located beneath the zygomatic arch. The donor area is closed primarily.

The musculocutaneous unit is rarely indicated, since it includes hairbearing scalp.

Precautions

- The temporal and zygomatic branches of the facial nerve are in close relationship to the zygomatic arch.
- Following flap rotation, there is a depression in the temporal fossa of the skull. This is a noticeable deformity in the bald patient.

HEAD AND NECK

Temporalis

Facial reconstruction

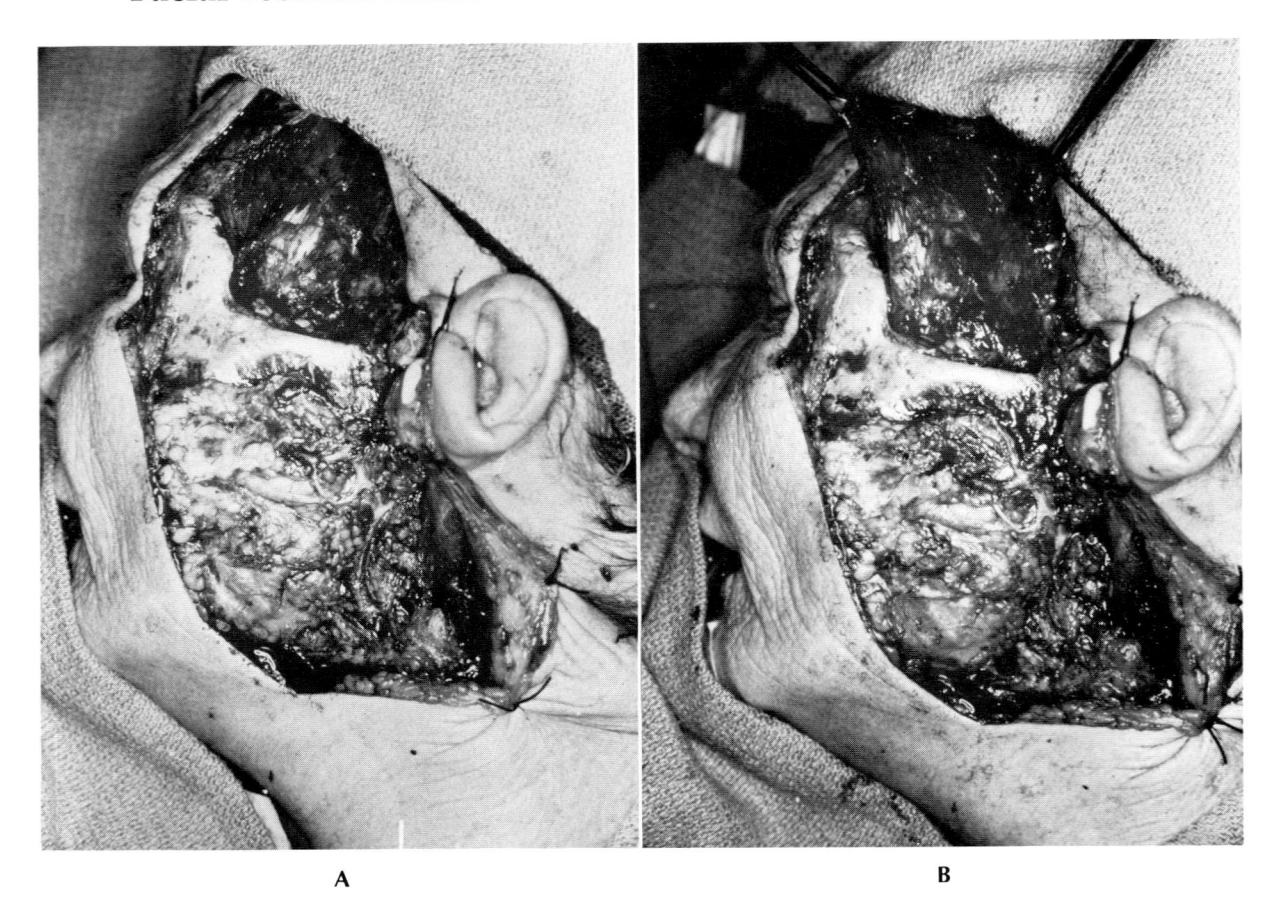

A Surgical defect following excision of recurrent squamous cell carcinoma. Exposed zygomatic arch and facial nerve branches.

B Detachment of temporalis muscle from origin in temporal fossa.

HEAD AND NECK

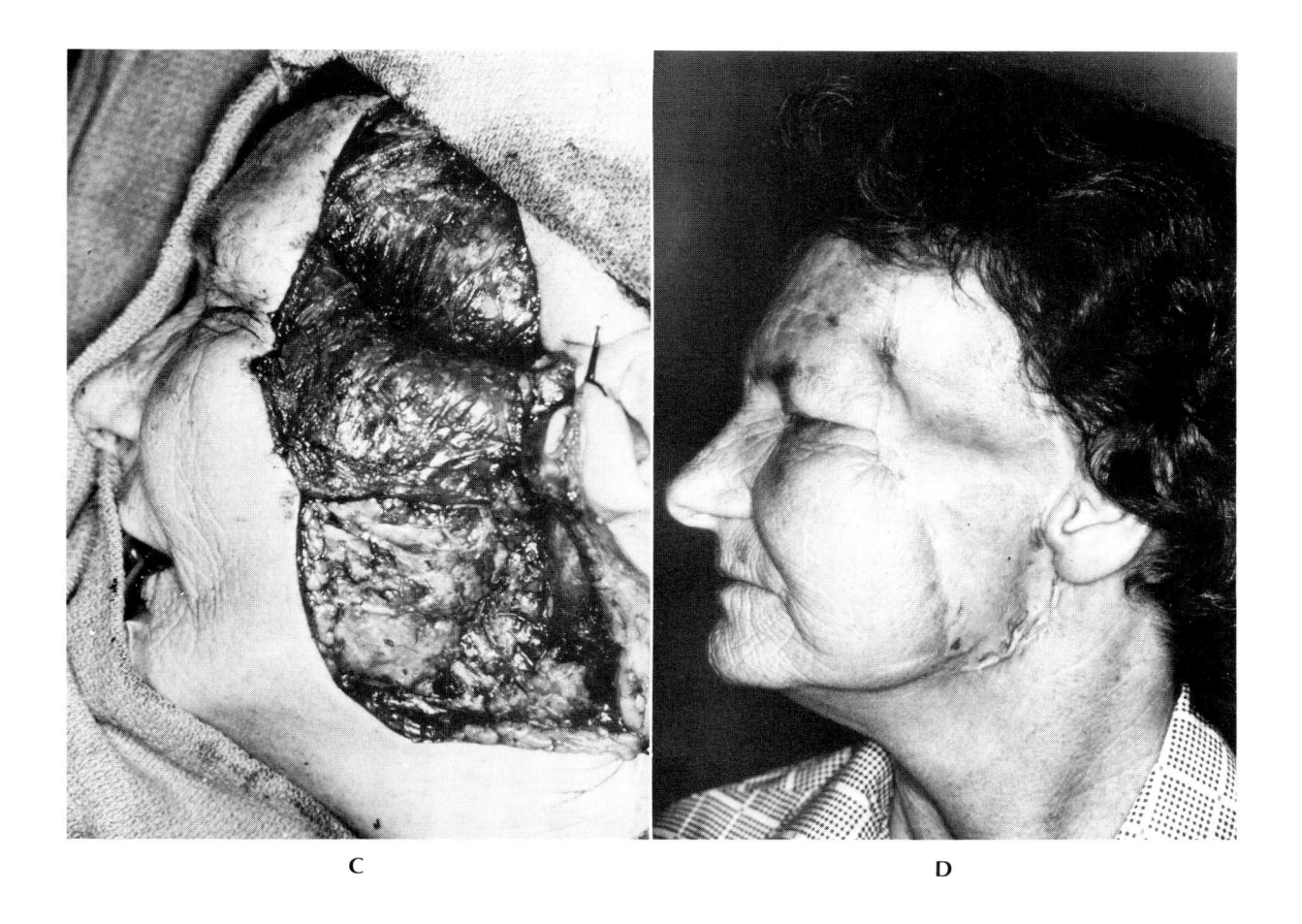

C   Muscle flap transposition over exposed facial bones and nerves.

D   Split-thickness skin graft over transposed muscle flap. Postoperative single-stage reconstruction.

APPENDIX

Applications of muscle and musculocutaneous flaps based on reconstructive problems

Based on our experience both in the laboratory and in reconstructive surgery, each muscle listed in this special index will provide soft-tissue coverage or reconstruction of the problem. Since each individual surgeon will have a preference as to the appropriate muscle or musculocutaneous flap for a specific problem, the muscles are placed in alphabetical order. This list is included primarily to assist the surgeon considering the use of muscle and musculocutaneous flaps in locating the discussion on appropriate muscles in the text.

Abdominal wall, coverage and reconstruction of

Gracilis
Latissimus dorsi
Rectus abdominis
Rectus femoris
Tensor fascia lata
Vastus lateralis

Acetabular fossa, reconstruction of

Vastus lateralis

Anal musculature, reconstruction of

Gracilis
Tensor fascia lata

Antecubital fossa

Brachioradialis
Flexor carpi ulnaris
Latissimus dorsi

Arm, coverage of

Biceps brachii
Brachioradialis
Flexor carpi ulnaris
Latissimus dorsi
Tensor fascia lata

Axilla, coverage of

Biceps brachii
Latissimus dorsi
Pectoralis major
Serratus anterior
Trapezius

Back, coverage of

Latissimus dorsi
Tensor fascia lata
Tapezius

Breast implant, coverage of

Latissimus dorsi
Pectoralis major
Serratus anterior

APPENDIX

Breast, reconstruction of
    Latissimus dorsi
    Pectoralis major
    Rectus abdominis
    Serratus anterior
    Trapezius

Buttocks, coverage of
    Biceps femoris
    Gluteus maximus
    Gracilis
    Semimembranosus
    Semitendinosus
    Tensor fascia lata
    Vastus lateralis

Chest wall, reconstruction of
    Latissimus dorsi
    Pectoralis major
    Rectus abdominis
    Serratus anterior
    Trapezius

Distal third of leg, coverage of (proximal portion of distal third of leg)
    Extensor digitorum longus
    Extensor hallucis longus
    Flexor digitorum longus
    Flexor hallucis longus
    Soleus (distally based)

Distant coverage (microsurgical free transfer)
    Brachioradialis
    Extensor digitorum longus
    Extensor hallucis longus
    Flexor digitorum brevis
    Gastrocnemius
    Gracilis
    Latissimus dorsi
    Pectoralis major
    Rectus abdominis
    Rectus femoris
    Serratus anterior
    Tensor fascia lata
    Trapezius

Face, coverage of (including oral cavity)
    Latissimus dorsi
    Pectoralis major
    Sternocleidomastoid
    Temporalis
    Trapezius

Femoral vessels in groin, coverage of
    Gracilis
    Rectus femoris
    Sartorius
    Tensor fascia lata

First and second metacarpal, coverage of
    First dorsal interosseous

Groin, coverage of
    Gracilis
    Rectus abdominis
    Rectus femoris
    Sartorius
    Tensor fascia lata
    Vastus lateralis

Heel, coverage of
    Abductor digiti minimi
    Abductor hallucis
    Flexor digitorum brevis

Inguinal hernia, reconstruction of
    Tensor fascia lata

Ischium, coverage of
    Biceps femoris
    Gluteus maximus
    Gracilis
    Rectus femoris
    Semimembranosus
    Semitendinosus
    Tensor fascia lata
    Vastus lateralis

APPENDIX

Knee, coverage of

    Gastrocnemius
    Gracilis (distally based with delay)
    Sartorius
    Semimembranosus (distally based)

Lateral ankle, coverage of

    Abductor digiti minimi

Medial ankle, coverage of

    Abductor hallucis

Middle third of leg, coverage of

    Peroneus brevis
    Peroneus longus
    Soleus
    Tibialis anterior

Myelomeningocele, coverage of

    Gluteus maximus
    Latissimus dorsi
    Trapezius

Neck, coverage of

    Pectoralis major
    Serratus anterior
    Sternocleidomastoid
    Trapezius

Penis, reconstruction of

    Gracilis

Perineum, coverage of

    Biceps femoris
    Gracilis
    Rectus femoris
    Semimembranosus
    Semitendinosus
    Tensor fascia lata

Posterior scalp and neck, coverage of

    Latissimus dorsi
    Sternocleidomastoid
    Temporalis
    Trapezius

Sacrum, coverage of

    Gluteus maximus
    Semimembranosus
    Tensor fascia lata

Shoulder, coverage of

    Latissimus dorsi
    Pectoralis
    Trapezius

Sternum (see thorax)

Thorax, coverage of

    Latissimus dorsi
    Pectoralis major
    Rectus abdominis
    Serratus anterior
    Trapezius

Trochanter, coverage of

    Biceps femoris
    Gluteus maximus
    Rectus femoris
    Semimembranosus
    Tensor fascia lata
    Vastus lateralis

Upper arm, coverage of

    Biceps brachii
    Latissimus dorsi
    Trapezius

Upper third of leg, coverage of

    Gastrocnemius

Vagina, reconstruction of

    Gracilis
    Tensor fascia lata

Vulva, reconstruction of

    Gracilis
    Tensor fascia lata

Wrist, coverage of

    Abductor digiti minimi
    Abductor pollicis brevis
    Rectus abdominis
    Tensor fascia lata

Suggested readings

Anatomy

Brash, J.: Neurovascular hila of limb muscles, London, 1955, E. & S. Livingstone.

Quiring, D., and Warfel, J.: The extremities, Philadelphia, 1958, Lea & Febiger.

Warwick, R., and Williams, P. L., editors: Gray's anatomy, ed. 35, Philadelphia, 1973, W. B. Saunders Co.

Clinical applications

Bakamjian, V.: A technique for primary reconstruction of the palate after radical maxillectomy for cancer, Plast. Reconstr. Surg. 31:103, 1963.

Baker, D. C., Barton, E. F., and Converse, J. M.: A combined biceps and semitendinosus muscle flap in the repair of ischial sores, Br. J. Plast. Surg. 31:26, 1978.

Barford, B., and Pers, M.: Gastrocnemius-plasty for primary closure of compound injuries of the knee, J. Bone Joint Surg. (Br.) 52:124, 1970.

Bhagwat, B. M., Pearl, R. M., and Laub, D. R.: Uses of the rectus femoris myocutaneous flap, Plast. Reconstr. Surg. 62:698, 1978.

Bors, E., and Carmarr, A. E.: Ischial decubitus ulcer, Surgery, 24:680, 1948.

Bostwick, J., III: Reconstruction of the heel pad by muscle transposition and split skin graft, Surg. Gynecol. Obstet. 143:972, 1976.

Bostwick, J., III, Nahai, F., Wallace, J. G., and Vasconez, L. O.: Sixty latissimus dorsi flaps, Plast. Reconstr. Surg. 63:31, 1979.

Bostwick, J., III, Vasconez, L. O., and Jurkiewicz, M. J.: Breast reconstruction after a radical mastectomy, Plast. Reconstr. Surg. 61:682, 1978.

Brantigan, O. C.: Evaluation of Hutchins' modification of radical mastectomy for cancer of the breast, Am. Surg. 40:86, 1979.

Brown, R. G., Fleming, W. H., and Jurkiewicz, M. J.: An island flap of the pectoralis major muscle, Br. J. Plast. Surg. 30:161, 1977.

Brown, R. G., Vasconez, L. O., and Jurkiewicz, M. J.: Transverse abdominal flap and the deep epigastric arcade, Plast. Reconstr. Surg. 55:416, 1975.

Campbell, D. A.: Reconstruction of the anterior thoracic wall, J. Thorac. Surg. 19:456, 1950.

Desprez, J. D., Kiehn, C. L., Eckstein, W.: Closure of large meningomyelocele defects by composite skin-muscle flaps, Plast. Reconstr. Surg. 47:234, 1971.

SUGGESTED READINGS

D'Este, S.: La technique de l'amputation de la mamelle pour carcinome mammaire, Rev. Chir. (Paris) **45**:164, 1912.

Feldman, J. J., Cohen, B. E., and May, J. W.: The medial gastrocnemius myocutaneous flap, Plast. Reconstr. Surg. **61**:531, 1978.

Franklin, E. W., Bostwick, J., III, Burrell, M. O., and Powell, J. L.: Reconstructive techniques in radical pelvic surgery. Am. J. Obstet. Gynecol. **129**:285, 1977.

Ger, R.: The coverage of vascular repairs by muscle transposition, J. Trauma **16**:974, 1976.

Ger, R.: The management of chronic ulcers of the foot by muscle transposition and free skin grafting, Br. J. Plast. Surg. **29**:199, 1976.

Ger, R.: The surgical management of ulcers of the heel, Surg. Gynecol. Obstet. **140**:909, 1975.

Ger, R.: The management of pretibial skin loss, Surgery **63**:757, 1968.

Ger, R.: The operative treatment of advanced stasis ulcer: A preliminary communication, Am. J. Surg. **111**:659, 1966.

Harii, K., Ohmori, K., and Sekiguchi, J.: The free musculocutaneous flap, Plast. Reconstr. Surg. **57**:294, 1976.

Hester, T. R., Hill, H. L., and Jurkiewicz, M. J.: One-stage reconstruction of the penis, Br. J. Plast. Surg. **31**:279, 1978.

Hill, H. L., Brown, R. G., and Jurkiewicz, M. J.: The transverse lumbosacral back flap, Plast. Reconstr. Surg. **62**:177, 1978.

Hill, H. L., Nahai, F., and Vasconez, L. O.: The tensor fascia lata myocutaneous free flap, Plast. Reconstr. Surg. **61**:517, 1978.

Mathes, S. J., and Bostwick, J., III: A rectus abdominis myocutaneous flap to reconstruct abdominal wall defects, Br. J. Plast. Surg. **30**:282, 1977.

Mathes, S. J., McCraw, J. B., and Vasconez, L. O.: Muscle transposition flaps for coverage of lower extremity defects: anatomic considerations, Surg. Clin. North Am. **54**:1337, 1974.

Mathes, S. J., Nahai, F., and Vasconez, L. O.: Myocutaneous free flap transfer: anatomical and experimental considerations, Plast. Reconstr. Surg. **62**:162, 1978.

Mathes, S. J., and Vasconez, L. O.: The cervicohumeral flap, Plast. Reconstr. Surg. **61**:7, 1978.

Mathes, S. J., Vasconez, L. O., and Jurkiewicz, M. J.: Extensions and further applications of muscle flap transposition, Plast. Reconstr. Surg. **60**:6, 1977.

McCraw, J. B., and Dibbell, D. G.: Experimental definition of independent myocutaneous vascular territories, Plast. Reconstr. Surg. **60**:212, 1977.

McCraw, J. B., Dibbell, D. G., and Carraway, J. H.: Clinical definition of independent myocutaneous vascular territories, Plast Reconstr. Surg. **60**:341, 1977.

McCraw, J. B., Fishman, J. M., and Sharzer, L. A.: The versatile gastrocnemius myocutaneous flap, Plast. Reconstr. Surg. **62**:15, 1978.

McCraw, J. B., Massey, F. M., Shanklin, K. D., and Horton, C. E.: Vaginal reconstruction with gracilis myocutaneous flaps, Plast. Reconstr. Surg. **58**:176, 1976.

SUGGESTED READINGS

McCraw, J. B., Penix, J. O., and Baker, J. W.: Repair of major defects of the chest wall and spine with latissimus dorsi myocutaneous flap, Plast. Reconstr. Surg. 62:197, 1978.

McHugh, M., and Prendiville, J. B.: Muscle flaps in the repair of skin defects over the exposed tibia, Br. J. Plast. Surg. 28:205, 1975.

Milward, T. M., Scott, W. G., and Kleinert, H. E.: The abductor digiti minimus muscle flap, Hand 9:82, 1977.

Minami, R. T., Hentz, V. R., and Vistnes, L. M.: Use of vastus lateralis muscle flap for repair of trochanteric pressure sores, Plast. Reconstr. Surg. 60:364, 1977.

Minami, R. T., Mills, R., and Pardee, R.: Gluteus maximus myocutaneous flaps for repair of pressure sores, Plast. Reconstr. Surg. 60:242, 1977.

Nahai, F., Brown, R. G., and Vasconez, L. O.: Blood supply of the abdominal wall as related to planning abdominal incisions, Am. Surg. 42:691, 1976.

Nahai, F., Silverton, J. S., Hill, H. L., and Vasconez, L. O.: The tensor fascia lata musculocutaneous flap, Ann. Plast. Surg. 1:372, 1978.

Olivari, N.: The latissimus flap, Br. J. Plast. Surg. 29:126, 1976.

Orticochea, M.: The musculo-cutaneous flap method: an immediate and heroic substitute for the method of delay, Br. J. Plast. Surg. 25:106, 1972.

Orticochea, M.: A new method of total reconstruction of the penis, Br. J. Plast. Surg. 25:347, 1972.

Owens, N. A.: Compound neck pedicle designed for the repair of massive facial defects: formation, development and application, Plast. Reconstr. Surg. 15:369, 1955.

Pers, M., and Medgyesi, S.: Pedicle muscle flaps and their application in the surgery of repair, Br. J. Plast. Surg. 26:313, 1973.

Pickrell, K., Broadbent, T., Masters, F. W., and Metzger, J. T.: Construction of a rectal sphincter and restoration of anal continence by transplanting the gracilis muscle, Ann. Surg. 135:853, 1952.

Reese, A. G., and Jones, I. S.: Exenteration of the orbit and repair by transplantation of the temporalis muscle, Am. J. Ophthalmol. 51:217, 1961.

Schneider, W. J., Hill, H. L., and Brown, R. G.: Latissimus dorsi myocutaneous flap for breast reconstruction, Br. J. Plast. Surg. 30:277, 1977.

Schottstaedt, E. R., Larson, L. J., and Bost, F. C.: Complete muscle transposition, J. Bone Joint Surg. (Am.) 37:897, 1955.

Silverton, J. S., Nahai, F., and Jurkiewicz, M. J.: The latissimus dorsi myocutaneous flap to replace a defect on the upper arm, Br. J. Plast. Surg. 31:29, 1978.

Stallings, J. O., Delgado, J. P., and Converse, J. M.: Turnover island flap of gluteus maximus muscle for the repair of sacral decubitus ulcer, Plast. Reconstr. Surg. 54:52, 1974.

Townsend, P. L. G.: An inferiorly based soleus muscle flap, Br. J. Plast. Surg. 31:210, 1978.

Vasconez, L. O., Bostwick, J., III, and McCraw, J.: Coverage of exposed bone by muscle transposition and skin grafting, Plast. Reconstr. Surg. 53:526, 1974.

SUGGESTED READINGS

Vasconez, L. O., Schneider, W. J., and Jurkiewicz, M. J.: Pressure sores, Curr. Probl. Surg. 24:1-62, 1977.

Wangensteen, O. H.: Repair of recurrent and difficult hernias and other large defects of the abdominal wall employing the iliotibial tract of fascia lata as a pedicle flap, Surg. Gynecol. Obstet. 59:766, 1934.

Webster, J. P.: Temporalis muscle transplants for defects following orbital exenteration. Transactions of the International Society of Plastic Surgeons, First Congress, 1955, Baltimore, 1957, The Williams & Wilkins Co., p. 291.

Woods, J. E., Irons, G. B., and Masson, J. K.: Use of musculocutaneous and omental flaps to reconstruct difficult defects, Plast. Reconstr. Surg. 59:191, 1977.

Index

A

Abdominal wall
  gracilis and, 13-31; *see also* Gracilis
  latissimus dorsi and, 369-391; *see also* Latissimus dorsi
  rectus abdominis and, 347-361; *see also* Rectus abdominis
    case studies of, 358-359, 360
  rectus femoris and, 41-50
  tensor fascia lata and, 63-85; *see also* Tensor fascia lata
    case study of, 76
  vastus lateralis and, 51-62
Abductor digiti minimi
  in foot, 265, 291-299
    anatomy of, 292-295
    arc of rotation and, 296-297
    blood supply of, 293-295
    elevation of flap and, 298
    function of, 293
    nerve supply to, 293
    origin and insertion of, 292-293
    precautions and, 299
  in hand, 465-469
Abductor hallucis, 265, 269-277
  anatomy of, 270-273
  arc of rotation and, 274-275
  blood supply to, 271-273
  elevation of flap and, 276
  function of, 271
  nerve supply of, 271
  origin and insertion of, 270
  precautions and, 277
Abductor pollicis brevis, 459-463
Accessory nerves, spinal
  sternocleidomastoid and, 477
  trapezius and, 395, 406
Acetabular fossa, 51-62
Achilles tendon
  abductor digiti minimi and, 296, 298
  flexor digitorum brevis and, 289

Achilles tendon—cont'd
  gastrocnemius and, 142, 144
  soleus and, 158, 166, 174-175
Adductor longus, 16, 17
Adductor magnus, 17
Anal musculature
  gracilis and, 13-31; *see also* Gracilis
  tensor fascia lata and, 63-85; *see also* Tensor fascia lata
Ankle
  lateral, abductor digiti minimi and, 264, 265, 291-299
  medial, abductor hallucis and, 264, 265, 269-277
Antecubital fossa
  brachioradialis and, 433-439
  flexor carpi ulnaris and, 441-447
  latissimus dorsi and, 369-391; *see also* Latissimus dorsi
Anterior chest; *see* Chest wall
Anterior thigh; *see* Thigh, anterior
Anterior tibial artery; *see* Tibial artery, anterior
Anterior tibial nerve, 209, 243; *see also* Tibial nerve
Anterior trunk; *see* Trunk, anterior
Arc of rotation
  of abductor digiti minimi
    in foot, 296-297
    in hand, 468-469
  of abductor hallucis, 274-275
  of abductor pollicis brevis, 462
  of biceps brachii, 430-431
  of biceps femoris, 110-111
  of brachioradialis, 438-439
  of dorsal interosseous, 456
  of extensor digitorum brevis, 306
  of extensor digitorum longus, 222-223
  of extensor hallucis longus, 232-233
  of flexor carpi ulnaris, 446-447
  of flexor digitorum brevis, 284-285
  of flexor digitorum longus, 137, 184-185

INDEX

Arc of rotation—cont'd
  of flexor hallucis longus, 137, 194-195
  of gastrocnemius, 137, 146-148
  of gluteus maximus, 96-97
  of gracilis, 18-21
  of latissimus dorsi, 376-381
  motor nerve and, 4
  of pectoralis major, 324-325
  of peroneus brevis, 256-257
  of peroneus longus, 246-247
  of rectus abdominis, 354-355
  of rectus femoris, 46-47
  of sartorius, 36-37
  of semimembranosus, 128-129
  of semitendinosus, 120
  of serratus anterior, 342-343
  of soleus, 137, 162-165
  of sternocleidomastoid, 480-481
  of temporalis, 492, 493
  of tensor fascia lata, 68-69
  of tibialis anterior, 212-213
  of trapezius, 400-403
  of vastus lateralis, 56-57
Arm; *see also* Extremities, upper
  biceps brachii and, 425-432
  brachioradialis and, 433-439
  flexor carpi ulnaris and, 441-447
  latissimus dorsi and, 369-391; *see also* Latissimus dorsi
  tensor fascia lata and, 63-85; *see also* Tensor fascia lata
  trapezius and, 393-419; *see also* Trapezius
Arteries; *see also* Blood supply
  auricular, 478
  axillary, 312-313
    latissimus dorsi and, 372, 384
    pectoralis major and, 320
    posterior trunk and, 366, 367
    serratus anterior and, 340
  brachial, 424
    biceps brachii and, 426, 428, 432
    flexor carpi ulnaris and, 444
  carotid, 473, 490
  cervical
    superficial, 398, 404
    transverse, 366, 367, 395, 398, 399, 400, 404, 405
  dorsalis pedis, 266-267
    extensor digitorum brevis and, 304, 306
    extensor hallucis longus and, 228
  epigastric
    inferior, 312, 313, 315, 352-353, 357
    superior, 313, 314, 315, 351, 356
  femoral; *see* Femoral arteries

Arteries—cont'd
  gluteal, 90, 94
  hypogastric, 90, 94
  iliac, external, 315
    rectus abdominis and, 350, 352
  intercostal, 323, 375, 399
  lumbar, 375
  mammary, 312, 314, 315
    pectoralis major and, 319, 322-323, 328
    rectus abdominis and, 350, 351
  maxillary, 473, 490, 493
  musculophrenic, 314, 351
  occipital, 473
    sternocleidomastoid and, 477, 482
    trapezius and, 406
  palmar, 451, 455
  perforating; *see* Perforating vessels
  peroneal, 138, 139
    flexor hallucis longus and, 192, 194, 196
    peroneus brevis and, 238, 254
    peroneus longus and, 238, 244
    soleus and, 159, 160, 161, 162
  plantar; *see* Plantar arteries
  popliteal; *see* Popliteal artery
  princeps pollicis, 451
  profunda femoris; *see* Profunda femoris artery
  radial; *see* Radial artery
  scapular
    circumflex, 313, 314, 399, 404
    posterior, 399, 404
  subclavian; *see* Subclavian artery
  subscapular
    anterior trunk and, 312-313, 314
    latissimus dorsi and, 371, 372, 384
    posterior trunk and, 366, 367
  tarsal, lateral, 267, 298, 304, 306
  temporal, 473, 490, 493
  thoracic
    lateral, 312, 313-314
    long, 340, 342, 344
  thoracoacromial, 312, 313, 314
    pectoralis major and, 319-320
  thoracodorsal, 313, 314
    latissimus dorsi and, 372, 384
    serratus anterior and, 340, 342
  tibial; *see* Tibial artery
  ulnar; *see* Ulnar artery
Auricular artery, 478
Auricular nerve, greater, 483
Axilla
  biceps brachii and, 425-432
  latissimus dorsi and, 369-391; *see also* Latissimus dorsi

INDEX

Axilla—cont'd
  pectoralis major and, 317-335; *see also* Pectoralis major
  serratus anterior and, 337-345
  trapezius and, 393-419; *see also* Trapezius
Axillary artery, 312-313
  latissimus dorsi and, 372, 384
  pectoralis major and, 320
  posterior trunk and, 366, 367
  serratus anterior and, 340
Axillary vein, 328

B

Back
  latissimus dorsi and, 369-391; *see also* Latissimus dorsi
  tensor fascia lata and, 63-85; *see also* Tensor fascia lata
  trapezius and, 393-419; *see also* Trapezius
Biceps brachii, 422, 423, 425-432
  anatomy of, 426-429
  arc of rotation and, 430-431
  blood supply to, 427-429
  elevation of flap and, 432
  function of, 427
  nerve supply to, 427
  origin and insertion of, 426
  precautions and, 432
Biceps femoris, 105-114
  anatomy of, 106-109
  arc of rotation and, 110-111
  blood supply to, 107-109
  elevation of flap and, 112-113
  function of, 107
  nerve supply to, 107
  origin and insertion of, 106
  precautions and, 114
Blood supply, 3
  to abductor digiti minimi
    in foot, 293-295
    in hand, 466-467
  to abductor hallucis, 271-273
  to abductor pollicis brevis, 460-461
  to arm, 424
  to biceps brachii, 427-429
  to biceps femoris, 107-109
  to brachioradialis, 435-437
  to dorsal interosseous, 454-455
  to extensor digitorum brevis, 303-305
  to extensor digitorum longus, 219-221
  to extensor hallucis longus, 229-231
  to flexor carpi ulnaris, 443-445
  to flexor digitorum brevis, 281-283
  to flexor digitorum longus, 181-183
  to flexor hallucis longus, 191-193

Blood supply—cont'd
  to foot, 266-267
  to gastrocnemius, 143-145
  to gluteus maximus, 93-95
  to gracilis, 16-17
  to hand, 451
  to head and neck, 473
  to latissimus dorsi, 371-375
  leg and
    anterior lateral muscles and, 204-205
    medial, 138-139
    posterior lateral muscles and, 238
  to pectoralis major, 319-323
  to peroneus brevis, 253-255
  to peroneus longus, 243-245
  to rectus abdominis, 349-353
  to rectus femoris, 44-45
  to sartorius, 35
  to semimembranosus, 125-127
  to semitendinosus, 117-119
  to serratus anterior, 339-341
  to soleus, 159-161
  to sternocleidomastoid, 477-479
  to temporalis, 489-491
  to tensor fascia lata, 65-67
  thigh and
    anterior, 10-11
    posterior, 90
  to tibialis anterior, 209-211
  to trapezius, 396-399
  trunk and
    anterior, 312-315
    posterior, 366-367
  to upper extremity, 424
  to vastus lateralis, 53-55
Bones
  flap transposition or free transfer and, 5
  tensor fascia lata flaps and, 72
Brachial artery, 424
  biceps brachii and, 426, 428, 432
  flexor carpi ulnaris and, 444
Brachial plexus, 319, 371
Brachioradialis, 422, 423, 433-439
  anatomy of, 434-437
  arc of rotation and, 438-439
  blood supply to, 435-437
  elevation of flap and, 439
  function of, 435
  nerve supply to, 435
  origin and insertion of, 434
  precaution for, 439
Breast
  latissimus dorsi and, 364, 369-391; *see also* Latissimus dorsi
  case studies of, 386-391

INDEX

Breast—cont'd
  pectoralis major and, 317-335; *see also* Pectoralis major
  rectus abdominis and, 347-361; *see also* Rectus abdominis
    case study of, 361
  serratus anterior and, 337-345
  trapezius and, 364, 393-419; *see also* Trapezius
  tumor of, 361
Breast implant; *see also* Breast
  latissimus dorsi and, 369-391; *see also* Latissimus dorsi
  pectoralis major and, 317-335; *see also* Pectoralis major
  serratus anterior and, 337-345
Buttocks
  biceps femoris and, 105-114
  gracilis and, 13-31; *see also* Gracilis
  semimembranosus and, 123-132
  semitendinosus and, 114-122
  tensor fascia lata and, 63-85; *see also* Tensor fascia lata
  vastus lateralis and, 51-62
Bypass graft, femero-femoral, 78

C

Calcaneus, 284, 286, 288
Carcinoma
  of face, 494
  of mouth, 334
  of neck, 409
Carotid arteries, 473, 490
Cervical arteries
  superficial, 398, 404
  transverse, 366, 367
    trapezius and, 395, 398, 399, 400, 404, 405
Cervical humeral flap, 398; *see also* Trapezius
Cervical nerves, 395, 477
Cervical vertebrae, 378
Cheek deformity, 332
Chest wall
  latissimus dorsi and, 369-391; *see also* Latissimus dorsi
  pectoralis major and, 317-335; *see also* Pectoralis major
  rectus abdominis and, 347-361; *see also* Rectus abdominis
    case study of, 361
  serratus anterior and, 337-345
  trapezius and, 393-419; *see also* Trapezius
Chin
  pectoralis major and, 334-335
  trapezius and, 410-413, 418-419
Clavicle segment with flap, 482

Cross leg flap, 141-156; *see also* Gastrocnemius
Cutaneous nerves
  antebrachial, 435, 443
  anterior, 43
  femoral, 65, 73
  posterior, 107

D

Deformities
  cheek, 332
  drooping shoulder, 395
  of skull, temporalis flaps and, 493
  winged scapula, 339, 344
Dialysis access shunts, 430, 432
Distal third of leg
  extensor digitorum longus and, 217-226
  extensor hallucis longus and, 227-236
  flexor digitorum longus and, 179-188
  flexor hallucis longus and, 189-198
  soleus and, 157-178; *see also* Soleus
Distant coverage
  brachioradialis and, 433-439
  extensor digitorum longus and, 217-226
  extensor hallucis longus and, 227-236
  flexor digitorum brevis and, 279-289
  gastrocnemius and, 141-156; *see also* Gastrocnemius
  gracilis and, 13-31; *see also* Gracilis
  latissimus dorsi and, 369-391; *see also* Latissimus dorsi
  pectoralis major and, 317-335; *see also* Pectoralis major
  rectus abdominis and, 347-361; *see also* Rectus abdominis
  rectus femoris and, 41-50
  serratus anterior and, 337-345
  tensor fascia lata and, 63-85; *see also* Tensor fascia lata
  trapezius and, 393-419; *see also* Trapezius
Dorsal interosseous, first, 453-457
Dorsalis pedis artery, 266-267
  extensor digitorum brevis and, 304, 306
  extensor hallucis longus and, 228
Drooping shoulder deformity, 395

E

Elbow flexion, 317-335; *see also* Pectoralis major
Elevation of flap, 5
  of abductor digiti minimi
    in foot, 298
    in hand, 469
  of abductor hallucis, 276
  of abductor pollicis brevis, 463
  of biceps brachii, 432

507

INDEX

Elevation of flap—cont'd
  of biceps femoris, 112-113
  of brachioradialis, 439
  of dorsal interosseous, 457
  of extensor digitorum brevis, 306
  of extensor digitorum longus, 224
  of extensor hallucis longus, 234
  of flexor carpi ulnaris, 447
  of flexor digitorum brevis, 286
  of flexor digitorum longus, 186-187
  of flexor hallucis longus, 196-197
  of gastrocnemius, 149-150
  of gluteus maximus, 98
  of gracilis, 22-23
  of latissimus dorsi, 382-385
  of pectoralis major, 328
  of peroneus brevis, 258
  of peroneus longus, 248
  of rectus abdominis, 356-357
  of rectus femoris, 48
  of sartorius, 38
  of semimembranosus, 130
  of semitendinosus, 121
  of serratus anterior, 344
  of soleus, 166-167
  of sternocleidomastoid, 482
  of temporalis, 492-493
  of tensor fascia lata, 72
  of tibialis anterior, 214-215
  of trapezius, 404-407
  of vastus lateralis, 58
Epigastric arteries
  inferior, 312, 313, 315
    rectus abdominis and, 352-353, 357
  superior, 313, 314, 315
    rectus abdominis and, 351, 356
Epimysium, 2
Extensor digitorum brevis, 301-307
  anatomy of, 302-305
  arc of rotation and, 306
  blood supply to, 303-305
  elevation of flap and, 306, 307
  function of, 303
  nerve supply to, 4, 303
  origin and insertion of, 302
Extensor digitorum longus, 202-203, 217-226
  anatomy of, 218-221
  arc of rotation and, 222-223
  blood supply of, 219-221
  elevation of flap and, 224
  function of, 219
  nerve supply to, 219
  origin and insertion of, 218
  precautions and, 225

Extensor hallucis longus, 202-203, 227-236
  anatomy of, 228-231
  arc of rotation and, 232-233
  blood supply to, 229-231
  elevation of flap and, 234, 235
  function of, 229
  nerve supply to, 229
  origin and insertion of, 228
  precautions and, 234
Extremities
  lower; *see* Foot; Leg; Thigh
  upper, 421-447; *see also* Hand
    biceps brachii in, 422, 423, 425-432
    blood supply to, 424
    brachioradialis and, 422, 423, 433-439
    choice of flaps for, 422
    flexor carpi ulnaris and, 422, 423, 441-447
    latissimus dorsi and, 369-391; *see also* Latissimus dorsi
    posterior trunk muscles and, 364
    trapezius and, 393-419; *see also* Trapezius

F
Face
  carcinoma of, 494
  choice of flap for, 472
  latissimus dorsi and, 369-391; *see also* Latissimus dorsi
  pectoralis major and, 317-335; *see also* Pectoralis major
  sternocleidomastoid and, 475-485
    case study of, 484-485
  temporalis and, 487-495
    case study of, 494-495
  trapezius and, 393-419; *see also* Trapezius
Facial nerve, coverage of, 487-495
Femero-femoral bypass graft, 78
Femoral arteries, 10-11; *see also* Femoral vessels
  gracilis and, 16-17
  rectus femoris and, 44
  superficial, 35
    posterior thigh and, 90
    semimembranosus and, 126, 130
  tensor fascia lata and, 67, 74
  vastus lateralis and, 55, 58
Femoral nerves
  rectus femoris and, 43
  sartorius and, 34
  tensor fascia lata and, 65, 73
  vastus lateralis and, 53
Femoral vessels; *see also* Femoral arteries
  gracilis and, 13-31; *see also* Gracilis
  rectus femoris and, 41-50

INDEX

Femoral vessels—cont'd
  sartorius and, 33-40
  tensor fascia lata and, 63-85; *see also* Tensor fascia lata
Fibula
  abductor digiti minimi and, 296
  excision of, 167
  peroneus brevis and, 260-261
First dorsal interosseous, 453-457
First and second metacarpals, 453-457
Flaps
  distant coverage with; *see* Distant coverage
  free; *see* Free flap
  island; *see* Island flap
  kinking of, 406
  muscle; *see* Muscle flaps
  musculocutaneous; *see* Musculocutaneous flaps
  musculofascial; *see* musculofascial flap
  neurosensory, 63-85; *see also* Tensor fascia lata
  osseous-musculocutaneous, 63-85; *see also* Tensor fascia lata
  tubed pedicle, 84-85
Flexor carpi ulnaris, 422, 423, 441-447
  anatomy of, 442-445
  arc of rotation and, 446-447
  blood supply to, 443-445
  elevation of flap and, 447
  function of, 443
  nerve supply to, 443
  origin and insertion of, 442
  precautions and, 447
Flexor digitorum brevis, 265, 279-289
  anatomy of, 280-283
  arc of rotation and, 284-285
  blood supply to, 281-283
  case study and, 288-289
  elevation of flap and, 286, 287
  function of, 281
  nerve supply of, 281
  origin and insertion of, 280
  precautions and, 286
Flexor digitorum longus, 179-188
  anatomy of, 180-183
  arc of rotation and, 184-185
  blood supply to, 181-183
  elevation of flap and, 186-187
  function of, 181
  nerve supply to, 180
  origin and insertion of, 180
  precautions and, 186
Flexor hallucis longus, 189-198
  anatomy of, 190-193

Flexor hallucis longus—cont'd
  arc of rotation and, 194-195
  blood supply to, 191-193
  elevation of flap and, 196-197
  function of, 191
  nerve supply to, 191
  origin and insertion of, 190
  precautions for, 196
Fluorescein injection, 5
Foot, 263-307
  abductor digiti minimi in, 265, 291-299
  abductor hallucis in, 265, 269-277
  blood supply to, 266-267
  extensor digitorum brevis and, 301-307
  flexor digitorum brevis and, 265, 279-289
Forearm
  brachioradialis and, 433
  choice of flaps for, 422
  flexor carpi ulnaris and, 441-447
  rectus abdominis and, 355, 360
Forehead coverage, 480; *see also* Head
Free flap
  bone and, 5
  brachioradialis and, 433-439
  extensor digitorum brevis and, 301-307
  extensor digitorum longus and, 217-226
  extensor hallucis longus and, 227-236
  flexor digitorum brevis and, 279-289
  flexor digitorum longus and, 185
  flexor hallucis longus and, 194
  gastrocnemius and, 141-156; *see also* Gastrocnemius
  gracilis and, 13-31; *see also* Gracilis
  latissimus dorsi and, 369-391; *see also* Latissimus dorsi
  pectoralis major and, 317-335; *see also* Pectoralis major
  rectus abdominis and, 347-361; *see also* Rectus abdominis
  rectus femoris and, 41-50
  serratus anterior in, 337-345
  tensor fascia lata and, 63-85; *see also* Tensor fascia lata
  trapezius and, 393-419; *see also* Trapezius
  vascular pedicle and, 3
Free transfer, microsurgical; *see* Distant coverage
Functional muscle transfer
  brachioradialis and, 433-439
  extensor digitorum brevis and, 301-307
  extensor digitorum longus and, 217-226
  extensor hallucis longus and, 227-236
  flexor digitorum brevis and, 279-289
  gracilis and, 13-31; *see also* Gracilis

INDEX

Functional muscle transfer—cont'd
  latissimus dorsi and, 369-391; *see also* Latissimus dorsi
  pectoralis major and, 317-335; *see also* Pectoralis major
  rectus femoris and, 41-50
  tensor fascia lata and, 63-85; *see also* Tensor fascia lata
  trapezius and, 393-419; *see also* Trapezius

G

Gastrocnemius, 141-156
  anatomy of, 142-145
  arc of rotation and, 146-148
  blood supply to, 143-145
  case studies and, 152-155
  elevation of flap and, 149-150
  function of, 143
  nerve supply to, 143
  origin and insertion of, 142
  precautions and, 150
Gluteal artery, inferior, 90, 94
Gluteal nerve
  inferior, 93
  superior, 65
Gluteus maximus, 91-104
  anatomy of, 92-95
  arc of rotation and, 96-97
  blood supply to, 93-95
  elevation of flap and, 4, 98
  function of, 4, 93
  nerve supply to, 93
  origin and insertion of, 92
  precautions and, 98
Gracilis, 13-31
  anatomy of, 14-17
  arc of rotation of, 18-21
  blood supply to, 10-11, 16-17
  elevation of flap and, 22-23
  function of, 15
  ischial coverage and, 26-27
  nerve supply of, 15
  origin and insertion of, 14
  penis reconstruction and, 30-31
  precautions and, 24
  vaginal reconstruction and, 28-29
Grafts
  femero-femoral bypass, 78
  skin; *see* Skin grafts
  vein, aneurysm of, 172-173
Groin
  gracilis and, 13-31; *see also* Gracilis
  rectus abdominis and, 347-361; *see also* Rectus abdominis
  rectus femoris and, 41-50

Groin—cont'd
  sartorius and, 33-40
  tensor fascia lata and, 63-85; *see also* Tensor fascia lata
    case study of, 77-79
  vastus lateralis and, 51-62

H

Hamstrings; *see* Thigh, posterior
Hand, 449-469
  abductor digiti minimi in, 465-469
  abductor pollicis brevis in, 459-463
  blood supply to, 451
  choice of flaps for, 422, 450
  first dorsal interosseous in, 453-457
  rectus abdominis and, 355
  tensor fascia lata and, 84-85
Head, 471-495
  anterior trunk muscles and, 310
  blood supply to, 473
  choice of flap and, 472
  latissimus dorsi and, 369-391; *see also* Latissimus dorsi
  pectoralis major and, 317-335; *see also* Pectoralis major
    case study of, 332-335
  posterior trunk muscles and, 364
  sternocleidomastoid and, 472, 475-485
  temporalis and, 472, 487-495
  trapezius and, 393-419; *see also* Trapezius
    case studies and, 408-413, 418-419
Heel
  abductor digiti minimi and, 264-265, 291-299
  abductor hallucis and, 264-265, 269-277
  flexor digitorum brevis and, 264-265, 279-289
  melanoma of, 288
Hernias
  inguinal, 63-85; *see also* Tensor fascia lata
  from rectus abdominis flaps, 357
Hip disarticulation, 56, 58, 60-61
Hypogastric artery, 90, 94

I

Iliac arteries, external, 315
  rectus abdominis and, 350, 352
Iliac crest bone, 72
Incisions
  in abductor digiti minimi
    in foot, 298
    in hand, 469
  in abductor hallucis, 276
  in abductor pollicis brevis, 463
  in biceps brachii, 432
  in biceps femoris, 112

INDEX

Incisions—cont'd
  in brachioradialis, 439
  in dorsal interosseous, 457
  in extensor digitorum brevis, 306, 307
  in extensor digitorum longus, 224
  in extensor hallucis longus, 234, 235
  in flexor carpi ulnaris, 447
  in flexor digitorum brevis, 286, 287
  in flexor digitorum longus, 186
  in flexor hallucis longus, 196
  in gastrocnemius, 149
  in gluteus maximus, 98, 99
  in gracilis, 22
  in lateral anterior leg muscles, 200
  in latissimus dorsi, 383, 384, 385
  in pectoralis major, 324, 325, 328
  in peroneus brevis, 258, 259
  in peroneus longus, 248, 249
  in rectus abdominis, 356, 357
  in rectus femoris, 48
  in semimembranosus, 130
  in semitendinosus, 121
  in serratus anterior, 344, 345
  in soleus, 166-167
  in sternocleidomastoid, 482
  in temporalis, 492
  in tibialis anterior, 214
  in trapezius, 404
  in vastus lateralis, 58-59
Inguinal hernia, 63-85; *see also* Tensor fascia lata
Intercostal arteries, 323, 375, 399
Intercostal nerves, 349
Intercostobrachial nerve, 339
Interosseous, first dorsal, 453-457
Ischium
  biceps femoris and, 89, 105-114
  gluteus maximus and, 91-104
    case study of, 102-103
  gracilis and, 26-27; *see also* Gracilis
  rectus femoris and, 41-50
  semimembranosus and, 89, 123-132
  semitendinosus and, 89, 114-122
  tensor fascia lata and, 63-85; *see also* Tensor fascia lata
    case study of, 80-81
  vastus lateralis and, 51-62
Island flap, 3
  biceps femoris and, 113
  gluteus maximus and, 100, 102-103
  latissimus dorsi and, 382, 383, 384, 386, 389
  pectoralis major and, 325, 328, 329, 332, 333, 335
  rectus abdominis flap and, 356, 357, 359
  serratus anterior and, 344

Island flaps—cont'd
  sternocleidomastoid and, 482
  tensor fascia lata and, 79, 83
  trapezius and, 400, 404

J
Jugular vein, internal, 483

K
Knee
  gastrocnemius and, 136, 141-156; *see also* Gastrocnemius
  gracilis and, 13-31; *see also* Gracilis
  sartorius and, 33-40
  semimembranosus and, 123-132

L
Lateral ankle, 264, 265, 291-299
Lateral leg, 199-262
  anterior muscles in, 199-236
    approach to, 200
    blood supply to, 204-205
    extensor digitorum longus and, 202-203, 217-226
    extensor hallucis longus and, 202-203, 227-236
    tibialis anterior and, 202-203, 207-216
  posterior muscles in, 237-262
    blood supply of, 238
    peroneus brevis and, 251-262
    peroneus longus and, 241-250
Latissimus dorsi, 364, 365, 369-391
  anatomy of, 370-375
  arc of rotation and, 376-381
  blood supply to, 371-375
  case studies and, 386-391
  elevation of flap and, 382-385
  function of, 371
  insertion of, 371
  nerve supply of, 371
  origin of, 370
  precautions and, 384
Leg
  distal third of
    extensor digitorum longus and, 217-226
    extensor hallucis longus and, 227-236
    flexor digitorum longus and, 179-188
    flexor hallucis longus and, 189-198
    soleus and, 157-178; *see also* Soleus
  lateral; *see* Lateral leg
  medial, 133-198
    anatomy of, 134-138
    blood supply to, 138-139
    flexor digitorum longus and, 179-188
    flexor hallucis longus and, 189-198

INDEX

Leg—cont'd
  medial—cont'd
    gastrocnemius and, 141-156; see also Gastrocnemius
    soleus and, 157-178; see also Soleus
  middle third of
    muscles of, 238, 239
    peroneus brevis and, 251-262
    peroneus longus and, 241-250
    soleus and, 157-178; see also Soleus
    tibialis anterior and, 207-216
  upper third of, 141-156; see also Gastrocnemius
Lingual nerve, 478
Lower motor neuron disease, 2
Lower third of leg; see Leg, distal third of
Lumbar arteries, 375
Lumbar vertebrae, 378

M

Malleolus
  lateral, 264, 265, 291-299
  medial, 264, 265, 269-277
Mammary arteries, internal, 312, 314, 315
  pectoralis major and, 319, 322-323, 328
  rectus abdominis and, 350, 351
Mandible
  sternocleidomastoid and, 475-485
  trapezius and, 410-413
  trigeminal nerve and, 489
Mastectomy, 386-391; see also Breast
Mastoid
  sternocleidomastoid and, 480
  temporalis and, 487-495
Maxillary arteries, 473, 490, 493
Medial ankle, 264, 265, 269-277
Medial leg; see Leg, medial
Median nerve
  abductor pollicis brevis and, 460
  biceps brachii and, 426, 432
Melanoma of heel, 288
Metacarpals, first and second, 453-457
Microsurgical free transfer; see Distant coverage
Middle third of leg
  muscles of, 238, 239
  peroneus brevis and, 251-262
  peroneus longus and, 241-250
  soleus and, 157-178; see also Soleus
  tibialis anterior and, 207-216
Motor nerves, 4; see also Nerves
Motor neuron disease, 2
Mouth
  pectoralis major and, 334-335

Mouth—cont'd
  sternocleidomastoid and, 475-485
  trapezius and, 410-413, 418-419
Muscle belly size, 2
Muscle flaps, 1, 2-6
  and abductor digiti minimi
    of foot, 291-299
    of hand, 465-469
  abductor hallucis and, 269-277
  abductor pollicis brevis and, 459-463
  biceps brachii and, 425-432
  biceps femoris and, 105-114
  brachioradialis and, 433-439
  dorsal interosseous and, 453-457
  elevation of; see Elevation of flap
  extensor digitorum longus and, 217-226
  extensor hallucis longus and, 227-236
  flexor carpi ulnaris and, 441-447
  flexor digitorum brevis and, 279-289
  flexor digitorum longus and, 179-188
  flexor hallucis longus and, 189-198
  free transfer of; see Free flap
  gastrocnemius and, 141-156; see also Gastrocnemius
  gluteus maximus and, 91-104
  gracilis and, 13-31; see also Gracilis
  latissimus dorsi and, 369-391; see also Latissimus dorsi
  pectoralis major and, 317-335; see also Pectoralis major
  peroneus brevis and, 251-262
  peroneus longus and, 241-250
  precautions for, 5
  rectus abdominis and, 347-361; see also Rectus abdominis
  rectus femoris and, 41-50
  sartorius and, 33-40
  semimembranosus and, 123-132
  semitendinosus and, 114-122
  serratus anterior and, 337-345
  soleus and, 157-178; see also Soleus
  sternocleidomastoid and, 475-485
  temporalis and, 487-495
  tensor fascia lata and, 63-85; see also Tensor fascia lata
  tibialis anterior and, 207-216
  trapezius and, 393-419; see also Trapezius
  vastus lateralis and, 51-62
Muscles; see also Muscle flaps; Musculocutaneous flaps
  accessibility of, 3
  arc of rotation of; see Arc of rotation
  epimysium removed from, 2
  preservation of function of, 4
  sensory innervation and, 4; see also Nerves

INDEX

Muscle flaps—cont'd
  suture of edge of, to edge of cutaneous territory, 5
Musculocutaneous flaps, 1
  and abductor digiti minimi
    of foot, 291-299
    of hand, 469
  abductor pollicis brevis and, 463
  biceps femoris and, 105-114
  brachioradialis and, 433-439
  flexor carpi ulnaris and, 441-447
  free transfer of; *see* Free flap
  gastrocnemius and, 141-156; *see also* Gastrocnemius
    case study of, 152-155
  gluteus maximus and, 91-104
    case study of, 102-103
  gracilis and, 13-31; *see also* Gracilis
  latissimus dorsi and, 369-391; *see also* Latissimus dorsi
  pectoralis major and, 317-335; *see also* Pectoralis major
  rectus abdominis and, 347-361; *see also* Rectus abdominis
  rectus femoris and, 41-50
  semimembranosus and, 123-132
  semitendinosus and, 114-122
  sensory innervation and, 4
  serratus anterior and, 337-345
  sternocleidomastoid and, 475-485
  temporalis and, 487-495
  tensor fascia lata and, 63-85; *see also* Tensor fascia lata
    case study of, 76
  tibialis anterior and, 207-216
  trapezius and, 393-419; *see also* Trapezius
Musculocutaneous nerve, 427, 432
Musculocutaneous perforating vessels; *see* Perforating vessels
Musculofascial flap
  latissimus dorsi and, 369-391; *see also* Latissimus dorsi
  tensor fascia lata and, 63-85; *see also* Tensor fascia lata
    technique for, 74-75
Musculophrenic arteries, 314, 351
Myelomeningocele
  gluteus maximus and, 91-104
  latissimus dorsi and, 369-391; *see also* Latissimus dorsi
  trapezius and, 393-419; *see also* Trapezius

N

Neck, 471-495
  anterior trunk muscles and, 310

Neck—cont'd
  blood supply to, 473
  carcinoma of, 409
  choice of flap for, 472
  latissimus dorsi and, 369-391; *see also* Latissimus dorsi
  pectoralis major and, 317-335; *see also* Pectoralis major
    case study of, 332-335
  posterior trunk muscles and, 364
  radiation necrosis of, 414-417
  serratus anterior and, 337-345
  sternocleidomastoid and, 472, 475-485
  temporalis and, 472, 487-495
  trapezius and, 393-419; *see also* Trapezius
    case studies of, 408-419
Necrosis, radiation, 330, 414-417
Nerves
  and abductor digiti minimi
    of foot, 293
    of hand, 466
  abductor hallucis and, 271
  abductor pollicis brevis and, 460
  auricular, greater, 483
  biceps brachii and, 427
  biceps femoris and, 107
  brachioradialis and, 435
  cervical, 395, 477
  cutaneous; *see* Cutaneous nerves
  dorsal interosseous and, 454
  extensor digitorum brevis and, 303
  extensor digitorum longus and, 219
  extensor hallucis longus and, 229
  facial, coverage of, 487-495
  femoral; *see* Femoral nerves
  flexor carpi ulnaris and, 443
  flexor digitorum brevis and, 281
  flexor digitorum longus and, 180
  flexor hallucis longus and, 191
  gastrocnemius and, 143
  gluteal
    inferior, 93
    superior, 65
  gluteus maximus and, 93
  gracilis and, 15
  intercostal, 349
  intercostobrachial, 339
  latissimus dorsi and, 371
  lingual, 478
  median
    abductor pollicis brevis and, 460
    biceps brachii and, 426, 432
  motor, 4
  musculocutaneous, 427, 432
  obturator, 15

INDEX

Nerves—cont'd
  pectoral, 319
  pectoralis major and, 319
  peroneal; see Peroneal nerve
  peroneus brevis and, 253
  peroneus longus and, 243
  plantar; see Plantar nerves
  radial
    brachioradialis and, 434, 435, 436, 439
    first dorsal interosseous and, 457
    superficial, brachioradialis and, 436, 438, 439
  rectus abdominis and, 349
  rectus femoris and, 43
  sartorius and, 34
  sciatic
    biceps femoris and, 106, 107, 112, 114
    gluteus maximus and, 94, 98
    semimembranosus and, 125, 131
    semitendinosus and, 117, 121
  semimembranosus and, 125
  semitendinosus and, 117
  sensory, musculocutaneous flap and, 4
  serratus anterior and, 339
  soleus and, 159
  spinal accessory
    sternocleidomastoid and, 477
    trapezius and, 395, 406
  sternocleidomastoid and, 477
  sural, 293, 298
  temporalis and, 489
  tensor fascia lata and, 65, 73
  thoracic
    cutaneous branch of, 65, 73
    long, 339
  thoracodorsal, 314, 371, 385
  tibial; see Tibial nerve
  tibialis anterior and, 209
  trapezius and, 395
  trigeminal, 489
  ulnar; see Ulnar nerve
  vastus lateralis and, 53
Neurosensory flap, 63-85; see also Tensor fascia lata

O

Obturator nerve, 15
Occipital artery, 473
  sternocleidomastoid and, 478, 482
  trapezius and, 406
Oral cavity
  pectoralis major and, 334-335
  sternocleidomastoid and, 475-485
  trapezius and, 410-413, 418-419
Orbital cavity coverage, 487-495

Osseous-musculocutaneous flap, 63-85; see also Tensor fascia lata
Osseous territory, 5

P

Palmar arch, 451, 455
Palmar arteries, 451, 455
Pectoral nerves, 319
Pectoralis major, 310, 311, 317-335
  anatomy of, 318-323
  arc of rotation and, 324-325
  blood supply to, 319-323
  case studies and, 326-327, 330-335
  elevation of flap and, 328
  function of, 319
  nerve supply to, 319
  origin and insertion of, 318
  precautions and, 328
Pedicle flap, tubed, 84-85
Penis, reconstruction of, 30-31; see also Gracilis
Perforating vessels
  biceps femoris and, 108, 114
  gastrocnemius and, 144, 151
  intercostal, 323, 328
  latissimus dorsi and, 374, 375
  pectoralis major and, 323, 328
  in posterior trunk, 364
  rectus abdominis and, 352, 360
  tensor fascia lata and, 70
  trapezius and, 398, 399, 404, 406
Perineum
  biceps femoris and, 105-114
  gracilis and, 13-31; see also Gracilis
  rectus femoris and, 41-50
  semimembranosus and, 123-132
  semitendinosus and, 114-122
  tensor fascia lata and, 63-85; see also Tensor fascia lata
Peroneal arteries, 138, 139
  flexor hallucis longus and, 192, 194, 196
  peroneus brevis and, 238, 254
  peroneus longus and, 238, 244
  soleus and, 159, 160, 161, 162
Peroneal nerve
  anterior tibial artery and, 204
  common
    biceps femoris and, 106, 112, 114
    peroneus longus and, 243
  deep
    extensor digitorum brevis and, 303, 306
    extensor digitorum longus and, 219
    extensor hallucis longus and, 228, 229, 234
    peroneus longus and, 243
  gastrocnemius and, 149, 150

INDEX

Peroneal nerve—cont'd
  superficial
    extensor digitorum longus and, 225
    peroneus brevis and, 253
    peroneus longus and, 243
    tibialis anterior and, 208, 209, 214
Peroneus brevis, 238, 239, 251-262
  anatomy of, 252-255
  arc of rotation and, 256-257
  blood supply to, 253-255
  case study and, 260-261
  elevation of flap and, 258
  function of, 253
  nerve supply of, 253
  origin and insertion of, 252
  precautions and, 258
Peroneus longus, 238, 239, 241-250
  anatomy of, 242-245
  arc of rotation and, 246-247
  blood supply of, 243-245
  elevation of flap and, 248
  function of, 243
  nerve supply of, 243
  origin and insertion of, 242
  precautions and, 248
Plantar arch, 266
Plantar arteries
  lateral, 266, 267
    abductor digiti minimi and, 294, 299
    extensor digitorum brevis and, 306
    flexor digitorum brevis and, 282, 286
  medial, 266, 267
    abductor hallucis and, 272, 276
    flexor digitorum brevis and, 282, 284, 286
Plantar nerves
  lateral, abductor digiti minimi and, 293
  medial
    abductor hallucis and, 271, 277
    flexor digitorum brevis and, 281, 286
Popliteal artery
  anterior tibial artery and, 204
  flexor hallucis longus and, 192
  peroneus longus and, 244
  soleus and, 160
  sural branches of, 138, 139
    gastrocnemius and, 144, 149, 150
Popliteal fossa, defects on, 21
Posterior scalp and neck; *see also* Neck
  latissimus dorsi and, 369-391; *see also* Latissimus dorsi
  sternocleidomastoid and, 475-485
  temporalis and, 487-495
  trapezius and, 393-419; *see also* Trapezius
Posterior thigh; *see* Thigh, posterior
Posterior trunk; *see* Trunk, posterior

Postmastectomy defect, 386-391; *see also* Breast
Princeps pollicis artery, 451
Profunda femoris artery
  anterior thigh and, 10-11
  biceps femoris and, 108
  gracilis and, 16-17
  posterior thigh and, 90
  semimembranosus and, 126
  semitendinosus and, 118, 121
  tensor fascia lata and, 67
Pubis
  gracilis and, 13-31; *see also* Gracilis
  rectus femoris and, 46

R

Radial artery, 424
  abductor pollicis brevis and, 461
  brachioradialis and, 434, 436, 438, 439
  first dorsal interosseous and, 455, 457
  hand and, 451
  recurrent, 436
Radial nerve
  brachioradialis and, 434, 435, 436, 439
  first dorsal interosseous and, 457
  superficial, 436, 438, 439
Radiation necrosis, 330, 414-417
Rectus abdominis, 310, 311, 347-361
  anatomy of, 348-353
  arc of rotation and, 354-355
  blood supply to, 349-353
  case studies and, 358-361
  elevation of flap and, 356-357
  function of, 349
  nerve supply to, 349
  origin and insertion of, 348
  precautions and, 357
Rectus femoris, 41-50
  anatomy of, 42-45
  arc of rotation of, 46-47
  blood supply to, 10-11, 44-45
  elevation of flap and, 48
  function of, 43
  nerve supply to, 43
  origin and insertion of, 42
  precautions and, 48
Rib, transfer of, 344
Rotation, arc of; *see* Arc of rotation

S

Sacrum
  gluteus maximus and, 89, 91-104
  semimembranosus and, 89, 123-132
  tensor fascia lata and, 63-85; *see also* Tensor fascia lata

INDEX

Sartorius, 33-40
  anatomy of, 34-35
  arc of rotation and, 36-37
  blood supply to, 10-11, 35
  elevation of flap and, 38
  function of, 34
  nerve supply of, 34
  origin and insertion of, 34
  precautions and, 38
Scalp, posterior
  latissimus dorsi and, 369-391; *see also* Latissimus dorsi
  sternocleidomastoid and, 475-485
  temporalis and, 487-495
  trapezius and, 393-419; *see also* Trapezius
Scapula deformity, winged, 339, 344
Scapular artery
  circumflex, 313, 314
    latissimus dorsi and, 372, 384
  posterior, 399, 404
Sciatic nerve
  biceps femoris and, 106, 107, 112, 114
  gluteus maximus and, 94, 98
  semimembranosus and, 125, 131
  semitendinosus and, 117, 121
Second metacarpal, 453-457
Semimembranosus, 123-132
  anatomy of, 124-127
  arc of rotation and, 128-129
  blood supply to, 125-127
  elevation of flap and, 130
  function of, 125
  nerve supply of, 125
  origin and insertion of, 124
  precautions and, 131
Semitendinosus, 114-122
  anatomy of, 116-119
  arc of rotation and, 120
  blood supply to, 117-119
  elevation of flap and, 121
  function of, 117
  nerve supply to, 117
  origin and insertion of, 116
  precautions and, 121
Sensory innervation, 4; *see also* Nerves
Serratus anterior, 310, 311, 337-345
  anatomy of, 338-341
  arc of rotation and, 342-343
  blood supply of, 339-341
  elevation of flap and, 344
  function of, 339
  nerve supply to, 339
  origin and insertion of, 338
  precautions for, 344

Shoulder
  drooping deformity of, 395
  latissimus dorsi and, 369-391; *see also* Latissimus dorsi
  pectoralis major and, 317-335; *see also* Pectoralis major
  trapezius and, 393-419; *see also* Trapezius
Skin grafts, 5
  abductor digiti minimi and, 298, 299
  flexor digitorum brevis and, 289
  latissimus dorsi and, 384
  pectoralis major and, 325
  rectus abdominis and, 356, 357
  rectus femoris and, 48
  soleus and, 169, 175
  temporalis and, 495
  tensor fascia lata and, 72, 76
  trapezius and, 406, 411, 418
Skin island; *see* Island flap
Skull
  choice of flap and, 472
  sternocleidomastoid and, 480
  temporalis and, 487-495
  trapezius and, 402-403
Soleus, 157-178
  anatomy of, 158-161
  arc of rotation of, 162-165
  blood supply to, 159-161
  case studies and, 168-177
  elevation of flap and, 166-167
  function of, 159
  nerve supply to, 159
  origin and insertion of, 158
  precautions and, 167
  retractions of, for flexor hallucis longus flap, 196-197
Spinal accessory nerves
  sternocleidomastoid and, 477
  trapezius and, 395, 406
Sternocleidomastoid, 472, 475-485
  anatomy of, 476-479
  arc of rotation and, 480-481
  blood supply to, 477-479
  case study and, 484-485
  elevation of flap and, 482
  function of, 477
  nerve supply to, 477
  origin and insertion of, 477
  precautions and, 483
Sternum; *see also* Thorax
  pectoralis major and, 317-335; *see also* Pectoralis major
    case study of, 326-327, 330-331
  rectus abdominis and, 354

INDEX

Subclavian artery
  anterior trunk and, 314
  head and neck and, 473
  pectoralis major and, 322
  posterior trunk and, 366
  trapezius and, 396, 399
Subscapular arteries
  anterior trunk and, 312-313, 314
  latissimus dorsi and, 371, 372, 384
  posterior trunk and, 366, 367
Sural nerve, 293, 298
Suture of muscle to cutaneous territory edge, 5

T

Tarsal artery, lateral, 267, 298, 304, 306
Temporal arteries, 473, 490, 493
Temporalis, 472, 487-495
  anatomy of, 488-491
  arc of rotation and, 492, 493
  blood supply to, 489-491
  case study and, 494-495
  elevation of flap and, 492-493
  function of, 489
  nerve supply to, 489
  origin and insertion of, 488
  precautions for, 493
Tensor fascia lata, 63-85
  anatomy of, 64-67
  arc of rotation of, 68-69
  blood supply to, 10-11, 65-67
  case studies and, 76-85
  elevation of flap and, 72
  function of, 65
  nerve supply to, 65, 73
  origin and insertion of, 64
  precautions and, 74
  sensory distribution and, 65, 73
TFL; *see* Tensor fascia lata
Thigh
  anterior, 7-85
    anatomy of, 8-9
    blood supply to, 10-11
    gracilis and, 13-31; *see also* Gracilis
    rectus femoris and, 41-50
    sartorius and, 33-40
    tensor fascia lata and, 63-85; *see also* Tensor fascia lata
    vastus lateralis and, 51-62
  inferior, 129
  lateral, 58
  posterior, 87-132
    anatomy of, 88-89
    biceps femoris and, 105-114
    blood supply to, 90

Thigh—cont'd
  posterior—cont'd
    gluteus maximus and, 91-104
    semimembranosus and, 123-132
    semitendinosus and, 115-122
Thoracic arteries
  lateral, 312, 313-314
  long, 340, 342, 344
Thoracic nerve
  cutaneous branch of, 65, 73
  long, 339
Thoracic vertebrae, 378
Thoracoacromial arteries, 312, 313, 314
  pectoralis major and, 319-320
Thoracodorsal artery, 313, 314
  latissimus dorsi and, 372, 384
  serratus anterior and, 340, 342
Thoracodorsal nerve, 314, 371, 385
Thorax
  latissimus dorsi and, 369-391; *see also* Latissimus dorsi
  pectoralis major and, 317-335; *see also* Pectoralis major
  rectus abdominis and, 347-361; *see also* Rectus abdominis
  serratus anterior and, 337-345
  trapezius and, 393-419; *see also* Trapezius
Tibia
  extensor digitorum longus and, 217-226
  gastrocnemius and, 141-156; *see also* Gastrocnemius
    case study of, 152-153
  medial leg muscles and, 136
  soleus and, 168-171, 176-177
  tibialis anterior flap and, 207-216
Tibial artery
  anterior, 204-205
    extensor digitorum longus and, 220, 224
    extensor hallucis longus and, 228, 234
    tibialis anterior and, 208, 210, 214
  posterior
    flexor digitorum longus and, 182, 185, 186
    flexor hallucis longus and, 196
    foot and, 266, 267
    medial leg and, 138, 139
    soleus and, 159, 160, 161, 162, 164, 166, 167
Tibial nerve
  anterior, 209
    peroneus longus and, 243
  flexor digitorum longus and, 180, 186
  gastrocnemius and, 143, 149, 150
  posterior, 191, 196
  soleus and, 159, 160, 166, 167

INDEX

Tibialis anterior, 202-203, 207-216
   anatomy of, 208-211
   arc of rotation and, 212-213
   blood supply to, 209-211
   elevation of flap and, 214-215
   function of, 209
   nerve supply of, 209
   origin and insertion of, 208
   precautions and, 214
Tourniquet, 5
Transfers; *see also* Muscle flaps; Musculocutaneous flaps
   functional muscle; *see* Functional muscle transfer
   microsurgical free; *see* Distant coverage
Trapezius, 364, 365, 393-419
   anatomy of, 394-399
   arc of rotation and, 400-403
   blood supply to, 396-399
   case studies and, 408-419
   elevation of flap and, 404-407
   function of, 395
   nerve supply to, 395
   origin and insertion of, 394
   precautions and, 406
Trigeminal nerve, 489
Trochanter
   biceps femoris and, 105-114
   gluteus maximus and, 91-104
   rectus femoris and, 41-50
   semimembranosus and, 123-132
   tensor fascia lata and, 63-85; *see also* Tensor fascia lata
      case study of, 80-83
   vastus lateralis and, 51-62
Trunk, 309-419
   anterior, 309-361
      blood supply to, 312-315
      pectoralis major in, 217-335; *see also* Pectoralis major
      rectus abdominis and, 347-361; *see also* Rectus abdominis
      serratus anterior and, 337-345
   posterior, 363-419
      blood supply to, 366-367
      latissimus dorsi and, 364, 365, 369-391; *see also* Latissimus dorsi
      trapezius and, 364, 365, 393-419; *see also* Trapezius
Tubed pedicle flap, 84-85

U

Ulcers, radiation necrosis, 330, 414-417
Ulnar artery, 424, 451

Ulnar artery—cont'd
   abductor digiti minimi and, 467
   flexor carpi ulnaris and, 442, 444, 447
Ulnar nerve
   abductor digiti minimi and, 466, 467, 469
   biceps brachii and, 432
   first dorsal interosseous and, 454
   flexor carpi ulnaris and, 442, 443, 444, 447
Upper arm; *see also* Arm
   biceps brachii and, 425-432
   latissimus dorsi and, 369-391; *see also* Latissimus dorsi
   trapezius and, 393-419; *see also* Trapezius
Upper extremity, 421-447; *see also* Arm; Hand
   biceps brachii and, 422, 423, 425-432
   blood supply to, 424
   brachioradialis in, 422, 423, 433-439
   choice of flaps for, 422
   flexor carpi ulnaris and, 422, 423, 441-447
   latissimus dorsi and, 369-391; *see also* Latissimus dorsi
   posterior trunk muscles and, 364
   trapezius and, 393-419; *see also* Trapezius
Upper motor neuron disease, 2
Upper third of leg, 141-156; *see also* Gastrocnemius; Leg

V

Vagina
   gracilis and, 28-29; *see also* Gracilis
   tensor fascia lata and, 63-85; *see also* Tensor fascia lata
Vascular anatomy, 3; *see also* Blood supply
Vascular graft, coverage of exposed, 172-173
Vascular pedicle, 3; *see also* Elevation of flap
Vastus lateralis, 51-62
   anatomy of, 52-55
   arc of rotation of, 56-57
   blood supply to, 10-11, 53-55
   elevation of flap and, 58
   function of, 53
   nerve supply to, 53
   origin and insertion of, 52
   precautions and, 58
Vein graft aneurysm, 172-173
Veins; *see also* Blood supply
   axillary, 328
   extensor digitorum brevis flaps and, 306
   jugular, 482
Vertebrae, 378
Vulva
   gracilis and, 13-31; *see also* Gracilis
   tensor fascia lata and, 63-85; *see also* Tensor fascia lata

INDEX

W

Winged scapula deformity, 339, 344

Wrist
   abductor digiti minimi and, 465-469

Wrist—cont'd
   abductor pollicis brevis and, 459-463
   tensor fascia lata and, 63-85; *see also* Tensor fascia lata